VOLUME FIVE HUNDRED AND FORTY SIX

# METHODS IN ENZYMOLOGY

## The Use of CRISPR/Cas9, ZFNs, and TALENs in Generating Site-Specific Genome Alterations

# METHODS IN ENZYMOLOGY

*Editors-in-Chief*

**JOHN N. ABELSON and MELVIN I. SIMON**
*Division of Biology*
*California Institute of Technology*
*Pasadena, California*

**ANNA MARIE PYLE**
*Departments of Molecular, Cellular and Developmental Biology and Department of Chemistry Investigator*
*Howard Hughes Medical Institute*
*Yale University*

*Founding Editors*

**SIDNEY P. COLOWICK and NATHAN O. KAPLAN**

VOLUME FIVE HUNDRED AND FORTY SIX

# METHODS IN ENZYMOLOGY

## The Use of CRISPR/Cas9, ZFNs, and TALENs in Generating Site-Specific Genome Alterations

Edited by

**JENNIFER A. DOUDNA**

*Howard Hughes Medical Institute Investigator,*
*Li Ka Shing Chancellor's Chair in Biomedical and Health Sciences and Professor of Biochemistry,*
*Biophysics and Structural Biology,*
*Faculty Affiliate, Physical Biosciences Division,*
*Lawrence Berkeley National Laboratory,*
*UC Berkeley, Berkeley, CA USA*

**ERIK J. SONTHEIMER**

*RNA Therapeutics Institute,*
*University of Massachusetts Medical School,*
*Worcester, MA USA*

AMSTERDAM • BOSTON • HEIDELBERG • LONDON
NEW YORK • OXFORD • PARIS • SAN DIEGO
SAN FRANCISCO • SINGAPORE • SYDNEY • TOKYO
Academic Press is an imprint of Elsevier

Academic Press is an imprint of Elsevier
525 B Street, Suite 1800, San Diego, CA 92101-4495, USA
225 Wyman Street, Waltham, MA 02451, USA
32 Jamestown Road, London NW1 7BY, UK
The Boulevard, Langford Lane, Kidlington, Oxford, OX5 1GB, UK

First edition 2014



**Notices**

Knowledge and best practice in this field are constantly changing. As new research and experience broaden our understanding, changes in research methods, professional practices, or medical treatment may become necessary.

Practitioners and researchers must always rely on their own experience and knowledge in evaluating and using any information, methods, compounds, or experiments described herein. In using such information or methods they should be mindful of their own safety and the safety of others, including parties for whom they have a professional responsibility.

To the fullest extent of the law, neither the Publisher nor the authors, contributors, or editors, assume any liability for any injury and/or damage to persons or property as a matter of products liability, negligence or otherwise, or from any use or operation of any methods, products, instructions, or ideas contained in the material herein.

ISBN: 978-0-12-801185-0
ISSN: 0076-6879

# CONTENTS

# CONTRIBUTORS

**Carolin Anders**
Department of Biochemistry, University of Zurich, Zurich, Switzerland

**Carlos F. Barbas III**
Department of Chemistry, and Department of Cell and Molecular Biology, The Skaggs Institute for Chemical Biology, The Scripps Research Institute, La Jolla, California, USA

**Ira L. Blitz**
Department of Developmental and Cell Biology, University of California, Irvine, California, USA

**Erika Brunet**
Museum National d'Histoire Naturelle, INSERM U1154, CNRS 7196, Paris, France

**Susan M. Byrne**
Department of Genetics, Harvard Medical School, Boston, Massachusetts, USA

**Jamie H.D. Cate**
Energy Biosciences Institute; Department of Molecular and Cell Biology; Department of Chemistry, University of California, and Physical Biosciences Division, Lawrence Berkeley National Laboratory, Berkeley, California, USA

**Regina Cencic**
Department of Biochemistry, McGill University, Montreal, Quebec, Canada

**Baohui Chen**
Department of Pharmaceutical Chemistry, University of California, San Francisco, California, USA

**Ken W.Y. Cho**
Department of Developmental and Cell Biology, University of California, Irvine, California, USA

**George M. Church**
Department of Genetics, Harvard Medical School, Boston, Massachusetts, USA

**Chad A. Cowan**
Department of Stem Cell and Regenerative Biology, Harvard University; Harvard Stem Cell Institute, Sherman Fairchild Biochemistry, Cambridge, and Center for Regenerative Medicine, Massachusetts General Hospital, Boston, Massachusetts, USA

**D. Dambournet**
Department of Molecular and Cell Biology, University of California, Berkeley, California, USA

**D.G. Drubin**
Department of Molecular and Cell Biology, University of California, Berkeley, California, USA

**Leonardo M.R. Ferreira**
Department of Stem Cell and Regenerative Biology, Harvard University, Cambridge, Massachusetts, USA

**Margaret B. Fish**
Department of Developmental and Cell Biology, University of California, Irvine, California, USA

**Yanfang Fu**
Molecular Pathology Unit, Center for Computational and Integrative Biology, and Center for Cancer Research, Massachusetts General Hospital, Charlestown, and Department of Pathology, Harvard Medical School, Boston, Massachusetts, USA

**Yoshitaka Fujihara**
Research Institute for Microbial Diseases, Osaka University, Suita, Japan

**Thomas Gaj**
Department of Chemistry, and Department of Cell and Molecular Biology, The Skaggs Institute for Chemical Biology, The Scripps Research Institute, La Jolla, California, USA

**Hind Ghezraoui**
Museum National d'Histoire Naturelle, INSERM U1154, CNRS 7196, Paris, France

**Andrew P.W. Gonzales**
Cardiovascular Research Center, Massachusetts General Hospital, Charlestown, and Department of Medicine, Harvard Medical School, Boston, Massachusetts, USA

**Federico González**
Developmental Biology Program, Sloan-Kettering Institute, New York, USA

**Robert M. Grainger**
Department of Biology, University of Virginia, Charlottesville, Virginia, USA

**A. Grassart**
Department of Molecular and Cell Biology, University of California, Berkeley, California, USA

**Yuting Guan**
Shanghai Key Laboratory of Regulatory Biology, Institute of Biomedical Sciences and School of Life Sciences, East China Normal University, Shanghai, China

**John P. Guilinger**
Department of Chemistry & Chemical Biology, and Howard Hughes Medical Institute, Harvard University, Cambridge, Massachusetts, USA

**S.H. Hong**
Department of Molecular and Cell Biology, University of California, Berkeley, California, USA

**Benjamin E. Housden**
Department of Genetics, Harvard Medical School, Boston, Massachusetts, USA

**Bo Huang**
Department of Pharmaceutical Chemistry, and Department of Biochemistry and Biophysics, University of California, San Francisco, California, USA

**Danwei Huangfu**
Developmental Biology Program, Sloan-Kettering Institute, New York, USA

**Masahito Ikawa**
Research Institute for Microbial Diseases, Osaka University, Suita, Japan

**Rayelle Itoua Maïga**
Department of Biochemistry, McGill University, Montreal, Quebec, Canada

**Maria Jasin**
Developmental Biology Program, Memorial Sloan-Kettering Cancer Center, New York, USA

**Martin Jinek**
Department of Biochemistry, University of Zurich, Zurich, Switzerland

**J. Keith Joung**
Molecular Pathology Unit, Center for Computational and Integrative Biology, and Center for Cancer Research, Massachusetts General Hospital, Charlestown, and Department of Pathology, Harvard Medical School, Boston, Massachusetts, USA

**Alexandra Katigbak**
Department of Biochemistry, McGill University, Montreal, Quebec, Canada

**Hyongbum Kim**
Graduate School of Biomedical Science and Engineering, College of Medicine, Hanyang University, Seoul, South Korea

**Jin-Soo Kim**
Center for Genome Engineering, Institute for Basic Science, and Department of Chemistry, Seoul National University, Seoul, South Korea

**Young-Hoon Kim**
Graduate School of Biomedical Science and Engineering, College of Medicine, Hanyang University, Seoul, South Korea

**Przemek M. Krawczyk**
Developmental Biology Program, Memorial Sloan-Kettering Cancer Center, New York, USA, and Department of Cell Biology and Histology, Academic Medical Center, University of Amsterdam, Amsterdam, The Netherlands

**Dali Li**
Shanghai Key Laboratory of Regulatory Biology, Institute of Biomedical Sciences and School of Life Sciences, East China Normal University, Shanghai, China

**Jian-Feng Li**
Department of Molecular Biology, Centre for Computational and Integrative Biology, Massachusetts General Hospital, and Department of Genetics, Harvard Medical School, Boston, Massachusetts, USA

**Shuailiang Lin**
Department of Genetics, Harvard Medical School, Boston, Massachusetts, USA

**David R. Liu**
Department of Chemistry & Chemical Biology, and Howard Hughes Medical Institute, Harvard University, Cambridge, Massachusetts, USA

**Mingyao Liu**
Shanghai Key Laboratory of Regulatory Biology, Institute of Biomedical Sciences and School of Life Sciences, East China Normal University, Shanghai, China

**Prashant Mali**
Department of Genetics, Harvard Medical School, Boston, Massachusetts, USA

**Abba Malina**
Department of Biochemistry, McGill University, Montreal, Quebec, Canada

**Pankaj K. Mandal**
Department of Stem Cell and Regenerative Biology, Harvard University, Cambridge, and Program in Cellular and Molecular Medicine, Division of Hematology/Oncology, Boston Children's Hospital, Boston, Massachusetts, USA

**Sumanth Manohar**
Department of Biology, University of Virginia, Charlottesville, Virginia, USA

**Torsten B. Meissner**
Department of Stem Cell and Regenerative Biology, Harvard University, Cambridge, Massachusetts, USA

**Hisashi Miura**
Department of Biochemistry, McGill University, Montreal, Quebec, Canada

**Dana C. Nadler**
Department of Chemical and Biomolecular Engineering, University of California, Berkeley, California, USA

**Takuya Nakayama**
Department of Biology, University of Virginia, Charlottesville, Virginia, USA

**Benjamin L. Oakes**
Department of Molecular & Cell Biology, University of California, Berkeley, California, USA

**Akinleye O. Odeleye**
Department of Biology, University of Virginia, Charlottesville, Virginia, USA

**Vikram Pattanayak**
Department of Pathology, Massachusetts General Hospital, Boston, Massachusetts, USA

**Jerry Pelletier**
Department of Biochemistry; Department of Oncology, and The Rosalind and Morris Goodman Cancer Research Center, McGill University, Montreal, Quebec, Canada

**Norbert Perrimon**
Department of Genetics, and Howard Hughes Medical Institute, Harvard Medical School, Boston, Massachusetts, USA

**Marion Piganeau**
Museum National d'Histoire Naturelle, INSERM U1154, CNRS 7196, Paris, France

**Benjamin Renouf**
Museum National d'Histoire Naturelle, INSERM U1154, CNRS 7196, Paris, France

**Deepak Reyon**
Molecular Pathology Unit, Center for Computational and Integrative Biology, and Center for Cancer Research, Massachusetts General Hospital, Charlestown, and Department of Pathology, Harvard Medical School, Boston, Massachusetts, USA

**Francis Robert**
Department of Biochemistry, McGill University, Montreal, Quebec, Canada

**Derrick J. Rossi**
Department of Stem Cell and Regenerative Biology, Harvard University, Cambridge; Program in Cellular and Molecular Medicine, Division of Hematology/Oncology, Boston Children's Hospital; Department of Pediatrics, Harvard Medical School, Boston, and Harvard Stem Cell Institute, Sherman Fairchild Biochemistry, Cambridge, Massachusetts, USA

**Owen W. Ryan**
Energy Biosciences Institute, University of California, Berkeley, California, USA

**David F. Savage**
Department of Molecular & Cell Biology; Department of Chemistry, and Energy Biosciences Institute, University of California, Berkeley, California, USA

**Hillel T. Schwartz**
Division of Biology and Biological Engineering, California Institute of Technology, and Howard Hughes Medical Institute, Pasadena, California, USA

**Yanjiao Shao**
Shanghai Key Laboratory of Regulatory Biology, Institute of Biomedical Sciences and School of Life Sciences, East China Normal University, Shanghai, China

**Jen Sheen**
Department of Molecular Biology, Centre for Computational and Integrative Biology, Massachusetts General Hospital, and Department of Genetics, Harvard Medical School, Boston, Massachusetts, USA

**Minjung Song**
Graduate School of Biomedical Science and Engineering, College of Medicine, Hanyang University, Seoul, South Korea

**Paul W. Sternberg**
Division of Biology and Biological Engineering, California Institute of Technology, and Howard Hughes Medical Institute, Pasadena, California, USA

**Alexandro E. Trevino**
Broad Institute of MIT and Harvard, 7 Cambridge Center; McGovern Institute for Brain Research; Department of Brain and Cognitive Sciences, and Department of Biological Engineering, Massachusetts Institute of Technology, Cambridge, Massachusetts, USA

**Lianne E.M. Vriend**
Developmental Biology Program, Memorial Sloan-Kettering Cancer Center, New York, USA, and Department of Cell Biology and Histology, Academic Medical Center, University of Amsterdam, Amsterdam, The Netherlands

**Jing-Ruey Joanna Yeh**
Cardiovascular Research Center, Massachusetts General Hospital, Charlestown, and Department of Medicine, Harvard Medical School, Boston, Massachusetts, USA

**Dandan Zhang**
Department of Molecular Biology, Centre for Computational and Integrative Biology, Massachusetts General Hospital, and Department of Genetics, Harvard Medical School, Boston, Massachusetts, USA

**Feng Zhang**
Broad Institute of MIT and Harvard, 7 Cambridge Center; McGovern Institute for Brain Research; Department of Brain and Cognitive Sciences, and Department of Biological Engineering, Massachusetts Institute of Technology, Cambridge, Massachusetts, USA

**Zengrong Zhu**
Developmental Biology Program, Sloan-Kettering Institute, New York, USA

# PREFACE

The availability of whole-genome sequencing data for large numbers and types of organisms including humans holds exciting promise for advancing both research and healthcare. A major challenge has been understanding and using genomic data through targeted manipulation of cellular DNA. Ever since the discovery of DNA structure in the 1950s, researchers and clinicians have been contemplating the possibility of making site-specific changes to the genomes of cells and organisms. Many of the earliest approaches to what has become known as genome editing relied on the principle of site-specific recognition of DNA sequences. The study of natural DNA repair pathways in bacteria and yeast, as well as the mechanisms of DNA recombination, showed that cells have endogenous machinery to repair DNA double-strand breaks that would otherwise be lethal. Thus, methods for introducing precise breaks in the DNA at desired editing sites were recognized as a valuable strategy for targeted genomic engineering.

Although some isolated successes were achieved using oligonucleotides or small molecules to localize DNA-cleaving activities to specific sequences, proteins that could be programmed to bind and cleave particular DNA sites proved more broadly useful. Modular DNA recognition proteins, when coupled to the sequence-independent nuclease domain of the restriction enzyme FokI, could function as site-specific nucleases. When designed to recognize a chromosomal sequence, such zinc-finger nucleases and TAL effector nucleases can be effective at inducing genomic sequence changes in both animal and plant cells. Difficulties of protein design, synthesis, and validation have limited widespread adoption of these engineered nucleases for routine use, paving the way for the CRISPR/Cas9 system in which a natural protein with double-stranded DNA-cleaving activity can be programmed with a short RNA sequence to recognize and cut DNA sites of interest. The ease of use, efficiency, and multiplexing capabilities of this technology have enabled rapid adoption for many different genome engineering applications.

In this volume, we provide readers with a collection of protocols for the major protein-based genome editing techniques, with a particular emphasis on the more recently developed CRISPR/Cas9 approaches. As these systems are used more widely and for ever-increasing types of projects, we anticipate that the facility of genome manipulation for application in human health and biotechnology will continue to expand.

Jennifer A. Doudna
Erik J. Sontheimer

CHAPTER ONE

# *In Vitro* Enzymology of Cas9

**Carolin Anders, Martin Jinek[1]**
Department of Biochemistry, University of Zurich, Zurich, Switzerland
[1]Corresponding author: e-mail address: jinek@bioc.uzh.ch

## Contents



## Abstract

Cas9 is a bacterial RNA-guided endonuclease that uses base pairing to recognize and cleave target DNAs with complementarity to the guide RNA. The programmable sequence specificity of Cas9 has been harnessed for genome editing and gene expression control in many organisms. Here, we describe protocols for the heterologous expression and purification of recombinant Cas9 protein and for *in vitro* transcription of guide RNAs. We describe *in vitro* reconstitution of the Cas9–guide RNA ribonucleoprotein complex and its use in endonuclease activity assays. The methods outlined here enable mechanistic characterization of the RNA-guided DNA cleavage activity of Cas9 and may assist in further development of the enzyme for genetic engineering applications.

## 1. INTRODUCTION

The clusters of regularly interspaced short palindromic repeat (CRISPR)-associated protein Cas9 is an RNA-guided endonuclease that generates double-strand DNA breaks (DSBs) (reviewed in Hsu, Lander, & Zhang, 2014; Mali, Esvelt, & Church, 2013). Found in type II CRISPR systems, Cas9 functions in conjunction with CRISPR RNAs (crRNAs) and a transactivating crRNA (tracrRNA) to mediate sequence-specific immunity against bacteriophages and other mobile genetic elements (Barrangou et al., 2007; Deltcheva et al., 2011; Garneau et al., 2010). Cas9 associates with

*Methods in Enzymology*, Volume 546
ISSN 0076-6879
http://dx.doi.org/10.1016/B978-0-12-801185-0.00001-5

a partially base-paired crRNA–tracrRNA guide structure and the resulting ribonucleoprotein complex recognizes and cleaves DNA molecules containing sequences complementary to a 20-nucleotide guide segment in the crRNA (Gasiunas, Barrangou, Horvath, & Siksnys, 2012; Jinek et al., 2012; Karvelis et al., 2013).

Due to its programmability, Cas9 has been developed into a versatile molecular tool for genome editing in numerous organisms and cell types (reviewed extensively in Hsu et al., 2014; Mali, Esvelt, et al., 2013; Sander & Joung, 2014), including human cells (Cong et al., 2013; Jinek et al., 2013; Mali, Yang, et al., 2013), mice (Wang et al., 2013; Yang et al., 2013), zebrafish (Hwang et al., 2013), *Drosophila melanogaster* (Bassett & Liu, 2014; Gratz et al., 2013), *Caenorhabditis elegans* (Cho, Lee, Carroll, Kim, & Lee, 2013; Friedland et al., 2013; Katic & Grosshans, 2013; Lo et al., 2013), and plants (Li et al., 2013; Nekrasov, Staskawicz, Weigel, Jones, & Kamoun, 2013; Shan et al., 2013; Xie & Yang, 2013). The sequence specificity of Cas9 permits the targeting of unique loci in a typical eukaryotic genome and can be readily altered *in vitro* and *in vivo* by supplying artificially designed guide RNAs either in the naturally occurring dual-RNA form or as single-molecule guide RNAs (sgRNAs) (Cong et al., 2013; Jinek et al., 2012, 2013; Mali, Yang, et al., 2013). Cas9 thus provides a superior alternative to existing protein-based approaches such as zinc finger nucleases and transcription activator-like effector nucleases. In eukaryotic cells, Cas9-generated DSBs are repaired by nonhomologous end joining or homologous recombination, which can be exploited to engineer insertions, deletions, and substitutions in the vicinity of the DSB. Furthermore, a catalytically inactive variant of Cas9 (the D10A/H840A mutant of *Streptococcus pyogenes* Cas9, referred to as dCas9) has been employed as an RNA-programmable DNA-binding protein for transcriptional regulation (Gilbert et al., 2013; Mali, Aach, et al., 2013; Qi et al., 2013). Variants of the basic targeting approach, including paired nickases (Mali, Aach, et al., 2013; Ran et al., 2013), dCas9-FokI fusion nucleases (Guilinger, Thompson, & Liu, 2014; Tsai et al., 2014), and 5′-truncated sgRNAs (Fu, Sander, Reyon, Cascio, & Joung, 2014) have emerged recently to address the issue of off-targeting and to further improve Cas9 specificity.

Extensive biochemical and structural studies have illuminated many aspects of the molecular mechanism of Cas9. The two nuclease domains found in Cas9, HNH and RuvC domains, catalyze the cleavage of the complementary and noncomplementary DNA strands, respectively (Chen, Choi, & Bailey, 2014; Gasiunas et al., 2012; Jinek et al., 2012). Target

DNA recognition is strictly dependent on the presence of a short protospacer adjacent motif (PAM) immediately downstream of the DNA region base-paired to the guide RNA (Gasiunas et al., 2012; Jinek et al., 2012). An 8–12 nt PAM-proximal "seed" region in the guide RNA–target DNA heteroduplex is critical for target binding by Cas9 (Jinek et al., 2012; Nishimasu et al., 2014). While seed region interactions are sufficient for target binding, DNA cleavage requires more extensive guide–target interactions (Wu et al., 2014). Nevertheless, Cas9 tolerates mismatches within the guide–target heteroduplex, which is the principal cause of off-target activity (Fu et al., 2013; Hsu et al., 2013; Mali, Aach, et al., 2013; Pattanayak et al., 2013). Recent crystal structures and electron microscopic reconstructions of Cas9 in its free and nucleic-acid-bound states have revealed that Cas9 undergoes a striking RNA-driven conformational rearrangement that results in the formation of the DNA-binding site (Anders, Niewoehner, Duerst, & Jinek, 2014; Jinek et al., 2014; Nishimasu et al., 2014). Additionally, single-molecule and ensemble biophysical studies of target recognition by the Cas9–guide RNA complex have indicated that target DNA binding is dependent on an initial recognition of the PAM, followed by local unwinding of the adjacent DNA duplex and directional formation of the guide RNA–target DNA heteroduplex (Sternberg, Redding, Jinek, Greene, & Doudna, 2014).

In this Chapter, we provide detailed protocols for the heterologous expression and purification of *S. pyogenes* Cas9, preparation of guide RNAs by *in vitro* transcription, and for the use of these reagents in endonuclease cleavage assays *in vitro*. The assays described here can be used to validate guide RNAs and target sites for *in vivo* gene targeting applications or to test the *in vitro* efficacy of new guide RNA structures and designs. Moreover, the described procedures can be implemented to utilize Cas9 as a programmable restriction enzyme for DNA manipulations *in vitro*. Although *S. pyogenes* Cas9 has been the mainstay of genome editing applications so far, the protocols are readily adaptable for Cas9 proteins and guide RNAs from other bacterial species and may aid in the rational design of novel Cas9 variants with altered specificity or PAM requirements.

## 2. EXPRESSION AND PURIFICATION OF Cas9

Cas9 from *S. pyogenes* (hereafter referred to as SpyCas9) is expressed from a pET-based T7 promoter-containing plasmid (pMJ806, available from Addgene, www.addgene.org) in the *E. coli* strain Rosetta 2 DE3. The expressed fusion protein construct contains an N-terminal $His_6$-tag, followed by maltose-binding protein (MBP) polypeptide sequence, a tobacco etch virus (TEV) protease cleavage site, and the SpyCas9 sequence spanning residues 1–1368. We found that expression in the Rosetta 2 strain was necessary to overcome the unfavorable codon bias in the *S. pyogenes* genomic DNA sequence, while inclusion of the MBP tag further boosted expression levels. The purification protocol includes three chromatography steps: immobilized metal ion affinity chromatography (IMAC), followed by cation exchange chromatography (IEX), and a final purification by size exclusion chromatography (SEC). The protocol is generally based on previously published procedures, with minor modifications (Jinek et al., 2012, 2014; Sternberg et al., 2014). The procedure can be used for the expression and purification of mutant SpyCas9 proteins and can be adapted for the expression of Cas9 orthologs from other bacterial species.

**Day 1: Cell transformation**

1. Transform chemically competent Rosetta 2 DE3 cells (Novagen, Merck Millipore) according to the protocol supplied with the cells. Briefly, add ~200 ng of plasmid DNA (pMJ806) to 50 μl of freshly thawed competent cells and incubate on ice for 15 min. Heat-shock cells by incubation at 42 °C for 45 s, then place cells on ice for further 3 min. Add 500 μl of LB (Luria Broth) medium to the cells and incubate the culture at 37 °C for 1 h in a shaking incubator. Plate 100 μl of culture out on LB agar containing 50 $\mu g\ ml^{-1}$ kanamycin and 33 $\mu g\ ml^{-1}$ chloramphenicol. Incubate plates overnight at 37 °C.

**Day 2: Culture growth and induction**

2. Pick one colony from the agar plate to inoculate 50 ml LB medium containing 50 $\mu g\ ml^{-1}$ kanamycin and 33 $\mu g\ ml^{-1}$ chloramphenicol. Incubate the preculture at 37 °C in a shaking incubator (250 rpm) for a minimum of 4–5 h or overnight.
3. Use 7.5 ml of the preculture to inoculate 750 ml prewarmed LB medium supplemented with 50 $\mu g\ ml^{-1}$ kanamycin and 33 $\mu g\ ml^{-1}$ chloramphenicol in a 2 l baffled flask. We typically express 6 × 750 ml total culture volume at a time. Incubate the cultures at 37 °C in a shaking incubator (90 rpm) while monitoring the cell growth by measuring optical density at 600 nm ($OD_{600}$). Reduce the temperature to 18 °C at an OD of ~0.8 and continue shaking for additional 30 min. Induce protein expression by the addition of 150 μl 1 *M* isopropyl-β-D-1-thiogalactopyranoside to each flask (200 μ*M* final concentration). Continue shaking at 18 °C overnight for another 12–16 h.

**Day 3: Cas9 purification by IMAC**

4. Harvest cells by centrifugation for 15 min at 3500 rpm (~2700 × *g*) in a swing-bucket rotor in 1 l bottles. Decant the supernatant and resuspend the cell pellets using ~15 ml ice-chilled lysis buffer (20 m*M* Tris–Cl, pH 8.0, 250 m*M* NaCl, 5 m*M* imidazole, pH 8.0, 1 m*M* phenylmethylsulfonyl fluoride) per cell pellet from 1 l culture. The resuspended cell pellets can either be used directly for further purification or flash frozen in liquid nitrogen and stored at −80 °C for several months without loss of Cas9 enzymatic activity.
5. Lyse the resuspended cell pellets using a cell homogenizer (Avestin Emulsiflex). Pass the cell suspension through the homogenizer three times at ~1000 bar to ensure complete lysis. The lysate should be cooled on ice between passes.
6. Clarify the lysate by centrifugation in 50 ml Nalgene Oak Ridge tubes at 18,000 rpm (~30,000 × *g*) in an SS-34 rotor (or equivalent) for 30 min at 4 °C. Collect the supernatant.
7. All chromatographic steps should be performed at 4 °C. Equilibrate 10 ml of His-Select Ni resin (Sigma–Aldrich) packed in a XK 16/20 column housing (GE Healthcare) with 20 ml lysis buffer. Load the cleared lysate on the column using a peristaltic pump at ~1.5 ml min$^{-1}$. Attach the column with bound protein to an FPLC system equilibrated in wash buffer (20 m*M* Tris–Cl, pH 8.0, 250 m*M* NaCl, 10 m*M* imidazole, pH 8.0).
8. Wash with ~50 ml wash buffer at 1.5 ml min$^{-1}$ until the absorbance nearly reaches the baseline again. Elute with ~50 ml elution buffer (20 m*M* Tris–Cl, pH 8.0, 250 m*M* NaCl, 250 m*M* imidazole, pH 8.0) and collect in 2 ml fractions. Analyze the peak fractions using SDS-PAGE and pool those containing Cas9 protein.
9. Estimate protein concentration by measuring the absorbance at 280 nm (use elution buffer as blank). Add 0.5 mg TEV protease per 50 mg of protein. Dilute the Cas9 sample to ~1 mg ml$^{-1}$ with dialysis buffer (20 m*M* HEPES–KOH, pH 7.5, 150 m*M* KCl, 10% (v/v) glycerol, 1 m*M* dithiothreitol (DTT), 1 m*M* EDTA) and dialyze the sample in dialysis tubing with a molecular weight cut off (MWCO) of 12–14 kDa against 2 l dialysis buffer at 4 °C overnight. Dialysis buffer (without DTT and glycerol) can be prepared as a 10× stock, but DTT should only be added immediately prior to use.

**Day 4: IEX and SEC chromatographic steps**

**10.** Recover the dialyzed sample from the dialysis tubing. Typically, slight precipitation occurs after successful TEV protease cleavage. Centrifuge the sample at 3900 rpm (~3200 × *g*) for 5 min at 4 °C to remove the precipitate. Check the extent of TEV protease cleavage using SDS-PAGE.

**11.** Equilibrate a 5-ml HiTrap SP FF column (GE Healthcare) with IEX buffer A (20 m*M* HEPES–KOH, pH 7.5, 100 m*M* KCl) and load the cleaved protein onto the column using a peristaltic pump or a sample loading superloop at a flow rate of ~2 ml $min^{-1}$. Attach the column to the FPLC system. Set the flow rate to 2 ml $min^{-1}$ and the pressure limit to 0.3 MPa for further steps using the HiTrap column. Collect 2 ml fractions throughout. Wash the column with 10 ml IEX buffer A and elute bound protein by applying a gradient from 0% to 50% IEX buffer B (20 m*M* HEPES–KOH, pH 7.5, 1 *M* KCl) over 60 ml. Cas9 typically elutes in two peaks with different ratios of the absorbances at 260 and 280 nm ($A_{260}/A_{280}$). The first peak starts eluting at ~15% IEX buffer B with its maximum at ~20%, the second peak elutes at ~25–40%, with a maximum at ~30%. Analyze all peak fractions for the presence of Cas9 using SDS-PAGE. Pool Cas9-containing fractions that have an $A_{260}/A_{280}$ of less than ~0.6. The pooled sample can be stored at 4 °C overnight or can be flash frozen in liquid nitrogen and stored at −80 °C.

**12.** Exchange the buffer to SEC buffer (20 m*M* HEPES–KOH, pH 7.5, 500 m*M* KCl, 1 m*M* DTT) while concentrating the protein to <1.5 ml volume using a 30,000 MWCO centrifugal concentrator (Amicon) at 3900 rpm. The buffer exchange helps circumvent precipitation in the centrifugal concentrator. Recover concentrate in a 1.5 ml Eppendorf tube and centrifuge for 10 min at 14,000 rpm (16,900 × *g*) at 4 °C to remove any precipitated material.

**13.** In the meantime, equilibrate a HiLoad 16/600 Superdex 200 PG gel filtration column (GE Healthcare) with the SEC buffer. Inject concentrated Cas9 onto column using a 2 ml sample loop. Elute with 120 ml SEC buffer at a flow rate of 1 ml $min^{-1}$, collecting 2 ml fractions. Cas9 typically elutes at a volume of ~66 ml. Analyze the peak fractions using SDS-PAGE and pool those containing Cas9 protein.

**Day 5: Concentration and storage**

**14.** Concentrate eluted Cas9 using a 30,000 MWCO centrifugal concentrator to a concentration required for further experiments. In the described SEC Buffer, Cas9 can be concentrated up to ~30 mg ml$^{-1}$ (189.3 μ*M*) without precipitation. The concentration is determined based on the assumption that 1 mg ml$^{-1}$ has an absorbance at 280 nm of 0.76 (based on a calculated extinction coefficient of 120,450 M$^{-1}$ cm$^{-1}$).

**15.** Divide concentrated protein sample into 50 μl aliquots and flash freeze in liquid nitrogen. Frozen Cas9 can be stored at −80 °C for several months without loss of activity. Prior to use, thaw an aliquot and dilute to the required concentration with SEC buffer. We typically dilute Cas9 to 15 μ*M* for endonuclease activity assays. To avoid unnecessary freeze–thaw cycles that can result in loss of enzymatic activity, Cas9 can be stored on ice or at 4 °C for at least 2 days without loss of activity if used for several experiments. However, it is recommended to check the stored sample periodically for the absence of any precipitated material and to monitor protein integrity by SDS-PAGE.

## 3. PREPARATION OF GUIDE RNAs

For sequence-specific DNA cleavage, Cas9 can be programmed either with a custom crRNA that is partially base-paired to an invariant tracrRNA molecule (i.e., the dual-RNA guide) or with chimeric sgRNAs that combine the essential parts of the crRNA and tracrRNA molecules in a single oligonucleotide chain (Jinek et al., 2012). When using dual-RNA guides, the crRNA guide is composed of a 5′-terminal 20-nt guide sequence, followed by an invariant 22-nt repeat-derived sequence at the 3′ end (5′-XXXXXXXXXXXXXXXXXXXX-GUUUUAGAGCUAUGCUGUUUUG-3′) that will ensure base pairing to the tracrRNA (Fig. 1.1A). The tracrRNA sequence remains identical for all guide crRNAs and corresponds to the sequence of the mature processed *S. pyogenes* tracrRNA (5′-AAACAGCAUAGCAAGUUAAAAUAAGGCUAGUCCGUUAUCAACUUGAAAAAGUGGCACCGAGUCGGUGCUU-3′) (Fig. 1.1A). Alternatively, a sgRNA can be used for Cas9 programming. The chimeric RNA is essentially composed of the desired 20-nt guide sequence at its 5′ end, followed by a segment corresponding to the 3′-terminal invariant sequence of the crRNA, and fused to a tracrRNA fragment with a GAAA tetraloop. The tracrRNA-derived part of the sgRNA consists of a region complementary to the repeat-derived part of the crRNA and three additional stem–loops (SL1–3) at the 3′ end (Fig. 1.1B). Although 3′-terminally

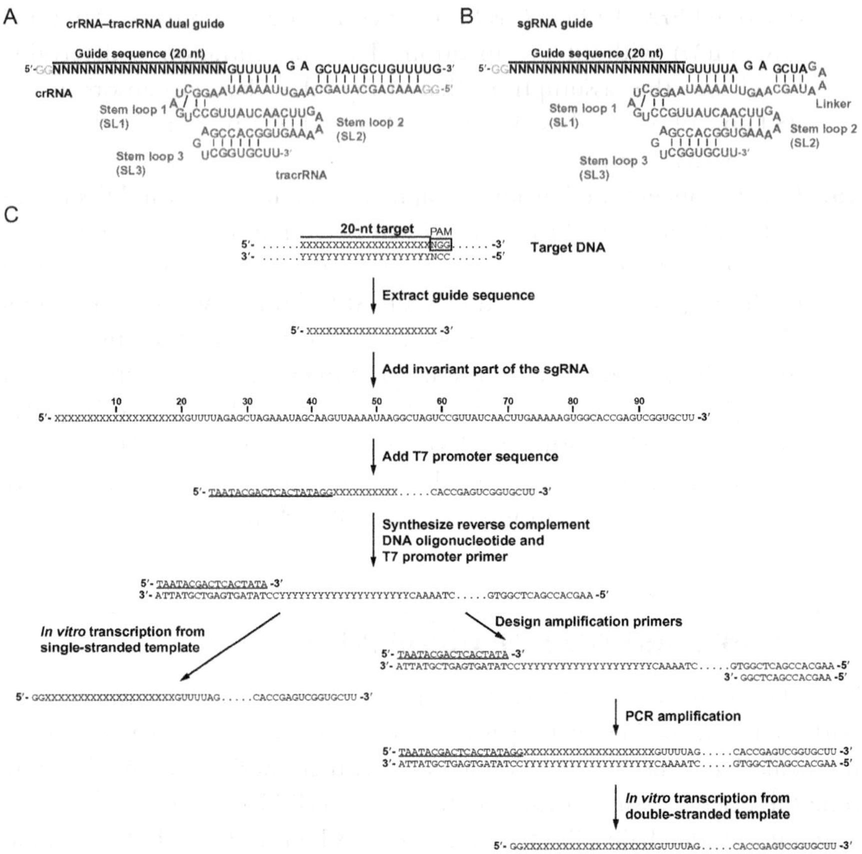


**Figure 1.1** (A) Schematic representation of the dual-RNA guide structure for programming SpyCas9. Blue: crRNA containing a 20-nt guide sequence and a 22-nt invariant sequence. The GG dinucleotide at the 5′ end is appended during *in vitro* transcription. Red: tracrRNA base-pairing with the invariant sequence of the crRNA. tracrRNA contains three stem-loops (SL1, SL2, and SL3) at its 3′ end. Although SL1 is sufficient for Cas9-mediated cleavage *in vitro* (Jinek et al., 2012), inclusion of both SL2 and SL3 increases cleavage efficiency by increasing the stability of the Cas9–crRNA–tracrRNA complex (Hsu et al., 2013; Nishimasu et al., 2014). (B) Schematic representation of chimeric single-guide RNA (sgRNA). crRNA- and tracrRNA-derived sequences are connected by a 5′-GAAA-3′ tetraloop linker. (C) Outline of the procedure for sgRNA guide design and preparation by *in vitro* transcription. (See the color plate.)

truncated sgRNAs containing just SL1 are functional *in vitro* (Jinek et al., 2012), inclusion of SL2 and SL3 increases the stability of the Cas9–crRNA–tracrRNA complex and enhances cleavage activity (Hsu et al., 2013; Nishimasu et al., 2014).

Custom 42-nt crRNAs can be obtained as synthetic oligonucleotides, whereas tracrRNA and sgRNAs need to be prepared by *in vitro* transcription using T7 RNA polymerase and subsequently purified by denaturing polyacrylamide gel electrophoresis. The RNAs are sufficiently short to be transcribed using synthetic DNA oligonucleotides as transcription templates (Milligan & Uhlenbeck, 1989), without the need to clone the transcribed sequence into a DNA plasmid (Fig. 1.1C). The RNAs are transcribed as run-off products, so that no transcriptional terminator is required. Note that an optimal T7 promoter contains two G nucleotides that are required for efficient transcription and that will be appended to the 5′ end of the transcribed RNA upstream of the 20-nt guide sequence. Addition of the 5′-terminal GG dinucleotide to the guide RNA has little effect on Cas9 loading and enzymatic activity (Jinek et al., 2012; Sternberg et al., 2014). T7 polymerase can transcribe a single-stranded DNA template but requires a double-stranded promoter region for efficient template binding (Milligan, Groebe, Witherell, & Uhlenbeck, 1987). Such partially duplexed template can be prepared by annealing a T7 promoter oligonucleotide to a synthetic oligonucleotide consisting of the antisense sequence of the desired RNA followed by the reverse complement of the T7 promoter sequence (Fig. 1.1C). While this method is generally efficient and produces hundreds of micrograms of RNA from a 1 ml transcription reaction, the RNA yield can be low in some cases. Transcription efficiency can be improved by converting the template to fully double-stranded DNA by PCR amplification (Fig. 1.1C).

The following protocol is used to prepare the sgRNA used in the endonuclease activity assays described in Section 4. The protocol describes preparation of a fully double-stranded transcription templates in step 1; to prepare partially duplexed templates by annealing synthetic oligonucleotides, start at step 2. We recommend PAGE-purification of DNA oligonucleotides prior to annealing (Lopez-Gomollon & Nicolas, 2013). We typically perform *in vitro* transcription in a 5 ml reaction to obtain ~0.5–1 mg of pure RNA, but the reaction can be scaled down accordingly. A control transcription reaction in a total volume of 100 μl should be carried out first and analyzed by denaturing PAGE before scaling up to large volumes. T7 RNA polymerase is available from a number of commercial sources. Alternatively, several published protocols describe the expression and purification of recombinant T7 RNA polymerase (Ellinger & Ehricht, 1998; He et al., 1997; Li, Wang, & Wang, 1999; Rio, 2013).

*Note*: Working with RNA requires an RNase-free environment. Wear gloves and use RNase-free plasticware and filter pipette tips. Prepare all reagents using nuclease-free or DEPC-treated water (DEPC–$H_2O$). Transcription reactions can be supplemented with RNase inhibitors if necessary.

**Day 1: Preparation of transcription template**

1. A double-stranded transcription template is prepared by amplifying a single-stranded oligonucleotide by PCR amplification of the antisense template oligonucleotide with the T7 promoter sequence as forward primer and the 3′ end of the antisense template as reverse primer (Fig. 1.1C). Mix reagents for PCR according to Table 1.1 and split into 50 μl aliquots into a 96-well plate suitable for PCR. Perform PCR cycling according to Table 1.2. After PCR, combine all reactions into 500 μl fractions in 2 ml tubes. Precipitate the PCR product by adding 50 μl 3 *M* sodium acetate, pH 5.2 and 1.45 ml ice-chilled 100% ethanol to each 500 μl fraction and incubate for 1 h at −20 °C. Centrifuge the

**Table 1.1** PCR reaction for preparing double-stranded template for *in vitro* transcription

| | Stock concentration | Final concentration | Volume (μl)[a] |
|---|---|---|---|
| DEPC–$H_2O$ | – | – | 3810 |
| Phusion buffer | 5 × | 1 × | 1000 |
| dNTP mix | 10 m*M* each | 200 μ*M* each | 100 |
| Primer forward | 100 μ*M* | 0.5 μ*M* | 25 |
| Primer reverse | 100 μ*M* | 0.5 μ*M* | 25 |
| Template | 1 μ*M* | 4 n*M* | 20 |
| Phusion polymerase | 2 U $\mu l^{-1}$ | 0.02 U $\mu l^{-1}$ | 20 |
| Total volume | | | 5000 |

[a]Master mix for a 96-well PCR plate. Aliquot 50 μl into each well. Scale down as appropriate.

**Table 1.2** PCR cycling program

| | | |
|---|---|---|
| 98 °C | 30 s | |
| 98 °C | 5 s | Repeat 34 × |
| 42 °C | 20 s | |
| 72 °C | 10 s | |
| 72 °C | 1 min | |
| 4 °C | ∞ | |

precipitated PCR product at 14,000 rpm (16,900 × *g*) for 40 min at 4 °C. Remove supernatant and wash the DNA pellet with 200 μl ice-chilled 70% ethanol. Centrifuge for 5 min, remove supernatant and air-dry the DNA pellet for 30 min. Dissolve the DNA pellet in 100 μl water. Measure the absorbance at 260 nm and calculate the concentration of the double-stranded DNA template. Extinction coefficients can be calculated using the OligoCalc server (Kibbe, 2007). Dilute the double-stranded template with 5 × transcription buffer (150 m*M* Tris–Cl, pH 8.1, 125 m*M* $MgCl_2$, 0.05% Triton X-100, 10 m*M* spermidine) and water to a final concentration of 10 μ*M* in a total volume of 1 ml. Incubate mixture at 75 °C for 5 min and allow it to cool down slowly to room temperature. Proceed with the transcription reaction described in step 3.

2. To prepare a partially single-stranded DNA template, mix 100 μl of 100 μ*M* template oligonucleotide stock with 150 μl 100 μ*M* (1.5-fold molar excess) complementary T7 promoter oligonucleotide (5′-TAATACGACTCACTATAGG-3′) and 200 μl 5 × transcription buffer. Add water to bring the total volume to 1 ml. Anneal the oligonucleotides at 75 °C for 5 min and let the mixture slowly cool down to room temperature.

**Day 2: *In vitro* transcription and gel purification**

3. *Transcription reaction*: Set up a transcription reaction by mixing the reagents according to Table 1.3. A 5 ml reaction is usually sufficient for ~0.5–1 mg of pure sgRNA. The reaction can be scaled down as needed.

**Table 1.3** *In vitro* transcription reaction

| | Stock concentration | Final concentration | Volume (ml) |
|---|---|---|---|
| DEPC–$H_2O$ | – | – | 1.95 |
| Transcription buffer | 5 × | 1 × | 1.00 |
| ATP | 100 m*M* | 5 m*M* | 0.25 |
| UTP | 100 mM | 5 m*M* | 0.25 |
| GTP | 100 m*M* | 5 m*M* | 0.25 |
| CTP | 100 m*M* | 5 m*M* | 0.25 |
| DTT | 1 *M* | 10 m*M* | 0.05 |
| Hybridized template | 10 μ*M* | 1 μ*M* | 0.50 |
| T7 RNA polymerase | 1 mg $ml^{-1}$ | 0.1 mg $ml^{-1}$ | 0.50 |
| Total volume | | | 5.00 |

Incubate the reaction at 37 °C for 1.5 h. During transcription, magnesium pyrophosphate precipitates from the reaction due to production of inorganic pyrophosphate upon NTP incorporation into nascent transcripts. To restore magnesium ion levels, add 25 μl 1 *M* $MgCl_2$ after 1.5 h and continue with incubation at 37 °C for another 1.5 h. The optimal reaction times may vary as a function of length and RNA sequence of the transcribed RNA and should be determined empirically.

4. Add 25 μl (25 U) of RNase-free RQ1 DNase (Promega) to the transcription reaction and incubate 15 min at 37 °C in order to remove the template DNA. Centrifuge the mixture for 5 min at 3900 rpm (~3200 × *g*) and 4 °C to pellet magnesium pyrophosphate and remove the supernatant containing transcribed RNA. The supernatant can be stored at −20 °C at this point.
5. Add 5 ml 2 × RNA loading dye (5% glycerol, 2.5 m*M* EDTA, pH 8.0, 90% formamide, trace bromophenol blue) to the sample. Separate the RNA by electrophoresis on a prewarmed 8% polyacrylamide, 7 *M* urea denaturing gel in 0.5 × TBE (44.5 m*M* Tris, 44.5 m*M* boric acid, 1 m*M* EDTA, pH 8.0) until the bromophenol blue dye reaches the lower quarter of the gel. We typically use gels with the dimensions of 400 mm × 360 mm × 4 mm (h × w × d) and run them at 50 W for ~4–6 h. Visualize RNA by UV-shadowing over a silica glass TLC plate and excise the band corresponding to the correct RNA product using a disposable scalpel. Crush the gel slices in a 50-ml tube using a clean spatula or a plastic serological pipette and add water to 50 ml total volume. Incubate overnight at 4 °C while rocking.

**Day 3: Gel purification—continued**

6. Centrifuge the crushed gel suspension for 5 min at 3900 rpm (~3200 × *g*) and 4 °C and collect the supernatant. Filter supernatant through a 0.22-μm disposable filter (Steriflip, Millipore). Concentrate the eluted RNA in 3000 Da MWCO centrifugal concentrator to a final volume of ~2 ml. Aliquot the concentrate into 500 μl fractions in 2-ml tubes. Add 50 μl 3 *M* sodium acetate, pH 5.2 and 1.5 ml ice-chilled 100% ethanol to each 500 μl fraction and incubate for 1 h at −20 °C. Centrifuge the precipitated PCR product at 14,000 rpm (16,900 × *g*) for 40 min at 4 °C. Remove supernatant and wash the DNA pellet with 200 μl ice-chilled 70% ethanol. Air-dry the pellet and dissolve in 100 μl water. Determine RNA concentration by measuring absorbance at 260 nm.

## 4. ENDONUCLEASE CLEAVAGE ASSAYS

Endonuclease cleavage assays can be used to characterize the activity of purified Cas9 or test the *in vitro* efficacy of a particular guide RNA or a target DNA site. In these assays, the target DNA site, including its PAM motif, is either inserted into a plasmid or provided in the form of an oligonucleotide duplex. Cleavage of plasmid substrates is monitored by agarose gel electrophoresis and staining with an intercalating dye, whereas oligonucleotide substrates typically require labeling (with a radioisotope or a fluorophore) of one or both target DNA strands. Although we previously used 5′-radiolabeling with $^{32}P$ phosphate, we recently switched to ATTO532-labeled oligonucleotides, choosing the fluorophore for its superior quantum yield and photostability. Custom 5′-ATTO532-labeled oligonucleotides are readily available from commercial sources.

In both plasmid and oligonucleotide cleavage assays, Cas9 and guide RNA are preincubated in a 1:1 molar ratio in the cleavage buffer to reconstitute the Cas9–guide RNA complex prior to the addition of target DNA. Preincubation is not strictly required for cleavage (Jinek et al., 2012). The cleavage reaction is started by the addition of DNA to the binary complex. As Cas9 is a single turnover enzyme (Sternberg et al., 2014), it is important to maintain the protein–RNA complex in excess over the DNA substrate. We recommend a molar ratio of 5:1 or higher to ensure complete cleavage. The reaction mixture is sampled at different time points and analyzed by gel electrophoresis. Since the reaction rate can strongly vary as a function of DNA source and length (oligonucleotide vs. plasmid, supercoiled circular vs. linear), optimal enzyme and substrate concentrations, and also reaction time points need to be determined empirically. In both assay types, product formation can be quantified by densitometry using a fluorescence scanner and curve fitting to extract pseudo-first-order rate constants.

For endonuclease cleavage assays shown in Fig. 1.2, SpyCas9 was programmed with chimeric sgRNA guides to cleave either linearized plasmid DNA or a short oligonucleotide duplex substrate. The sgRNA guide included stem loops SL1 and SL2 in its 3′-terminal region (Fig. 1.2A), which were sufficient for Cas9 loading and robust DNA cleavage activity. Plasmid cleavage was carried out with a pUC19-derived plasmid in which the target site (including a 5′-NGG-3′ PAM) was inserted between the EcoRI and BamHI sites (Fig. 1.2B). The oligonucleotide duplex substrate contained a 5′-ATTO532-labeled target strand containing an 8-nt linker at the 5′

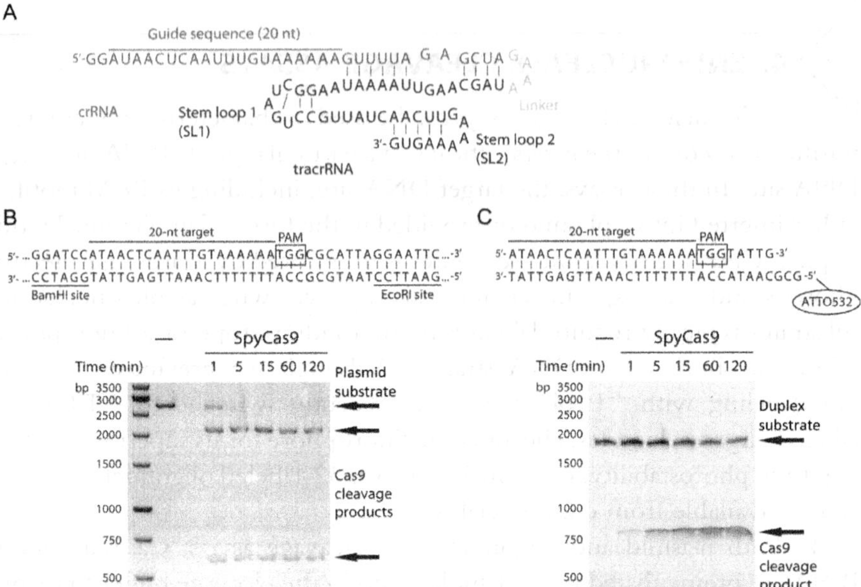


**Figure 1.2** (A) sgRNA used in endonuclease activity assays in panels B and C. The sgRNA contains stem-loops SL1 and SL2 but lacks SL3 (total length of sgRNA is 83 nt). (B) Endonuclease activity assay of SpyCas9 using SspI-linearized plasmid (2702 bp). Samples were taken at indicated time points. Cleavage products (2104 and 598 bp) were resolved on a 1% agarose gel and stained with GelRed. The sequence of the target site in the plasmid substrate is shown above the gel image. (C) Endonuclease activity assay of SpyCas9 using a double-stranded oligonucleotide target. The oligonucleotide duplex substrate is shown above the gel image. The cleavage reaction was sampled the indicated time points and analyzed by electrophoresis on a denaturing (7 *M* urea) polyacrylamide gel. ATTO532 fluorescence was detected using a FLA9500 laser scanner.

end, followed by the complementary PAM sequence (5′-CCN-3′) and the 20-nt target sequence. The target strand was annealed to a complementary unlabeled nontarget strand containing the 5′-NGG′3-PAM (Fig. 1.2C). A 250-fold excess of Cas9–guide RNA complex over DNA substrate was used. The following protocol describes the technical details of the two assays.

**Substrate preparation**

1. For plasmid substrates, linearize the circular DNA by a restriction digest. Choose a restriction enzyme that cuts at a unique site away from the Cas9 target site so that Cas9-mediate cleavage of the linear DNA produces two well-separated fragments. For pUC19-based plasmids, use 50 units of SspI-HF (New England Biolabs) and 5 μg plasmid in 1 × CutSmart™ buffer in a total volume of 50 μl. Incubate the reaction

for 1 h at 37 °C. Heat-inactivate SspI-HF by incubating for 20 min at 65 °C. Analyze plasmid cleavage by electrophoresis on a 1% agarose gel in 1 × TAE buffer (40 m*M* Tris, pH 8.0, 20 m*M* glacial acetic acid, 1 m*M* EDTA, pH 8.0) stained with GelRed (Biotium) or similar nucleic acid stain. Complete digestion of pUC19-derived vectors yields a single band of ~2700 bp.

2. For oligonucleotide duplex substrate, the 5′-ATTO532-labeled target strand oligonucleotide should be PAGE-purified. To generate the oligonucleotide duplex, anneal target and nontarget strand by mixing the oligonucleotides in a molar ratio of 1:1.5. Prepare 100 μ*M* stock solutions of the target and nontarget strand oligonucleotides. Mix 1.0 μl of the target strand with 1.5 μl of the nontarget strand and add water to a total volume of 25 μl. Heat the mixture to 75 °C for 5 min and cool slowly to room temperature (Table 1.5). Dilute the mixture with 225 μl DEPC–H2O to obtain 400 n*M* stock solution of the oligonucleotide duplex substrate.

**Cleavage assay**

3. Prepare 5 × cleavage buffer (100 m*M* HEPES, pH 7.5, 500 m*M* KCl, 25% glycerol, 5 m*M* DTT, 2.5 m*M* EDTA, pH 8.0, 10 m*M* $MgCl_2$).
4. Dilute Cas9 to 15 μ*M* with SEC Buffer (see Section 2). Dilute *in vitro* transcribed sgRNA to 15 μ*M* with water.
5. Anneal the guide RNA by heating to 90 °C for 5 min and slowly cooling to room temperature.
6. Set up reaction mixes according to Table 1.4 (plasmid DNA substrates) or Table 1.5 (oligonucleotide DNA substrates). Add equimolar

**Table 1.4** Endonuclease activity assay using linearized plasmid DNA substrate

| | Stock concentration | Final concentration | Volume (μl) |
|---|---|---|---|
| DEPC–$H_2O$ | – | – | 22.0 |
| Cleavage buffer | 5 × | 1 × | 11.0 |
| SpyCas9 | 15 μ*M* | 1.5 μ*M* | 5.5 |
| Guide RNA | 15 μ*M* | 1.5 μ*M* | 5.5 |
| Incubate for 10–15 min at room temperature. Then add the DNA. | | | |
| Plasmid DNA | 100 ng $\mu l^{-1}$ | 10 ng $\mu l^{-1}$ | 5.5 |
| Total volume | | | 55.0 |

**Table 1.5** Endonuclease activity assay using oligonucleotide duplex substrate

| | Stock | Final concentration | Volume (μl) |
|---|---|---|---|
| DEPC–$H_2O$ | – | – | 1.5 |
| Cleavage buffer | 5 × | 1 × | 21.0 |
| SpyCas9 | 15 μ*M* | 5 μ*M* | 35.0 |
| Guide RNA | 15 μ*M* | 5 μ*M* | 35.0 |
| Incubate for 10–15 min at room temperature. Then add annealed DNA duplex. | | | |
| Duplex substrate[a] | 400 n*M* | 20 n*M* | 12.5 |
| Total volume | | | 105.0 |

[a]The target and nontarget DNA are annealed prior addition to the reaction. See step 2 for the annealing procedure.

amounts of Cas9 protein and guide RNA in the reaction mix without DNA. Incubate for 10–15 min at room temperature.

7. Start the cleavage reaction by adding the DNA target to the reaction mix and incubate at 37 °C immediately.
8. Remove 10 μl aliquots at different time points and quench by adding to 1.0 μl of 500 m*M* EDTA, pH 8.0 (final concentration 50 n*M*) in a separate 1.5 ml tube. Mix by pipetting up and down and store at −80 °C until all time points are collected.
9. Thaw the samples and add 1.0 μl Proteinase K (20 mg $ml^{-1}$) to digest DNA-bound Cas9. Incubate for 20 min at room temperature. Add 6 × plasmid DNA-loading buffer (10 m*M* Tris, pH 7.6, 60 m*M* EDTA, 60% glycerol, 0.03% bromophenol blue, 0.03% xylene cyanol FF) or 2 × oligonucleotide-loading buffer (90% formamide, 10% glycerol) to each sample.
10. *Plasmid cleavage assay*: Analyze each plasmid sample on a 1% agarose gel in 1 × TAE stained with GelRed (Biotium) or compatible nucleic acid stains. For visualization with a Typhoon FLA 9500 scanner, loading 2 μl of sample was sufficient.
11. *Oligonucleotide cleavage assay*: Resolve oligonucleotide duplex samples on a 16% polyacrylamide, 7 *M* urea denaturing gel (300 mm × 180 mm; 1 mm thick) in 0.5 × TBE buffer. Load 12.5 μl of each sample and run at 25 W for ~2 h until the bromophenol blue dye front has reached the bottom half of the gel. Scan the gel using a laser gel scanner (e.g., Typhoon FLA9500, GE Healthcare) with the appropriate excitation and emission wavelength settings (532 and 553 nm for ATTO532, respectively).

**Interpretation of cleavage assays**

12. *Plasmid-based assays*: In the absence of Cas9-mediated cleavage, linearized plasmid yields a single band at 2702 bp. Cleavage by Cas9 leads to gradual emergence of two cleavage products at 2104 and 598 bp.
13. *Oligonucleotide-based assays*: Cleavage of the oligonucleotide substrate by Cas9 leads to a mobility shift of the fluorescent signal. The 5′-labeled target strand of the substrate has a length of 31 nt, while the cleaved product runs as a 14-nt band.

## 5. CONCLUDING REMARKS

Thanks to its specificity and easy programmability, Cas9 represents a revolutionary advance in genetic engineering technologies for basic research, biomedicine, and biotechnology. In this chapter, we have outlined procedures for generating recombinant Cas9 and guide RNAs and for performing endonuclease cleavage assays with the reconstituted Cas9–guide ribonucleoprotein complex. These protocols provide the experimental framework for further *in vitro* mechanistic studies of Cas9 that will facilitate ongoing development of Cas9-based genetic engineering technologies. A number of questions pertaining to the molecular mechanism of Cas9 remain unanswered, in particular the nature of the conformational rearrangements that activate the Cas9 nuclease domains prior to target cleavage. In the context of genome editing and regulation in eukaryotic cells, major outstanding questions concern the off-target activity of Cas9 and the effect of chromatin structure on Cas9 targeting and DNA cleavage. Therefore, further *in vitro* studies will be required to define the functional constraints of Cas9. Finally, although SpyCas9 has been extensively characterized, relatively little is known about the biochemical properties of other Cas9 orthologs. The protocols provided here can be readily applied to Cas9 proteins from other bacterial species or novel rationally designed Cas9 variants in order to expand the molecular toolbox for genome engineering.

## ACKNOWLEDGMENTS

We thank Ole Niewoehner and Alessia Duerst for technical assistance. This work was supported by the University of Zurich and the European Research Council (ERC) Starting Grant ANTIVIRNA (337284).

## REFERENCES

Anders, C., Niewoehner, O., Duerst, A., & Jinek, M. (2014). Structural basis of PAM-dependent target DNA recognition by the Cas9 endonuclease. *Nature, 513*, 569–573. http://dx.doi.org/10.1038/nature13579.

Barrangou, R., Fremaux, C., Deveau, H., Richards, M., Boyaval, P., Moineau, S., et al. (2007). CRISPR provides acquired resistance against viruses in prokaryotes. *Science, 315*, 1709–1712. http://dx.doi.org/10.1126/science.1138140.

Bassett, A. R., & Liu, J.-L. (2014). CRISPR/Cas9 and genome editing in Drosophila. *Journal of Genetics and Genomics, 41*, 7–19. http://dx.doi.org/10.1016/j.jgg.2013.12.004.

Chen, H., Choi, J., & Bailey, S. (2014). Cut site selection by the two nuclease domains of the Cas9 RNA-guided endonuclease. *Journal of Biological Chemistry, 289*, 13284–13294. http://dx.doi.org/10.1074/jbc.M113.539726.

Cho, S. W., Lee, J., Carroll, D., Kim, J.-S., & Lee, J. (2013). Heritable gene knockout in caenorhabditis elegans by direct injection of Cas9-sgRNA ribonucleoproteins. *Genetics, 195*, 1177–1180. http://dx.doi.org/10.1534/genetics.113.155853.

Cong, L., Ran, F. A., Cox, D., Lin, S., Barretto, R., Habib, N., et al. (2013). Multiplex genome engineering using CRISPR/Cas systems. *Science, 339*, 819–823. http://dx.doi.org/10.1126/science.1231143.

Deltcheva, E., Chylinski, K., Sharma, C. M., Gonzales, K., Chao, Y., Pirzada, Z. A., et al. (2011). CRISPR RNA maturation by trans-encoded small RNA and host factor RNase III. *Nature, 471*, 602–607. http://dx.doi.org/10.1038/nature09886.

Ellinger, T., & Ehricht, R. (1998). Single-step purification of T7 RNA polymerase with a 6-histidine tag. *BioTechniques, 24*, 718–720.

Friedland, A. E., Tzur, Y. B., Esvelt, K. M., Colaiácovo, M. P., Church, G. M., & Calarco, J. A. (2013). Heritable genome editing in C. elegans via a CRISPR-Cas9 system. *Nature Methods, 10*, 741–743. http://dx.doi.org/10.1038/nmeth.2532.

Fu, Y., Foden, J. A., Khayter, C., Maeder, M. L., Reyon, D., Joung, J. K., et al. (2013). High-frequency off-target mutagenesis induced by CrIsPr-Cas nucleases in human cells. *Nature Biotechnology, 31*, 822–826. http://dx.doi.org/10.1038/nbt.2623.

Fu, Y., Sander, J. D., Reyon, D., Cascio, V. M., & Joung, J. K. (2014). Improving CRISPR-Cas nuclease specificity using truncated guide RNAs. *Nature Biotechnology, 32*, 279–284. http://dx.doi.org/10.1038/nbt.2808.

Garneau, J. E., Dupuis, M.-È., Villion, M., Romero, D. A., Barrangou, R., Boyaval, P., et al. (2010). The CRISPR/Cas bacterial immune system cleaves bacteriophage and plasmid DNA. *Nature, 468*, 67–71. http://dx.doi.org/10.1038/nature09523.

Gasiunas, G., Barrangou, R., Horvath, P., & Siksnys, V. (2012). Cas9-crRNA ribonucleoprotein complex mediates specific DNA cleavage for adaptive immunity in bacteria. *Proceedings of the National Academy of Sciences of the United States of America, 109*, E2579–E2586. http://dx.doi.org/10.1073/pnas.1208507109.

Gilbert, L. A., Larson, M. H., Morsut, L., Liu, Z., Brar, G. A., Torres, S. E., et al. (2013). CRISPR-mediated modular RNA-guided regulation of transcription in eukaryotes. *Cell, 154*, 442–451. http://dx.doi.org/10.1016/j.cell.2013.06.044.

Gratz, S. J., Cummings, A. M., Nguyen, J. N., Hamm, D. C., Donohue, L. K., Harrison, M. M., et al. (2013). Genome engineering of drosophila with the CRISPR RNA-guided Cas9 nuclease. *Genetics, 194*, 1029–1035. http://dx.doi.org/10.1534/genetics.113.152710.

Guilinger, J. P., Thompson, D. B., & Liu, D. R. (2014). Fusion of catalytically inactive Cas9 to FokI nuclease improves the specificity of genome modification. *Nature Biotechnology, 32*, 577–582. http://dx.doi.org/10.1038/nbt.2909.

He, B., Rong, M., Lyakhov, D., Gartenstein, H., Diaz, G., Castagna, R., et al. (1997). Rapid mutagenesis and purification of phage RNA polymerases. *Protein Expression and Purification, 9*, 142–151. http://dx.doi.org/10.1006/prep.1996.0663.

Hsu, P. D., Lander, E. S., & Zhang, F. (2014). Development and applications of CRISPR-Cas9 for genome engineering. *Cell*, *157*, 1262–1278. http://dx.doi.org/10.1016/j.cell.2014.05.010.

Hsu, P. D., Scott, D. A., Weinstein, J. A., Ran, F. A., Konermann, S., Agarwala, V., et al. (2013). DNA targeting specificity of RNA-guided Cas9 nucleases. *Nature Biotechnology*, *31*(9), 827–832. http://dx.doi.org/10.1038/nbt.2647.

Hwang, W. Y., Fu, Y., Reyon, D., Maeder, M. L., Tsai, S. Q., Sander, J. D., et al. (2013). Efficient genome editing in zebrafish using a CRISPR-Cas system. *Nature Biotechnology*, *31*, 227–229. http://dx.doi.org/10.1038/nbt.2501.

Jinek, M., Chylinski, K., Fonfara, I., Hauer, M., Doudna, J. A., & Charpentier, E. (2012). A programmable dual-RNA-guided DNA endonuclease in adaptive bacterial immunity. *Science*, *337*, 816–821. http://dx.doi.org/10.1126/science.1225829.

Jinek, M., East, A., Cheng, A., Lin, S., Ma, E., & Doudna, J. (2013). RNA-programmed genome editing in human cells. *eLife*, *2*, e00471. http://dx.doi.org/10.7554/eLife.00471.

Jinek, M., Jiang, F., Taylor, D. W., Sternberg, S. H., Kaya, E., Ma, E., et al. (2014). Structures of Cas9 endonucleases reveal RNA-mediated conformational activation. *Science*, *343*, 1247997. http://dx.doi.org/10.1126/science.1247997.

Karvelis, T., Gasiunas, G., Miksys, A., Barrangou, R., Horvath, P., & Siksnys, V. (2013). crRNA and tracrRNA guide Cas9-mediated DNA interference in Streptococcus thermophilus. *RNA Biology*, *10*, 841–851.

Katic, I., & Grosshans, H. (2013). Targeted heritable mutation and gene conversion by Cas9-CRISPR in Caenorhabditis elegans. *Genetics*, *195*(3), 1173 1176. http://dx.doi.org/10.1534/genetics.113.155754.

Kibbe, W. A. (2007). OligoCalc: an online oligonucleotide properties calculator. Nucleic Acids Res, 35 (Web Server issue):W43-6. http://dx.doi.org/10.1093/nar/gkm234.

Li, J.-F., Norville, J. E., Aach, J., McCormack, M., Zhang, D., Bush, J., et al. (2013). Multiplex and homologous recombination-mediated genome editing in Arabidopsis and Nicotiana benthamiana using guide RNA and Cas9. *Nature Biotechnology*, *31*, 688–691. http://dx.doi.org/10.1038/nbt.2654.

Li, Y., Wang, E., & Wang, Y. (1999). A modified procedure for fast purification of T7 RNA polymerase. *Protein Expression and Purification*, *16*, 355–358. http://dx.doi.org/10.1006/prep.1999.1083.

Lo, T.-W., Pickle, C. S., Lin, S., Ralston, E. J., Gurling, M., Schartner, C. M., et al. (2013). Precise and heritable genome editing in evolutionarily diverse nematodes using TALENs and CRISPR/Cas9 to engineer insertions and deletions. *Genetics*, *195*, 331–348. http://dx.doi.org/10.1534/genetics.113.155382.

Lopez-Gomollon, S., & Nicolas, F. E. (2013). Purification of DNA Oligos by denaturing polyacrylamide gel electrophoresis (PAGE). *Methods in Enzymology*, *529*, 65–83. http://dx.doi.org/10.1016/B978-0-12-418687-3.00006-9.

Mali, P., Aach, J., Stranges, P. B., Esvelt, K. M., Moosburner, M., Kosuri, S., et al. (2013). CAS9 transcriptional activators for target specificity screening and paired nickases for cooperative genome engineering. *Nature Biotechnology*, *31*, 833–838. http://dx.doi.org/10.1038/nbt.2675.

Mali, P., Esvelt, K. M., & Church, G. M. (2013). Cas9 as a versatile tool for engineering biology. *Nature Methods*, *10*, 957–963. http://dx.doi.org/10.1038/nmeth.2649.

Mali, P., Yang, L., Esvelt, K. M., Aach, J., Guell, M., Dicarlo, J. E., et al. (2013). RNA-guided human genome engineering via Cas9. *Science*, *339*, 823–826. http://dx.doi.org/10.1126/science.1232033.

Milligan, J. F., Groebe, D. R., Witherell, G. W., & Uhlenbeck, O. C. (1987). Oligoribonucleotide synthesis using T7 RNA polymerase and synthetic DNA templates. *Nucleic Acids Research*, *15*, 8783–8798.

Milligan, J. F., & Uhlenbeck, O. C. (1989). Synthesis of small RNAs using T7 RNA polymerase. *Methods in Enzymology*, *180*, 51–62.

Nekrasov, V., Staskawicz, B., Weigel, D., Jones, J. D. G., & Kamoun, S. (2013). Targeted mutagenesis in the model plant Nicotiana benthamiana using Cas9 RNA-guided endonuclease. *Nature Biotechnology*, *31*, 691–693. http://dx.doi.org/10.1038/nbt.2655.

Nishimasu, H., Ran, F. A., Hsu, P. D., Konermann, S., Shehata, S. I., Dohmae, N., et al. (2014). Crystal structure of Cas9 in complex with guide RNA and target DNA. *Cell*, *156*(5), 935–949. http://dx.doi.org/10.1016/j.cell.2014.02.001.

Pattanayak, V., Lin, S., Guilinger, J. P., Ma, E., Doudna, J. A., & Liu, D. R. (2013). High-throughput profiling of off-target DNA cleavage reveals RNA-programmed Cas9 nuclease specificity. *Nature Biotechnology*, *31*, 839–843. http://dx.doi.org/10.1038/nbt.2673.

Qi, L. S., Larson, M. H., Gilbert, L. A., Doudna, J. A., Weissman, J. S., Arkin, A. P., et al. (2013). Repurposing CRISPR as an RNA-guided platform for sequence-specific control of gene expression. *Cell*, *152*, 1173–1183. http://dx.doi.org/10.1016/j.cell.2013.02.022.

Ran, F. A., Hsu, P. D., Lin, C.-Y., Gootenberg, J. S., Konermann, S., Trevino, A. E., et al. (2013). Double nicking by RNA-guided CRISPR Cas9 for enhanced genome editing specificity. *Cell*, *154*, 1380–1389. http://dx.doi.org/10.1016/j.cell.2013.08.021.

Rio, D. C. (2013). Expression and purification of active recombinant T7 RNA polymerase from E. coli. *Cold Spring Harbor Protocols*. http://dx.doi.org/10.1101/pdb.prot078527.

Sander, J. D., & Joung, J. K. (2014). CRISPR-Cas systems for editing, regulating and targeting genomes. *Nature Biotechnology*, *32*, 347–355. http://dx.doi.org/10.1038/nbt.2842.

Shan, Q., Wang, Y., Li, J., Zhang, Y., Chen, K., Liang, Z., et al. (2013). Targeted genome modification of crop plants using a CRISPR-Cas system. *Nature Biotechnology*, *31*, 686–688. http://dx.doi.org/10.1038/nbt.2650.

Sternberg, S. H., Redding, S., Jinek, M., Greene, E. C., & Doudna, J. A. (2014). DNA interrogation by the CRISPR RNA-guided endonuclease Cas9. *Nature*, *507*, 62–67. http://dx.doi.org/10.1038/nature13011.

Tsai, S. Q., Wyvekens, N., Khayter, C., Foden, J. A., Thapar, V., Reyon, D., et al. (2014). Dimeric CRISPR RNA-guided FokI nucleases for highly specific genome editing. *Nature Biotechnology*, *32*, 569–576. http://dx.doi.org/10.1038/nbt.2908.

Wang, H., Yang, H., Shivalila, C. S., Dawlaty, M. M., Cheng, A. W., Zhang, F., et al. (2013). One-step generation of mice carrying mutations in multiple genes by CRISPR/Cas-mediated genome engineering. *Cell*, *153*, 910–918. http://dx.doi.org/10.1016/j.cell.2013.04.025.

Wu, X., Scott, D. A., Kriz, A. J., Chiu, A. C., Hsu, P. D., Dadon, D. B., et al. (2014). Genome-wide binding of the CRISPR endonuclease Cas9 in mammalian cells. *Nature Biotechnology*, *32*, 670–676. http://dx.doi.org/10.1038/nbt.2889.

Xie, K., & Yang, Y. (2013). RNA-guided genome editing in plants using a CRISPR-Cas system. *Molecular Plant*, *6*, 1975–1983. http://dx.doi.org/10.1093/mp/sst119.

Yang, H., Wang, H., Shivalila, C. S., Cheng, A. W., Shi, L., & Jaenisch, R. (2013). One-step generation of mice carrying reporter and conditional alleles by CRISPR/Cas-mediated genome engineering. *Cell*, *154*, 1370–1379. http://dx.doi.org/10.1016/j.cell.2013.08.022.

CHAPTER TWO

# Targeted Genome Editing in Human Cells Using CRISPR/Cas Nucleases and Truncated Guide RNAs

**Yanfang Fu*,†, Deepak Reyon*,†,1, J. Keith Joung*,†,2**

*Molecular Pathology Unit, Center for Computational and Integrative Biology, and Center for Cancer Research, Massachusetts General Hospital, Charlestown, Massachusetts, USA

†Department of Pathology, Harvard Medical School, Boston, Massachusetts, USA

[1]Current address: Genetically Engineered Models Center, Biogen Idec, Cambridge, MA 02142, USA.

[2]Corresponding author: e-mail address: jjoung@mgh.harvard.edu

## Contents



## Abstract

CRISPR RNA-guided nucleases have recently emerged as a robust genome-editing platform that functions in a wide range of organisms. To reduce off-target effects of these nucleases, we developed and validated a modified system that uses truncated guide RNAs (tru-gRNAs). The use of tru-gRNAs leads to decreases in off-target effects and does not generally compromise the on-target efficiencies of these genome-editing nucleases. In this chapter, we describe guidelines for identifying potential tru-gRNA target sites and protocols for measuring the on-target efficiencies of CRISPR RNA-guided nucleases in human cells.

## 1. INTRODUCTION

Methods to edit genome sequence in living cells provide a powerful and versatile approach for elucidating gene function and could also potentially be useful for therapy of inherited diseases. Over the past few decades,

*Methods in Enzymology*, Volume 546
ISSN 0076-6879
http://dx.doi.org/10.1016/B978-0-12-801185-0.00002-7

successful genome modification has relied on various technologies including transposons, lentiviral vectors, and recombinases. However, each of these platforms has certain limitations. For example, transposable elements and lentiviral vectors integrate in a semi-random fashion, and recombinases are limited by their lack of programmability. In the last decade, highly efficient and programmable genome-editing nucleases such as zinc finger nucleases (ZFNs), transcription activator-like effector nucleases (TALENs), and clustered regularly interspaced short palindromic repeat (CRISPR) RNA-guided nucleases have rapidly emerged and been shown to work in a wide range of model organisms. These customizable nucleases mediate genome editing by introducing a double-stranded break (DSB) in a target DNA sequence, which in turn can lead to the efficient generation of insertions or deletion mutations (indels) by nonhomologous end-joining repair. Alternatively, in the presence of an appropriately designed, homologous donor DNA template (which can be either single- or double-stranded), precise alterations can be created by homology-directed repair of the DSB.

ZFNs and TALENs are each composed of a customizable DNA-binding domain fused to the nonspecific cleavage domain of the *Fok*I endonuclease. Both types of nucleases have been used successfully to modify genome sequences in a large number of different cell types and organisms (Joung & Sander, 2013; Urnov, Rebar, Holmes, Zhang, & Gregory, 2010). Engineered zinc finger arrays with novel DNA-binding specificities can be challenging to construct if one accounts for the context-dependent activities of individual zinc finger domains within an array (Wolfe, Nekludova, & Pabo, 2000). By contrast, the activities of individual transcription activator-like effector (TALE) repeat domains are quite modular in their activities (Reyon, Tsai, et al., 2012). As a result, a very high percentage of TALE repeat arrays can bind to their intended target sites in human and other cell types (Reyon, Tsai, et al., 2012). Although TALEN-encoding constructs can be very rapidly assembled, the highly repetitive nature of TALE repeat arrays has required the use of nonstandard molecular biology methods to speed up the process of assembling DNA constructs encoding these proteins (Joung & Sander, 2013). In addition, the highly repetitive nature of these TALE repeat-encoding sequences has led to challenges in packaging them into certain virus-based delivery systems (Holkers et al., 2012).

CRISPR RNA (crRNA)-guided nucleases provide a simpler genome-editing alternative to ZFNs and TALENs. The initial version of this platform was based on components derived from the *Streptococcus pyogenes* type

II CRISPR immune system (Jinek et al., 2012), which forms an adaptive system responsible for silencing of invading plasmids and viral DNA in many bacteria (Wiedenheft, Sternberg, & Doudna, 2012). The double-stranded DNA cleavage activity of Cas9 can be programmed by a RNA duplex of crRNAs and trans-activating crRNAs (tracrRNAs) to cleave 20 bp target sites that lie next to a protospacer adjacent motif (PAM) sequence of the form NGG. Charpentier, Doudna, and colleagues first showed that a chimeric "guide RNA" (sgRNA) consisting of parts of the crRNA and tracrRNA can also direct Cas9 to cleave specific target DNA sites by altering the first 20 nts of this chimeric sgRNA (Jinek et al., 2012) (Fig. 2.1A). The results of this study opened the door to use of the CRISPR/Cas9 system as a programmable genome-editing tool with initial studies showing its use in bacteria (Jiang, Bikard, Cox, Zhang, & Marraffini, 2013), zebrafish (Hwang et al., 2013), and human cells (Cho, Kim, Kim, & Kim, 2013; Cong et al., 2013; Jinek et al., 2013; Mali, Yang, et al., 2013). Subsequently, a large number of studies have shown the successful use of the CRISPR/Cas9 system for genome editing in a variety of organisms (Sander & Joung, 2014).

A number of groups have studied the specificities of crRNA-guided nucleases and demonstrated that off-target effects can be observed in plants, zebrafish, mouse, rat, and cultured human cells. A study from our group first showed that RNA-guided Cas9 could induce high-frequency off-target mutations in human cells (Fu et al., 2013). In this work, we screened ~60 computationally identified candidate off-target sites for six sgRNAs targeted to four endogenous genes using a T7 Endonuclease I (T7EI) genotyping assay that can detect indels at frequencies of 2–5% or higher. Surprisingly, we found that it was relatively easy to identify off-target sites for four of the six sgRNAs and that the rates of mutagenesis observed at these off-target sites were comparable to (or, in some cases, higher than) those observed at the on-target site. Importantly, some of the off-target sites we identified differed from the off-target sites by as many as five mismatches and many of these off-target mutations could be identified in three different human cell lines. At least four subsequent studies have observed similar findings in human cells (Cradick, Fine, Antico, & Bao, 2013; Hsu et al., 2013; Pattanayak et al., 2013), in rice (Xie & Yang, 2013), and in model organisms such as zebrafish (Auer, Duroure, De Cian, Concordet, & Del Bene, 2014; Jao, Wente, & Chen, 2013), mouse (Yang et al., 2013), and rat (Ma, Shen, et al., 2014; Ma, Zhang, et al., 2014). The results of various studies that have identified off-target mutations are summarized in Table 2.1.

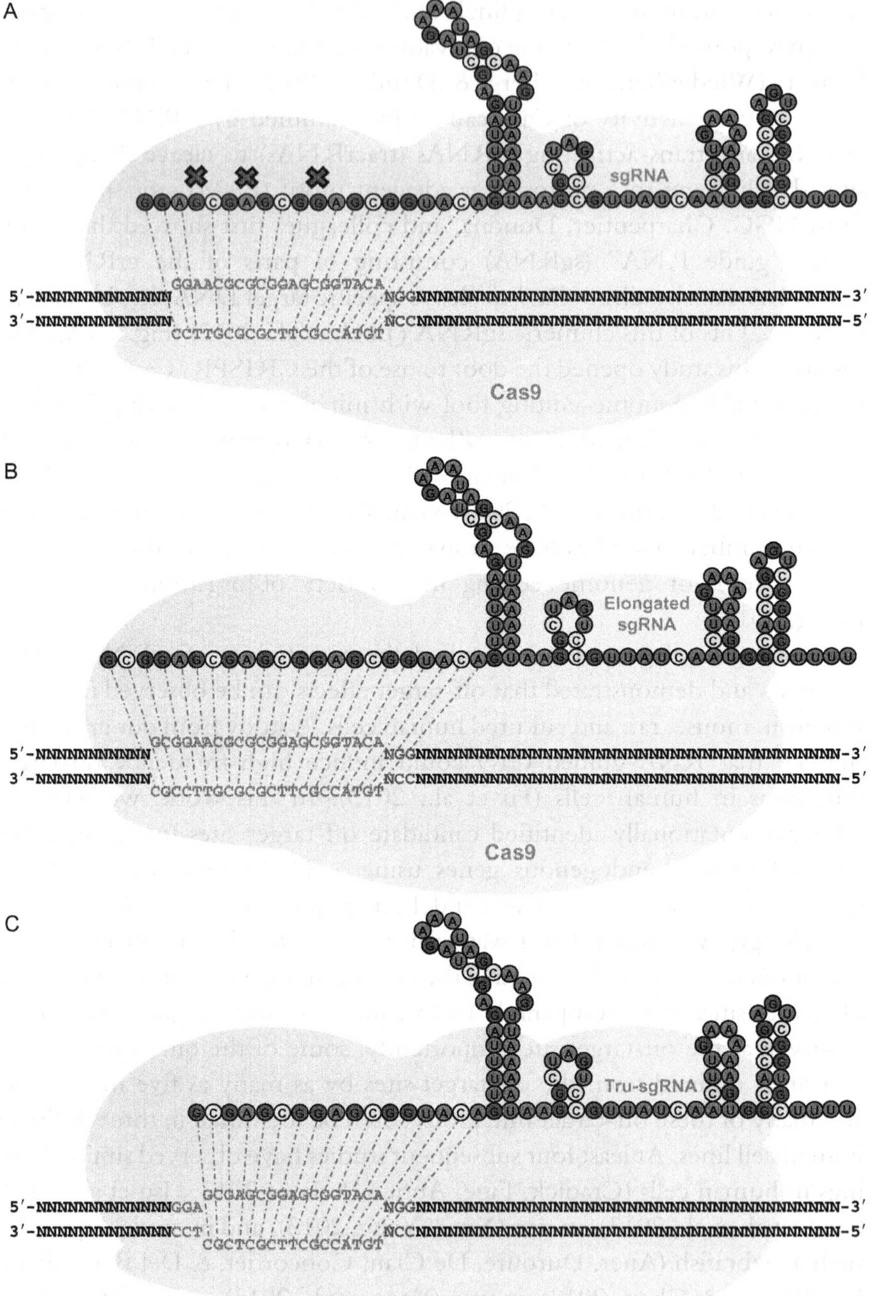
A
sgRNA
GGAACGCGCGGAGCGGTACA
CCTTGCGCGCTTCGCCATGT
Cas9
B
Elongated sgRNA
GCGGAACGCGCGGAGCGGTACA
CGCCTTGCGCGCTTCGCCATGT
Cas9
C
Tru-sgRNA
GCGAGCGGAGCGGTACA
CGCTCGCTTCGCCATGT
Cas9

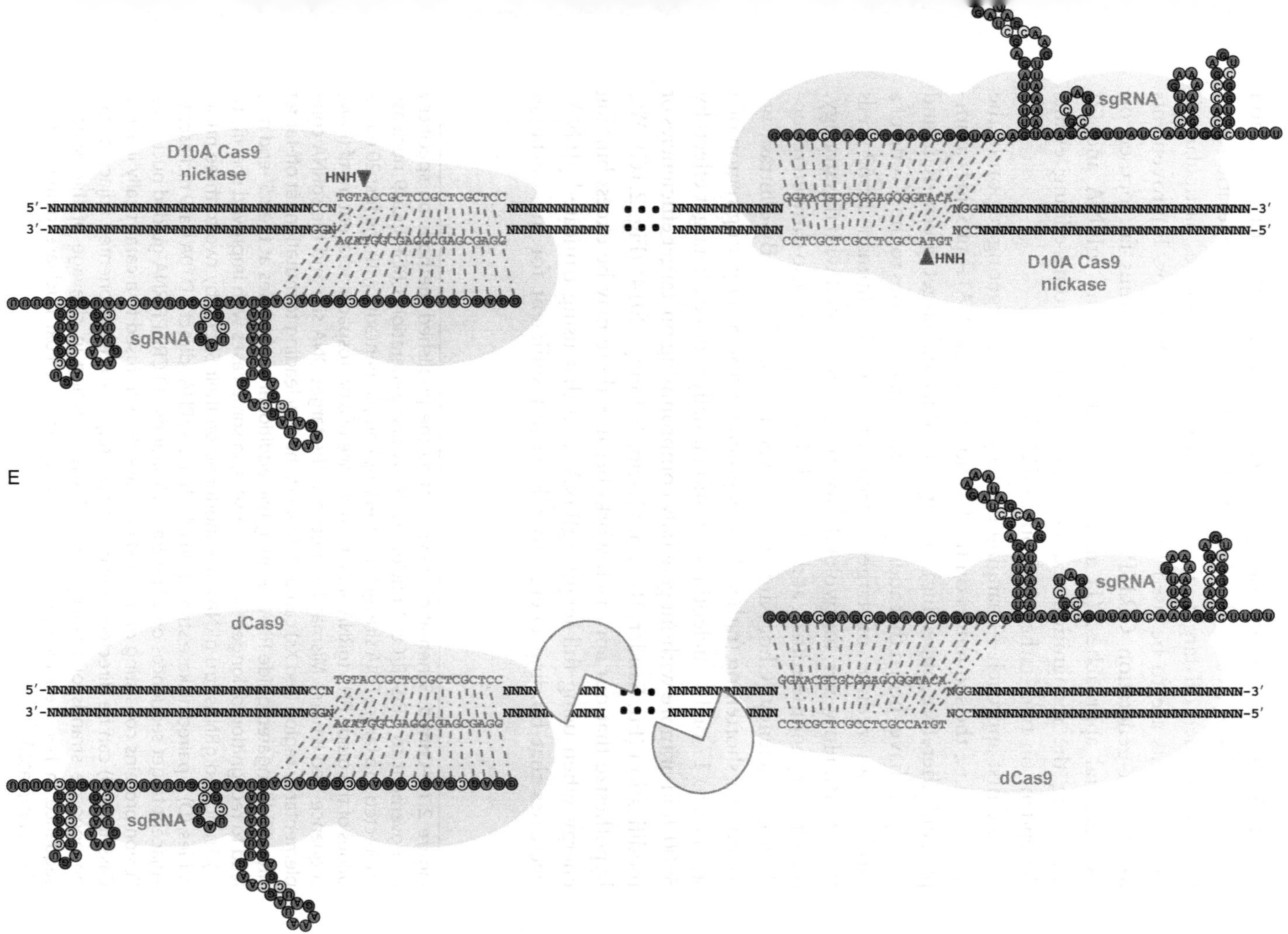

E

**Figure 2.1** (See figure legend on next page.)

Given the limitations of specificity observed with first-generation CRISPR/Cas nucleases, the development of newer, next-generation platforms to reduce off-target effects is of utmost importance, particularly if these reagents are to be used for therapeutic applications. Improvements to the first-generation CRISPR/Cas platform have thus far focused on two general approaches: (1) altering the length of the sgRNA and (2) increasing the recognition sequence by making genome-editing events dependent on two, rather than one, sgRNAs.

The first approach—changing the length of the targeting region of the sgRNA (i.e., the sequences on the 5′-end of the sgRNA)—provides a simple (and, therefore, appealing) strategy to reduce off-target effects. Kim and colleagues have shown that adding two additional guanines to the 5′-end of a sgRNA can reduce off-target effects of Cas9 in human K562 cells (Fig. 2.1B). However, for two of the four target sites tested with this strategy, the rates of genome editing at the intended on-target site also showed reductions in efficiency (Cho, Kim, Kim, Kweon, et al., 2013). By contrast, we found that shortening (rather than lengthening) the 5′-targeting region of a sgRNA to 17 or 18 nucleotides can substantially reduce off-target effects by 5000-fold or more without generally compromising on-target efficiencies of modification (Fu, Sander, Reyon, Cascio, & Joung, 2014) (Fig. 2.1C). We hypothesize that our approach works because there may be excess binding energy when using full-length sgRNAs and that using truncated gRNAs decreases that binding energy to a level just sufficient for full on-target

**Figure 2.1—Cont'd** Schematic overview of all the published methods on specificity improvement to CRISPR/Cas9 nucleases. (A) A first-generation CRISPR/Cas9 nuclease is directed to a target DNA site (green letters) by complementarity with the first 20 nucleotides of the guide RNA (gRNA) and the presence of a protospacer adjacent motif (PAM) sequence (red letters). Mismatches between the target DNA site and the sgRNA complementarity region (red Xs) can be tolerated, thereby leading to mutations at off-target sites. (B) Elongated guide RNAs bearing two additional nucleotides at their 5′-end may lead to recognition of longer target DNA sites, in some cases with improved specificity. (C) Truncated gRNAs (tru-gRNAs) are shortened on their 5′-end by two to three nucleotides. (D) A paired nickase strategy in which two sgRNAs direct a D10A Cas9 nickases to adjacent target sequences on opposite DNA strands. (E) CRISPR RNA-guided nucleases. Fusion proteins consisting of the *FokI* nuclease domain fused to a catalytically inactive Cas9 (dCas9) can be directed to two appropriately spaced and oriented adjacent sites on opposite strands of DNA by two sgRNAs, resulting in cleavage of the "spacer" sequence in between by the dimerization-dependent *FokI* nuclease domains. (See the color plate.)

**Table 2.1** Published examples of off-target mutations induced by RGNs in human cells and model organisms

| Reference | Target locus | Sequences | Indel mutation frequencies (%) | | | % GC content of on-target sequences | Number of mismatches | Cell type or organism | Detection methods used |
|---|---|---|---|---|---|---|---|---|---|
| | | | U2OS.EGFP | K562 | HEK293 | | | | |
| Fu et al. (2013) | *VEGFA* site 1 | GGGTGGGGGGAGTTTGCTCCtGG | 26 | 10.5 | 3.3 | 70 | 0 | Human cells | T7EI and Sanger sequencing |
| | | GG**A**TGG**A**GGGAGTTTGCTCCtGG | 25.7 | 18.9 | 2.9 | | 2 | | |
| | | GGG**A**GGG**T**GGAGTTTGCTCCtGG | 9.2 | 8.3 | N.D. | | 2 | | |
| | | **C**GG**G**GG**A**GGGAGTTTGCTCCtGG | 5.3 | 3.7 | N.D. | | 3 | | |
| | | GGG**GA**GGGG**A**AGTTTGCTCCtGG | 17.1 | 8.5 | N.D. | | 3 | | |
| | *VEGFA* site 2 | GACCCCCTCCACCCCGCCTCcGG | 50.2 | 38.6 | 15 | 80 | 0 | | |
| | | GACCCCC**C**CCACCCCGCC**C**CcGG | 14.4 | 33.6 | 4.1 | | 2 | | |
| | | G**GG**CCCCTCCACCCCGCCTCtGG | 20 | 15.6 | 3 | | 2 | | |
| | | **CTA**CCCCTCCACCCCGCCTCcGG | 8.2 | 15 | 5.2 | | 3 | | |
| | | G**C**CCCC**AC**CCACCCCGCCTCtGG | 50.7 | 30.7 | 7.1 | | 3 | | |
| | | **T**ACCCCC**CA**CACCCCGCCTCtGG | 9.7 | 6.97 | 1.3 | | 3 | | |
| | | **ACA**CCCC**C**CCACCCCGCCTCaGG | 14 | 12.3 | 1.8 | | 4 | | |
| | | **ATT**CCCC**C**CCACCCCGCCTCaGG | 17 | 19.4 | N.D. | | 4 | | |
| | | **CC**CC**A**CC**C**CCACCCCGCCTCaGG | 6.1 | N.D. | N.D. | | 4 | | |
| | | **CG**CCC**T**C**C**CCACCCCGCCTCcGG | 44.4 | 28.7 | 4.2 | | 4 | | |
| | | **CT**CCCC**AC**CCACCCCGCCTCaGG | 62.8 | 29.8 | 21.1 | | 4 | | |
| | | **TG**CCCC**TC**CCACCCCGCCTCtGG | 13.8 | N.D. | N.D. | | 4 | | |
| | | **AGG**CCCC**CA**CACCCCGCCTCaGG | 2.8 | N.D. | N.D. | | 5 | | |

*Continued*

**Table 2.1** Published examples of off-target mutations induced by RGNs in human cells and model organisms—cont'd

| Reference | Target locus | Sequences | Indel mutation frequencies (%) | | | % GC content of on-target sequences | Number of mismatches | Cell type or organism | Detection methods used |
|---|---|---|---|---|---|---|---|---|---|
| | | | U2OS.EGFP | K562 | HEK293 | | | | |
| | *VEGFA* site 3 | GGTGAGTGAGTGTGTGCGTGtGG | 49.4 | 35.7 | 28 | 60 | 0 | | |
| | | GGTGAGTGAGTGTGTG**T**GTGaGG | 7.4 | 8.97 | N.D. | | 1 | | |
| | | **A**GTGAGTGAGTGTGTG**T**GTGgGG | 24.3 | 23.9 | 8.9 | | 2 | | |
| | | G**C**TGAGTGAGTGT**A**TGCGTGtGG | 20.9 | 11.2 | N.D. | | 2 | | |
| | | GGTGAGTGAGTG**C**GTGCG**G**GtGG | 3.2 | 2.34 | N.D. | | 2 | | |
| | | G**T**TGAGTGA**A**TGTGTGCGTGaGG | 2.9 | 1.27 | N.D. | | 2 | | |
| | | **T**GTG**G**GTGAGTGTGTGCGTGaGG | 13.4 | 12.1 | 2.4 | | 2 | | |
| | | **A**G**A**GAGTGAGTGTGTGC**A**TGaGG | 16.7 | 7.64 | 1.2 | | 3 | | |
| | *EMX1* | GAGTCCGAGCAGAAGAAGAAgGG | 42.1 | 26 | 10.7 | 50 | 0 | | |
| | | GAGT**TA**GAGCAGAAGAAGAAaGG | 16.8 | 8.43 | 2.5 | | 2 | | |
| Hsu et al. (2013) | *EMX1* target 1 | GTCACCTCCAATGACTAGGGtGG | ~22 | | | 55 | 0 | Human 293FT cells | Deep sequencing |
| | | G**CT**ACCTCCA**G**TGACTAGGGtGG | ~4 | | | | 3 | | |
| | *EMX1* target 3 | GAGTCCGAGCAGAAGAAGAAgGG | ~48 | | | 50 | 0 | | |
| | | GAG**G**CCGAGCAGAAGAA**AG**AgGG | ~1 | | | | 3 | | |
| | | GAGTCC**T**AGCAGA**G**GAAGAAgAG | ~8 | | | | 2 | | |
| | | GAGTC**TA**AGCAGAAGAAGAAgAG | ~18 | | | | 2 | | |

| | | | | | | | |
|---|---|---|---|---|---|---|---|
| Pattanayak et al. (2013) | *CLTA* | GCAGATGTAGTGTTTCCACAgGG | 76 | 45 | 0 | Human HEK293T cells | Deep sequencing |
| | | **A**CA**T**ATGTAGT**A**TTTCCACAgGG | 24 | | 3 | | |
| | | **C**CAGATGTAGT**A**TT**C**CCACAgGG | 0.46 | | 3 | | |
| | | **CT**AGATG**A**AGTG**C**TTCCACAtGG | 0.73 | | 4 | | |
| Cradick et al. (2013) | *HBB* R01 | GTGAACGTGGATGAAGTTGGtGG | 54 | 50 | 0 | Human HEK-293T cells | T7EI |
| | | GTGAACGTGGATG**C**AGTTGGtGG | 27 | | 1 | | |
| | *HBB* R02 | **C**TTGCCCCACAGGGCAGTAAcGG | 66 | | 1 | | |
| | | **TCA**GCCCCACAGGGCAGTAAcGG | 33 | | 3 | | |
| | *HBB* R03 | **C**ACGTTCACCTTGCCCCACAgGG | 55 | | 1 | | |
| | | **C**ACGTTCAC**T**TTGCCCCACAgGG | 58 | | 2 | | |
| | *HBB* R04 | **C**CACGTTCACCTTGCCCCACaGG | 53 | | 1 | | |
| | | **C**CACGTTCACtTTGCCCCACaGG | 12 | | 1 | | |
| | *HBB* R05 | **A**GTCTGCCGTTACTGCCCTGtGG | 51 | | 1 | | |
| | *HBB* R06 | **C**GTTACTGCCCTGTGGGGCAaGG | 59 | | 1 | | |
| | *HBB* R07 | **A**AGGTGAACGTGGATGAAGTtGG | 61 | | 1 | | |
| | | **A**AGGTGAACGTGGATG**C**AGTtGG | 7 | | 2 | | |
| | *HBB* R08 | **C**CTGTGGGGCAAGGTGAACGtGG | 36 | | 1 | | |
| | | **C**CTGTGGGGCAA**A**GTGAACGtGG | 48 | | 2 | | |
| | *HBB* R30 | GTAGAGCGGAGGCAGGAGGCgGG | 21 | 70 | 0 | | |
| | | GTAGAGCGGAGGCAGGAG**TT**gGG | 5 | | 2 | | |

*Continued*

**Table 2.1** Published examples of off-target mutations induced by RGNs in human cells and model organisms—cont'd

| Reference | Target locus | Sequences | Indel mutation frequencies (%) | | | % GC content of on-target sequences | Number of mismatches | Cell type or organism | Detection methods used |
|---|---|---|---|---|---|---|---|---|---|
| | | | U2OS·EGFP | K562 | HEK293 | | | | |
| Wang, Wei, Sabatini, & Lander (2014) | *AAVS1* | GGGGCCACTAGGGACAGGATtGG | 96.9 | | | 65 | 0 | Mouse Cas9-KBM7 cells | Deep sequencing |
| | | GGGGC**TT**CTA**A**GGACAGGATtGG | 29.5 | | | | 3 | | |
| | | GGGGC**A**ACTAG**A**GACAGGA**A**tGG | 2.46 | | | | 3 | | |
| | | GGGGCC**C**CT**G**GGGACAG**A**ATtGG | 1.36 | | | | 3 | | |
| | | GG**T**GCCAC**C**AGGGA**G**AGGATtGG | 0.1 | | | | 3 | | |
| Shalem et al. (2014) | *MED12*-sg1 | GTTGTGCTCAGTACTGACTTtGG | 100 | | | 45 | 0 | A375 cells | Deep sequencing |
| | | **A**T**CC**TGCTC**T**GTACTGACTTgAG | ~3 | | | | 4 | | |
| | *NF2*-sg2 | ATTCCACGGGAAGGAGATCTtGG | 100 | | | 50 | 0 | | |
| | | **G**TT**G**CAC**A**G**A**AAGGAGATCTtGG | ~15 | | | | 4 | | |
| | *NF2*-sg4 | GTACTGCAGTCCAAAGAACCaGG | 100 | | | 50 | 0 | | |
| | | **TA**ACT**A**CAGTCCAAAGAACCaGG | ~50 | | | | 3 | | |
| | *MED12*-sg2 | CGTCAGCTTCAATCCTGCCAaGG | 100 | | | 55 | 0 | | |
| | | **A**GTCAGCTTCA**G**TCCTGCCAcGG | ~30 | | | | 2 | | |
| | | **CAG**CAGCTTCAATCCTGCCAgGG | ~95 | | | | 3 | | |
| | | **CAG**CAGCTTCAATCCTGCCAgGG | ~50 | | | | 3 | | |
| Xie and Yang (2013) | *Tet1* | GTCTACATCGCCACGGAGCTCAtGG | 8.2 | | | 55 | 0 | Rice | Sanger sequencing |
| | | GTCTA-A**C**CGC-ACGGAGCTCAtGG | 1.6 | | | | 3 | | |

| | | | | | | | |
|---|---|---|---|---|---|---|---|
| Auer et al. (2014) | *GFP* | GGCGAGGGCGATGCCACCTAcGG | 66 | 70 | 0 | Zebrafish | T7EI and Sanger sequencing |
| | | GG**T**GAGGGC**A**ATGC**A**A**TA**TAcGG | <3% but T7E1 detectable | | 5 | | |
| | | GGC**C**AGGGCGA**G**G**G**CACC**GC**cGG | <3% but T7E1 detectable | | 5 | | |
| Jao et al. (2013) | *GFP* | GGGCACGGGCAGCTTGCCGGtGG | ~80 | 80 | 0 | Zebrafish | T7EI |
| | | GGGCA**T**GG**A**CAGCTTGCCGGtGG | ~80 | | 2 | | |
| Yang et al. (2013) | *Nanog* | CGTAAGTCTCATATTTCACCtGG | Unknown | 40 | 0 | Mouse | Sanger sequencing |
| | | **T**GTAAGTCTCATATTTCACCtGG | 10 | | 1 | | |
| | *Oct4* | GCTCAGTGATGCTGTTGATCaGG | Unknown | 50 | 0 | | |
| | | G**T**TCAGTGATGCTGTTGATCaGG | 67 | | 1 | | |
| | *Mecp2* | AGGAGTGAGGTCTAGTACTTGGG | Unknown | 45 | 0 | | |
| | | **T**GGAGTGAGGTCT**T**GTACTTGGG | 10 | | 1 | | |
| Ma, Shen, et al. (2014) | *Dnmt1* | GGCGAGGGGCGGACCGATGcGG | Unknown | 80 | 0 | Rat | T7EI and Sanger sequencing |
| | | GG**G**GAGGGGC**A**GGACC**A**ATGcGG | 58.3 | | 3 | | |
| | *Dnmt3b* | GGTAGCTGGGGCACATGGTGaGG | Unknown | 65 | 0 | | |
| | | GGTA**C**CTGGGGCACATGGTGtGG | 30 | | 1 | | |
| Ma, Zhang, et al. (2014) | *Prkdc* | CGAGCTGTTCAGAAACACCAaGG | 100 | 50 | 0 | Rat | T7EI |
| | | **AA**AGCTGT**A**CAGAAACACCAaGG | 57.5 | | 3 | | |

activity but poised to be more sensitive to mismatches at the sgRNA/target DNA site interface.

The second approach of making the genome-editing activities of CRISPR/Cas systems dependent on pairs of sgRNAs targeted adjacent sequences has been implemented in two different ways. With the paired Cas9 nickase approach, two sgRNAs localize a Cas9 nuclease variant that nicks DNA to opposite strands of DNA at a target site of interest (Fig. 2.1D) (Mali, Aach, et al., 2013). This approach has been shown to reduce off-target effects associated with single sgRNAs (Ran et al., 2013). However, one potential limitation of this approach is that it is not a truly dimeric system: the two Cas9 nickase molecules recruited to target sites are enzymatically active as monomers and therefore capable of inducing mutations at monomeric binding sites elsewhere in the genome. An alternative approach is to create fusions of the dimerization-dependent *Fok*I nuclease domain (used in ZFNs and TALENs) to catalytically inactive versions of Cas9 (so-called "dead Cas9" or dCas9) that can still be recruited to specific target sites by associated sgRNAs. In this configuration, two *Fok*I–dCas9 fusion proteins are recruited to adjacent sites by two sgRNAs with resulting cleavage by the *Fok*I domains in the sequence between the two sgRNA target sites (Fig. 2.1E). *Fok*I–dCas9 fusions function robustly in human cells for genome editing, and direct comparisons show that these proteins are generally less active for mutagenic activities as monomers than Cas9 nickases (Guilinger, Thompson, & Liu, 2014; Tsai et al., 2014).

A significant advantage of the truncated gRNA platform is that it provides a simple strategy for improving the specificities of CRISPR/Cas nucleases that does not require the expression of multiple sgRNAs or larger fusion proteins. Here we describe how to identify potential target sites for Cas9 directed by truncated gRNAs. We also detail how to introduce these components into cultured human cells and how to quantify their activities using a simple T7EI-based genotyping assay.

# 2. METHODS

## 2.1. Identification of target sites using ZiFiT

The latest version of our publicly available Web-based server ZiFiT Targeter has been upgraded to include functionality that enables users to pick target sites for both standard full-length and tru-gRNAs. Users can query ZiFiT Targeter with either in single-sequence or in batch mode (up to 96 sequences in FASTA format). When used in batch mode, ZiFiT Targeter will identify one target site per query sequence; whereas when used in single-sequence mode, it will identify all potential target sites within the query sequence. ZiFiT Targeter will first attempt to identify target sites that span a user-specified bracketed single nucleotide of interest, but if no sites are identified, the brackets are ignored, and the search is attempted again. ZiFiT Targeter allows users to change several design constraints including length of the target site and choice of promoter used to express the potential sgRNAs identified (for example, use of a U6 promoter requires a 5′-G on the sgRNA whereas use of a T7 RNA polymerase promoter requires two Gs at the 5′-end of the sgRNA).

ZiFiT Targeter returns the list of target sites along with additional information regarding oligonucleotides required for construction of the sgRNA expression vectors. Users can save these outputs into a comma-separated value (CSV) file for subsequent use. In addition to the simple identification of target sites, ZiFiT Targeter can also identify potential off-target sites present in genomes of interest. Users can query ZiFiT Targeter to determine the orthogonality of a given target site against the genomes of several model organisms including, human, rat, mouse, zebrafish, *C. elegans*, mosquito, and *E. coli*.

**Required materials**

1. ZiFiT available at http://zifit.partners.org.
2. Repeat Masker available at http://www.repeatmasker.org
3. Genomic sequence of the locus to be targeted
4. Computer with Internet connection

**Ensure query sequence is valid**

1. Ensure the query sequence entered is the genomic sequence and not cDNA sequence. This distinction is important because target sites that span intron–exon junctions will not work on genomic DNA. Users are also encouraged to sequence the region of interest in the actual cells to be modified to ensure that there are no any unexpected polymorphisms.
2. Use Repeat Masker (http://www.repeatmasker.org) to determine if the query sequence contains any repeat elements. Load the Website and input your sequence of interest into the text box labeled "sequence" and click "submit." Repeat Masker will scan the sequence and change all repeat sequences to "N"s. Download this masked sequence and use it as the input for ZiFiT Targeter. This is a critically important step because the user should not target highly repetitive regions of the genome.

**Design target sites**

1. Using any Web browser, open ZiFiT Targeter by entering the following URL: http://ZiFiT.partners.org.
2. Click on "ZiFiT" on the menu along the top of the Web page, this will direct users to a disclaimer. Clicking on "Proceed to ZiFiT" will open the main menu. Note that ZiFiT Targeter also provides functionality to design target sites for ZFNs and TALENs as previously described (Reyon, Khayter, Regan, Joung, & Sander, 2012; Reyon et al., 2013; Sander, Maeder, & Joung, 2011).
3. From the main menu, click the appropriate link under "Design Genome Editing Nucleases/Nickases": *For Tru-gRNAs: click "CRISPR/Cas Nucleases."*
4. Paste up to 96 sequences in FASTA format in the text box.

   *Note*: While querying in batch mode, only one target site will be returned per sequence.

   *Note*: If a nucleotide of interest is marked by surrounding it with brackets, ZiFiT Targeter will attempt to identify target sites that span it.

   *Note*: All characters other than A, C, G, T, and N will be ignored.
5. Set the promoter type.

   **5.1.** Choose the U6 promoter if expression will be from a plasmid-based vector using this promoter.

   *Note*: The U6 promoter will include a guanine at the 5′-end of the transcribed RNA, so the target sites returned by ZiFiT Targeter will all start with a G.

   **5.2.** Choose the T7 promoter if the sgRNA will be transcribed *in vitro* using this promoter.

   *Note*: The T7 promoter will include two guanines at the 5′-end of the sgRNA being transcribed, so ZiFiT Targeter will identify sites that start with a pair G's. RNA yield after in vitro transcription will be lower if a single G is present and extremely low in the absence of any G at the 5′-end.

   **5.3.** Choose the "None" option if you would like to relax the 5′ "G" constraint.

   *Note*: If no appropriate standard tru-gRNA site can be identified in your target gene of interest, there are two potential solutions:

   1. Relax the 5′-G constraint in your target sites but then use a G at the 5′-position of your sgRNA (this will cause a mismatch at this position). If using tru-gRNAs, we have found that this

strategy will work for 18 nt target sites but not with 17 nt target sites.

2. Relax the 5′-G constraint in your target sites and use a non-G nucleotide at the 5′-position. Although the U6 promoter has optimal activity with a 5′-G, we have found that use of a non-G at the 5′-end can in some cases still lead to functional tru-gRNAs (data not shown).
3. Relax the 5′-G constraint in your target sites and express your sgRNAs using a recently described multiplex expression system that does not require a G to be present at the 5′-end of sgRNAs (Tsai et al., 2014).

6. Set the desired length of the sgRNA complementarity region to 17 or 18 nt.

   *Note*: If there is no available 17 nt tru-gRNA site in the DNA sequence of interest, then we suggest the use of 18 nt sites.

7. Click "Identify target sites" and ZiFiT Targeter will return the list of target sites identified in a table that can be saved as a CSVs file.

   The table contains four columns: (i) sequence name, (ii) target site, (iii) first oligonucleotide required for cloning, and (iv) second oligonucleotide required for cloning.

8. Identify potential off-target sites based on similarity to the on-target site. Enable the option to check the orthogonality of on-target sites relative to a genome of choice.

   8.1. Attest that the query sequence has been repeat masked.

   *Note*: This is extremely important because ZiFiT Targeter will return all potential off-target sites that have up to three mismatches relative to the on-target site. If any of the identified on-target sites contain repeat elements, typically a 1000 or more sites will be returned causing the system to slow down or crash.

   8.2. Choose the organism against which the orthogonality will be checked. Users can choose from several model organisms including, including, human, rat, mouse, zebrafish, *C. elegans*, mosquito, and *E. coli*.

   8.3. Click "Identify potential off-targets." ZiFiT Targeter will scan all the previously identified target sites against the selected genome and return potential off-target sites in a table that users can save in a CSV format.

9. Analyze orthogonality data. For each potential, off-target site identified the following information is displayed:

**I.** A unique identifier
**II.** Chromosomal location
**III.** DNA strand
**IV.** Genomic coordinate
**V.** Number of times that a potential off-target site occurs within the genome of interest
**VI.** Position and nature of the mismatches (up to three mismatches)

*Note*: For sgRNAs with 17 nt complementarity regions, potential off-target sites with only up to two mismatches are identified because our previous work has shown that a single mismatch tends to disrupt the activity of these tru-gRNAs.

*Note*: Mismatches are labeled from 5′ to 3′, with the first nt of the target site is designated position 'N' (where N = length of the complementarity region in the sgRNA) and the last nt of the target site, which is closest to PAM is designated as position "1."

*Note*: In considering which sgRNAs to use for experiments, one needs to consider multiple factors including the observation that off-target sites bearing more mismatches are less likely to show off-target effects, that not all potential off-target sites will actually show evidence of mutation, and the concept that off-target sites in coding sequences are more likely to be problematic than those that fall within introns. At present there are no perfect guidelines for choosing sgRNAs so each user must ultimately make their own choices.

## 2.2. Construction of tru-gRNA expression plasmids

Construction of a plasmid that expresses a specific sgRNA from a U6 promoter is simple and rapid. It involves the cloning of a pair of annealed oligonucleotides (sequences provided by the ZiFiT Targeter program) into *Bsm*BI-digested plasmid pMLM3636 (Fig. 2.2).

### *2.2.1 Reagents*

Plasmid pMLM3636 (Addgene: http://www.addgene.org/43860/)
Oligonucleotides encoding the sgRNA complementarity region (order from any standard vendor, sequences of oligonucleotides provided by ZiFiT Targeter as described earlier)
*Bsm*BI restriction enzyme (New England Biolabs, cat. no. R0580S)
T4 DNA ligase (New England Biolabs, cat. no. M0202S)
10× Restriction enzyme buffer 3.1 (New England Biolabs)
2× Quick Ligase Buffer (New England Biolabs, cat. no. M2200S)

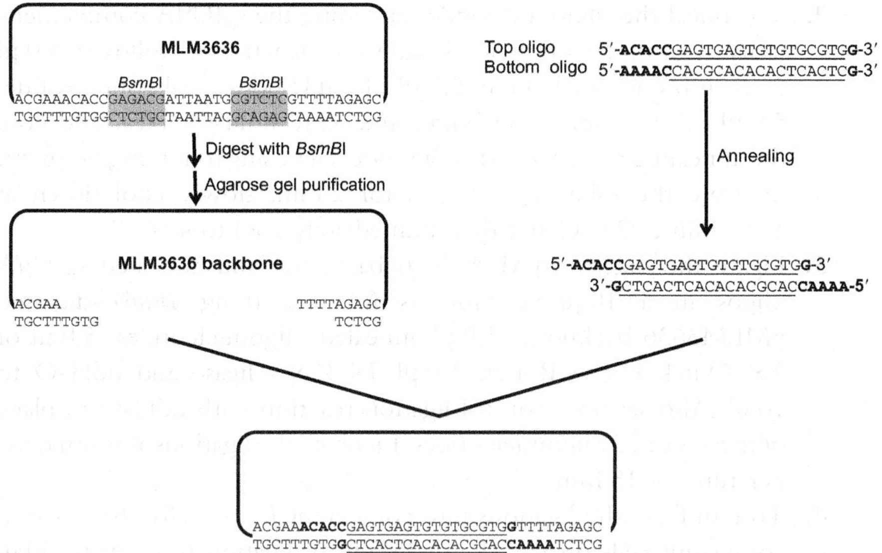


**Figure 2.2** Schematic overview of tru-gRNA expression vector construction. The left panel shows how to digest pMLM3636 with *Bsm*BI restriction enzyme to generate a linearized pMLM3636 vector backbone. The right panel shows annealing of two oligos. Ligation of pMLM3636 backbone and annealed oligos results in the desired tru-gRNA Expression vector. "ACACC" is part of U6 promoter and "GTTT" is the beginning of tracrRNA sequence.

LB medium (Difco, cat. no. 244620)
LB agar medium (Difco, cat. no. 244520)
LB/Carb plates (LB agar supplemented with 100 μg/ml carbenicillin)
QIAquick Gel Extraction Kit (Qiagen, cat. no. 28704)
QIAprep Spin Miniprep Kit (Qiagen, cat. no. 27106)
Chemically competent *recA-E. coli* cells (e.g., Top10 cells or equivalent)
10 × Annealing buffer: 0.4 *M* Tris (pH 8), 0.2 *M* $MgCl_2$, 0.5 *M* NaCl, 0.01 *M* EDTA (pH 8).

### *2.2.2 Protocol*

**Day 1**

1. To digest pMLM3636, set up the following 50 μl reaction in a microcentrifuge tube: 1.0 μg pMLM3636 plasmid, 5.0 μl NEB buffer 3.1, 5.0 μl *Bsm*BI, and $ddH_2O$ up to 50 μl. Digest at 55 °C for 2 h, run the digested products on a 2% agarose gel, and then purify the digested vector backbone from the gel using a QIAquick Gel Extraction Kit according to the manufacturer's instructions.

2. To anneal the oligonucleotides encoding the sgRNA complementarity region, set up a 50 μl annealing reaction as follows: 5.0 μl 1 μ*M* top oligonucleotide, 5.0 μl of 1 μ*M* bottom oligonucleotide, 5.0 μl of 10 × annealing buffer and 35 μl dd$H_2O$. Heat and cool the annealing reaction with a thermocycler using following program: incubate the mixture at 95 °C for 2 min, slowly cool down at 1 °C/min to 25 °C and then immediately cool to 4 °C.
3. Ligate the digested pMLM3636 backbone and annealed sgRNA oligos in a 10 μl reaction as follows: 10 ng *Bsm*BI-digested pMLM3636 backbone, 1.0 μl annealed oligonucleotides, 5.0 μl of 2 × Quick Ligase Buffer, 1.0 μl T4 DNA ligase and dd$H_2O$ to 10 μl. Also set up a control ligation reaction with dd$H_2O$ in place of the annealed oligonucleotides. Incubate the ligations at room temperature for 15 min.
4. To transform the ligations into competent *E. coli* cells: thaw frozen competent cells on ice, then add 5 μl of ligation reaction to 50 μl chemically competent Top10 cells, incubate on ice for 5–10 min, heat shock at 42 °C for 1 min, then return to ice for 2 min, add 350 μl LB medium and recover the cells at 37 °C for 1 h. Plate half the volume of each transformation on a LB/Carb plate. Incubate the plates for 12–16 h at 37 °C.

**Day 2**

Ensure that there are at least 10-fold more colonies on the actual ligation transformation plate than the control transformation plate. Pick to two to four colonies for each construct and inoculate each into 5 ml of LB/Carb medium. Incubate with shaking overnight at 37 °C.

**Day 3**

Isolate plasmids DNA using a QIAprep Spin Miniprep Kit following the manufacturer's instructions and verify the sequences of the inserted oligonucleotides by Sanger sequencing using the following primer: 5′-AGGGAATAAGGGCGACACGGAAAT-3′.

## 2.3. Transfection of sgRNA and Cas9 expression plasmids into human cells

Here, we describe transfection of plasmids encoding sgRNA and Cas9 into human U2OS cells using Nucleofection. Performing similar experiments in other cell lines may require optimization and the use of different methods of transfection.

### 2.3.1 Reagents

*U2OS cell culture medium*: advanced DMEM (Life Technologies, cat. no. 12491–023) supplemented with 10% FBS (Life Technologies, cat. no. 16140–071), 2 m*M* GlutaMax (Life Technologies, cat. no. 35050–061), penicillin/streptomycin (Life Technologies, 15070–063), and 400 μg/ml G418 (Life Technologies, cat. no. 35050–061)

*U2OS cell transfection medium*: advanced DMEM (Life Technologies, cat. no. 12491–023) supplemented with 10% FBS (Life Technologies, 16140–071), 2 m*M* GlutaMax (Life Technologies, cat. no. 35050–061)

SE Cell Line 4D-Nucleofector® X Kit S (Lonza, cat. no. V4XC-1032)

4D-Nucleofector machine (Lonza)

Sterile 50-ml conical tube (Corning, cat. no. 430290)

Agencourt DNAdvance kit (Beckman, cat. no. A48705)

### 2.3.2 Protocol

#### 2.3.2.1 Prior to Day 1

Maintain human U2OS cells in U2OS cell culture medium, passaging cells every 2–3 days, never allowing the cells reach a confluence of >90%. Perform testing for mycoplasma contamination every 4 weeks.

**Day 1**

1. Add 0.5 ml U2OS·EGFP cell transfection medium to each well of 24-well plate and prewarm the plate at 37 °C incubator for ~1 h.
2. Trypsinize U2OS·EGFP cells from a confluent plate and resuspend the cells in cell transfection medium.
3. Count the number of cells using a hemocytometer. 20,000 cells will be used for each 4D nucleofection. Calculate the cell density and then spin down appropriate number of cells in a sterile 50-ml conical tube.
4. Resuspend the cells in SE solution with supplementary reagent at a density of 20,000 cells/20 μl. Co-transfect 250 ng sequence-verified sgRNA expression plasmid, 750 ng of a Cas9 expression plasmid (e.g., plasmid pJDS246; Fu et al., 2014), and 5 ng Td-tomato plasmids (used as a transfection control) into 20,000 cells using the program DN100.

   *Note*: It is important to perform a transfection optimization experiment whenever working on a new cell type using Cell Line Optimization Nucleofector™ Kit for the 4D-Nucleofector™ System (Lonza, cat. no. V4XC-9064). Most of the cell lines we have tested show a better transfection efficiency and lower toxicity using an optimized protocol rather than the protocol recommended by the manufacturer.

5. Allow the transfected cells to sit at room temperature for 10–15 min, then add 100 μl cell transfection medium into each sample and transfer into a single well of a 24-well plate.

   *Note*: The 10–15 min incubation is critical for 4D Nucleofection. Also, using transfection medium after Nucleofection without antibiotics is much less toxic than using cell culture medium.

**Day 2**

Check cell viability and Td-tomato expression with a fluorescent microscope to make sure all the samples have been successfully transfected (as judged by Td-tomato fluorescence) and that the cells do not show gross evidence of toxicity.

**Day 3**

Isolate genomic DNA from transfected cells using the Agencourt DNAdvance kit following manufacturer's instructions.

## 2.4. Quantitative T7EI assays to assess frequencies of targeted genome editing

The T7EI assay is a simple and reproducible assay that can be routinely used to quantify mutation frequencies in a population of cells. In this assay, target loci are amplified from genomic DNA. The resulting amplicons are then denatured and then re-annealed allowing the formation of heteroduplexes between mutant and wild-type alleles. Heteroduplex fragments are then specifically cleaved at the site of mismatch by T7EI enzyme. Digested and undigested PCR fragments can then be analyzed using either a capillary electrophoresis system (e.g., the QIAxcel) or gel-based electrophoresis. The relative amounts of digested and undigested PCR fragments can then be used to calculate the original frequency of mutated alleles from the cell population (Fig. 2.3).

### *2.4.1 Reagents*

Phusion® High-Fidelity DNA Polymerase (New England Biolabs, cat. no. M0530L)

5 × Phusion HF buffer (New England Biolabs, included with Phusion® High-Fidelity DNA Polymerase)

T7 Endonuclease I (New England Biolabs, cat. no. M0302L)

10 × NEB buffer 2.1 (New England Biolabs, included with T7 Endonuclease I)

***VEGFA* Ctrl**

Expected amplicon size: 488bp

Actual amplicon size: 486bp

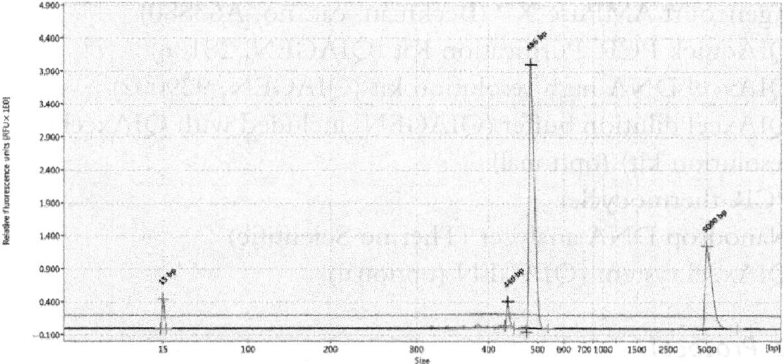


***VEGFA* sample**

Expected cleavage peak: 184bp and 304bp

Actual cleavage peak: 191bp and 304bp

Fraction cleaved: 70.09%

Gene modification frequency: 45.3%

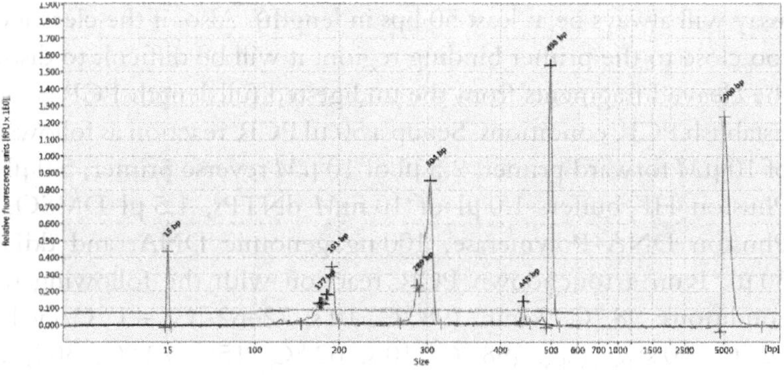

**Overlay traces of *VEGFA* Ctrl and VEGFA sample**

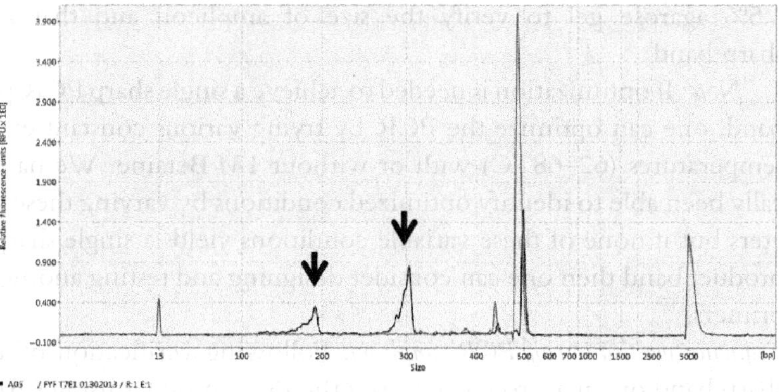


**Figure 2.3** Representative capillary electrophoresis traces from a T7EI experiment. Top panel: electrophoretic trace from negative control cells. Middle panel: trace from cells transfected with gRNA and Cas9 expression plasmids. Bottom panel: overlay of traces from the top (blue-colored line; dark gray in print version) and middle panels (red-colored line; light gray in print version). T7EI cleavage fragments of the expected sizes are indicated with red-colored arrows (dark gray in print version).

Agencourt AMPure XP (Beckman, cat. no. A63880)
QIAquick PCR Purification Kit (QIAGEN, 28106)
QIAxcel DNA high resolution kit (QIAGEN, 929002)
QIAxcel dilution buffer (QIAGEN, included with QIAxcel DNA high resolution kit) (optional)
PCR thermocycler
Nanodrop DNA analyzer (Thermo Scientific)
QIAxcel system (QIAGEN (optional)

### *2.4.2 Protocol*

**Before Day 1**

1. Design primers to amplify nuclease target site of interest. Always design primers that are located at least 50 bps away from the 5′ and 3′-ends of the sgRNA target site (this ensures that digestion products of the T7EI assay will always be at least 50 bps in length). Also, if the cleavage site is too close to the primer binding region, it will be difficult to distinguish the cleaved fragments from the undigested full-length PCR product.
2. Establish PCR conditions. Set up a 50 μl PCR reaction as follows: 2.5 μl of 10 μ*M* forward primer, 2.5 μl of 10 μ*M* reverse primer, 5.0 μl of 5 × Phusion HF buffer, 1.0 μl of 10 m*M* dNTPs, 1.5 μl DMSO, 0.5 μl Phusion DNA Polymerase, 100 ng genomic DNA, and $ddH_2O$ to 50 μl. Run a touchdown PCR reaction with the following reaction conditions: 98 °C, 3 min, (98 °C, 10 s; 72–62 °C, −1 °C/cycle, 15 s; 72 °C, 30 s)$_{10\ \text{cycles}}$, (98 °C, 10 s; 62 °C, 15 s; 72 °C, 30 s)$_{25\ \text{cycles}}$, 62 °C, 5 min, 4 °C, and hold. Run 5.0 μl of this PCR product on 1.5% agarose gel to verify the size of amplicon and that a single sharp band.

   *Note*: If optimization is needed to achieve a single sharp PCR product band, one can optimize the PCR by trying various constant extension temperatures (62–68 °C) with or without 1*M* Betaine. We have typically been able to identify optimized conditions by varying these parameters but if none of these variable conditions yields a single sharp PCR product band then one can consider designing and testing another set of primers.
3. *Sequence verification of PCR amplicon*: Following verification of a single sharp band on an agarose gel, purify the PCR product using QIAquick PCR Purification Kit and then sequence-verify the product using either the forward primer or reverse primer.

   *Note*: If a polymorphism exists in the target amplicon, it will be interfere with the T7E1 assay.

**Day 1**

1. Set up PCR reactions with genomic DNA from cells in which nucleases have been expressed using the conditions established above. Also set up the following control PCR reactions: (1) a reaction containing genomic DNA from unmodified cells and (2) a negative control lacking any genomic DNA. Run 5 μl of PCR products on 1.5% agarose gel to verify successful amplification.
2. Purify PCR products using the Agencourt AMPure XP kit following manufacturer's instructions.
3. Quantify the purified PCR products using a Nanodrop or other equivalent DNA analyzer.
4. Denaturation and re-annealing of PCR products. Set up a 19-μl denaturation/re-annealing reaction as follows: 200 ng purified PCR product, 2.0 μl of 10 × NEB buffer 2.1, add $ddH_2O$ up to 19 μl. Denature and re-anneal the purified PCR product to form heteroduplexes in a thermocycler using the program: 95 °C, 5 min; 95–85 °C at −2 °C/s; 85–25 °C at −0.1 °C/s; hold at 4 °C.
5. *T7EI reaction*: Following denaturation/re-annealing, briefly spin down reactions and add 1 μl T7E1 enzyme to make a 20 μl reaction, mix by pipetting up and down several times, incubate at 37 °C for 15 min, and then add 0.25 μ*M* EDTA to stop the T7E1 digestion. Purify the digested PCR product using Agencourt AMPure XP.
6. Quantify T7E1 digested product on a QIAxcel machine as follows: Mix 10 μl QIAxcel dilution buffer with 10 μl of purified T7EI-digested fragment and run with program OM500.
7. *Data analysis*: To estimate gene modification frequency, analyze data using QIAxcel BioCalculator Software following the vendor's instructions. Check for the following two parameters: (a) verify that the sizes of the T7EI-digested products are correct and present only in the experimental sample and not in the negative control and (b) quantify the relative amounts of the digested and the undigested PCR fragments. Mutation frequencies can then be calculated using the following formula as previously described (Guschin et al., 2010):

$$\%\ \text{gene modification} = 100 \times (1 - (1 - \text{fraction cleaved})1/2)$$

*Note*: T7EI reactions can also be analyzed by agarose gel electrophoresis or polyacrylamide gel electrophoresis. Digital gel images can be generated using a gel documentation station and the cleavage products can be analyzed using image quantification software.

## CONFLICT OF INTEREST

JKJ is a consultant for Horizon Discovery. JKJ has financial interests in Editas Medicine and Transposagen Biopharmaceuticals. JKJ's interests were reviewed and are managed by Massachusetts General Hospital and Partners HealthCare in accordance with their conflict of interest policies.

## REFERENCES

Auer, T. O., Duroure, K., De Cian, A., Concordet, J. P., & Del Bene, F. (2014). Highly efficient CRISPR/Cas9-mediated knock-in in zebrafish by homology-independent DNA repair. *Genome Research*, *24*(1), 142–153.

Cho, S. W., Kim, S., Kim, J. M., & Kim, J. S. (2013). Targeted genome engineering in human cells with the Cas9 RNA-guided endonuclease. *Nature Biotechnology*, *31*(3), 230–232.

Cho, S. W., Kim, S., Kim, Y., Kweon, J., Kim, H. S., Bae, S., et al. (2013). Analysis of off-target effects of CRISPR/Cas-derived RNA-guided endonucleases and nickases. *Genome Research*, *24*(1), 132–141.

Cong, L., Ran, F. A., Cox, D., Lin, S., Barretto, R., Habib, N., et al. (2013). Multiplex genome engineering using CRISPR/Cas systems. *Science*, *339*(6121), 819–823.

Cradick, T. J., Fine, E. J., Antico, C. J., & Bao, G. (2013). CRISPR/Cas9 systems targeting beta-globin and CCR5 genes have substantial off-target activity. *Nucleic Acids Research*, *41*(20), 9584–9592.

Fu, Y., Foden, J. A., Khayter, C., Maeder, M. L., Reyon, D., Joung, J. K., et al. (2013). High-frequency off-target mutagenesis induced by CRISPR-Cas nucleases in human cells. *Nature Biotechnology*, *31*(9), 822–826.

Fu, Y., Sander, J. D., Reyon, D., Cascio, V. M., & Joung, J. K. (2014). Improving CRISPR-Cas nuclease specificity using truncated guide RNAs. *Nature Biotechnology*, *32*(3), 279–284.

Guilinger, J. P., Thompson, D. B., & Liu, D. R. (2014). Fusion of catalytically inactive Cas9 to FokI nuclease improves the specificity of genome modification. *Nature Biotechnology*, *32*(6), 577–582.

Guschin, D. Y., Waite, A. J., Katibah, G. E., Miller, J. C., Holmes, M. C., & Rebar, E. J. (2010). A rapid and general assay for monitoring endogenous gene modification. *Methods in Molecular Biology*, *649*, 247–256.

Holkers, M., Maggio, I., Liu, J., Janssen, J. M., Miselli, F., Mussolino, C., et al. (2012). Differential integrity of TALE nuclease genes following adenoviral and lentiviral vector gene transfer into human cells. *Nucleic Acids Research*, *41*(5), e63.

Hsu, P. D., Scott, D. A., Weinstein, J. A., Ran, F. A., Konermann, S., Agarwala, V., et al. (2013). DNA targeting specificity of RNA-guided Cas9 nucleases. *Nature Biotechnology*, *31*(9), 827–832.

Hwang, W. Y., Fu, Y., Reyon, D., Maeder, M. L., Tsai, S. Q., Sander, J. D., et al. (2013). Efficient genome editing in zebrafish using a CRISPR-Cas system. *Nature Biotechnology*, *31*(3), 227–229.

Jao, L. E., Wente, S. R., & Chen, W. (2013). Efficient multiplex biallelic zebrafish genome editing using a CRISPR nuclease system. *Proceedings of the National Academy of Sciences of the United States of America*, *110*(34), 13904–13909.

Jiang, W., Bikard, D., Cox, D., Zhang, F., & Marraffini, L. A. (2013). RNA-guided editing of bacterial genomes using CRISPR-Cas systems. *Nature Biotechnology*, *31*(3), 233–239.

Jinek, M., Chylinski, K., Fonfara, I., Hauer, M., Doudna, J. A., & Charpentier, E. (2012). A programmable dual-RNA-guided DNA endonuclease in adaptive bacterial immunity. *Science*, *337*(6096), 816–821.

Jinek, M., East, A., Cheng, A., Lin, S., Ma, E., & Doudna, J. (2013). RNA-programmed genome editing in human cells. *Elife*, *2*, e00471.

Joung, J. K., & Sander, J. D. (2013). TALENs: A widely applicable technology for targeted genome editing. *Nature Reviews Molecular Cell Biology*, *14*(1), 49–55.

Ma, Y., Shen, B., Zhang, X., Lu, Y., Chen, W., Ma, J., et al. (2014). Heritable multiplex genetic engineering in rats using CRISPR/Cas9. *PLoS One*, *9*(3), e89413.

Ma, Y., Zhang, X., Shen, B., Lu, Y., Chen, W., Ma, J., et al. (2014). Generating rats with conditional alleles using CRISPR/Cas9. *Cell Research*, *24*(1), 122–125.

Mali, P., Aach, J., Stranges, P. B., Esvelt, K. M., Moosburner, M., Kosuri, S., et al. (2013). CAS9 transcriptional activators for target specificity screening and paired nickases for cooperative genome engineering. *Nature Biotechnology*, *31*(9), 833–838.

Mali, P., Yang, L., Esvelt, K. M., Aach, J., Guell, M., DiCarlo, J. E., et al. (2013). RNA-guided human genome engineering via Cas9. *Science*, *339*(6121), 823–826.

Pattanayak, V., Lin, S., Guilinger, J. P., Ma, E., Doudna, J. A., & Liu, D. R. (2013). High-throughput profiling of off-target DNA cleavage reveals RNA-programmed Cas9 nuclease specificity. *Nature Biotechnology*, *31*(9), 839–843.

Ran, F. A., Hsu, P. D., Lin, C. Y., Gootenberg, J. S., Konermann, S., Trevino, A. E., et al. (2013). Double nicking by RNA-guided CRISPR Cas9 for enhanced genome editing specificity. *Cell*, *154*(6), 1380–1389.

Reyon, D., Khayter, C., Regan, M. R., Joung, J. K., & Sander, J. D. (2012). Engineering designer transcription activator-like effector nucleases (TALENs) by REAL or REAL-fast assembly. *Current Protocols in Molecular Biology*, (Chapter 12, Unit 12.15).

Reyon, D., Maeder, M. L., Khayter, C., Tsai, S. Q., Foley, J. E., Sander, J. D., et al. (2013). Engineering customized TALE nucleases (TALENs) and TALE transcription factors by fast ligation-based automatable solid-phase high-throughput (FLASH) assembly. *Current Protocols in Molecular Biology*, (Chapter 12, Unit 12.16).

Reyon, D., Tsai, S. Q., Khayter, C., Foden, J. A., Sander, J. D., & Joung, J. K. (2012). FLASH assembly of TALENs for high-throughput genome editing. *Nature Biotechnology*, *30*(5), 460–465. http://dx.doi.org/10.1038/nbt.2170.

Sander, J. D., & Joung, J. K. (2014). CRISPR-Cas systems for editing, regulating and targeting genomes. *Nature Biotechnology*, *32*(4), 347–355.

Sander, J. D., Maeder, M. L., & Joung, J. K. (2011). Engineering designer nucleases with customized cleavage specificities. *Current Protocols in Molecular Biology*, (Chapter 12, Unit 12.13).

Shalem, O., Sanjana, N. E., Hartenian, E., Shi, X., Scott, D. A., Mikkelsen, T. S., et al. (2014). Genome-scale CRISPR-Cas9 knockout screening in human cells. *Science*, *343*(6166), 84–87.

Tsai, S. Q., Wyvekens, N., Khayter, C., Foden, J. A., Thapar, V., Reyon, D., et al. (2014). Dimeric CRISPR RNA-guided FokI nucleases for highly specific genome editing. *Nature Biotechnology*, *32*(6), 569–576.

Urnov, F. D., Rebar, E. J., Holmes, M. C., Zhang, H. S., & Gregory, P. D. (2010). Genome editing with engineered zinc finger nucleases. *Nature Reviews Genetics*, *11*(9), 636–646.

Wang, T., Wei, J. J., Sabatini, D. M., & Lander, E. S. (2014). Genetic screens in human cells using the CRISPR-Cas9 system. *Science*, *343*(6166), 80–84.

Wiedenheft, B., Sternberg, S. H., & Doudna, J. A. (2012). RNA-guided genetic silencing systems in bacteria and archaea. *Nature*, *482*(7385), 331–338.

Wolfe, S. A., Nekludova, L., & Pabo, C. O. (2000). DNA recognition by Cys2His2 zinc finger proteins. *Annual Review of Biophysics and Biomolecular Structure*, *29*, 183–212.

Xie, K., & Yang, Y. (2013). RNA-guided genome editing in plants using a CRISPR-Cas system. *Molecular Plant*, *6*(6), 1975–1983.

Yang, H., Wang, H., Shivalila, C. S., Cheng, A. W., Shi, L., & Jaenisch, R. (2013). One-step generation of mice carrying reporter and conditional alleles by CRISPR/Cas-mediated genome engineering. *Cell*, *154*(6), 1370–1379.

CHAPTER THREE

# Determining the Specificities of TALENs, Cas9, and Other Genome-Editing Enzymes

**Vikram Pattanayak**[*,1], **John P. Guilinger**[†,‡,1], **David R. Liu**[†,‡,2]

[*]Department of Pathology, Massachusetts General Hospital, Boston, Massachusetts, USA
[†]Department of Chemistry & Chemical Biology, Harvard University, Cambridge, Massachusetts, USA
[‡]Howard Hughes Medical Institute, Harvard University, Cambridge, Massachusetts, USA
[1]These authors contributed equally to this work.
[2]Corresponding author: e-mail address: drliu@fas.harvard.edu

## Contents



## Abstract

The rapid development of programmable site-specific endonucleases has led to a dramatic increase in genome engineering activities for research and therapeutic purposes. Specific loci of interest in the genomes of a wide range of organisms including mammals can now be modified using zinc-finger nucleases, transcription activator-like effectornucleases, and CRISPR-associated Cas9 endonucleases in a site-specific manner, in some cases requiring relatively modest effort for endonuclease design, construction, and application. While these technologies have made genome engineering widely accessible, the ability of programmable nucleases to cleave off-target sequences can limit their applicability and raise concerns about therapeutic safety. In this chapter, we review methods to evaluate and improve the DNA cleavage activity of programmable site-specific endonucleases and describe a procedure for a comprehensive off-target profiling method based on the *in vitro* selection of very large ($\sim10^{12}$-membered) libraries of potential nuclease substrates.

*Methods in Enzymology*, Volume 546
ISSN 0076-6879
http://dx.doi.org/10.1016/B978-0-12-801185-0.00003-9

# 1. INTRODUCTION

## 1.1. Introduction to programmable nucleases for genome editing

Programmable site-specific nucleases such as zinc-finger nucleases (ZFNs), transcription activator-like effectornucleases (TALENs), and CRISPR-associated Cas9 nucleases can be designed to target any gene of interest and therefore are powerful research tools with significant therapeutic implications. In cells, a targeted double-strand break can lead to gene modification or insertion through homology-directed repair (HDR) with exogenous DNA or to gene knockout via nonhomologous end-joining (NHEJ). In the HDR pathway, the creation of a double-strand break at a chromosomal DNA locus by a sequence-specific endonuclease can increase the efficiency of insertion of an exogenous donor DNA template by several orders of magnitude (Choulika, Perrin, Dujon, & Nicolas, 1995). If no donor template is provided, endogenous NHEJ pathways that repair the break will often introduce missense mutations that abrogate production of functional protein product (Lukacsovich, Yang, & Waldman, 1994; Rouet, Smih, & Jasin, 1994). Programmable nucleases have been used to modify the genomes of a variety of organisms and human cell lines, as has been reviewed extensively (Carroll, 2011; Joung & Sander, 2013; Sander & Joung, 2014). In addition to engineering the genomes of cells or organisms for direct biological interrogation, genetic screens have recently been performed with these enzymes in human tissue culture to uncover genetic factors underlying specific cellular processes in an unbiased manner (Koike-Yusa, Li, Tan, Velasco-Herrera Mdel, & Yusa, 2014; Shalem et al., 2014; Wang, Wei, Sabatini, & Lander, 2014; Zhou et al., 2014).

These nucleases also serve as the promising basis of a new generation of human therapeutics. Clinical trials of two site-specific nucleases are currently underway as potential treatments for HIV and glioblastoma. Researchers have completed and are conducting clinical trials using a ZFN, developed by Sangamo BioSciences, that targets a sequence in the *CCR5* gene (Tebas et al., 2014). CCR5 is a co-receptor used by HIV in early-stage infection (Scarlatti et al., 1997), and mutation of CCR5 (*CCR5Δ32*) is known to confer resistance to HIV infection (Huang et al., 1996; Liu et al., 1996; Samson et al., 1996).

The second ZFN in clinical trials, also developed by Sangamo BioSciences, disrupts the gene for the glucocorticoid receptor (Reik et al.,

2008) as part of a potential treatment for glioblastoma. The target cells for the ZFN are T cells modified by other methods to express a cell-surface receptor that specifically recognizes malignant glioblastoma cells (Kahlon et al., 2004). The therapeutic cells, however, are rendered inactive by glucocorticoids, which are often also a component of therapy. ZFN-mediated modification of the glucocorticoid receptor in the therapeutic cells confers resistance to glucocorticoid treatment, while maintaining anti-glioblastoma activity, allowing the cells to recognize their malignant targets. These and other examples demonstrate that, in addition to serving as powerful research tools, programmable nucleases are promising platforms for clinically relevant genetic manipulation.

## 1.2. Overview of methods to study specificity of genome-editing agents

Specificity is a crucial feature of programmable endonucleases, and a high (though currently undefined) level of specificity is desired for the vast majority of therapeutic applications. Until recently, however, few methods existed to study the DNA cleavage specificity of active, site-specific nucleases. An ideal study of off-target activities of site-specific endonucleases would measure nuclease activity against each of the $>10^9$ potential off-target sites for every target site in the human genome. While whole-exome sequencing has been used in studies of site-specific endonuclease specificity (Cho et al., 2014; Ding et al., 2013; Li et al., 2011b), sequencing offers limited sensitivity in detecting rare off-target events, and exomes represent only a small fraction of genomic DNA containing potential off-target sites. Therefore, the general study of off-target activities of site-specific endonucleases has relied on the experimental identification of likely off-target sites. Off-target studies have taken one of three general forms: discrete off-target site testing, genome-wide selections, and minimally biased *in vitro* selections (Fig. 3.1).

### *1.2.1 Discrete off-target site testing*

Perhaps, the most obvious approach to evaluating the sequence specificity of nucleases is by assaying discrete potential off-target substrates, either in a low- or high-throughput format. While the methods summarized below are not a comprehensive list of such efforts, they are representative examples of this strategy.

Homing endonucleases such as I-*Sce*I were the subjects of some of the earliest studies of the specificity of nucleases that recognize sites sufficiently long to be unique in the human genome, even though the presence of

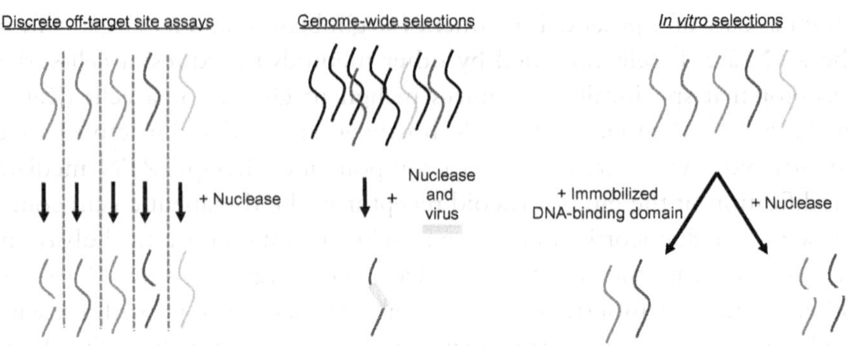


**Figure 3.1** Overview of methods to study the specificity of nucleases. Potential substrate sequences of interest (colored strands) are subjected to nuclease cleavage to identify cleaved sequences (broken red and orange strands). In discrete off-target site assays, sequences are individually subjected to nuclease cleavage in a low- or high-throughput manner. In genome-wide selections, a few potential off-target sites are cleaved within predominantly uncleaved genomic DNA (black strands) and detected by viral integration. Using *in vitro* selection, many potential off-target sites in a vast DNA library are selected for their ability to be bound or cleaved by site-specific nucleases *in vitro*. (See the color plate.)

integrated binding and cleavage domains complicates engineering homing endonucleases with tailor-made specificities (Chen, Wen, Sun, & Zhao, 2009; Chen & Zhao, 2005; Doyon, Pattanayak, Meyer, & Liu, 2006; Gimble, Moure, & Posey, 2003). In early studies of I-*Sce*I homing endonuclease specificity, Dujon and coworkers interrogated a subset of the 54 potential single-mutant individual off-target sequences of the 18 base pair target site (Colleaux, D'Auriol, Galibert, & Dujon, 1988).

The throughput of this approach was increased in the multi-target ELISA method, initially applied to zinc-fingers, developed by Barbas and coworkers (Segal, Dreier, Beerli, & Barbas, 1999), in which 96 biotinylated oligonucleotides or oligonucleotide pools are plated individually in the streptavidin-coated wells of a 96-well plate. Fusions to maltose-binding protein of a DNA-binding domain of interest are incubated with the oligonucleotides in the wells. After a wash step to remove unbound protein, the wells are incubated with a primary antibody that recognizes maltose-binding protein, followed by a secondary antibody that allows visualization of wells containing bound protein.

Church and coworkers (Bulyk, Huang, Choo, & Church, 2001) have used a microarray approach to study zinc-finger DNA-binding specificity. They prepared DNA microarrays containing all 64 possible three-base pair subsequences within a longer target site. The microarrays were incubated with M13 phage displaying the DNA-binding domain of interest, washed, and visualized with primary and secondary antibody staining to reveal DNA-binding specificities. A variant of this method, developed by Bulyk and coworkers (Philippakis, Qureshi, Berger, & Bulyk, 2008), has been extended to profiling 10-base pair subsequences of transcription factor binding sites. Another microarray-based method (Carlson et al., 2010) has also been used to profile the DNA-binding specificity of engineered zinc fingers.

More recently, discrete testing of potential single- and double-mutant off-target sequences has been used in human cells to study the sequence preferences of Cas9. In these methods, a single target site in human cells is assayed for its ability to be modified through NHEJ by a set of endonucleases that are targeted to cleave either the target site or discrete single- or multiple-mutant variants of the target site. At least two separate studies have used this strategy. In one study by Joung and coworkers, an eGFP reporter is the target of a collection of Cas9:guide RNA complexes containing mutant (mismatched) guide RNAs (Fu et al., 2013). In this approach, off-target endonuclease activity leads to the loss of cellular GFP expression. A second study, developed by Zhang and coworkers (Hsu et al., 2013), assayed the ability of a set of Cas9:guide RNA complexes to cleave the *EMX1* gene. Cleavage activity was detected as NHEJ events at the *EMX1* locus using high-throughput sequencing. Although if Cas9 cleaved with perfect specificity the site would not be modified by Cas9:guide RNA complexes containing mutated guide RNAs, many of the mutated guide RNAs resulted in NHEJ, thereby demonstrating off-target activity. In both methods, other potential genomic off-target sites are extrapolated from the small set of off-target sites directly screened. The results of this approach applied to Cas9 are summarized in Section 1.5 and further demonstrate the utility of simple, discrete-off-target site testing to identify genomic off-target sites.

### 1.2.2 Genome-wide selections

In contrast to discrete screening assays of potential off-target sequences to be cleaved by a nuclease of interest, genome-wide selections have also been used to identify those sequences in a population of human cells that can bind to or are cleaved by a nuclease of interest. In assessments of genome-wide binding of Cas9, Adli, Sharp, Zhang, and their respective colleagues (Kuscu, Arslan, Singh, Thorpe, & Adli, 2014; Wu et al., 2014) used chromatin immunoprecipitation followed by sequencing (ChIP-seq) to study the ability of inactive Cas9 to bind off-target sequences in the genome. In this method, hemagglutinin-tagged, catalytically inactive Cas9 is expressed in human cells. A crosslinking step covalently attaches the tagged Cas9 to any DNA target sites it is bound to in the cell. The bound DNA is then fractionated, the crosslinks are reversed, and high-throughput sequencing of the resulting DNA reveals the genomic sequences bound by Cas9. While these studies show that Cas9 is capable of extensively binding off-target sites, they also suggest that most of the off-target sites bound are not modified.

Genome-wide selections for DNA cleavage, rather than binding alone, have been achieved by exploiting the tendency of certain viruses to preferentially integrate at sites of double-strand breaks. The endonuclease of interest is expressed in cultured human cells, creating double-strand breaks at cleaved genomic sites. Cells are then exposed to a virus that preferentially integrates at double-strand breaks. Genomic DNA sequences containing integrated virus are then identified through selection or direct DNA sequencing. In a selection method developed by Miller and coworkers, adeno-associated virus packaged with antibiotic resistance markers and an *E. coli* plasmid origin is used as an integration marker (Petek, Russell, & Miller, 2010). Any on-target and off-target substrates in the genome containing the integration marker would therefore contain a plasmid origin and antibiotic resistance markers. Genomic DNA is then isolated from infected cells, fragmented with a cocktail of restriction enzymes, circularized, and transformed into *E. coli*. Only fragments containing integrated adeno-associated virus have an *E. coli* origin of replication and the appropriate antibiotic resistant markers, and therefore only fragments containing integrated virus survive. Sequencing of the plasmid reveals the viral-chromosomal junctions, which contain the off-target sites of the endonuclease. Subsequent studies by von Kalle, Tolar, and their respective coworkers to study the specificities of ZFNs and TALENs have extended this approach, using instead integrase-deficient lentiviral vectors (IDLVs) and read-out of integration sites using high-throughput sequencing (Gabriel et al., 2011; Osborn et al., 2013).

Advantages of viral integration methods include their abilities to study specificity directly in the context of the target genome and the unbiased nature of the selection, allowing for identification of off-target sites that are not highly similar in sequence to the on-target site. Results from these methods should be interpreted carefully, however, as integration can occur at double-strand breaks that arise naturally, independent of nuclease activity. As with whole-genome sequencing, viral integration methods may not be sufficiently sensitive to detect low-level off-target modification. In addition, abstraction of general properties of endonuclease specificity could be complicated by cellular factors such as DNA accessibility, which varies from site to site and between cell types (Maeder et al., 2008; Wu et al., 2014).

### *1.2.3 Minimally biased selections* in vitro *and in cells*

The most general method to determine site-specific endonuclease specificity would test the activity of a given endonuclease against each potential off-target sequence. Since therapeutic endonucleases target long sequences ($\geq \sim$20 base pairs) to ensure uniqueness in the genome, a truly comprehensive specificity study would require an assay with at least $4^{20}(\sim 1 \times 10^{12})$ different substrates. Since libraries of this size are challenging to generate and process even using *in vitro* methods, selections to determine site-specific endonuclease specificity either rely on the use of "minimally biased" libraries or focus on smaller subsets of the DNA substrate to be studied. Minimally biased libraries are randomized across the nucleotide positions being studied, but the composition of nucleotides at each position is biased toward the target sequence rather than fully randomized. For example, if a particular target site of a three-base pair specific endonuclease is ATG, a fully randomized library would contain equal proportions of all sequences (NNN). A minimally biased library contains higher proportions of sequences that are similar to the target site. In this example, the most common sequence in the library would be ATG, followed by the single-mutant sequences (cTG, gTG, tTG, AaG, AcG, AgG, ATa, ATc, ATt), the double-mutant sequences, and then triple-mutant sequences, which are the rarest in the library. Biasing is accomplished through the incorporation of mixtures of nucleoside phosphoramidites at each position during DNA synthesis, such that the on-target base is incorporated at a higher frequency than the other off-target bases (Fig. 3.2A). In other variants of this approach, portions of the sequence are fixed, while subsets are fully randomized (for example, nTG, AnG, or ATn).

Using minimally biased libraries, several methods have studied the binding specificities of monomeric zinc-finger domains, in the absence of cleavage domains and dimeric binding partners. In the bacterial one-hybrid system developed by Wolfe and coworkers (Meng, Brodsky, & Wolfe, 2005), a DNA target site library is placed upstream of a selectable marker on a plasmid. The DNA-binding domain of interest is expressed in *E. coli* as a fusion to the α-subunit of RNA polymerase. In each individual bacterium, which each contains only one member of the target site library, RNA

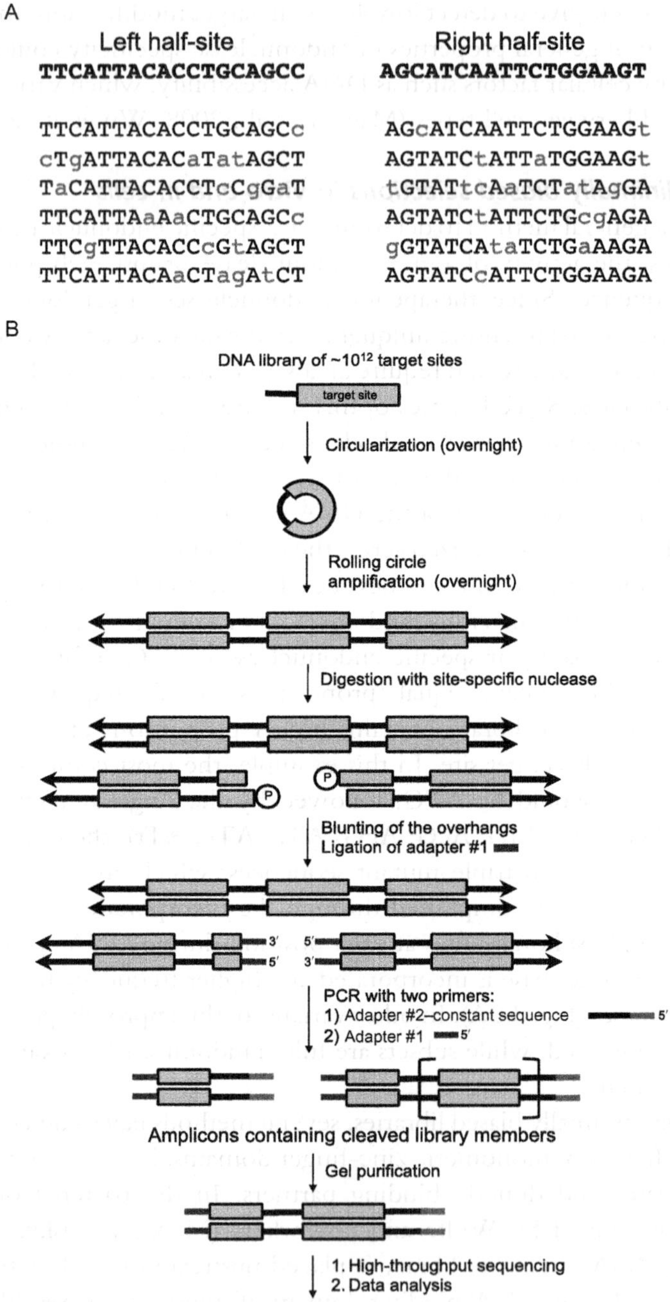

A

| Left half-site | Right half-site |
|---|---|
| **TTCATTACACCTGCAGCC** | **AGCATCAATTCTGGAAGT** |
| TTCATTACACCTGCAGCc | AGcATCAATTCTGGAAGt |
| cTgATTACACaTatAGCT | AGTATCtATTaTGGAAGt |
| TaCATTACACCTcCgGaT | tGTATtcAaTCTatAgGA |
| TTCATTAaAaCTGCAGCc | AGTATCtATTCTGcgAGA |
| TTCgTTACACCcGtAGCT | gGTATCAtaTCTGaAAGA |
| TTCATTACAaCTagAtCT | AGTATCcATTCTGGAAGA |

B

DNA library of ~$10^{12}$ target sites

target site

Circularization (overnight)

Rolling circle amplification (overnight)

Digestion with site-specific nuclease

P

P

Blunting of the overhangs
Ligation of adapter #1

3′ 5′

5′ 3′

PCR with two primers:
1) Adapter #2–constant sequence 5′
2) Adapter #1 5′

Amplicons containing cleaved library members

Gel purification

1. High-throughput sequencing
2. Data analysis

**Figure 3.2** (See figure legend on next page.)

polymerase is recruited to the promoter of the selectable marker if the DNA-binding domain is able to bind to the library DNA sequence present. Only cells that have target sites that can be bound by the DNA-binding domain will express the selectable marker and survive. Wolfe and coworkers have used this approach to assay libraries of up to $10^8$ molecules for DNA binding (Meng, Thibodeau-Beganny, Jiang, Joung, & Wolfe, 2007). A computational structure-based approach developed by Bradley and colleagues (Yanover & Bradley, 2011) using the Rosetta algorithm has also been used to study monomeric zinc-finger domain specificity and has accurately predicted DNA-binding profiles that were obtained by the bacterial one-hybrid system.

Church and coworkers recently used a variant of the bacterial one-hybrid approach to study the specificity of Cas9 in human cells (Mali et al., 2013). In this method, a library of target sites was placed upstream of a reporter gene in a plasmid. Instead of using active Cas9 as an endonuclease in the selection, an inactive variant was expressed as a DNA-binding domain alone, fused to the VP64 activation domain. Therefore, any inactive Cas9 that could bind to a library member caused expression of the reporter gene. The results of this study are summarized in Section 1.5.

Larger libraries, covering more potential off-target sites, have been used to evaluate DNA-binding domain specificity *in vitro*. Applying an *in vitro* SELEX approach (Oliphant, Brandl, & Struhl, 1989) to large ($\sim10^{14}$-membered) libraries of randomized target site DNA (Miller et al., 2011), Struhl and coworkers, and later several other groups, enriched DNA sequences that can bind a given DNA-binding domain of interest

---

**Figure 3.2** *In vitro* selection scheme for profiling the specificity of site-specific nucleases. (A) Example sequences biased toward a target sequence for both the left- and right-half sites of TALEN targeting the human *CCR5* gene. The on-target sequences are in bold and below are examples of variant sequences from minimally biased libraries. (B) A single-stranded library of DNA oligonucleotides containing partially randomized target sites (gray box) and constant region (thick black line) is circularized, then transformed into concatemeric repeats by rolling-circle amplification. The concatemeric repeats of double-stranded DNA (double arrows) target site variants are incubated *in vitro* with a site-specific nuclease of interest. The resulting cleaved ends are blunted and then are ligated to adapter #1. The ligation products are amplified by PCR using one primer consisting of adapter #1 and the other primer consisting of adapter #2—constant sequence, which anneals to the constant regions of the library. From the resulting ladder of amplicons containing 0.5, 1.5, 2.5, ... repeats of a target site, amplicons corresponding to 1.5 target-sites in length are isolated by gel purification and subjected to high-throughput DNA sequencing and computational analysis.

(Thiesen & Bach, 1990; Zykovich, Korf, & Segal, 2009). In this approach, the DNA-binding domain is immobilized and incubated with a randomized target site library. After washing steps to remove unbound DNA, the bound DNA is eluted, amplified, and cycled through the procedure several times before being sequenced.

The bacterial one-hybrid and SELEX methods described above study DNA-binding domains alone, outside of the context of catalysis. Since site-specific endonucleases involve DNA cleavage in addition to DNA binding, and since DNA-binding specificities may not exactly predict DNA cleavage specificities, methods to study the specificity of DNA cleavage reactions are desirable. Monnat and coworkers developed a gel electrophoresis-based method (Argast, Stephens, Emond, & Monnat, 1998) in which an active endonuclease is incubated *in vitro* with a target site library that was cloned into a circular plasmid. Cleavage of library members results in linearization of the plasmid, and the pool of cleavable, linearized DNA sequences is separated from uncleaved, circular DNA through agarose gel electrophoresis and gel purification. The linear DNAs containing bona fide substrate sequences are ligated back into circles and amplified in *E. coli*. After several rounds of enrichment of a pre-selection library with a theoretical complexity of $10^8$ to $10^9$ members (constrained by the need to introduce library members into *E. coli*), the post-selection library is sequenced and analyzed.

To combine the benefits of both large library sizes and the context of cleavage selection, Liu and coworkers developed a fully *in vitro* selection strategy to profile the DNA cleavage specificity of ZFNs, TALENs, and Cas9 using libraries of $10^{11}$ to $10^{12}$ potential off-target sites (Guilinger et al., 2014; Pattanayak et al., 2013; Pattanayak, Ramirez, Joung, & Liu, 2011). In this strategy, library construction is performed entirely *in vitro* and therefore is not bottlenecked by cell transformation efficiency. This method, which is described in detail in Section 2, uses the generation of 5′ phosphates upon DNA cleavage to selectively tag and amplify library members that are cleaved by nucleases. These cleaved library members are then revealed by high-throughput DNA sequencing (Fig. 3.2B).

When applied to ZFNs, this *in vitro* DNA cleavage specificity profiling strategy demonstrated that a SELEX study on the specificity of individual DNA-binding domains, in the absence of dimerization and cleavage, did not detect some genomic off-target sites. Analysis of hundreds of thousands of off-target sites cleaved *in vitro* suggested that interactions between ZFN monomers affect DNA cleavage specificity and explain differences with the SELEX study (Pattanayak et al., 2011). For the CCR5-targeting

ZFN described above, the *in vitro* cleavage selection also identified more genomic off-target sites than a genome-wide selection method on the same ZFN using IDLVs reported by von Kalle and coworkers (Gabriel et al., 2011). However, each method identified off-target sites that were missed by the other method. As was also demonstrated for SELEX results (Perez et al., 2008), additional computational analysis of *in vitro* selection results improves the sensitivity of the *in vitro* cleavage selection method to determine off-target sites. Joung and coworkers (Sander et al., 2013) applied a machine-learning classifier algorithm to *in vitro* cleavage selection results for the CCR5-targeting ZFN, and identified 26 more off-target sites than had previously been identified, including all of the previously determined off-target sites. These studies collectively demonstrate how *in vitro* selection methods and genome-wide selection methods can serve as complementary tools in the determination of gene-editing nuclease specificities.

## 1.3. Insights and improvements from ZFN specificity studies

ZFNs (Kim, Cha, & Chandrasegaran, 1996) are dimeric fusions of the non-specific *Fok*I restriction endonuclease cleavage domain (Hirsch, Wah, Dorner, Schildkraut, & Aggarwal, 1997) with zinc-finger DNA-binding domains (Fig. 3.3A). The *Fok*I cleavage domain must dimerize to be active; therefore, ZFNs can cleave DNA only after dimerizing and bridging two half-sites (Vanamee, Santagata, & Aggarwal, 2001) that are separated by an unspecified DNA spacer sequence. Target site specificity is therefore determined by two zinc-finger DNA-binding domains, each of which consists of three or more tandem repeats of individual zinc fingers. Each individual zinc finger recognizes three base pairs (Beerli, Segal, Dreier, & Barbas, 1998), and a zinc-finger DNA-binding domain in total recognizes at least nine base pairs. Therefore, in total, ZFNs recognize sites that are at least 18 bp long (not including the spacer).

The DNA-binding specificity of ZFNs is programmed by the composite individual zinc fingers. Each individual zinc finger consists of a compact ββα fold with a hydrophobic core stabilized by a zinc ion coordinated by two cysteines and two histidines. While a great deal of progress has been reported in the design of zinc fingers that can target any DNA triplet, primarily by Barbas, Joung, Klug, Pabo, and their respective coworkers (Beerli et al., 1998; Choo, Sanchez-Garcia, & Klug, 1994; Dreier, Beerli, Segal, Flippin, & Barbas, 2001; Dreier, Segal, & Barbas, 2000; Dreier et al., 2005; Maeder et al., 2008; Rebar & Pabo, 1994; Sander et al., 2011;

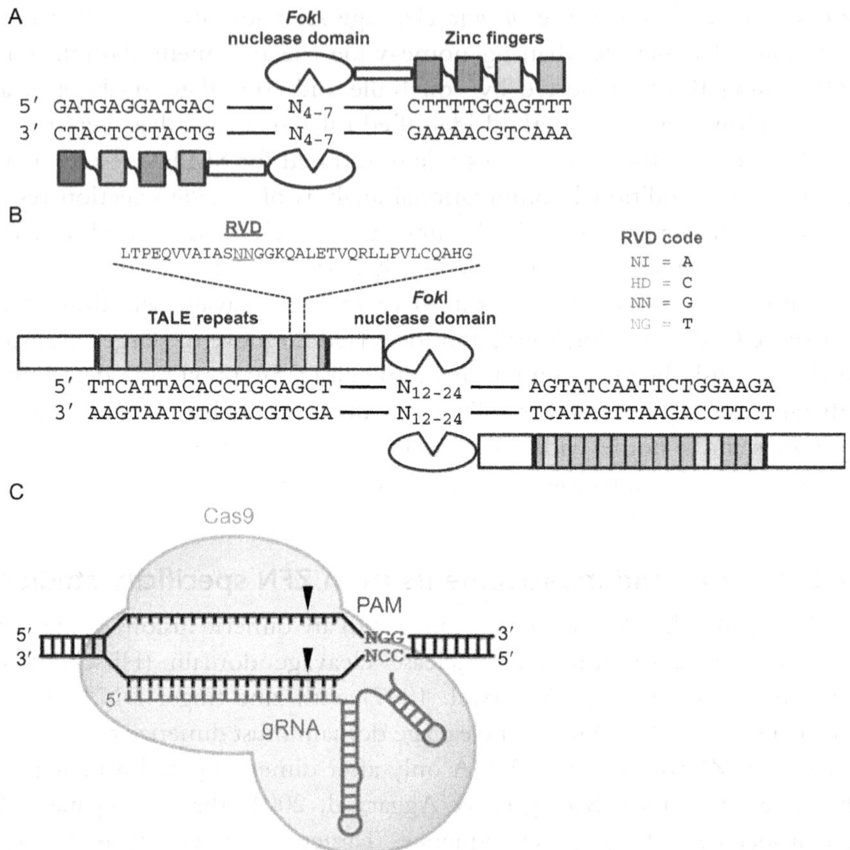


**Figure 3.3** Architecture of ZFN, TALEN, and Cas9 programmable nucleases. (A) A ZFN monomer is a fusion of a *FokI* nuclease cleavage domain (purple; dark gray in print version) to a set of adjoined zinc fingers (four, in this example) each targeting three base pairs (for a total of 12 base pairs recognized, in this example). Two different ZFN monomers bind their corresponding half-sites, allowing *FokI* dimerization and DNA cleavage between the half-sites. (B) A TALEN monomer contains an N-terminal domain followed by an array of TALE repeats (filled boxes), a C-terminal domain, and a *FokI* nuclease cleavage domain (purple; dark gray in print version). The 12th and 13th amino acids (the RVD, red; dark gray in print version) of each TALE repeat recognize a specific DNA base pair. Two different TALEN monomers bind their corresponding half-sites, allowing *FokI* dimerization and DNA cleavage between the half-sites. (C) Cas9 protein (yellow; light gray in print version) binds to target DNA in complex with a single guide RNA (sgRNA, green; light gray in print version). The *S. pyogenes* Cas9 protein and sgRNA complex recognizes the PAM sequence NGG (blue; light gray in print version). Black triangles indicate the cleavage points in the target DNA three bases from the PAM on both DNA strands.

Wu, Yang, & Barbas, 1995), designing the multi-finger domains of a ZFN often requires selection (Greisman & Pabo, 1997; Isalan, Klug, & Choo, 2001; Maeder et al., 2008) or computational approaches (Sander et al., 2011) such as those described by Joung and coworkers.

Initial studies of ZFN specificity using SELEX on zinc-finger binding domains alone (Perez et al., 2008) suggested that ZFNs are highly specific, especially when heterodimeric versions, first developed by Rebar and Cathomen, are used (Miller et al., 2007; Szczepek et al., 2007). The heterodimeric ZFNs have mutations in the *Fok*I cleavage domain that only allow dimerization between different ZFN monomers (Doyon et al., 2011; Miller et al., 2007; Szczepek et al., 2007). While many CCR5 ZFN off-target sites have been identified (Gabriel et al., 2011; Pattanayak et al., 2011, Sander et al., 2013), to date, no toxicity has been reported in clinical trials (Tebas et al., 2014).

In addition to identifying genomic off-target sites, *in vitro* selections on two different ZFNs by Liu and coworkers also illuminated several general properties of ZFN specificity (Pattanayak et al., 2011). Like other enzymes, ZFNs exhibit concentration-dependent specificity, such that a larger set of off-target sites can be cut when the ZFN is at higher concentration. In general, ZFN off-target sites with a small number of mutations (for example, for the CCR5 ZFN, three or fewer mutations out of 24 target base pairs) can be recognized and cleaved. Although no sequence preference in the spacer region between half-sites was observed, sites with disfavored four- and seven-base pair spacers were generally recognized with greater specificity than sites with the more favored five- and six-base pair spacers. Finally, off-target sites with several mutations in one half-site likely contain few to no mutations in the other half-site. All of these observations are consistent with a model in which ZFN:DNA-binding energy must meet a minimum threshold for cleavage to take place, and that off-target cleavage activity arises from excess binding energy between a ZFN and DNA that can tolerate the energetic penalty incurred by protein–DNA mismatches.

## 1.4. Insights and improvements from TALEN specificity studies

Like ZFNs, TALENs are engineered fusions of DNA-binding domains with *Fok*I nuclease domains (Fig. 3.1B). In the case of TALENs, the DNA-binding domains consist of TALE repeat arrays (Christian et al., 2010; Li et al., 2011a, 2011b; Miller et al., 2011). TALE repeats are naturally found in the plant pathogen *Xanthomonas* and are part of transcriptional activator proteins that lead to gene expression upon binding to specific promoter elements in the plant host cell (Gu et al., 2005; Kay, Hahn, Marois, Hause, & Bonas, 2007; Yang, Sugio, & White, 2006). Canonical TALE repeats are 34-amino acid sequences that each recognize one base pair of DNA. The DNA-binding specificity of each repeat is determined by two amino acids referred to as the repeat-variable di-residue (RVD) (Boch et al., 2009; Moscou & Bogdanove, 2009). Examples of RVDs that recognize each of the four DNA base pairs are known. The only known sequence constraint on TALE repeat domains is a requirement for the 5′ end of the target site to contain deoxythymidine (T). Beyond this requirement, TALEs can be designed to target virtually any DNA sequence, and have been successfully used to manipulate genomes in a variety of organisms (Cermak et al., 2011; Moore et al., 2012; Tesson et al., 2011; Wood et al., 2011) and cell lines (Hockemeyer et al., 2011; Mussolino et al., 2011; Reyon et al., 2012).

Multiple studies, using genome-wide studies and minimally biased selections, have demonstrated that TALEN-mediated genome modification can be accompanied by very rare off-target effects. Whole-genome sequencing of TALEN-treated yeast strains (Li et al., 2011b) and whole-exome sequencing of human cell lines derived from TALEN-treated cells (Ding et al., 2013) revealed no evidence of TALEN-induced genomic off-target mutations. However, whole-genome sequencing may not be sensitive to detecting rare mutations in the absence of sequencing the genomic DNA from an impractically large number of treated cells.

Discrete DNA cleavage studies using homology to on-target sequences to predict potential off-target sites found no TALEN-induced modification of potential off-target sites in *Xenopus* (Lei et al., 2012) and human cell lines (Kim et al., 2013). Several groups have studied the specificity of the TALE repeat DNA-binding domains in isolation, in the absence of cleavage domains. Initial minimally biased selection experiments using SELEX and TALE activator binding (Hockemeyer et al., 2011; Mali et al., 2013; Miller et al., 2011; Tesson et al., 2011) on monomeric TALE repeat array domains demonstrated strong preferences for the intended target base pair at each position in the binding site, and a study by Duchateau and coworkers using a cellular GFP reporter assay found that relatively few mismatches can be accommodated (Juillerat et al., 2014).

Several studies, in human cell lines (Mussolino et al., 2011), zebrafish (Dahlem et al., 2012), and rats (Tesson et al., 2011) have demonstrated TALEN-mediated off-target modification of multiple genomic sites that differ from the on-target site at two to six base pairs. The detection of these sites is not thought to be a general problem of TALEN specificity, since for many applications a TALEN on-target site (up to 36 bp long) can be chosen to be at least seven mutations from any other site in the human genome. However, at least three studies have uncovered off-target sites modified in cells with more than seven mutations from the target site. In one study, Jaenisch and coworkers used DNA-binding SELEX results on TALE repeat domains in isolation to computationally predict potential genomic off-target sites of a fully active heterodimeric TALEN. Of the 19 predicted sites assayed, two off-target sites containing 9 or 10 mutations relative to the on-target site, were modified in cultured human cells (Hockemeyer et al., 2011). Tolar and coworkers used genome-wide selection with IDLVs (Gabriel et al., 2011; Osborn et al., 2013) to capture off-target double-strand break sites in cells, resulting in the identification of three off-target sites in the genome with up to 12 mutations from the target sequence.

Finally, Liu and colleagues applied the *in vitro* cleavage selection method described above to reveal 16 sites confirmed to be off-target sites in human cells with modification efficiencies ranging from 0.03% to 2.3% (Guilinger et al., 2014). The 16 off-target sites contained 8 to 12 mutations compared to the on-target site, demonstrating that TALENs can have appreciable off-target activities in human cells even at loci that are quite distant from the on-target sequence. Similar to the model developed to describe ZFN specificity, the *in vitro* cleavage results of Liu and coworkers suggested that reducing the cationic charge of the canonical 63-aa TALE C-terminal domain or the canonical N-terminal TALE domain could improve specificity by reducing nonspecific DNA-binding energy. Consistent with this hypothesis, the ability of off-target sites to survive the *in vitro* selection decreased as these cationic residues were mutated to neutral amino acids. Many of these charge-engineered TALENs demonstrated improved specificity across all positions in the target site. Specificity profiles generated using the *in vitro* selection method applied to charge-engineered TALENs indeed showed ~10 to ~100-fold improved specificity from assays of on-target and off-target activity both *in vitro* and in cells.

## 1.5. Insights and improvements from Cas9 specificity studies

In contrast to ZFNs and TALENs, RNA-guided Cas9 nucleases (referred to below as Cas9) do not require the design of separate DNA-binding domains for each new target site (Fig. 3.3C). Cas9 is a member of the CRISPR/Cas family of proteins that naturally defend bacterial genomes through endonuclease activity against foreign DNA sequences. In contrast to ZFNs and TALENs, the target DNA specificity of Cas9 is programmed by hybridization of the target DNA to a Cas9-bound guide RNA sequence (sgRNA) (Jinek et al., 2012). Similar to TALENs, Cas9 target sequences are constrained at one end. All Cas9-targeted sequences require a sequence motif called a protospacer adjacent motif (PAM), the identity of which depends on the species of the Cas9 protein. For example, the most commonly used Cas9, from *S. pyogenes*, cleaves most efficiently target sequences containing an NGG PAM. Unlike ZFNs and TALENs, Cas9 target sites described to date consist of at most 20 base pairs, not including the PAM sequence.

Early studies of Cas9 specificity in nature by Siksyns, Severinov, Maraffini, and their respective coworkers (Cong et al., 2013; Jiang, Bikard, Cox, Zhang, & Marraffini, 2013; Jinek et al., 2012; Sapranauskas et al., 2011; Semenova et al., 2011) suggested that specific recognition of target DNA by Cas9 was limited to a 7–12 base pair subsequence adjacent to the PAM end of the target site. Further *in vitro* study using discrete off-target site testing by Doudna, Charpentier, and colleagues (Jinek et al., 2012) also supported the model. In this model of Cas9 specificity, mismatches were thought to be tolerated at the non-PAM end of the molecule. This model would suggest that Cas9 could not be used for specific genome modification, since a 12-base pair sequence plus two base pair PAM is not long enough to specify a unique sequence in the human genome. Several studies had shown, however, that Cas9 could be used for genome modification in several organisms without adverse effects; for example, Joung and coworkers reported that Cas9-mediated gene modification in zebrafish embryos exhibited a similar rate of off-target toxicity as ZFNs and TALENs (Hwang et al., 2013).

Four subsequent studies, two using discrete off-target testing in human cell culture by Joung (Fu et al., 2013), Zhang (Hsu et al., 2013), and their respective coworkers, one using a minimally biased selection in cells by Church and coworkers (Mali et al., 2013), and one using a minimally biased selection *in vitro* by Liu and coworkers (Pattanayak et al., 2013), investigated Cas9 specificity and showed that while Cas9 specificity is sufficient for at least some genome-editing applications, several off-target cleavage sites could be detected for most Cas9 target sites tested. While the magnitude of off-target activity varied in the four studies, Joung and coworkers observed that some off-target sites could be modified at similar frequencies to the on-target site.

All four studies showed that Cas9 specificity extended past the 7–12 base pair subsequence near the PAM, and that specificity is decreased at the end of the target site farthest from the PAM. The subsequence near the PAM, while highly specified, tolerates certain single-base pair mismatches in an unpredictable fashion depending on the target site. These functional observations of cleavage specificity have been supported both by a molecular dynamics study of Cas9 by Doudna and colleagues (Sternberg, Redding, Jinek, Greene, & Doudna, 2014), as well as crystallographic models of Cas9 elucidated by Doudna, Nureki, and their respective colleagues (Jinek et al., 2014; Nishimasu et al., 2014). Genomic binding site profiling of inactive Cas9 by Adli, Sharp, Zhang, and their respective colleagues, in addition to confirming that the entire Cas9 target site is necessary for cleavage, suggests that Cas9 can bind many more sites in the genome than it actually cleaves (Kuscu et al., 2014; Wu et al., 2014).

Within the PAM itself, certain mismatches can also be tolerated. While the *S. pyogenes* Cas9 specifies an NGG PAM, studies in bacteria by Marraffini and coworkers (Jiang et al., 2013) using a fully randomized PAM library demonstrated a tolerance of NAG and NNGGN PAM sequences. Further observations by Church, Liu, Zhang, and their respective coworkers in the specificity studies described above showed that an NAG PAM can also be recognized, and *in vitro*, an NNG or an NGN PAM can be recognized with weak activity when the rest of the target sequence is fully complementary to the sgRNA (Pattanayak et al., 2013). The observation that an NAG PAM can be tolerated has also been supported by crystallographic studies by Jinek and coworkers (Anders, Niewoehner, Duerst, & Jinek, 2014).A more recent study on Cas9 specificity by Bao and colleagues (Lin et al., 2014) also suggests that Cas9 can tolerate single-base pair insertions or deletions in the target sequence relative to the sgRNA, though with reduced activity. In addition, several studies have established that specificity is dependent on Cas9 concentration (Fu et al., 2013; Hsu et al., 2013; Pattanayak et al., 2013) and guide RNA architecture (Fu, Sander, Reyon, Cascio, & Joung, 2014; Hsu et al., 2013; Pattanayak et al., 2013).

Given the significant off-target activity of Cas9 endonucleases, numerous groups have engineered Cas9 or guide RNA variants with enhanced specificity. Joung and coworkers improved the specificity of the Cas9: sgRNA complex by truncating the sgRNA to target less than the canonical 20-bp target sites (Fig. 3.4A) (Fu et al., 2014). By analogy to a study by Kim and colleagues on dimeric zinc-finger nickases (Kim et al., 2012), Church, Zhang, and their respective coworkers demonstrated that mutant Cas9 proteins that cleave only a single strand of dsDNA can be used to nick opposite strands of two nearby target sites, generating what is effectively a double-strand break with reduced off-target activity (Fig. 3.4B) (Ran et al., 2013; Cho et al., 2014; Mali et al., 2013).

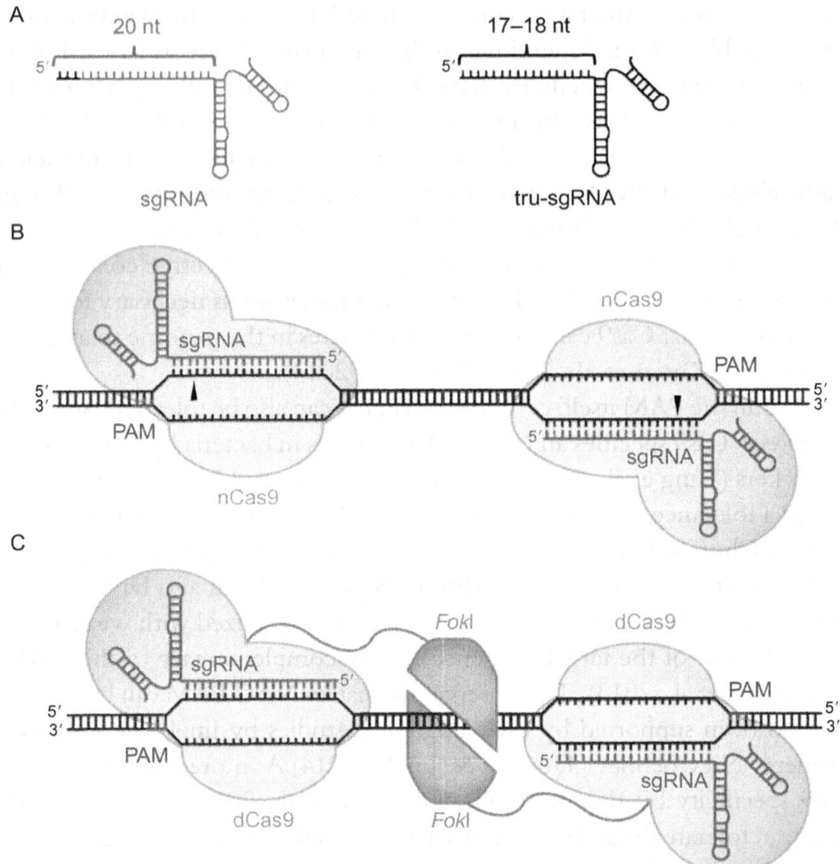


**Figure 3.4** Engineered Cas9 components with improved DNA cleavage specificity. (A) A truncated guide RNA (tru-gRNA, right), contains 17–18 base pairs of complementarity to its DNA target site, rather than 20 base pairs in a canonical sgRNA (left). The base pairs in the sgRNA that are not present in the tru-gRNA are colored black. (B) Mutant Cas9 proteins that cleave only a single strand of dsDNA (nCas9) can be targeted to opposite strands of adjacent sites as pairs to cause double-strand breaks. (C) Monomers of *Fok*I nuclease (red) fused to catalytically inactive Cas9 bind to separate sites within a target locus. Only adjacently bound *Fok*I–dCas9 monomers can assemble a catalytically active *Fok*I nuclease dimer, triggering dsDNA cleavage. (See the color plate.)

Nickases even when bound to off-target loci as monomers retain their ability to nick DNA, which can result in a low level of undesired genome modification (Cho et al., 2014; Fu et al., 2014; Ran et al., 2013), as has previously been described for single zinc-finger nickases (Ramirez et al., 2012; Wang et al., 2012). Therefore, Liu, Joung, and their respective coworkers developed engineered Cas9 variants that are only able to cleave DNA when two monomers are adjacently bound to a target locus by fusing a *Fok*I restriction endonuclease cleavage domain to a catalytically inactive Cas9 (dCas9) (Fig. 3.4C) (Guilinger, Thompson, & Liu, 2014; Tsai et al., 2014), analogous to dimeric ZFNs and TALENs. In discrete off-target studies, the *Fok*I–dCas9 fusions maintain substantial on-target DNA modification with a large reduction in off-target modification at known Cas9 off-target sites.

Collectively, studies that reveal in detail the DNA cleavage specificity of Cas9, together with the engineering of improved Cas9 variants, demonstrate the potential of Cas9 as an accessible and specific genome engineering tool.

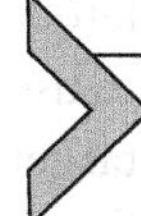

## 2. METHODS

### 2.1. Overview of *in vitro* selection-based nuclease specificity profiling

The *in vitro* selection method developed by our group to profile the DNA cleavage specificity of a nuclease comprises three major steps: pre-selection library construction, *in vitro* selection, and high-throughput sequencing and analysis. Briefly, synthetic 5′-phosphorylated oligonucleotides are converted into concatemeric repeats of a library of potential off-target sites through intramolecular circularization followed by rolling-circle amplification. The resulting pre-selection libraries are then incubated *in vitro* with the appropriate nuclease, either in purified form or used directly from *in vitro* translation systems. Cleaved library members, which contain free 5′ phosphates, are captured by adapter ligation enabling their separation from uncleaved pre-selection library members, which do not contain 5′ phosphates. Cleaved post-selection library members are then amplified by PCR prior to high-throughput DNA sequencing.

### 2.2. Pre-selection library design

While it would be ideal to use a pre-selection library that consists of all possible off-target sequences of a given length (for example, an $N_{22}$ library for Cas9, including PAM), the *in vitro* selection method has an upper limit of approximately $10^{12}$ sequences in the pre-selection library. Since an $N_{22}$ library would contain $4^{22}$ ($\sim 10^{13}$ sequences), a library biased in favor of sequences resembling the on-target recognition site is used instead. Library biasing is accomplished through the use of randomized nucleotide mixtures at all target-site base pairs during library construction. We and others (Argast et al., 1998; Doyon et al., 2006; Guilinger et al., 2014; Mali et al., 2013; Pattanayak et al., 2013, 2011) have had success using mixtures that contain 79% on-target base pair at each targeted position, with the remaining 21% of the mixture comprising of the three off-target base pairs. For Cas9, this approach results in a pre-selection library that in theory contains at least ten copies of each potential off-target sequence containing eight or fewer mutations relative to the on-target site. For a 36-base pair TALEN on-target site, the pre-selection library provides at least 10-fold coverage of all sequences with six or fewer mutations. The partially randomized on-target site is also flanked by fully randomized base pairs on each side to test for patterns of specificity beyond the canonical target site.

The concentration of nuclease used to digest the pre-selection library should be enough to produce sufficient cleaved sequences for robust detection but not enough to completely digest highly cleaved sequences. The use of high nuclease concentrations will augment the detection of rare off-target cleavage events and could result in lower apparent nuclease specificity. Therefore, careful consideration of the nuclease concentrations used, or at least reporting the percentage of on-target sequences cleaved under the assay conditions, is required when studying and describing specificity.

## 2.3. *In vitro* selection protocol

### 2.3.1 *Before Day 1: Design and synthesize pre-selection library oligonucleotides*

For a selection using a guide RNA (CLTA4) targeted to the human clathrin gene (*CLTA*) (Pattanayak et al., 2013), the library oligonucleotide was ordered from Integrated DNA Technologies and was of the form:

/5Phos/TTG TGTNNN NC*C* NT*G* T*G*G* A*A*A* C*A*C* T*A*C* A*T*C* T*G*C* NNN N**AC CTG CCG AG**T TGT GT

"/5Phos/" refers to a 5′ phosphate modification. The underlined sequence refers to the target site library, where each asterisk denotes a position that was ordered as a mixture of bases with 79% of the mixture corresponding to the base preceding the asterisk and 7% each corresponding to the other three bases. For Cas9, we found that this target site orientation (with the reverse complement of the PAM at the 5′ end of the oligonucleotide) yielded higher quality data than the reverse complement of this orientation. The italicized sequences denote a repeated six-base pair barcode that can be used to identify the target site used in the selection, if multiple selections are assayed at once. Other barcodes that we have used include *AAC ACA*, *TCT TCT*, and *AGA GAA*. Any barcodes can be used, although we recommend a minimum of two base pairs differing between each barcode. The sequence in bold denotes a constant region that remains the same for all selections. The constant sequence includes a *Bsp*MI restriction site that is used for pre-selection library preparation for high-throughput sequencing.

#### 2.3.2 Day 1: Circularize library oligonucleotides

Dilute library oligonucleotides to 10 μ*M* in 1 m*M* Tris, pH 8.0. In PCR tubes, add 1 μL of library oligo (10 pmol), 2 μL of 10× CircLigase II 10× Reaction Buffer, 1 μL of 50 m*M* $MnCl_2$, 15 μL water, and 1 μL CircLigase II ssDNA Ligase (100 U) (Epicentre #CL9021K). Incubate 16 h at 60 °C, followed by a 10 min inactivation step at 85 °C.

#### 2.3.3 Day 2: Confirm circularization of library oligonucleotides and perform rolling-circle amplification

On a 15% TBE-Urea polyacrylamide gel, load 2.5 pmol of uncircularized library oligo and 2.5 μL (1.25 pmol) of the CircLigase mixture without purification. Run for 75 min at 200 V. Stain the gel in 100 mL of 0.5× TBE containing 10 μL SYBR Gold (Invitrogen #11494) for 2 h. Rinse with water before imaging the gel. Under these conditions, the circularized oligonucleotides should migrate more slowly than the linear oligonucleotide control.

We use the Illustra TempliPhi Amplification Kit (GE Healthcare #25-6400-10) for the rolling-circle amplification reaction. Combine 5 μL (2.5 pmol) of each unpurified CircLigase reaction and 50 μL of TempliPhi sample buffer. Incubate 3 min at 95 °C. Cool at 0.5 °C/s to 4 °C. Add 50 μL TempliPhi reaction buffer and 1 μL TempliPhi enzyme mix. Incubate 16 h at 30 °C. Heat-inactivate for 10 min at 65 °C. The rolling-circle amplification step can be halved or doubled in scale, with the final pre-selection library size determined by the amount of CircLigase reaction that is used.

#### 2.3.4 Day 3: Quantify and digest pre-selection library

Since the pre-selection library is used in the selection without purification, a double-stranded DNA quantification reagent (Quant-it PicoGreen dsDNA Assay Kit, Invitrogen) is used to quantify the amplified double-stranded DNA. To quantify, add 1 μL rolling-circle amplified DNA or 10–200 ng of lambda DNA standard to 200 μL of 1 m*M* Tris, pH 8.0. Incubate 10 min at room temperature in the dark, before reading fluorescence in a plate reader (excitation wavelength ~ 480 nm, emission wavelength ~ 520 nm). Create a standard curve relating DNA concentration to fluorescence using, for example, pre-quantitated phage lambda DNA and calculate the concentration of the rolling-circle amplified pre-selection library.

To perform the *in vitro* selection, digest the pre-selection library with purified or *in vitro* translated site-specific endonuclease. For Cas9, 200 n*M* amplified pre-selection library was incubated with 100 n*M* Cas9 and

100 n*M* sgRNA or 1000 n*M* Cas9 and 1000 n*M* sgRNA in Cas9 cleavage buffer (20 m*M* HEPES, pH 7.5, 150 m*M* potassium chloride, 10 m*M* magnesium chloride, 0.1 m*M* EDTA, 0.5 m*M* dithiothreitol) for 10 min at 37 °C. Separately, the pre-selection library is also incubated with 2 U of *Bsp*MI restriction endonuclease (NEB) in NEBuffer 3 (100 m*M* NaCl, 50 m*M* Tris–HCl, 10 m*M* $MgCl_2$, 1 m*M* dithiothreitol, pH 7.9) for 1 h at 37 °C. Both nuclease-digested and restriction-digested libraries are purified with the QIAQuick PCR Purification Kit (Qiagen).

For site-specific nucleases that leave overhangs, such as ZFNs and TALENs, an additional step to convert the cut overhangs into blunt ends is performed before adapter ligation. In this step, 50 μL of purified, digested DNA is incubated with 3 μL of 10 m*M* dNTP mix (10 m*M* dATP, 10 m*M* dCTP, 10 m*M* dGTP, 10 m*M* dTTP) (NEB), 6 μL of 10 × NEBuffer 2, and 1 μL of 5 U/μL Klenow Fragment DNA Polymerase (NEB) for 30 min at room temperature. The blunt-ended mixture is purified with the QIAQuick PCR Purification Kit (Qiagen).

Once the cut ends have been made blunt, either by Cas9, or for TALENs by Klenow polymerase, sequencing adapters are ligated. For post-selection blunt libraries, adapter 1 (5′ AAT GAT ACG GCG ACC ACC GAG ATC TAC ACT CTT TCC CTA CAC GAC GCT CTT CCG ATC T*AA CA*) and adapter 2 (5′ *TGT T*AG ATC GGA AGA GCG TCG TGT AGG GAA AGA GTG TAG ATC TCG GTG G) are used to incorporate sequences for Illumina sequencing. The reverse complementary sequences in italics can be varied to barcode multiple reaction conditions. The rest of the adapter sequences can also be appropriately varied to be used with other high-throughput sequencing platforms. In the ligation step, 10 pmol each of adapter 1 and adapter 2 are incubated with the blunt-ended post-selection library and 1000 U T4 DNA Ligase (NEB) in NEB T4 DNA Ligase Reaction Buffer (50 m*M* Tris–HCl, pH 7.5, 10 m*M* magnesium chloride, 1 m*M* ATP, 10 m*M* dithiothreitol) overnight at room temperature. For the restriction-digested pre-selection library, the ligation protocol is the same, with the exception of the use of lib adapter 1 (5′ GAC GGC ATA CGA GAT) and lib adapter 2 (5′ *TTG T*AT CTC GTA TGC CGT CTT CTG CTT G). Of note, the first four bases of lib adapter 2 (in italics) must match the first four bases of the library oligonucleotide barcode (see Section 2.3.1), since *Bsp*MI digestion will leave an overhang that is specific to the barcode used. Therefore, if multiple target sites are tested in the same selection run, multiple lib adapter 2's must be used.

### *2.3.5 Day 4: PCR of pre- and post-selection libraries*

The PCR amplification step prior to high-throughput sequencing must be well controlled to minimize the potential effects of PCR bias on the final sequencing results. Prior to PCR, the adapter ligation mixtures from Day 3 are purified with the QIAQuick PCR Purification Kit (Qiagen) and eluted with 50 μL 1 m*M* Tris, pH 8.0. We use Phusion Hot Start Flex DNA Polymerase (NEB) in Buffer HF with an annealing temperature of 60 °C and an extension temperature of 72 °C for 1 min per cycle. For the nuclease-digested post-selection PCR, primers sel PCR (5′ CAA GCA GAA GAC GGC ATA CGA GAT AC*A CAA* CTC GGC AGG T) and PE2 short (5′ AAT GAT ACG GCG ACC ACC GA) are used. For the restriction-digested pre-selection library PCR, use the same PCR cycling conditions with primers lib fwd PCR (5′ CAA GCA GAA GAC GGC ATA CGA GAT) and lib seq PCR (5′ AAT GAT ACG GCG ACC ACC GAG ATC TAC ACT CTT TCC CTA CAC GAC GCT CTT CCG ATC TNN NNA CCT ACC TGC CGA *GTT GTG T*). Four Ns are included in the lib seq PCR to provide a randomized initiation sequence to maintain compatibility with Illumina sequencing requirements. Of note, if multiple target sites are used in the same selection run, multiple sel PCR primers must be used, with the four base sequence in sel PCR and the six base sequence in lib seq PCR listed in italics should be modified to maintain complementarity to the original library oligonucleotide backbone (see Section 2.3.1).

Before PCR of the full volume of post-selection library, we suggest performing a test PCR or test qPCR with 1 μL of purified post-selection library under the PCR conditions listed above to determine the number of cycles required to reach saturation. If doing a test PCR, remove aliquots every 5 cycles for 35 cycles and visualize the reaction products. Determine the point at which the PCR amplification saturates and subtract an appropriate number of cycles when the PCR is scaled up. For example, if scaling from 1 to 32 μL of purified post-selection library, subtract five cycles ($2^5 = 32$).

After PCR, there may be a ladder of products, corresponding to amplified post-selection library members that contain 0.5, 1.5, 2.5, etc., repeats of a given library member (Fig. 3.5). The variation in PCR product size results from the concatemeric nature of the pre-selection library. During PCR, the sel PCR primer can also anneal to one of multiple repeats, also leading a distribution of PCR product sizes. To standardize the analysis, only those PCR products that contain 1.5 repeats are analyzed. Therefore, before high-throughput sequencing, a final gel purification step is used to enrich the

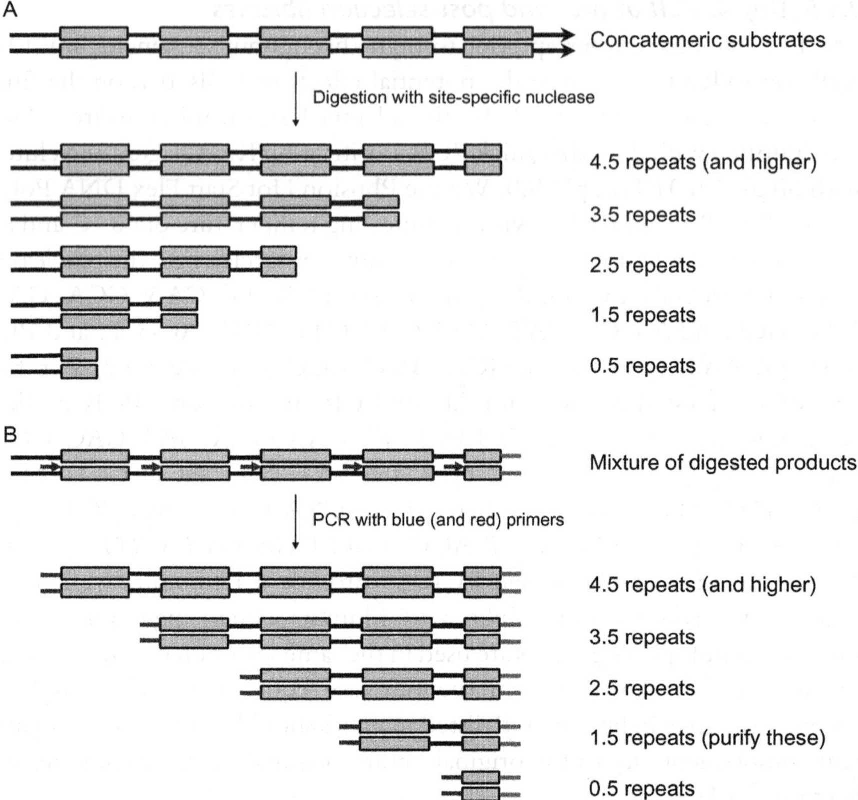


**Figure 3.5** Sample processing during *in vitro* selection-based nuclease specificity profiling. (A) Pre-selection DNA consisting of many repeats of a library member (gray boxes) becomes smaller in size due to nuclease digestion, depending on which target site along the pre-selection DNA is cleaved. (B) During post-selection library amplification, the PCR primer (blue arrows; dark gray in print version) can anneal to any one of the repeats, leading to a set of smaller PCR products. To simplify analysis, only PCR products with 1.5 repeats are purified and analyzed.

amplified post-selection library members that contain exactly 1.5 repeats and to remove any remaining free adapters and primers.

### 2.3.6 Day 5: High-throughput sequencing and analysis

The gel-purified post-selection and pre-selection libraries can be quantified using the KAPA Library Quantification Kit-Illumina (KAPA Biosystems) before high-throughput sequencing. We used single-read sequencing on an Illumina HiSeq or Illumina MiSeq, though the selection should be compatible with any high-throughput sequencing platform as long as the adapter sequences and PCR primers are modified appropriately. For Cas9, a minimum of 66 bases must be sequenced to capture the entire library member. If using a selection condition barcode (for example, *AACA* below), we recommend spiking in a PhiX library control at 25% with the sequencing run to provide appropriate initial base-calling diversity if using Illumina sequencing.

The sequencing output can be binned using a simple scripting language, such as C++ or Python. The components of the sequencing read are illustrated below:

*AACAcatgggtcgACACAAACACAA*CTCGGCAGGT**ACTTGCAGATGTAGTCTTTCCACATGGGTCG***ACACAAACACAA*CTCGGCAGGTATCTCGTATGCC

*AACA* is the four-base pair barcode for selection conditions. *catgggtcg* is the cut "half" of the library target site. *ACACAAACACAA* is the Cas9 target barcode. CTCGGCAGGT is the constant sequence (the reverse complement of the bold sequence in the Library Design section). **ACTTGCAGATGTAGTCTTTCCACATGGGTCG** is the full sequence of the post-selection library member. This sequence can be recognized and cut in the selection. The nonunderlined portion of the sequence consists of the eight random base pairs (four on each side) that flanked the target site library. Once sequences are binned, standard analyses can be performed on the set of target sites (bold and underlined). For example, specificity profiles can be represented as heat maps of specificity scores calculated as the enrichment level of each possible base pair at every position in the post-selection sequences relative to the pre-selection sequences, normalized to the maximum possible enrichment of that base pair (Fig. 3.6).

## 2.4. Confirmation of *in vitro*-identified genomic off-target sites

To identify genomic off-target sites, the set of target sites identified *in vitro* by selection and high-throughput sequencing can be searched for sequences that appear in the human genome. In addition to this simple comparison, a machine-learning algorithm can be trained on the *in vitro* dataset to assist the identification of potential genomic off-target sequences (Sander et al., 2013). The tested site-specific nuclease is then expressed in cultured human cells, along with a parallel experiment with a control, inactive form of the same site-specific nuclease. Genomic DNA is isolated, followed by PCR with primers specific to each potential off-target site. A primer design tool,

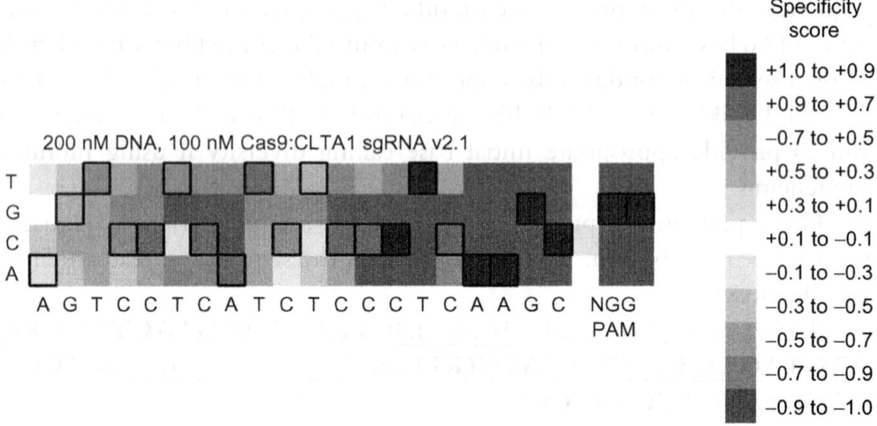


**Figure 3.6** An *in vitro* selection-derived specificity profile. The heat map shows the specificity profile resulting from a selection performed on Cas9:sgRNA targeting the human *CLTA* gene. Specificity scores of 1.0 (dark blue) and −1.0 (dark red) corresponds to 100% enrichment for and against, respectively, a particular base pair at a particular position. Black boxes denote the intended target nucleotides. (See the color plate.)

such as NCBI Primer-BLAST (http://www.ncbi.nlm.nih.gov/tools/primer-blast/), can be useful in the design of primers that lead to specific amplification of the target site of interest. Portions of the high-throughput adapter sequences can be incorporated into the primers (for example, for Illumina, 5′ ACA CTC TTT CCC TAC ACG ACG CTC TTC CGA TCT at the 5′ end of one primer and 5′ GTG ACT GGA GTT CAG ACG TGT GCT CTT CCG ATCT on the other) for the initial PCR. When assaying multiple off-target sites at once, PCRs can be pooled in equimolar ratios, purified, and then reamplified using primers PE1-barcode (5′ CAA GCA GAA GAC GGC ATA CGA GAT *ATA TCA GT*G TGA CTG GAG TTC AGA CGT GTG CT) and PE2-barcode (5′ AAT GAT ACG GCG ACC ACC GAG ATC TAC AC*A TTA CTC G*AC ACT CTT TCC CTA CAC GAC). The number of cycles of the re-amplification PCR should be minimized to avoid introducing significant PCR bias. The italicized bases in PE1-barcode and PE2-barcode correspond to barcodes that can be used in Illumina sequencing. Different barcodes should be used for PCR products derived from active-nuclease-treated DNA compared to inactive-nuclease-treated DNA.

Following high-throughput sequencing, nuclease-modified off-target sequences can be identified through sequence alignment or through computational methods. One algorithm for identifying modified-sequences involves searching for the 20 base pairs flanking each off-target site for each high-throughput sequencing read. For example:

5′ CAATCTCCCGCATGCGCTCAGTCCTCATCTCCCTCAAG CAGGCCCCGCTGGTGCACTGAAGAGCCA**CCCTGTGAAACAC TACATCTGC**AATATCTTAATCCTACTCAGTGAAGCTCTTCAC AGTCATTGGATTAATTATGTTGAGTTCTTTTGGACCAAACC

The flanking sequences (underlined) can be used to identify the off-target site being assayed (in bold). In the reference genome sequence, the sequence between the underlined flanking regions is 5′ CCCTGTGGAAACACTA CATCTGC. In this example, the sequence between the underlined flanking regions is 5′ **CCCTGT-GAAACACTACATCTGC**, where the dash indicates a one base pair deletion. For each potential off-target site tested, the fraction of sequences with insertions and deletions can be calculated and compared between active-nuclease and inactive-nuclease experiments. For target sites with high modification efficiencies, it may be necessary to use flanking sequences (the underlined sequences above) that are more distal to the target site, in case NHEJ leads to deletion of a region that is larger than the off-target site (the bold sequence).

## 3. CONCLUSION

Genome engineering in the last few years has become more facile through the use of programmable site-specific nucleases such as TALENs and Cas9, which can be designed to target nearly any DNA sequence. As the use of ZFNs, TALENs, and Cas9 in research and clinical settings continues to grow, efforts to reveal in depth the DNA cleavage specificity of programmable nucleases will become increasingly important. Efforts to characterize programmable nuclease specificity have ranged from discrete target-site assays to *in vitro* selections to genome-wide selections, all of which have been applied recently to study TALEN and Cas9 specificity. The findings from these methods will continue to deepen our understanding of the basis of the DNA cleavage specificity of these important proteins, inform the development of programmable nucleases with improved specificity, and perhaps eventually enable the broad application of these or other programmable nucleases to treat human genetic diseases.

## ACKNOWLEDGMENTS

V. P., J. P. G., and D. R. L. were supported by DARPA HR0011-11-2-0003, DARPA N66001-12-C-4207, NIH/NIGMS R01 GM095501 (D. R. L.), and the Howard Hughes Medical Institute. D. R. L. was supported as a HHMI Investigator. V. P. was supported by NIGMS T32GM007753. We are grateful to J. Keith Joung for helpful comments.

## REFERENCES

Anders, C., Niewoehner, O., Duerst, A., & Jinek, M. (2014). Structural basis of PAM-dependent target DNA recognition by the Cas9 endonuclease. *Nature, 513*(7519), 569–573.

Argast, G. M., Stephens, K. M., Emond, M. J., & Monnat, R. J., Jr. (1998). I-PpoI and I-CreI homing site sequence degeneracy determined by random mutagenesis and sequential in vitro enrichment. *Journal of Molecular Biology, 280*(3), 345–353.

Beerli, R. R., Segal, D. J., Dreier, B., & Barbas, C. F., 3rd. (1998). Toward controlling gene expression at will: Specific regulation of the erbB-2/HER-2 promoter by using polydactyl zinc finger proteins constructed from modular building blocks. *Proceedings of the National Academy of Sciences of the United States of America, 95*(25), 14628–14633.

Boch, J., et al. (2009). Breaking the code of DNA binding specificity of TAL-type III effectors. *Science, 326*(5959), 1509–1512.

Bulyk, M. L., Huang, X., Choo, Y., & Church, G. M. (2001). Exploring the DNA-binding specificities of zinc fingers with DNA microarrays. *Proceedings of the National Academy of Sciences of the United States of America, 98*(13), 7158–7163.

Carlson, C. D., et al. (2010). Specificity landscapes of DNA binding molecules elucidate biological function. *Proceedings of the National Academy of Sciences of the United States of America, 107*(10), 4544–4549.

Carroll, D. (2011). Genome engineering with zinc-finger nucleases. *Genetics, 188*(4), 773–782.

Cermak, T., et al. (2011). Efficient design and assembly of custom TALEN and other TAL effector-based constructs for DNA targeting. *Nucleic Acids Research, 39*(12), e82.

Chen, Z., Wen, F., Sun, N., & Zhao, H. (2009). Directed evolution of homing endonuclease I-SceI with altered sequence specificity. *Protein Engineering, Design and Selection, 22*(4), 249–256.

Chen, Z., & Zhao, H. (2005). A highly sensitive selection method for directed evolution of homing endonucleases. *Nucleic Acids Research, 33*(18), e154.

Cho, S. W., et al. (2014). Analysis of off-target effects of CRISPR/Cas-derived RNA-guided endonucleases and nickases. *Genome Research, 24*(1), 132–141.

Choo, Y., Sanchez-Garcia, I., & Klug, A. (1994). In vivo repression by a site-specific DNA-binding protein designed against an oncogenic sequence. *Nature, 372*(6507), 642–645.

Choulika, A., Perrin, A., Dujon, B., & Nicolas, J. F. (1995). Induction of homologous recombination in mammalian chromosomes by using the I-SceI system of Saccharomyces cerevisiae. *Molecular and Cellular Biology, 15*(4), 1968–1973.

Christian, M., et al. (2010). Targeting DNA double-strand breaks with TAL effector nucleases. *Genetics, 186*(2), 757–761.

Colleaux, L., D'Auriol, L., Galibert, F., & Dujon, B. (1988). Recognition and cleavage site of the intron-encoded omega transposase. *Proceedings of the National Academy of Sciences of the United States of America, 85*(16), 6022–6026.

Cong, L., et al. (2013). Multiplex genome engineering using CRISPR/Cas systems. *Science, 339*(6121), 819–823.

Dahlem, T. J., et al. (2012). Simple methods for generating and detecting locus-specific mutations induced with TALENs in the zebrafish genome. *PLoS Genetics, 8*(8), e1002861.

Ding, Q., et al. (2013). A TALEN genome-editing system for generating human stem cell-based disease models. *Cell Stem Cell, 12*(2), 238–251.

Doyon, J. B., Pattanayak, V., Meyer, C. B., & Liu, D. R. (2006). Directed evolution and substrate specificity profile of homing endonuclease I-SceI. *Journal of the American Chemical Society, 128*(7), 2477–2484.

Doyon, Y., et al. (2011). Enhancing zinc-finger-nuclease activity with improved obligate heterodimeric architectures. *Nature Methods, 8*(1), 74–79.

Dreier, B., Beerli, R. R., Segal, D. J., Flippin, J. D., & Barbas, C. F., 3rd. (2001). Development of zinc finger domains for recognition of the 5'-ANN-3' family of DNA sequences and their use in the construction of artificial transcription factors. *The Journal of Biological Chemistry*, *276*(31), 29466–29478.

Dreier, B., Segal, D. J., & Barbas, C. F., 3rd. (2000). Insights into the molecular recognition of the 5'-GNN-3' family of DNA sequences by zinc finger domains. *Journal of Molecular Biology*, *303*(4), 489–502.

Dreier, B., et al. (2005). Development of zinc finger domains for recognition of the 5'-CNN-3' family DNA sequences and their use in the construction of artificial transcription factors. *The Journal of Biological Chemistry*, *280*(42), 35588–35597.

Fu, Y., Sander, J. D., Reyon, D., Cascio, V. M., & Joung, J. K. (2014). Improving CRISPR-Cas nuclease specificity using truncated guide RNAs. *Nature Biotechnology*, *32*(3), 279–284.

Fu, Y., et al. (2013). High-frequency off-target mutagenesis induced by CRISPR-Cas nucleases in human cells. *Nature Biotechnology*, *31*(9), 822–826.

Gabriel, R., et al. (2011). An unbiased genome-wide analysis of zinc-finger nuclease specificity. *Nature Biotechnology*, *29*(9), 816–823.

Gimble, F. S., Moure, C. M., & Posey, K. L. (2003). Assessing the plasticity of DNA target site recognition of the PI-SceI homing endonuclease using a bacterial two-hybrid selection system. *Journal of Molecular Biology*, *334*(5), 993–1008.

Greisman, H. A., & Pabo, C. O. (1997). A general strategy for selecting high-affinity zinc finger proteins for diverse DNA target sites. *Science*, *275*(5300), 657–661.

Gu, K., et al. (2005). R gene expression induced by a type-III effector triggers disease resistance in rice. *Nature*, *435*(7045), 1122–1125.

Guilinger, J. P., Thompson, D. B., & Liu, D. R. (2014). Fusion of catalytically inactive Cas9 to FokI nuclease improves the specificity of genome modification. *Nature Biotechnology*, *32*(6), 577–582.

Guilinger, J. P., et al. (2014). Broad specificity profiling of TALENs results in engineered nucleases with improved DNA-cleavage specificity. *Nature Methods*, *11*(4), 429–435.

Hirsch, J. A., Wah, D. A., Dorner, L. F., Schildkraut, I., & Aggarwal, A. K. (1997). Crystallization and preliminary X-ray analysis of restriction endonuclease FokI bound to DNA. *FEBS Letters*, *403*(2), 136–138.

Hockemeyer, D., et al. (2011). Genetic engineering of human pluripotent cells using TALE nucleases. *Nature Biotechnology*, *29*(8), 731–734.

Hsu, P. D., et al. (2013). DNA targeting specificity of RNA-guided Cas9 nucleases. *Nature Biotechnology*, *31*(9), 827–832.

Huang, Y., et al. (1996). The role of a mutant CCR5 allele in HIV-1 transmission and disease progression. *Nature Medicine*, *2*(11), 1240–1243.

Hwang, W. Y., et al. (2013). Efficient genome editing in zebrafish using a CRISPR-Cas system. *Nature Biotechnology*, *31*(3), 227–229.

Isalan, M., Klug, A., & Choo, Y. (2001). A rapid, generally applicable method to engineer zinc fingers illustrated by targeting the HIV-1 promoter. *Nature Biotechnology*, *19*(7), 656–660.

Jiang, W., Bikard, D., Cox, D., Zhang, F., & Marraffini, L. A. (2013). RNA-guided editing of bacterial genomes using CRISPR-Cas systems. *Nature Biotechnology*, *31*(3), 233–239.

Jinek, M., Chylinski, K., Fonfara, I., Hauer, M., Doudna, J. A., & Charpentier, E. (2012). A programmable dual-RNA-guided DNA endonuclease in adaptive bacterial immunity. *Science*, *337*(6096), 816–821.

Jinek, M., et al. (2014). Structures of Cas9 endonucleases reveal RNA-mediated conformational activation. *Science*, *343*(6176), 1247997.

Joung, J. K., & Sander, J. D. (2013). TALENs: A widely applicable technology for targeted genome editing. *Nature Reviews Molecular Cell Biology*, *14*(1), 49–55.

Juillerat, A., et al. (2014). Comprehensive analysis of the specificity of transcription activator-like effector nucleases. *Nucleic Acids Research*, *42*(8), 5390–5402.

Kahlon, K. S., Brown, C., Cooper, L. J., Raubitschek, A., Forman, S. J., & Jensen, M. C. (2004). Specific recognition and killing of glioblastoma multiforme by interleukin 13-zetakine redirected cytolytic T cells. *Cancer Research*, *64*(24), 9160–9166.

Kay, S., Hahn, S., Marois, E., Hause, G., & Bonas, U. (2007). A bacterial effector acts as a plant transcription factor and induces a cell size regulator. *Science*, *318*(5850), 648–651.

Kim, Y. G., Cha, J., & Chandrasegaran, S. (1996). Hybrid restriction enzymes: Zinc finger fusions to Fok I cleavage domain. *Proceedings of the National Academy of Sciences of the United States of America*, *93*(3), 1156–1160.

Kim, E., Kim, S., Kim, D. H., Choi, B. S., Choi, I. Y., & Kim, J. S. (2012). Precision genome engineering with programmable DNA-nicking enzymes. *Genome Research*, *22*(7), 1327–1333.

Kim, Y., et al. (2013). A library of TAL effector nucleases spanning the human genome. *Nature Biotechnology*, *31*(3), 251–258.

Koike-Yusa, H., Li, Y., Tan, E. P., Velasco-Herrera Mdel, C., & Yusa, K. (2014). Genome-wide recessive genetic screening in mammalian cells with a lentiviral CRISPR-guide RNA library. *Nature Biotechnology*, *32*(3), 267–273.

Kuscu, C., Arslan, S., Singh, R., Thorpe, J., & Adli, M. (2014). Genome-wide analysis reveals characteristics of off-target sites bound by the Cas9 endonuclease. *Nature Biotechnology*, *32*(7), 677–683.

Lei, Y., et al. (2012). Efficient targeted gene disruption in Xenopus embryos using engineered transcription activator-like effector nucleases (TALENs). *Proceedings of the National Academy of Sciences of the United States of America*, *109*(43), 17484–17489.

Li, T., et al. (2011a). TAL nucleases (TALNs): Hybrid proteins composed of TAL effectors and FokI DNA-cleavage domain. *Nucleic Acids Research*, *39*(1), 359–372.

Li, T., et al. (2011b). Modularly assembled designer TAL effector nucleases for targeted gene knockout and gene replacement in eukaryotes. *Nucleic Acids Research*, *39*(14), 6315–6325.

Lin, Y., et al. (2014). CRISPR/Cas9 systems have off-target activity with insertions or deletions between target DNA and guide RNA sequences. *Nucleic Acids Research*, *42*(11), 7473–7485.

Liu, R., et al. (1996). Homozygous defect in HIV-1 coreceptor accounts for resistance of some multiply-exposed individuals to HIV-1 infection. *Cell*, *86*(3), 367–377.

Lukacsovich, T., Yang, D., & Waldman, A. S. (1994). Repair of a specific double-strand break generated within a mammalian chromosome by yeast endonuclease I-SceI. *Nucleic Acids Research*, *22*(25), 5649–5657.

Maeder, M. L., et al. (2008). Rapid "open-source" engineering of customized zinc-finger nucleases for highly efficient gene modification. *Molecular Cell*, *31*(2), 294–301.

Mali, P., et al. (2013). CAS9 transcriptional activators for target specificity screening and paired nickases for cooperative genome engineering. *Nature Biotechnology*, *31*(9), 833–838.

Meng, X., Brodsky, M. H., & Wolfe, S. A. (2005). A bacterial one-hybrid system for determining the DNA-binding specificity of transcription factors. *Nature Biotechnology*, *23*(8), 988–994.

Meng, X., Thibodeau-Beganny, S., Jiang, T., Joung, J. K., & Wolfe, S. A. (2007). Profiling the DNA-binding specificities of engineered Cys2His2 zinc finger domains using a rapid cell-based method. *Nucleic Acids Research*, *35*(11), e81.

Miller, J. C., et al. (2007). An improved zinc-finger nuclease architecture for highly specific genome editing. *Nature Biotechnology*, *25*(7), 778–785.

Miller, J. C., et al. (2011). A TALE nuclease architecture for efficient genome editing. *Nature Biotechnology*, *29*(2), 143–148.

Moore, F. E., et al. (2012). Improved somatic mutagenesis in zebrafish using transcription activator-like effector nucleases (TALENs). *PLoS One*, 7(5), e37877.

Moscou, M. J., & Bogdanove, A. J. (2009). A simple cipher governs DNA recognition by TAL effectors. *Science*, *326*(5959), 1501.

Mussolino, C., Morbitzer, R., Lutge, F., Dannemann, N., Lahaye, T., & Cathomen, T. (2011). A novel TALE nuclease scaffold enables high genome editing activity in combination with low toxicity. *Nucleic Acids Research*, *39*(21), 9283–9293.

Nishimasu, H., et al. (2014). Crystal structure of Cas9 in complex with guide RNA and target DNA. *Cell*, *156*(5), 935–949.

Oliphant, A. R., Brandl, C. J., & Struhl, K. (1989). Defining the sequence specificity of DNA-binding proteins by selecting binding sites from random-sequence oligonucleotides: Analysis of yeast GCN4 protein. *Molecular and Cellular Biology*, *9*(7), 2944–2949.

Osborn, M. J., et al. (2013). TALEN-based gene correction for epidermolysis bullosa. *Molecular Therapy*, *21*(6), 1151–1159.

Pattanayak, V., Lin, S., Guilinger, J. P., Ma, E., Doudna, J. A., & Liu, D. R. (2013). High-throughput profiling of off-target DNA cleavage reveals RNA-programmed Cas9 nuclease specificity. *Nature Biotechnology*, *31*(9), 839–843.

Pattanayak, V., Ramirez, C. L., Joung, J. K., & Liu, D. R. (2011). Revealing off-target cleavage specificities of zinc-finger nucleases by in vitro selection. *Nature Methods*, *8*(9), 765–770.

Perez, E. E., et al. (2008). Establishment of HIV-1 resistance in CD4+ T cells by genome editing using zinc-finger nucleases. *Nature Biotechnology*, *26*(7), 808–816.

Petek, L. M., Russell, D. W., & Miller, D. G. (2010). Frequent endonuclease cleavage at off-target locations in vivo. *Molecular Therapy*, *18*(5), 983–986.

Philippakis, A. A., Qureshi, A. M., Berger, M. F., & Bulyk, M. L. (2008). Design of compact, universal DNA microarrays for protein binding microarray experiments. *Journal of Computational Biology*, *15*(7), 655–665.

Ramirez, C. L., et al. (2012). Engineered zinc finger nickases induce homology-directed repair with reduced mutagenic effects. *Nucleic Acids Research*, *40*(12), 5560–5568.

Ran, F. A., et al. (2013). Double nicking by RNA-guided CRISPR Cas9 for enhanced genome editing specificity. *Cell*, *154*(6), 1380–1389.

Rebar, E. J., & Pabo, C. O. (1994). Zinc finger phage: Affinity selection of fingers with new DNA-binding specificities. *Science*, *263*(5147), 671–673.

Reik, A., et al. (2008). Zinc finger nucleases targeting the glucocorticoid receptor allow IL-13 zetakine transgenic CTLs to kill glioblastoma cells *in vivo* in the presence of immunosuppressing glucocorticoids. *Molecular Therapy*, *16*(Suppl. 1), S13–S14.

Reyon, D., Tsai, S. Q., Khayter, C., Foden, J. A., Sander, J. D., & Joung, J. K. (2012). FLASH assembly of TALENs for high-throughput genome editing. *Nature Biotechnology*, *30*(5), 460–465.

Rouet, P., Smih, F., & Jasin, M. (1994). Introduction of double-strand breaks into the genome of mouse cells by expression of a rare-cutting endonuclease. *Molecular and Cellular Biology*, *14*(12), 8096–8106.

Samson, M., et al. (1996). Resistance to HIV-1 infection in caucasian individuals bearing mutant alleles of the CCR-5 chemokine receptor gene. *Nature*, *382*(6593), 722–725.

Sander, J. D., & Joung, J. K. (2014). CRISPR-Cas systems for editing, regulating and targeting genomes. *Nature Biotechnology*, *32*(4), 347–355.

Sander, J. D., et al. (2011). Selection-free zinc-finger-nuclease engineering by context-dependent assembly (CoDA). *Nature Methods*, *8*(1), 67–69.

Sander, J. D., et al. (2013). In silico abstraction of zinc finger nuclease cleavage profiles reveals an expanded landscape of off-target sites. *Nucleic Acids Research*, *41*(19), e181.

Sapranauskas, R., Gasiunas, G., Fremaux, C., Barrangou, R., Horvath, P., & Siksnys, V. (2011). The Streptococcus thermophilus CRISPR/Cas system provides immunity in Escherichia coli. *Nucleic Acids Research*, *39*(21), 9275–9282.

Scarlatti, G., et al. (1997). In vivo evolution of HIV-1 co-receptor usage and sensitivity to chemokine-mediated suppression. *Nature Medicine*, *3*(11), 1259–1265.

Segal, D. J., Dreier, B., Beerli, R. R., & Barbas, C. F., 3rd. (1999). Toward controlling gene expression at will: Selection and design of zinc finger domains recognizing each of the 5'-GNN-3' DNA target sequences. *Proceedings of the National Academy of Sciences of the United States of America, 96*(6), 2758–2763.

Semenova, E., et al. (2011). Interference by clustered regularly interspaced short palindromic repeat (CRISPR) RNA is governed by a seed sequence. *Proceedings of the National Academy of Sciences of the United States of America, 108*(25), 10098–10103.

Shalem, O., et al. (2014). Genome-scale CRISPR-Cas9 knockout screening in human cells. *Science, 343*(6166), 84–87.

Sternberg, S. H., Redding, S., Jinek, M., Greene, E. C., & Doudna, J. A. (2014). DNA interrogation by the CRISPR RNA-guided endonuclease Cas9. *Nature, 507*(7490), 62–67.

Szczepek, M., Brondani, V., Buchel, J., Serrano, L., Segal, D. J., & Cathomen, T. (2007). Structure-based redesign of the dimerization interface reduces the toxicity of zinc-finger nucleases. *Nature Biotechnology, 25*(7), 786–793.

Tebas, P., et al. (2014). Gene editing of CCR5 in autologous CD4 T cells of persons infected with HIV. *The New England Journal of Medicine, 370*(10), 901–910.

Tesson, L., et al. (2011). Knockout rats generated by embryo microinjection of TALENs. *Nature Biotechnology, 29*(8), 695–696.

Thiesen, H. J., & Bach, C. (1990). Target detection assay (TDA): A versatile procedure to determine DNA binding sites as demonstrated on SP1 protein. *Nucleic Acids Research, 18*(11), 3203–3209.

Tsai, S. Q., et al. (2014). Dimeric CRISPR RNA-guided FokI nucleases for highly specific genome editing. *Nature Biotechnology, 32*(6), 569–576.

Vanamee, E. S., Santagata, S., & Aggarwal, A. K. (2001). FokI requires two specific DNA sites for cleavage. *Journal of Molecular Biology, 309*(1), 69–78.

Wang, T., Wei, J. J., Sabatini, D. M., & Lander, E. S. (2014). Genetic screens in human cells using the CRISPR-Cas9 system. *Science, 343*(6166), 80–84.

Wang, J., et al. (2012). Targeted gene addition to a predetermined site in the human genome using a ZFN-based nicking enzyme. *Genome Research, 22*(7), 1316–1326.

Wood, A. J., et al. (2011). Targeted genome editing across species using ZFNs and TALENs. *Science, 333*(6040), 307.

Wu, H., Yang, W. P., & Barbas, C. F., 3rd. (1995). Building zinc fingers by selection: Toward a therapeutic application. *Proceedings of the National Academy of Sciences of the United States of America, 92*(2), 344–348.

Wu, X., et al. (2014). Genome-wide binding of the CRISPR endonuclease Cas9 in mammalian cells. *Nature Biotechnology, 32*(7), 670–676.

Yang, B., Sugio, A., & White, F. F. (2006). Os8N3 is a host disease-susceptibility gene for bacterial blight of rice. *Proceedings of the National Academy of Sciences of the United States of America, 103*(27), 10503–10508.

Yanover, C., & Bradley, P. (2011). Extensive protein and DNA backbone sampling improves structure-based specificity prediction for C2H2 zinc fingers. *Nucleic Acids Research, 39*(11), 4564–4576.

Zhou, Y., et al. (2014). High-throughput screening of a CRISPR/Cas9 library for functional genomics in human cells. *Nature, 509*(7501), 487–491.

Zykovich, A., Korf, I., & Segal, D. J. (2009). Bind-n-Seq: High-throughput analysis of in vitro protein-DNA interactions using massively parallel sequencing. *Nucleic Acids Research, 37*(22), e151.

CHAPTER FOUR

# Genome Engineering with Custom Recombinases

**Thomas Gaj*,†,1,2, Carlos F. Barbas III*,†**

*Department of Chemistry, The Skaggs Institute for Chemical Biology, The Scripps Research Institute, La Jolla, California, USA

†Department of Cell and Molecular Biology, The Skaggs Institute for Chemical Biology, The Scripps Research Institute, La Jolla, California, USA

[1]Current address: University of California, Berkeley, California, USA.

[2]Corresponding author: e-mail address: gaj@berkeley.edu

## Contents



## Abstract

Site-specific recombinases are valuable tools for myriad basic research and genome engineering applications. In particular, hybrid recombinases consisting of catalytic domains from the resolvase/invertase family of serine recombinases fused to $Cys_2$–$His_2$ zinc-finger or TAL effector DNA-binding domains are capable of introducing targeted modifications into mammalian cells. Due to their inherent modularity, new recombinases with distinct targeting specificities can readily be generated and utilized in a "plug-and-play" manner. In this protocol, we provide detailed, step-by-step instructions for generating new hybrid recombinases with user-defined specificity, as well as methods for achieving site-specific integration into targeted genomic loci using these systems.

## 1. INTRODUCTION

Hybrid recombinases composed of catalytic domains derived from the resolvase/invertase family of serine recombinases fused to engineered

*Methods in Enzymology*, Volume 546
ISSN 0076-6879
http://dx.doi.org/10.1016/B978-0-12-801185-0.00004-0

$Cys_2$–$His_2$ zinc-finger (Akopian, He, Boocock, & Stark, 2003; Gordley, Smith, Graslund, & Barbas, 2007) or TAL effector DNA-binding domains (Mercer, Gaj, Fuller, & Barbas, 2012) are powerful tools for targeted genome engineering (Fig. 4.1A). Unlike classical site-specific recombination systems, such as Cre-*loxP*, Flp-*FRT*, and phiC31-*att*, hybrid recombinases are modular chimeric proteins that are capable of introducing genomic modifications at user-defined sites in human cells (Gaj, Sirk, & Barbas, 2014). Depending on the length of the custom DNA-binding domain, these enzymes can recognize target sites between 44- and 62-bp in length. In general, each target site consists of a central 20-bp core sequence recognized by the recombinase catalytic domain, flanked by two inverted

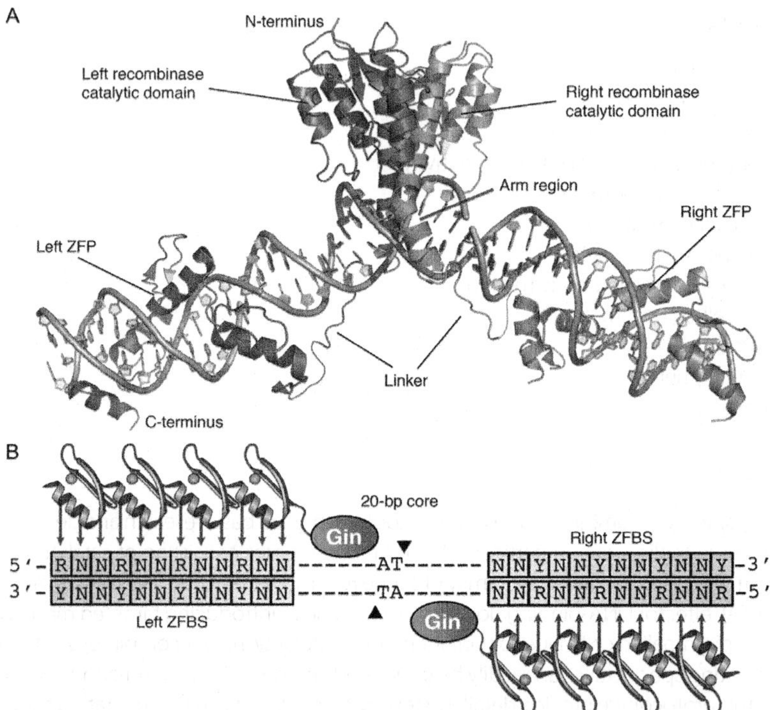


**Figure 4.1** Structure of the zinc-finger recombinase (ZFR) dimer bound to DNA. (A) Each ZFR monomer (blue or orange) consists of an activated serine recombinase catalytic domain fused to a custom-designed $Cys_2$-$His_2$ zinc-finger DNA-binding domain. (B) Cartoon of the ZFR dimer bound to DNA. ZFR target sites consist of two inverted zinc-finger binding sites flanking a central 20-bp core sequence recognized by the serine recombinase catalytic domain. Abbreviations are as follows: N = A, T, C, or G; R = G or A; and Y = C or T. *Adapted from Gaj, Mercer, et al. (2013).* (See the color plate.)

zinc-finger or TAL effector binding sites (Fig. 4.1B). These enzymes catalyze recombination via a concerted mechanism in which the recombinase catalytic domain cleaves all four DNA strands before ultimately promoting strand exchange and religation (Grindley, Whiteson, & Rice, 2006). For both zinc-finger and TAL effector recombinase platforms, enzyme specificity is the cooperative product of modular, site-specific DNA recognition, and sequence-dependent catalysis (Gordley, Gersbach, & Barbas, 2009). As such, advances in the design and synthesis of modular zinc-finger (Gersbach, Gaj, & Barbas, 2014) and TAL effector DNA-binding proteins (Joung & Sander, 2013) have enabled the possibility of generating new custom recombinases capable of recognizing a wide range of DNA sequences. In addition, new recombinase variants with distinct catalytic specificities can readily be generated via a "plug-and-play" manner using a collection of preselected recombinase catalytic domains derived from an activated mutant of the DNA invertase Gin from bacteriophage Mu (Gaj, Mercer, Gersbach, Gordley, & Barbas, 2011; Gaj, Mercer, Sirk, Smith, & Barbas, 2013; Gersbach, Gaj, Gordley, & Barbas, 2010; Klippel, Cloppenborg, & Kahmann, 1988; Proudfoot, McPherson, Kolb, & Stark, 2011). These catalytic domain variants, referred to here as Gin α, β, γ, δ, ε, and ζ, were generated by targeted saturation mutagenesis of the Gin recombinase C-terminal arm, a region of the enzyme that connects the catalytic and DNA-binding domains, and also contacts substrate DNA. These redesigned catalytic domains demonstrate high specificity for their intended DNA targets and can be used in a "mix and match" approach to recognize highly diverse DNA sequences. Indeed, our laboratory and others have demonstrated that hybrid recombinases composed of these reengineered catalytic domains are capable of targeted integration of therapeutic factors into endogenous genomic loci at specificities of >80% (Gaj, Sirk, Tingle, et al., 2014) and excision of transgenes with efficiencies of >15% in human cells (Gordley et al., 2007).

Here, we provide a step-by-step protocol for the design and validation of hybrid recombinases based on zinc-finger protein technology. We include assays for evaluating zinc-finger recombinase (ZFR) activity in mammalian cells and describe how to achieve site-specific integration into endogenous genomic loci using this technology.

## 2. TARGET IDENTIFICATION

Putative ZFR target sites can be identified using the consensus sequences provided in Fig. 4.2. Engineered ZFRs are fully modular enzymes

```
  +1  Left ZFBS            20-bp core             Right ZFBS  +44
5'-RNNRNNRNNRNNNNNNNAAABNWWNVTTTNNNNNNNYNNYNNYNNY-3'
3'-YNNYNNYNNYNNNNNNNTTTVNWWNBAAANNNNNNNRNNRNNRNNR-5'
```

**Figure 4.2** The consensus 44-bp target sequence used to identify potential ZFR target sites. Underlined bases indicate zinc-finger DNA-binding sites and core positions 3 and 2. Abbreviations are as follows: N = A, T, C, or G; R = A or G; Y = T or C; B = T, C, or G; V = A, C, or G; and W = A or T.

composed of an N-terminal recombinase catalytic domain fused to a C-terminal zinc-finger DNA-binding protein. Using the collection of preselected recombinase catalytic domains described in Fig. 4.3 and Table 4.1, we estimate that one potential ZFR target site can be identified for every ~160,000 bp of random sequence. The specificity profile of each reengineered Gin catalytic domain is presented in Fig. 4.3B. Currently, the presence of adenine at core positions 6, 5, and 4 is the only major targeting requirement for ZFRs containing the Gin recombinase catalytic domain. In cases that require recognition of nonadenine bases at these positions, ZFRs derived from the Sin and β recombinase catalytic domains (Sirk, Gaj, Jonsson, Mercer, & Barbas, 2014), which recognize both guanine and thymine at these sites, can be used. In general, 20-bp core sequences that display >50% sequence identity to the native recombinase target site typically show the highest level of activity. Following the identification of a specific target site, ZFRs with the intended complementary specificity can be generated from a preselected library of modular parts.

## 3. RECOMBINASE CONSTRUCTION

We have designed vectors to facilitate ZFR assembly and expression: ZFR construction requires the assembly of two zinc-finger protein assays that recognize the DNA sequences flanking the central 20-bp core sequence. We note that our laboratory (Gonzalez et al., 2010) and others (Carroll, Morton, Beumer, & Segal, 2006; Maeder, Thibodeau-Beganny, Sander, Voytas, & Joung, 2009; Wright et al., 2006) have described a number of protocols for zinc-finger assembly, which are described in detail elsewhere.

Gin catalytic domains engineered to recognize ~$10^7$ distinct core sequences can be obtained from the SuperZiF-compatible subcloning plasmids pBH-Gin-α, β, γ, δ, ε, and ζ (Gaj, Mercer, et al., 2013). These vectors are easily modified for compatibility with zinc-finger proteins constructed by OPEN (Maeder et al., 2008), CoDA (Sander et al., 2011), and the other

A

-10 -9 -8 -7 -6 -5 -4 -3 -2 -1

5′-T-C-C-A-A-A-A-C-C-A-T-G-G-T-T-T-A-C-A-G-3′

3′-A-G-G-T-T-T-T-G-G-T-A-C-C-A-A-A-T-G-T-C-5′

+1 +2 +3 +4 +5 +6 +7 +8 +9 +10

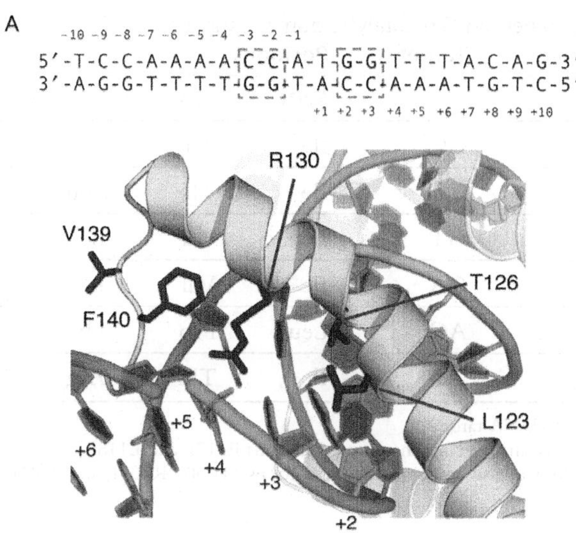

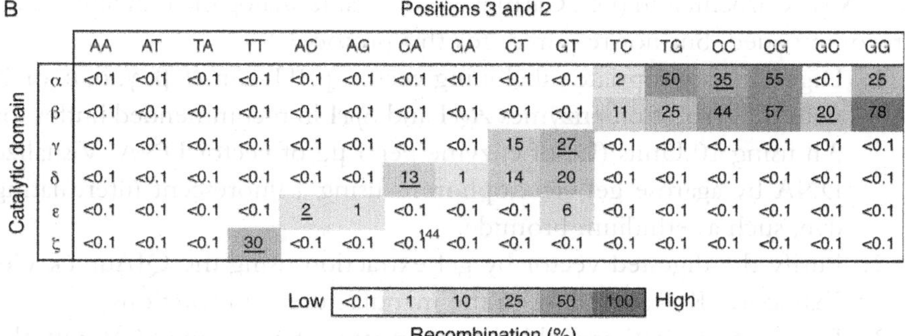

B

Positions 3 and 2

| Catalytic domain | AA | AT | TA | TT | AC | AG | CA | GA | CT | GT | TC | TG | CC | CG | GC | GG |
|---|---|---|---|---|---|---|---|---|---|---|---|---|---|---|---|---|
| α | <0.1 | <0.1 | <0.1 | <0.1 | <0.1 | <0.1 | <0.1 | <0.1 | <0.1 | <0.1 | 2 | 50 | 35 | 55 | <0.1 | 25 |
| β | <0.1 | <0.1 | <0.1 | <0.1 | <0.1 | <0.1 | <0.1 | <0.1 | <0.1 | <0.1 | 11 | 25 | 44 | 57 | 20 | 78 |
| γ | <0.1 | <0.1 | <0.1 | <0.1 | <0.1 | <0.1 | <0.1 | <0.1 | 15 | 27 | <0.1 | <0.1 | <0.1 | <0.1 | <0.1 | <0.1 |
| δ | <0.1 | <0.1 | <0.1 | <0.1 | <0.1 | <0.1 | 13 | 1 | 14 | 20 | <0.1 | <0.1 | <0.1 | <0.1 | <0.1 | <0.1 |
| ε | <0.1 | <0.1 | <0.1 | <0.1 | 2 | 1 | <0.1 | <0.1 | <0.1 | 6 | <0.1 | <0.1 | <0.1 | <0.1 | <0.1 | <0.1 |
| ζ | <0.1 | <0.1 | <0.1 | 30 | <0.1 | <0.1 | <0.1 | <0.1 | <0.1 | <0.1 | <0.1 | <0.1 | <0.1 | <0.1 | <0.1 | <0.1 |

Low <0.1 1 10 25 50 100 High

Recombination (%)

**Figure 4.3** Gin catalytic domain specificities. (A, top) The native 20-bp core sequence recognized by the Gin catalytic domain. All base positions within the core site are numbered. Positions 3 and 2 are boxed. (A, bottom) Structure of a serine recombinase catalytic domain in complex with DNA. Residues subject to reprogramming are shown as magenta (black color in the print version) sticks. (B) Recombination specificity of the Gin α, β, γ, δ, ε, and ζ catalytic domains for each possible two-base combination at positions 3 and 2. Intended DNA targets are underlined. *Adapted from Gaj, Mercer, et al. (2013).*

open-source assembly methods (Bhakta et al., 2013), as well as alternative serine recombinase catalytic domains. The following is a protocol for constructing ZFR heterodimers that recognize the user-defined target site identified in Section 2, using SuperZiF-assembled zinc-finger proteins. Importantly, ZFR targeting of asymmetric DNA sequences requires both the presence of "left"- and "right"-ZFR monomers. We note that the following protocol describes the generation of a single ZFR monomer.

**Table 4.1** Reengineered Gin catalytic domain substitutions

| Catalytic domain | Target | Positions | | | | |
|---|---|---|---|---|---|---|
| | | 120 | 123 | 127 | 136 | 137 |
| α | CC[a] | Ile | Thr | Leu | Ile | Gly |
| β | GC | Ile | Thr | Leu | Arg | Phe |
| γ | GT | Leu | Val | Ile | Arg | Trp |
| δ | CA | Ile | Val | Leu | Arg | Phe |
| ε[b] | AC | Leu | Pro | His | Arg | Phe |
| ζ[c] | TT | Ile | Thr | Arg | Ile | Phe |

[a]Indicates wild-type DNA target.
[b]The ε catalytic domain also contains the substitutions E117L and L118S.
[c]The ζ catalytic domain also contains the substitutions M124S, R131I, and P141R.

*Note*: Carbenicillin (i.e., Carb) is a more stable analog of ampicillin and is recommended, but not required, for this protocol.

1. Digest the appropriate subcloning vector (pBH-Gin-α, β, γ, δ, ε, or ζ) with the restriction enzymes *Age*I and *Spe*I in recommended buffer for 3 h using 10 Units (U) of enzyme per 1 μg of vector DNA. Visualize DNA by agarose gel electrophoresis using a fluorescent intercalating dye, such as ethidium bromide.
2. Purify the digested vector by gel extraction using the QIAquick Gel Extraction Kit, according to the manufacturer's instructions.
3. Release SuperZiF-assembled zinc-finger proteins from pSCV with the restriction enzymes *Xma*I and *Spe*I in appropriate buffer for 3 h using 10 U of enzyme per 1 μg of DNA and isolate via gel electrophoresis using the QIAquick Gel Extraction Kit.
4. Ligate the purified zinc-finger protein DNA into 25–50 ng of purified pBH-Gin-α, β, γ, δ, ε, or ζ vector using 1 U T4 DNA ligase for 1 h at room temperature. For best results, perform the ligation reaction using a 6:1 insert-to-vector ratio.
5. Transform 10–20 ng of ligated pBH-Gin-α, β, γ, δ, ε, or ζ into any competent laboratory strain of *Escherichia coli*, such as TOP10 or XL1 Blue, by electroporation and recover in 2 mL SOC for 1 h at 37 °C with shaking at 250 rpm.
6. Spread 100 μL of recovery culture on an LB agar plate with 100 μg/mL carbenicillin to determine ligation/transformation efficiency, and for possible use in Step 8. Inoculate remaining ~2 mL recovery culture

for overnight growth by adding 4 mL SB medium and 100 μg/mL carbenicillin. Incubate cultures for 16–24 h at 37 °C with shaking at 250 rpm.

7. The following day, purify plasmid from overnight culture with any commercially available Miniprep kit, according to the manufacturer's instructions.
8. Release ZFR from miniprepped pBH by restriction digestion with *Sfi*I for 3 h using 10 U of enzyme per 1 μg of DNA and visualize DNA by agarose gel electrophoresis. If unable to recover ZFRs, we recommend performing colony PCR using the plated recovery culture from Step 6 to screen for individual clones containing full-length ZFR gene inserts.
9. Purify the ZFR insert by gel extraction using the QIAquick Gel Extraction Kit.
10. Ligate the purified full-length ZFR insert into 25–50 ng of *Sfi*I-digested pcDNA 3.1 (Invitrogen). Transform 10–20 ng of the ligation reaction into any competent laboratory strain of *E. coli* and recover in 1–2 mL of SOC. After 1 h, spread 100 μL of cells on LB agar plates containing 100 μg/mL of carbenicillin and incubate overnight at 37 °C.
11. The following day, inoculate 6 mL of SB medium containing 100 μg/mL of carbenicillin with one colony from the LB agar carb plate and culture overnight at 37 °C with shaking at 250 rpm.
12. The next day, harvest the overnight culture and purify pcDNA-ZFR plasmid by Miniprep. Confirm ZFR identity by standard DNA sequencing using the primer T7 Universal (5′-TAATACGACTCACTATAGGG-3′).

## 4. MEASUREMENTS OF RECOMBINASE ACTIVITY

Our laboratory has developed a transient reporter assay for use in mammalian cells that links ZFR-mediated recombination to reduced luciferase expression (Gaj, Mercer, et al., 2013; Gaj, Sirk, Tingle, et al., 2014). To achieve this, recombinase target sites are introduced both up- and downstream a Simian vacuolating virus 40 (SV40) promoter that drives expression of a luciferase reporter gene. Active recombinases excise the promoter, resulting in reduced luciferase expression. The following protocol describes the generation and application of luciferase-based reporter vectors, but is also broadly adaptable to other types of reporter genes, including EGFP and β-galactosidase.

## 4.1. Reporter plasmid construction

1. The SV40 promoter sequence can be amplified from many different sources. In our studies, we used the primers SV40-ZFR-*Bgl*II-Fwd and SV40-ZFR-*Hind*III-Rev to PCR amplify the SV40 promoter from pGL3-Prm (Promega). These primers also encode flanking ZFR target sites.

   SV40-ZFR-*Bgl*II-Fwd:

   5′-TTAATTAAGAG*AGATCT*GCTGATGCAGATACAG
   AAACCAAGGTTTTCTTACTTGCTG CTGCGCGATCTGC
   ATCTCAATTAGTCAGC-3′

   SV40-ZFR-*Hind*III-Rev:

   5′-ACTGACCTAGAG*AAGCTT*GCAGCAGCAAGTAAG
   AAAACCTTGGTTTCTGTATCTGCA TCAGCTTTGC
   AAAAGCCTAGGCCTCCAAA-3′

   *Note*: ZFR target sites are underlined. Restriction sites are italicized.
2. Purify the PCR product by gel electrophoresis using the QIAquick Gel Extraction Kit, according to the manufacturer's instructions.
3. Digest the purified PCR product and pGL3-Prm with the restriction enzymes *Bgl*II and *Hind*III and purify both DNA fragments by gel electrophoresis.
4. Ligate the purified SV40-ZFR insert into 25–50 ng of purified pGL3 vector for 1 h at 25 °C, creating pGL3-target.
5. Transform 10–20 ng of ligated plasmid into *E. coli* by electroporation. Recover cells for 1 h in 1–2 mL of SOC, and spread 100 μL of cells on LB agar plates containing 100 μg/mL of carbenicillin. Incubate plates overnight at 37 °C.
6. The following day, inoculate 6 mL of SB medium containing 100 μg/mL of carbenicillin with one colony from the LB agar plate and culture overnight at 37 °C with shaking at 250 rpm.
7. Purify pGL3-target plasmid by Miniprep and confirm reporter plasmid identity by DNA sequencing using SV40-Mid-Prim-Fwd (5′-ACCATAGTCCCGCCCCTAACTCC-3′) and SV40-Mid-Prim-Rev (5′-GGAGTTAGGGGCGGGACTATGGT-3′).

## 4.2. Luciferase assay

8. Seed human embryonic kidney (HEK) 293 T cells in a 96-well plate at a density of $4 \times 10^4$ cells per well in Dulbecco's modified Eagle's medium (DMEM) containing 10% (v/v) fetal bovine serum (FBS; Gibco).

9. Transfect cells 24 h after seeding. For best results, cells should be 80–90% confluent at time of transfection.
   a. Use 25–50 ng of each pcDNA-ZFR monomer expression vector, 25 ng of pGL3-target reporter vector, and 1 ng of pRL-CMV (*Renilla* luciferase transfection control; Promega), according to the manufacturer's instructions. We typically transfect each well with 0.8 μL Lipofectamine 2000 (Invitrogen).
   b. To accurately assess recombinase activity, include the following samples:
      i. *Experimental*: reporter construct with ZFR expression vectors.
      ii. *Background luciferase activity*: reporter plasmid only.
      iii. *Negative control*: mock transfected cells that receive no reporter plasmid or ZFR expression vectors.
10. Evaluate the fold reduction in luminescence using a Microplate luminometer and dual-luciferase reporter assay 48 h after transfection, normalizing to cotransfected *Renilla* luciferase control.
11. Recombinases that induce a >20-fold decrease in luminescence are considered sufficiently active for endogenous genomic targeting studies. In our experience, recombinases that lead to a >60-fold reduction in luminescence yield the best results for downstream applications.

## 5. SITE-SPECIFIC INTEGRATION

Site-specific integration is one core application of ZFR technology. We have developed a donor plasmid backbone (pDonor) based on the pBABE vector system (Morgenstern & Land, 1990) that facilitates ZFR-mediated transgene insertion into the human genome (Gaj, Sirk, Tingle, et al., 2014; Gordley et al., 2009). This vector contains a constitutively expressed puromycin-resistance gene for enrichment of ZFR-modified cells, and a multiple-cloning site upstream of the cloned ZFR target site for gene of interest (GOI) placement. The following protocol describes a step-by-step procedure for constructing ZFR donor plasmids and achieving site-specific integration with ZFRs.

### 5.1. Donor plasmid construction

1. PCR amplify the cDNA sequence of the GOI using primers that encode a 5′ *Pst*I and 3′ *Bam*H1 restriction site. The 3′ primer must also encode the ZFR target site selected in Section 2 upstream of the *Bam*H1 restriction site. *Note*: pDonor does not contain a universal promoter or polyA

region, and therefore these components should be amplified with the GOI.

2. Gel purify the PCR product using QIAquick Gel Extraction Kit, according to the manufacturer's instructions.
3. Digest both the purified PCR product and pDonor (empty) with the restriction enzymes *Pst*I and *Bam*H1. Purify both DNA fragments by gel electrophoresis.
4. Ligate the purified GOI insert into 25–50 ng of purified pDonor vector for 1 h at room temperature.
5. Transform 10–20 ng of ligated plasmid into any competent laboratory strain of *E. coli* by electroporation. Recover cells for 1 h in 1–2 mL of SOC, and spread 100 μL of cells on LB agar plates containing 100 μg/mL of carbenicillin. Incubate plates overnight at 37 °C.
6. Purify plasmid DNA by Miniprep and confirm pDonor plasmid identity by DNA sequencing.

## 5.2. Cell culture methods

7. Seed HEK293 cells in a 24-well plate at a density of $2 \times 10^5$ cells per well in DMEM containing 10% FBS. *Note*: we test the ability of our recombinases to integrate donor plasmid in HEK293 cells; however, these enzymes should demonstrate high activity in most cell types.
8. At 24 h after seeding, transfect cells with 80 ng of pDonor plasmid, 10 ng each of pcDNA-ZFR-L and pcDNA-ZFR-R expression vectors and (optionally) 10 ng of pCMV-EGFP for transfection control using any desired transfection protocol. We note that the efficiency of ZFR-mediated integration is dependent on the amount of plasmid transfected and thus may require subsequent optimization, e.g., less specific recombinases may require up to 100 ng of each L- and R-ZFR plasmids.
9. At 72 h after transfection, perform the following evaluations and expand clonal populations.

### *5.2.1 PCR confirmation of integration*

i. At 72 h after transfection, harvest cells, and isolate bulk genomic DNA using Quick Extract DNA Extraction Reagent (Epicentre), according to manufacturer's instructions.
ii. Design primers complementary to the 5′ and 3′ junctions of the genomic region targeted and PCR amplify this region by nested PCR. As an

internal control, we recommend PCR amplifying the GAPDH gene from all harvested cell types.

*Note*: ZFR-mediated integration can occur in both the forward and reverse orientation. We thus recommend designing internal nested primers for detecting both the forward and reverse integration of the GOI.

**iii.** Gel purify the PCR product and confirm the identity of the donor vector by DNA sequencing.

### 5.2.2 Measurements of modification efficiency

**iv.** At 72 h after transfection, split cells into 6-well plates at a density of $1 \times 10^4$ cells per well and maintain in DMEM/FBS with or without 2 μg/mL puromycin.

**v.** At 14–18 days after seeding, stains cells with 0.2% crystal violet staining solution and calculate genome-wide integration rates by dividing the number of colonies in puromycin-containing media by the number of colonies in the absence of puromycin.

### 5.2.3 Isolation and expansion of modified clones

**vi.** At 72 h after transfection, split $1 \times 10^4$ cells into a 100 mm dish and maintain cells in DMEM/FBS with 2 μg/mL puromycin. Individual colonies can be isolated with 10 mm × 10 mm open-ended cloning cylinders with sterile silicone grease and expanded in 96-well plates in the presence of puromycin. Alternatively, modified cells can be isolated and expanded by harvesting 72 h after transfection and reseeding in 96-well plates at limiting dilution with 2 μg/mL puromycin. Sample gel illustrating positive integration events among expanded clones in the forward and reverse directions are shown in Fig. 4.4.

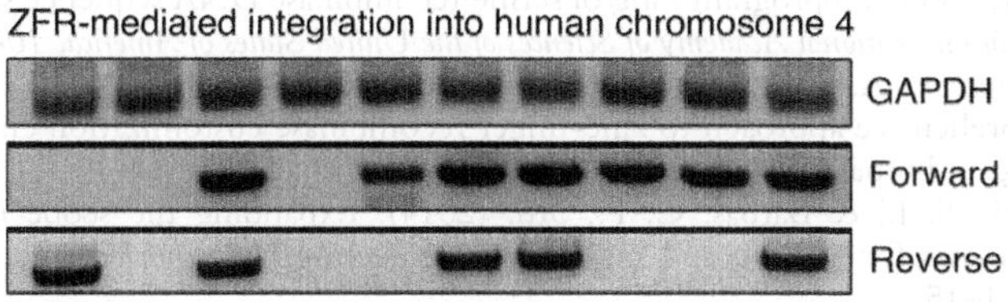


**Figure 4.4** ZFR-mediated integration into the human genome. Example clonal analysis of puromycin-resistant cells transfected with a pair of ZFR expression vectors (left and right) and pDonor plasmid.

## 6. CONCLUSIONS

The protocol provided here details the generation of engineered ZFRs capable of recognizing user-defined sites. The recombinases described here have the capacity to catalyze a diverse range of genome modification outcomes, including targeted gene integration, excision, and cassette exchange. These chimeric enzymes can also be used in tandem with other genome engineering technologies, including site-specific nucleases (Gaj, Gersbach, et al., 2013) and transposases (Gersbach, Gaj, Gordley, Mercer, & Barbas, 2011), for even more diverse editing outcomes. While this protocol does not include guidance for the generation of recently described custom TAL effector recombinases (Mercer et al., 2012), it should be easily adapted for the creation of any type of hybrid recombinases.

## ACKNOWLEDGMENTS

We thank S.J. Sirk for critical reading of the manuscript. Molecular graphics were generated using PyMol (http://pymol.org). This work was supported by the U.S. National Institutes for Health (DP1CA174426) and The Skaggs Institute for Chemical Biology.

## REFERENCES

Akopian, A., He, J., Boocock, M. R., & Stark, W. M. (2003). Chimeric recombinases with designed DNA sequence recognition. *Proceedings of the National Academy of Sciences of the United States of America*, *100*(15), 8688–8691.

Bhakta, M. S., Henry, I. M., Ousterout, D. G., Das, K. T., Lockwood, S. H., Meckler, J. F., et al. (2013). Highly active zinc-finger nucleases by extended modular assembly. *Genome Research*, *23*(3), 530–538.

Carroll, D., Morton, J. J., Beumer, K. J., & Segal, D. J. (2006). Design, construction and in vitro testing of zinc finger nucleases. *Nature Protocols*, *1*(3), 1329–1341.

Gaj, T., Gersbach, C. A., & Barbas, C. F., 3rd. (2013). ZFN, TALEN, and CRISPR/Cas-based methods for genome engineering. *Trends in Biotechnology*, *31*(7), 397–405.

Gaj, T., Mercer, A. C., Gersbach, C. A., Gordley, R. M., & Barbas, C. F., 3rd. (2011). Structure-guided reprogramming of serine recombinase DNA sequence specificity. *Proceedings of the National Academy of Sciences of the United States of America*, *108*(2), 498–503.

Gaj, T., Mercer, A. C., Sirk, S. J., Smith, H. L., & Barbas, C. F., 3rd. (2013). A comprehensive approach to zinc-finger recombinase customization enables genomic targeting in human cells. *Nucleic Acids Research*, *41*(6), 3937–3946.

Gaj, T., Sirk, S. J., & Barbas, C. F., 3rd. (2014). Expanding the scope of site-specific recombinases for genetic and metabolic engineering. *Biotechnology and Bioengineering*, *111*(1), 1–15.

Gaj, T., Sirk, S.J., Tingle, R. D., Mercer, A. C., Wallen, M. C., & Barbas, C. F., 3rd. (2014). Enhancing the specificity of recombinase-mediated genome engineering through dimer interface redesign. *Journal of the American Chemical Society*, *136*(13), 5047–5056.

Gersbach, C. A., Gaj, T., & Barbas, C. F., 3rd. (2014). Synthetic zinc finger proteins: The advent of targeted gene regulation and genome modification technologies. *Accounts of Chemical Research*, *47*, 2309–2318. http://dx.doi.org/10.1021/ar500039w.

Gersbach, C. A., Gaj, T., Gordley, R. M., & Barbas, C. F., 3rd. (2010). Directed evolution of recombinase specificity by split gene reassembly. *Nucleic Acids Research*, *38*(12), 4198–4206.

Gersbach, C. A., Gaj, T., Gordley, R. M., Mercer, A. C., & Barbas, C. F., 3rd. (2011). Targeted plasmid integration into the human genome by an engineered zinc-finger recombinase. *Nucleic Acids Research*, *39*(17), 7868–7878.

Gonzalez, B., Schwimmer, L. J., Fuller, R. P., Ye, Y., Asawapornmongkol, L., & Barbas, C. F., 3rd. (2010). Modular system for the construction of zinc-finger libraries and proteins. *Nature Protocols*, *5*(4), 791–810.

Gordley, R. M., Gersbach, C. A., & Barbas, C. F., 3rd. (2009). Synthesis of programmable integrases. *Proceedings of the National Academy of Sciences of the United States of America*, *106*(13), 5053–5058.

Gordley, R. M., Smith, J. D., Graslund, T., & Barbas, C. F., 3rd. (2007). Evolution of programmable zinc finger-recombinases with activity in human cells. *Journal of Molecular Biology*, *367*(3), 802–813.

Grindley, N. D., Whiteson, K. L., & Rice, P. A. (2006). Mechanisms of site-specific recombination. *Annual Review of Biochemistry*, *75*, 567–605.

Joung, J. K., & Sander, J. D. (2013). TALENs: A widely applicable technology for targeted genome editing. *Nature Reviews Molecular Cell Biology*, *14*(1), 49–55.

Klippel, A., Cloppenborg, K., & Kahmann, R. (1988). Isolation and characterization of unusual gin mutants. *EMBO Journal*, 7(12), 3983–3989.

Maeder, M. L., Thibodeau-Beganny, S., Osiak, A., Wright, D. A., Anthony, R. M., Eichtinger, M., et al. (2008). Rapid "open-source" engineering of customized zinc-finger nucleases for highly efficient gene modification. *Molecular Cell*, *31*(2), 294–301.

Maeder, M. L., Thibodeau-Beganny, S., Sander, J. D., Voytas, D. F., & Joung, J. K. (2009). Oligomerized pool engineering (OPEN): An 'open-source' protocol for making customized zinc-finger arrays. *Nature Protocols*, *4*(10), 1471–1501.

Mercer, A. C., Gaj, T., Fuller, R. P., & Barbas, C. F., 3rd. (2012). Chimeric TALE recombinases with programmable DNA sequence specificity. *Nucleic Acids Research*, *40*(21), 11163–11172.

Morgenstern, J. P., & Land, H. (1990). Advanced mammalian gene transfer: High titre retroviral vectors with multiple drug selection markers and a complementary helper-free packaging cell line. *Nucleic Acids Research*, *18*(12), 3587–3596.

Proudfoot, C., McPherson, A. L., Kolb, A. F., & Stark, W. M. (2011). Zinc finger recombinases with adaptable DNA sequence specificity. *PLoS One*, *6*(4), e19537.

Sander, J. D., Dahlborg, E. J., Goodwin, M. J., Cade, L., Zhang, F., Cifuentes, D., et al. (2011). Selection-free zinc-finger-nuclease engineering by context-dependent assembly (CoDA). *Nature Methods*, *8*(1), 67–69.

Sirk, S. J., Gaj, T., Jonsson, A., Mercer, A. C., & Barbas, C. F., 3rd. (2014). Expanding the zinc-finger recombinase repertoire: Directed evolution and mutational analysis of serine recombinase specificity determinants. *Nucleic Acids Research*, *42*(7), 4755–4766.

Wright, D. A., Thibodeau-Beganny, S., Sander, J. D., Winfrey, R. J., Hirsh, A. S., Eichtinger, M., et al. (2006). Standardized reagents and protocols for engineering zinc finger nucleases by modular assembly. *Nature Protocols*, *1*(3), 1637–1652.

CHAPTER FIVE

# Genome Engineering in Human Cells

**Minjung Song*, Young-Hoon Kim*, Jin-Soo Kim[†,‡,1], Hyongbum Kim*[,1]**

*Graduate School of Biomedical Science and Engineering, College of Medicine, Hanyang University, Seoul, South Korea
[†]Center for Genome Engineering, Institute for Basic Science, Seoul, South Korea
[‡]Department of Chemistry, Seoul National University, Seoul, South Korea
[1]Corresponding authors: e-mail address: jskim01@snu.ac.kr; hkim1@hanyang.ac.kr

## Contents



## Abstract

Genome editing in human cells is of great value in research, medicine, and biotechnology. Programmable nucleases including zinc-finger nucleases, transcription activator-like effector nucleases, and RNA-guided engineered nucleases recognize a specific target sequence and make a double-strand break at that site, which can result in gene disruption, gene insertion, gene correction, or chromosomal rearrangements. The target sequence complexities of these programmable nucleases are higher than 3.2 mega base

*Methods in Enzymology*, Volume 546
ISSN 0076-6879
http://dx.doi.org/10.1016/B978-0-12-801185-0.00005-2

pairs, the size of the haploid human genome. Here, we briefly introduce the structure of the human genome and the characteristics of each programmable nuclease, and review their applications in human cells including pluripotent stem cells. In addition, we discuss various delivery methods for nucleases, programmable nickases, and enrichment of gene-edited human cells, all of which facilitate efficient and precise genome editing in human cells.

## 1. INTRODUCTION

Genome engineering in human cells is of great value in research, medicine, and biotechnology. In research, one of the best ways to determine the function of a human gene or genetic element is to compare the phenotype of human cells containing a mutation in the gene or element of interest with that of isogenic normal human cells. This process is increasingly important given that a growing number of researchers are using human pluripotent stem cells as disease models to investigate disease pathophysiology and screen therapeutic drugs *in vitro* (Colman & Dreesen, 2009; Saha & Jaenisch, 2009). Furthermore, if reporter genes or peptide tags are inserted into endogenous genes through genome engineering, it becomes possible to monitor or trace those genes. In medicine, many genetic diseases could be prevented or treated if the genetic mutations that cause the disease were corrected, as has been done in cell or animal models (Li et al., 2011; Osborn et al., 2013; Schwank et al., 2013; Voit, Hendel, Pruett-Miller, & Porteus, 2014; Yin et al., 2014). This targeted genetic modification can potentially also be used to treat nongenetic diseases such as human immunodeficiency virus (HIV) infection, which has been tested in human patients (Holt et al., 2010; Tebas et al., 2014). In biotechnology, targeted genetic modification of human cells can also contribute to technical developments. For example, when human cells such as Chinese hamster ovary cells are used to produce specific proteins, genome engineering can improve yields and enhance the efficiency of this process.

Conventional gene targeting approaches based on homologous recombination (HR), which occurs in nature when sperms and eggs are generated, can be used to achieve targeted genetic modification in human cells (Smithies, Gregg, Boggs, Koralewski, & Kucherlapati, 1985; Song, Schwartz, Maeda, Smithies, & Kucherlapati, 1987). However, the efficiency of HR is extremely low, necessitating elaborate positive and negative selection to obtain cells that contain the desired modification.

Double-strand breaks (DSBs) at the target site can increase the efficiency of HR by at least two orders of magnitude (Rouet, Smih, & Jasin, 1994). Furthermore, error-prone repair of these DSBs through nonhomologous end joining (NHEJ) can lead to targeted mutagenesis (Bibikova, Golic, Golic, & Carroll, 2002). DSBs at specific genomic loci can be generated by specific sequence-recognizing programmable nucleases, which include zinc-finger nucleases (ZFNs), transcription activator-like effector nucleases (TALENs), and RNA-guided engineered nucleases (RGENs) (Kim & Kim, 2014).

In this chapter, we will first briefly review the structure of the human genome. We will then describe the three programmable nucleases, i.e., ZFNs, TALENs, and RGENs, and their applications in human cells, including their potential utilization for the treatment of both genetic and non-genetic diseases. We will also review various methods for delivering programmable nuclease into human cells as well as techniques for improving the efficiency of the editing process by using nickases or surrogate reporters in human cells.

## 2. STRUCTURE OF THE HUMAN GENOME

The Human Genome Project was initiated in 1990 and declared complete in 2003. The sequence was determined using a combination of high-throughput experiments and bioinformatics approaches (International Human Genome Sequencing, 2004; Lander et al., 2001; She et al., 2004; Venter et al., 2001). The genome in diploid (somatic) cells is composed of 22 pairs of autosomal chromosomes and two sex chromosome (XX in females, XY in males) (Fig. 5.1A and B). The haploid genome (contained in egg and sperm cells) has a total length of 3.2 billion base pairs (bp); the diploid genome is 6.4 billion bp in length. The genome includes approximately 20,000 protein-coding DNA genes, which represent about 1.5% of the genome, and noncoding DNA, which occupies the remaining fraction (approximately 98%) (International Human Genome Sequencing, 2004; Lander et al., 2001; Pennisi, 2012). Although the human genome sequence has been completely determined, the biological functions of the genes and noncoding elements are not fully understood. Experiments aimed at elucidating biological functions have been mainly conducted with coding sequences (Harrow et al., 2009). Noncoding DNA is composed of gene-related sequences [pseudogenes, gene fragments, introns, and untranslated regions (UTRs)], long interspersed elements (LINEs), short interspersed

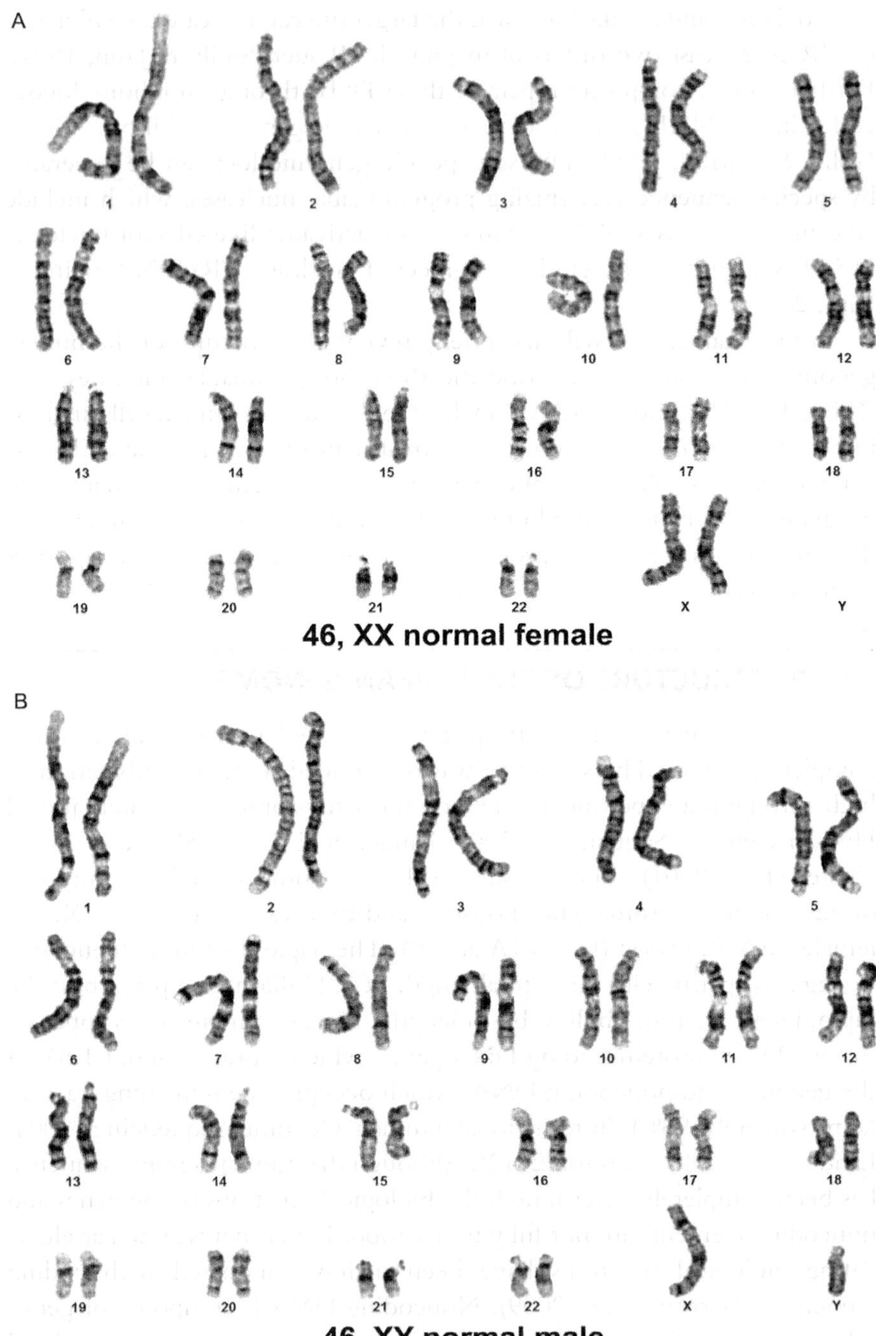


**Figure 5.1**—See legend on opposite page.

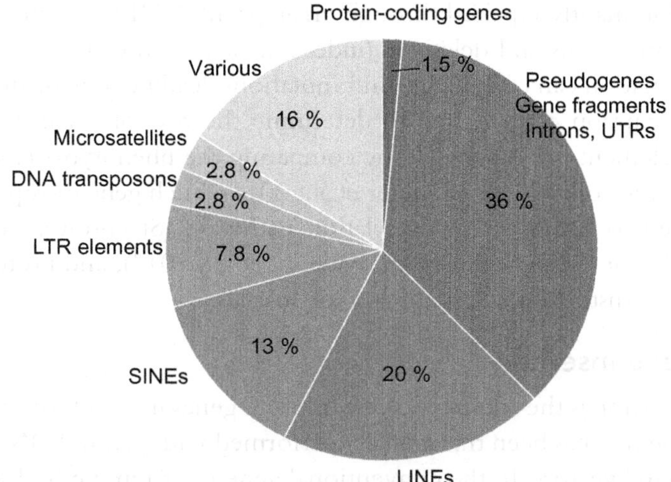


**Figure 5.2** Human genome organization. The human genome is composed of protein-coding genes and noncoding elements that include pseudogenes, gene fragments, introns, and untranslated regions (UTRs), long interspersed elements (LINEs), short interspersed elements (SINEs), long terminal repeats (LTRs), DNA transposons, microsatellites, and various elements of unknown function. The relative percentage of each component is calculated from the previously reported genome sequence data (Lander et al., 2001; Nussbaum, McInnes, & Willard, 2007).

elements (SINEs), long terminal repeat (LTR) elements, DNA transposons, microsatellites, and various other elements (Fig. 5.2).

## 3. SCOPE OF HUMAN GENE EDITING USING PROGRAMMABLE NUCLEASES

DSBs generated by programmable nucleases can lead to various genetic modifications including gene disruptions, gene insertions, gene corrections, point mutations, and chromosomal rearrangements.

### 3.1. Gene disruption

Gene disruption is the simplest form of genome editing that can be achieved using programmable nucleases. DSBs generated by programmable nucleases

**Figure 5.1** Human chromosomes. Human somatic cells contain two sets of chromosomes, one set given by each parent. Each set has 23 single chromosomes—22 autosomes and an "X" or "Y" sex chromosome. (A) A normal female karyotype is composed of 22 pairs of autosomes and two X chromosomes. (B) A normal male karyotype is composed of 22 pairs of autosomes, an X chromosome, and an Y chromosome.

are predominantly repaired through error-prone NHEJ, which often leads to small insertions and deletions (indels) at or near the cleavage site. This indel formation can cause frameshift mutations, leading to gene disruptions. Gene disruption can be used to determine the role of a specific gene or genetic element in human cells by comparing the phenotypes of knockout and isogenic control cells (Soldner et al., 2011). Such gene disruptions have been used to investigate glycosylation pathways (Steentoft et al., 2011), nuclear factor κB signaling (Kim, Kweon, et al., 2013), and protein methylation (Kernstock et al., 2012) in human cells.

### 3.2. Gene insertion

Gene insertion is the addition of one or more genes into a DNA sequence, a technique that has been traditionally performed with plasmid DNA or integrating viral vectors. In the conventional gene insertion method, the insertion site cannot be controlled. Uncontrolled integration of a transgene or its regulatory sequences into undesired sites can inactivate essential genes or activate proto-oncogenes (Dave, Jenkins, & Copeland, 2004; Hacein-Bey-Abina et al., 2003; McCormack & Rabbitts, 2004). Programmable nuclease-induced DSB generation at or near the desired gene insertion site drastically (by at least two orders of magnitude) enhances the efficiency of gene insertion through homology-directed repair (HDR). Several sites in the human genome have been proposed to be genomic safe harbors, where transgenes can be inserted and expressed without causing significant alternations in the expression of other genetic elements (Lombardo et al., 2011).

### 3.3. Gene correction

Gene correction or point mutagenesis can be achieved by delivering both programmable nucleases and correction/mutation templates such as target vectors (Bibikova, Beumer, Trautman, & Carroll, 2003) or single-stranded DNA oligonucleotides (ssODNs) (Chen et al., 2011). Both targeting vectors and ssODNs have homology arms, the sequence of which are identical or similar to that of regions neighboring the nuclease cleavage sites. Compared to targeting vectors, ssODNs are relatively easy and simple to prepare. ssODNs combined with ZFNs have been used to generate isogenic disease models using human pluripotent stem cells (Soldner et al., 2011).

### 3.4. Chromosomal rearrangement

If two DSBs are generated, chromosomal rearrangements, including deletions, duplications, inversions, and translocations, can be induced in human cells (Brunet et al., 2009; Lee, Kim, & Kim, 2010; Lee, Kweon, Kim, Kim, & Kim, 2012). Rearrangements of chromosomal segments of up to a few mega bp in size can be generated using ZFNs (Lee et al., 2010, 2012; Urnov, Rebar, Holmes, Zhang, & Gregory, 2010), TALENs (Joung & Sander, 2013), or RGENs (Sander & Joung, 2014). RGENs have an advantage over ZFNs and TALENs for inducing these rearrangements because RGEN-based multiplex genome editing is relatively easily achieved; a single protein (Cas9) is used in common by the RGEN targeting each site and only the addition of one guide RNA is required to generate cleavage at the additional site (Table 5.1).

# 4. PROGRAMMABLE NUCLEASES USED FOR GENOME EDITING IN HUMAN CELLS

## 4.1. ZFNs

ZFNs are chimeric programmable nucleases composed of a DNA-binding zinc-finger protein (ZFP) domain at the amino terminus and the *Fok*I nuclease cleavage domain at the carboxyl terminus. ZFNs work as heterodimers because *Fok*I must dimerize to cut the DNA (Bitinaite, Wah, Aggarwal, & Schildkraut, 1998). ZFPs contain a tandem array of Cys2His2 zinc fingers, each recognizing approximately 3 bp of DNA (Tupler, Perini, & Green, 2001; Wolfe, Nekludova, & Pabo, 2000). The binding specificity of the designed zinc-finger domain directs the ZFN to a specific genomic site. Compared with other programmable nucleases, the use of ZFNs is often limited by poor targeting density and relatively high levels of off-target effects, leading to cytotoxicity. However, ZFNs are also the smallest type of programmable nuclease, making it possible to express them using delivery vectors such as adeno-associated viral (AAV) vector (Li et al., 2011). Furthermore, ZFNs are the only type of programmable nuclease that has been tested in a completed clinical trial in human patients (Tebas et al., 2014).

## 4.2. TALENs

TALENs are composed of a *Fok*I nuclease domain and a customizable DNA-binding domain derived from transcription activator-like effectors of *Xanthomonas,* a plant pathogen (Miller et al., 2011). Within the TALE structure, the DNA sequence recognition domain is characterized by a repeating unit of 33–35 conserved amino acids. Each repeat is almost identical except for two highly variable amino acids at positions 12 and 13, which are called repeat variable diresidues (Boch et al., 2009; Moscou &

**Table 5.1** Examples of genome editing using programmable nucleases in human cells

| Engineered nucleases | Genes | Associated diseases | Types of genetic modification | References |
|---|---|---|---|---|
| ZFNs | Inserted *XIST* into the *DYRK1A* locus on chromosome 21 | Down's syndrome | Gene insertion by HDR | Jiang, Jing, et al. (2013) |
| | Inserted gp91phox minigene into the *AAS1* locus | X-linked chronic granulomatous | Gene insertion by HDR | Zou, Sweeney, et al. (2011) |
| | *HBB* | Sickle cell anemia | Gene correction by HDR | Zou, Mali, et al. (2011) and Sebastiano et al. (2011) |
| | $\alpha$-Synuclein | Parkinson disease | Disease modeling by HDR | Soldner et al. (2011) |
| | *PPP1R12C/p84* gene on chromosome 19 and the *IL2R$\gamma$* gene on X chromosome | Human tumors | Translocation | Brunet et al. (2009) |
| | *A1AT* | $\alpha_1$—Antitrypsin deficiency | Gene correction by HDR | Yusa et al. (2011) |
| | *CCR5* | HIV infection | Gene disruption by NHEJ | Holt et al. (2010), Maier et al. (2013), Perez et al. (2008), and Tebas et al. (2014) |
| | *HBV cccDNA* | HBV infection | Gene disruption | Cradick et al. (2010) |

| | | | | |
|---|---|---|---|---|
| TALENs | *HBB* | β-Thalassemia | Gene correction by HDR | Ma et al. (2013) |
| | *COL7A1* | Epidermolysis bullosa | Gene correction by HDR | Osborn et al. (2013) |
| | *DMD* | Duchenne muscular dystrophy | Gene disruption by NHEJ | Ousterout et al. (2013) |
| | *APOB* | HCV infection | Disease modeling by NHEJ | Ding, Lee, et al. (2013) |
| | *SORT1* | Dyslipidemia, Insulin resistance, Motor-neuron death | Disease modeling by NHEJ | Ding, Lee, et al. (2013) |
| | *AKT2* | Insulin resistance | Disease modeling by NHEJ | Ding, Lee, et al. (2013) |
| | | Hypoglycemia | Disease modeling by NHEJ | Ding, Lee, et al. (2013) |
| | | Hypoinsulinemia | Disease modeling by NHEJ | Ding, Lee, et al. (2013) |
| | *PLIN1* | Lipodystrophy | Disease modeling by NHEJ | Ding, Lee, et al. (2013) |
| | *HPRT* | Lesch-Nyhan syndrome | Disease modeling by NHEJ | Frank et al. (2013) |
| | HBV cccDNA | HBV infection | Gene disruption by NHEJ | Bloom et al. (2013) and Chen et al. (2014) |
| RGENs | *CFTR* | Cystic fibrosis | Gene correction by HDR | Schwank et al. (2013) |
| | *HTT* | Huntington's disease | Disease modeling by HDR | An et al. (2014) |

HDR, homology-directed repair; NHEJ, nonhomologous end joining.

Bogdanove, 2009). One repeat unit recognizes one base pair in the major groove of DNA (Deng et al., 2012; Mak, Bradley, Cernadas, Bogdanove, & Stoddard, 2012). Like ZFNs, TALENs work as heterodimers because the *Fok1* domain requires dimerization to function. Genome-wide TALEN libraries, targeting 18,740 human protein-coding genes (Kim, Kweon, et al., 2013) and 274 human miRNA-coding sequences (Kim, Wee, et al., 2013), are available.

## 4.3. RGENs

RGENs are composed of a guide RNA and the Cas9 nuclease. This programmable nuclease is derived from an adaptive immune system in bacteria and archaea called the clustered, regularly interspaced, short palindromic repeat (CRISPR) system (Barrangou et al., 2007; Makarova, Grishin, Shabalina, Wolf, & Koonin, 2006). Bacteria and archaea capture small (~20 bp) foreign DNA fragments from invading phages or viruses and insert these fragments into the genomic CRISPR locus. In type II systems, the fragments of foreign DNA, called protospacers, are transcribed as pre-CRISPR RNA (pre-crRNA) and processed to crRNA in the presence of transactivating crRNA (tracrRNA), which is also transcribed from the CRISPR locus (Deltcheva et al., 2011). The crRNA and tracrRNA are then complexed with CRISPR-associated protein 9 (Cas9) nuclease to construct an active sequence-specific endonuclease. Among several Cas9 nucleases, Cas9 derived from *Streptococcus pyogenes* has been widely used (Cho, Kim, Kim, & Kim, 2013; Cong et al., 2013; Hwang et al., 2013; Jiang, Bikard, Cox, Zhang, & Marraffini, 2013; Jinek et al., 2013; Mali, Yang, et al., 2013). The target sequence of an RGEN is 23 bp in length, which comprises the 20 bp guide sequence in the crRNA and a 3 bp (5′-NGG-3′) protospacer-adjacent motif sequence that is directly recognized by Cas9 (Mojica, Diez-Villasenor, Garcia-Martinez, & Almendros, 2009). The combination of crRNA and tracrRNA can be replaced with a single-chain guide RNA (sgRNA) (Jinek et al., 2012), simplifying RGENs to only two components.

RGENs have several advantages over ZFNs and TALENs. First, their design and preparation are easy and simple. Because the Cas9 protein is a common component, new RGENs that target a given sequence can be prepared by cloning a 20-bp guide sequence into a vector that expresses guide RNA. Alternatively, the guide RNA can be transcribed *in vitro* before delivery (Cho et al., 2013; Kim, Kim, Cho, Kim, & Kim, 2014; Ramakrishna, Kwaku Dad, et al., 2014), bypassing the cloning process. Second, multiplex genome editing is facilitated. To target one additional locus, one new pair of ZFNs or TALENs must be prepared, which requires an additional, relatively complex cloning process, whereas adding one more guide RNA suffices in the RGEN system. A more detailed comparison of the three programmable nucleases has recently been published (Kim & Kim, 2014).

Very recently, catalytically inactive Cas9 protein has been combined with the Fok1 nuclease domain to generate a highly specific programmable nuclease, called an RNA-guided Fok1 nuclease (RFN) (Guilinger, Thompson, & Liu, 2014; Tsai et al., 2014). Whereas RGENs function as monomers, limiting their complexity to $4^{22}$, RFNs function as heterodimers, similar to ZFNs and TALENs; thus, RFN complexity is $4^{44}$, making these nucleases more specific.

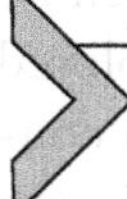

## 5. CORRECTION OF HUMAN GENETIC DISEASES USING PROGRAMMABLE NUCLEASES

Genetic diseases are primarily caused by genomic mutations and correction of these mutations using programmable nucleases can provide promising therapeutic modalities. Such corrections have been achieved *ex vivo* using patient-derived cells (Jiang, Jing, et al., 2013; Ma et al., 2013; Ousterout et al., 2013; Schwank et al., 2013; Sebastiano et al., 2011; Yusa et al., 2011; Zou, Mali, Huang, Dowey, & Cheng, 2011; Zou, Sweeney, et al., 2011) or *in vivo* in mouse cells containing human genetic sequences (Li et al., 2011). For *ex vivo* gene corrections, patient-derived induced pluripotent stem cells (iPSCs) (Jiang, Jing, et al., 2013; Ma et al., 2013; Sebastiano et al., 2011; Yusa et al., 2011; Zou, Mali, et al., 2011), adult stem cells (Schwank et al., 2013), somatic cells (Ousterout et al., 2013), and cancer cells (Sun, Liang, Abil, & Zhao, 2012; Voit et al., 2014) have been used. Transplantation of the genetically corrected human cells into animal model has been shown to lead to amelioration of the disease (Yusa et al., 2011).

ZFN-induced genome corrections have been mostly performed in patient-derived iPSCs, involving diseases that include sickle cell disease (Sebastiano et al., 2011; Zou, Mali, et al., 2011), X-linked chronic granulomatous disease (X-CGD) (Zou, Sweeney, et al., 2011), α1-antitrypsin deficiency (Yusa et al., 2011), and Down's syndrome (Jiang, Jing, et al., 2013). In an initial stage of ZFN-induced genome correction study, ZFNs were used to correct the single point mutation in the β-globin gene (*HBB*) that is responsible for sickle cell disease (Sebastiano et al., 2011; Zou, Mali, et al., 2011). iPSCs containing two mutated β-globin alleles were derived from a patient with sickle cell disease and the mutations were corrected with ZFNs *ex vivo*. When the corrected iPSCs were differentiated, 25–40% of the resulting cell population consisted of wild-type erythrocytes. ZFN-mediated safe harbor targeting enabled genetic correction of X-CGD, a defect of neutrophil microbicidal reactive oxygen species (ROS) production (Zou, Sweeney, et al., 2011). Corrected iPSCs were differentiated into neutrophils, which displayed restored ROS production. In another study, ZFNs were used to correct the point mutation in the α1-antitrypsin gene (*A1AT*) that is responsible for α1-antitrypsin deficiency (Yusa et al., 2011). Human iPSCs were differentiated into hepatocytes and engrafted into the mouse liver, where they replaced endogenous mouse hepatocytes. Cellular structures and functions were restored both *in vitro* and *in vivo*. Recently, ZFN-induced correction of Down's syndrome (trisomy 21) was explored (Jiang, Jing, et al., 2013). Human iPSCs from an individual with this condition were genetically edited with ZFNs to express a gene involved with dosage compensation, silencing one copy of chromosome 21. As a result, impaired cell proliferation and neurogenesis were rapidly recovered.

TALEN-induced gene corrections relevant to human disease have been performed in various cell types such as cancer cells for sickle cell disease and β-thalassemia (Sun et al., 2012; Voit et al., 2014), myoblasts/fibroblasts for Duchenne muscular dystrophy (DMD) (Ousterout et al., 2013), skin cells for recessive dystrophic epidermolysis bullosa (RDEB) (Osborn et al., 2013), and iPSCs for β-thalassemia (Ma et al., 2013). For sickle cell disease, TALENs were constructed to target the mutated human β-globin gene (*HBB*) locus and achieved high levels of targeted gene repair in HeLa (Sun et al., 2012) and K562 cells (Voit et al., 2014). In skeletal myoblasts and dermal fibroblasts from DMD patients, TALEN-induced genomic editing restored the expression of a functional dystrophin protein (Ousterout et al., 2013). Skin cells from an individual with RDEB with type VII collagen gene (*COL7A1*) defects showed normal protein expression after TALEN-based gene correction (Osborn et al., 2013). TALENs were also used in a study related to β-thalassemia, a life-threatening blood disorder associated with a β-globin gene (*HBB*) mutation. Corrected iPSCs were successfully differentiated into erythroblasts that expressed normal β-globin (Ma et al., 2013).

Genome editing with RGENs has been explored for treating cystic fibrosis (Schwank et al., 2013). This genome editing system was used to correct the cystic fibrosis transmembrane conductor receptor (*CFTR*) locus in intestinal stem cells from cystic fibrosis patients. Recently, the RGEN system was also used to correct the phenotype in a humanized mouse model of hereditary tyrosinemia, which is a fatal disease caused by mutation of the fumarylacetoacetate hydrolase (*FAH*) gene (Yin et al., 2014). Appropriate CRISPR/Cas9 components were directly delivered to the mice. As a result, several Fah-positive hepatocytes were generated, which ultimately expanded and rescued the body weight loss phenotype.

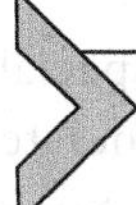

## 6. TREATMENT OF HUMAN NONGENETIC DISEASES USING PROGRAMMABLE NUCLEASES

Programmable nucleases can be used as novel therapeutic modalities for nongenetic diseases. So far, studies have focused on viral infectious diseases; therapeutic effects have been achieved by targeting the human receptor for the virus (Holt et al., 2010; Li et al., 2013; Maier et al., 2013; Perez et al., 2008; Tebas et al., 2014) or the virus genome itself (Bloom, Ely, Mussolino, Cathomen, & Arbuthnot, 2013; Chen et al., 2014; Cradick, Keck, Bradshaw, Jamieson, & McCaffrey, 2010; Schiffer et al., 2012).

Representative studies have been performed for HIV infection. HIV entry into human T cells is mediated by CD4 and CCR5 receptors. Roughly, 1% of people in Northern Europe have biallelic mutations in the *CCR5* gene and these people are resistant to HIV infection (Duncan, Scott, & Duncan, 2005). Furthermore, "cure" of HIV infection has been achieved by bone marrow transplantation using a donor with biallelic mutations in the *CCR5* gene (Hutter et al., 2009). ZFN-mediated disruption of the *CCR5* gene in human T cells prevented HIV infection into the T cells *in vitro* (Maier et al., 2013; Perez et al., 2008). Given that human T cells have limited self-renewal capacity, ZFN-mediated *CCR5* gene disruption has been pursued in human $CD34^+$ hematopoietic stem/progenitor cells, which significantly lowers HIV-1 levels upon transplantation into a humanized mouse model (Holt et al., 2010). Currently, three clinical trials involving ZFN-treated cells (identifier #: NCT00842634, NCT01044654, NCT01543152) are ongoing. The first results from the clinical trials (NCT00842634), which showed the safety and feasibility of *CCR5*-modified autologous $CD4^+$ T cell infusions in patients, were recently published. After transfusion, these cells successfully engrafted and persisted, leading to an increase in $CD4^+$ T cells and a decrease in HIV DNA levels in most patients (Tebas et al., 2014).

The use of programmable nucleases has also been investigated in human hepatitis B (HBV) infection. HBV is a DNA virus and targeting of HBV genes using ZFNs (Cradick et al., 2010) and TALENs (Bloom et al., 2013; Chen et al., 2014; Schiffer et al., 2012) resulted in decreases in HBV titer *in vitro* and *in vivo*.

## 7. GENOME ENGINEERING IN HUMAN PLURIPOTENT STEM CELLS

Genome engineering in human pluripotent stem cells is of special value because once such cells are engineered, they can be used to generate unlimited numbers of all types of cells, each genetically engineered. Such cells can serve as disease models (Colman & Dreesen, 2009; Saha & Jaenisch, 2009) and for transplantation (Yusa et al., 2011). These disease models include Parkinson disease (Soldner et al., 2011), Huntington's disease (An et al., 2014), Down's syndrome (Jiang, Jing, et al., 2013), Lesch-Nyhan-Syndrome (Frank, Skryabin, & Greber, 2013), dyslipidemia, insulin resistance, hypoglycemia, lipodystrophy, motor-neuron death, and hepatitis C infection (Ding, Lee, et al., 2013). In another case, ZFNs were used to induce chromosomal translocations in stem or precursor cells, which could be applied in modeling various human tumors (Brunet et al., 2009). In a different line of work, genetically corrected patient iPSCs were differentiated into hepatocyte-like cells and transplanted onto an $\alpha_1$ antitrypsin deficiency mouse model, where they successfully restored the structure and function of liver cells (Yusa et al., 2011).

Targeted genetic modification is possible using traditional HR, but the efficiency is extremely low (Zwaka & Thomson, 2003). Efficient targeted genetic modification using programmable nuclease in human pluripotent stem was achieved using ZFNs (Brunet et al., 2009; Hockemeyer et al., 2009, 2011) and TALENs (Ding, Lee, et al., 2013). The efficiency of genome engineering in pluripotent stem cells has been further improved using RGENs (Ding, Regan, et al., 2013).

For useful disease modeling, patient-derived pluripotent stem cells need to be coupled with a good control. Allogenic pluripotent stem cells derived from normal subjects can be controls (Brennand et al., 2011; Ebert et al., 2009; Itzhaki et al., 2011; Lee et al., 2009; Marchetto et al., 2010; Rashid et al., 2010), but their different genetic background can be a compounding factor for making comparisons. As a result, the investigation of disease pathophysiology and the screening of therapeutic small molecules are often inefficient unless the phenotypes of the control and patient-derived pluripotent stem cells are dramatically different (e.g., for early onset or metabolic diseases). Alternatively, good isogenic controls can be obtained from the patient-derived stem cells themselves if the pathogenic genes are corrected using programmable nucleases (Soldner et al., 2011; Fig. 5.3). Furthermore, *in vitro* disease models can be created by engineering the disease-linked mutations in normal pluripotent stem cells using programmable nucleases (Soldner et al., 2011). This generation of isogenic controls extends the pluripotent stem cell-mediated investigation of disease pathophysiology and/or drug screening to diseases that have subtle phenotypic changes, late age onset, or are slow progressing.

Besides facilitating the creation of disease models, genome editing in pluripotent stem cells has been used to obtain greater knowledge about stem cell biology. For example, ZFNs enabled the introduction of reporter genes into endogenous genomic loci, allowing pluripotency or cellular differentiation to be monitored (Collin & Lako, 2011).

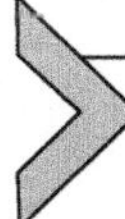

## 8. DELIVERY OF PROGRAMMABLE NUCLEASES TO HUMAN CELLS

For efficient genome editing, the successful delivery of programmable nuclease and/or homologous templates (e.g., targeting vectors or ssODNs) into target cells is essential. Plasmids have been widely used to deliver the genes encoding programmable nucleases into human cells (Kim et al., 2014; Porteus & Baltimore, 2003; Urnov et al., 2010). However, using plasmids requires the selection of optimized promoters and codons and is complicated with uncontrolled integration of the plasmid DNA into the host genome (Kim et al., 2014) and unwanted immune responses (Hemmi et al., 2000; Wagner, 2001).

Nonintegrating viral vectors have been used to deliver programmable nucleases both into human cells *in vitro* and into mouse cells that contain human genes *in vivo*. Such vectors include integrase-defective lentiviral vectors (IDLVs) (Lombardo et al., 2007, 2011), adenoviral vectors (Holkers et al., 2013), and AAVs (Handel et al., 2012). Because AAVs require a relatively small cargo size, only ZFNs, which are encoded by shorter sequences than the other programmable nucleases, have been delivered via AAVs into human cells *in vitro* (Handel et al., 2012) and into mice that contain human genetic material *in vivo* (Li et al., 2011). IDLVs successfully delivered ZFNs into various cells, including hard-to-transfect primary cells such as human

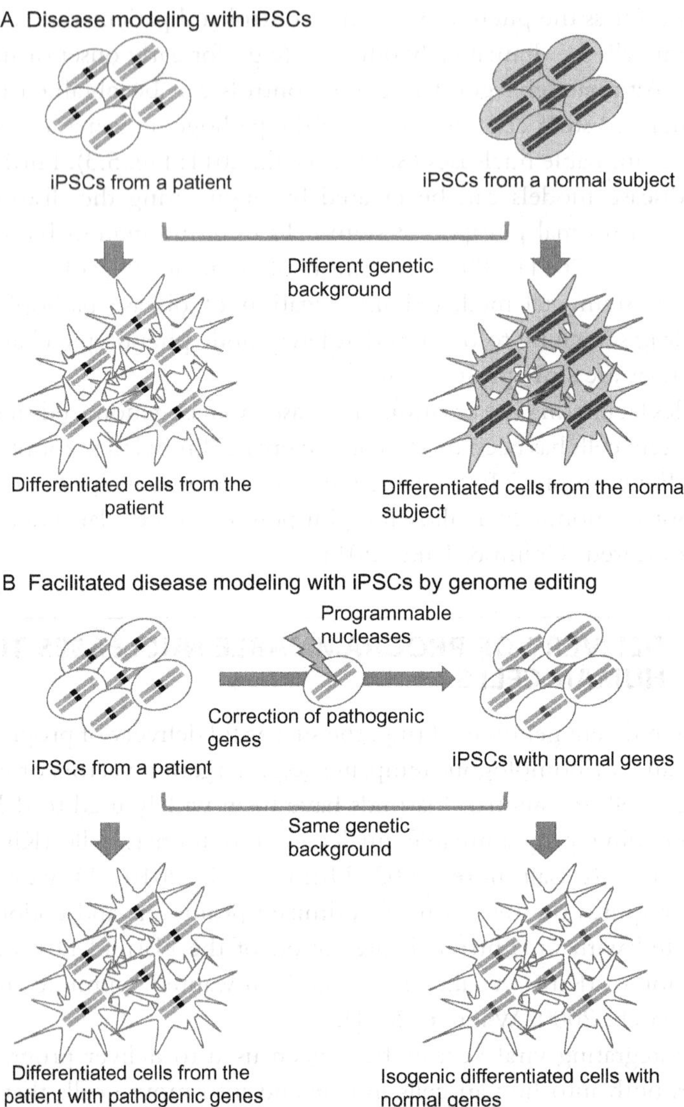


**Figure 5.3** Improved disease modeling in human pluripotent stem cells by programmable nuclease-mediated genome editing. (A) Conventional disease modeling using induced pluripotent stem cells (iPSCs). iPSCs are derived from a patient and a control normal individual. Both types of iPSCs are differentiated into the cell type of interest (e.g., neurons) and the phenotypes are compared. The different genetic backgrounds can be a compounding factor for such investigations. (B) Pathogenic genes in patient-derived iPSCs are corrected by genome editing using programmable nucleases. Both corrected and uncorrected iPSCs are differentiated into the cell type of interest and the phenotypes are compared. In this case, both cell types have the same genetic background, making the gene-corrected cells a good control for the study.

hematopoietic progenitor cells and embryonic stem cells (Lombardo et al., 2007). TALEN delivery is more challenging due to the large size and highly repetitive nature of the sequences that encode these nucleases. Because of size limitations, TALEN delivery with AAVs was unsuccessful (Asokan, Schaffer, & Samulski, 2012; Wu, Asokan, & Samulski, 2006). IDLVs are also incompatible with TALENs because of the highly homologous TALE repeats (Holkers et al., 2013). Integrating vectors such as lentiviral vectors have been used for continuous expression of Cas9 and sgRNAs in human cells (Shalem et al., 2014), which results in more efficient genome editing than transient delivery. However, this prolonged expression can aggravate off-target cleavage effects. Other drawbacks of viral vectors include the time required for their production, potential problematic immune responses, and safety concerns (Thomas, Ehrhardt, & Kay, 2003).

In contrast, direct delivery of either protein or RNA, which cannot integrate into the host genome, has relatively few safety concerns. ZFNs have a net positive charge and can be directly delivered into human cells without the need for cell-penetrating peptides (CPPs); when delivered in this manner ZFNs can cause specific gene disruption (Gaj, Guo, Kato, Sirk, & Barbas, 2012). CPP-conjugated TALENs can be directly delivered into human cells, likewise causing gene editing (Liu, Gaj, Patterson, Sirk, & Barbas, 2014). Direct RGEN delivery is slightly more complex because two components, i.e., protein and RNA, should be delivered together. When various human cells were treated with CPP-conjugated Cas9 and CPP-complexed guide RNA, efficient gene disruption was achieved, with reduced off-target mutations relative to plasmid delivery (Ramakrishna, Kwaku Dad, et al., 2014). Furthermore, the ribonucleoprotein Cas9–sgRNA complex can be directly delivered into human cells using electroporation, without the use of CPPs, causing highly efficient genome editing with reduced off-target effects (Kim & Kim, 2014). The reduction in off-target effects is attributable to the shorter working time available for the nucleases; proteins and RNA are rapidly degraded, whereas transfected plasmids continue to produce proteins and RNAs for longer periods (Gaj et al., 2012; Kim & Kim, 2014).

## 9. NICKASES FOR MODIFYING THE HUMAN GENOME

Even in the presence of targeting vectors or ssODNs, DSBs can be repaired through error-prone NHEJ, which often causes uncontrolled and unwanted indels at both target and off-target sites. To avoid these undesirable mutations, nickases—enzymes that generate DNA single-strand breaks (SSBs)—have been used as precise genome editing tools. SSBs can simulate HDR without activating the error-prone NHEJ pathway, essentially preventing unwanted indel formation (Davis & Maizels, 2011; McConnell Smith et al., 2009; Metzger, McConnell-Smith, Stoddard, & Miller, 2011). The first programmable nickases (termed ZFNickases) were derived from ZFNs through the addition of a mutation in the Fok1 cleavage domain; the resulting enzyme did not cause unwanted DSBs or indels at target or off-target sites in human cells (Kim et al., 2012; Ramirez et al., 2012; Wang et al., 2012). A different approach used CRISPR system components. Cas9 has two catalytic domains; inactivation of one by site-directed mutagenesis (D10A or H840A) resulted in RGENickases (Sapranauskas et al., 2011). Such enzymes produce site-specific nicks, leading to precise HDR-mediated genome editing with negligible NHEJ-driven mutations in human cells (Cong et al., 2013). Generation of two SSBs on different strands using two nickases can lead to genome editing similar to that mediated by nuclease-generated DSBs, but with much higher specificity (Cho et al., 2014; Kim et al., 2012; Mali, Esvelt, & Church, 2013; Ran et al., 2013). Increasing the specificity and precision of genomic engineering is especially important in human applications because uncontrolled genetic modifications can lead to adverse effects including tumor development.

## 10. ENRICHMENT OF GENE-EDITED HUMAN CELLS

Programmable nuclease activity and delivery are frequently limited, resulting in only a minor fraction of nuclease-treated human cells gaining the targeted genetic modification. For applications in research, medicine, and biotechnology, a means of selecting or isolating gene-edited cells is required. However, because genetically modified cells usually show indistinguishable phenotypes compared to unmodified cells, the genomic DNA must be evaluated, which is often a laborious and time-consuming process. Thus, simple methods for enriching cells containing nuclease-induced mutations would facilitate the application of this tool.

Gene-modified cells have been enriched by selecting cells transfected with the nuclease-encoding vector; this technique has some effectiveness because of the low delivery levels of programmable nucleases. For example, TALEN-modified human cells can be enriched by selecting transfected cells via flow cytometric sorting using vectors that express fluorescent proteins or through antibiotic resistance factors (Ding, Lee, et al., 2013; Frank et al., 2013).

Our group has described selection methods that overcome not only low levels of programmable nuclease delivery but also low levels of activity. We have developed various methods of selecting human cells containing mutations induced by ZFNs, TALENs, and RGENs using surrogate reporters (Kim & Kim, 2014; Kim, Kim, et al., 2013; Kim, Um, Cho, Jung, & Kim, 2011; Ramakrishna, Cho, et al., 2014). The surrogate reporter plasmids include a gene for a marker protein that is constitutively expressed, the nuclease target sequence, and gene(s) for other marker protein(s). These latter genes are not expressed in the absence of nuclease activity because they lack a dedicated promoter and are out of frame with the upstream marker. Furthermore, a stop codon is present upstream. The reporter plasmids and nuclease-encoding plasmids are codelivered into human cells. In cells with sufficient nuclease activity, DSBs are generated in the target sequence of the reporter plasmids (and, potentially, at the genomic site). NHEJ-mediated DSB repair can cause frameshift mutations, which can render the marker genes in frame with the upstream, expressed marker and inactivate the stop codon. Cells expressing the markers (GFP, H-2K$^{k}$, and hygromycin resistance) can be isolated using flow cytometry, magnetic separation, and hygromycin selection, respectively. This surrogate reporter-based selection can result in enrichment of cells containing nuclease-induced mutations in the genome by up to 92-fold (Kim et al., 2011).

## 11. CONCLUSION

Technologies for genome editing with programmable nucleases in human cells have been rapidly evolving. Four years ago, ZFNs were the only practical option. Now, genome editing can be achieved using ZFNs, TALENs, RGENs, RFNs, and nickases. Programmable nuclease efficiency and specificity are also improving at enormous speed. These technological improvements should render previously impossible tasks possible. They will facilitate disease pathophysiology studies, drug screening, and the development of next-generation gene therapy for genetic and nongenetic human diseases, thereby opening a new era of biomedical research and medicine.

## ACKNOWLEDGMENTS

We are grateful to Dr. Sung Han Shim (CHA University, Korea) for providing the pictures of normal human chromosomes. H. K. is supported in part by the National Research Foundation of Korea (2014R1A1A1A05006189, 2013M3A9B4076544, 2008-0062287). S.M. is supported in part by NRF-2013R1A1A1075992.

## REFERENCES


An, M. C., O'Brien, R. N., Zhang, N., Patra, B. N., De La Cruz, M., Ray, A., et al. (2014). Polyglutamine disease modeling: Epitope based screen for homologous recombination using CRISPR/Cas9 system. *PLoS Currents*, *6*, 1–19.

Asokan, A., Schaffer, D. V., & Samulski, R. J. (2012). The AAV vector toolkit: Poised at the clinical crossroads. *Molecular Therapy*, *20*, 699–708.

Barrangou, R., Fremaux, C., Deveau, H., Richards, M., Boyaval, P., Moineau, S., et al. (2007). CRISPR provides acquired resistance against viruses in prokaryotes. *Science*, *315*, 1709–1712.

Bibikova, M., Beumer, K., Trautman, J. K., & Carroll, D. (2003). Enhancing gene targeting with designed zinc finger nucleases. *Science*, *300*, 764.

Bibikova, M., Golic, M., Golic, K. G., & Carroll, D. (2002). Targeted chromosomal cleavage and mutagenesis in Drosophila using zinc-finger nucleases. *Genetics*, *161*, 1169–1175.

Bitinaite, J., Wah, D. A., Aggarwal, A. K., & Schildkraut, I. (1998). FokI dimerization is required for DNA cleavage. *Proceedings of the National Academy of Sciences of the United States of America*, *95*, 10570–10575.

Bloom, K., Ely, A., Mussolino, C., Cathomen, T., & Arbuthnot, P. (2013). Inactivation of hepatitis B virus replication in cultured cells and in vivo with engineered transcription activator-like effector nucleases. *Molecular Therapy*, *21*, 1889–1897.

Boch, J., Scholze, H., Schornack, S., Landgraf, A., Hahn, S., Kay, S., et al. (2009). Breaking the code of DNA binding specificity of TAL-type III effectors. *Science*, *326*, 1509–1512.

Brennand, K. J., Simone, A., Jou, J., Gelboin-Burkhart, C., Tran, N., Sangar, S., et al. (2011). Modelling schizophrenia using human induced pluripotent stem cells. *Nature*, *473*, 221–225.

Brunet, E., Simsek, D., Tomishima, M., DeKelver, R., Choi, V. M., Gregory, P., et al. (2009). Chromosomal translocations induced at specified loci in human stem cells. *Proceedings of the National Academy of Sciences of the United States of America*, *106*, 10620–10625.

Chen, F., Pruett-Miller, S. M., Huang, Y., Gjoka, M., Duda, K., Taunton, J., et al. (2011). High-frequency genome editing using ssDNA oligonucleotides with zinc-finger nucleases. *Nature Methods*, *8*, 753–755.

Chen, J., Zhang, W., Lin, J., Wang, F., Wu, M., Chen, C., et al. (2014). An efficient antiviral strategy for targeting hepatitis B virus genome using transcription activator-like effector nucleases. *Molecular Therapy*, *22*, 303–311.

Cho, S. W., Kim, S., Kim, J. M., & Kim, J. S. (2013). Targeted genome engineering in human cells with the Cas9 RNA-guided endonuclease. *Nature Biotechnology*, *31*, 230–232.

Cho, S. W., Kim, S., Kim, Y., Kweon, J., Kim, H. S., Bae, S., et al. (2014). Analysis of off-target effects of CRISPR/Cas-derived RNA-guided endonucleases and nickases. *Genome Research*, *24*, 132–141.

Collin, J., & Lako, M. (2011). Concise review: Putting a finger on stem cell biology: Zinc finger nuclease-driven targeted genetic editing in human pluripotent stem cells. *Stem Cells*, *29*, 1021–1033.

Colman, A., & Dreesen, O. (2009). Pluripotent stem cells and disease modeling. *Cell Stem Cell*, *5*, 244–247.

Cong, L., Ran, F. A., Cox, D., Lin, S., Barretto, R., Habib, N., et al. (2013). Multiplex genome engineering using CRISPR/Cas systems. *Science*, *339*, 819–823.

Cradick, T. J., Keck, K., Bradshaw, S., Jamieson, A. C., & McCaffrey, A. P. (2010). Zinc-finger nucleases as a novel therapeutic strategy for targeting hepatitis B virus DNAs. *Molecular Therapy*, *18*, 947–954.

Dave, U. P., Jenkins, N. A., & Copeland, N. G. (2004). Gene therapy insertional mutagenesis insights. *Science*, *303*, 333.

Davis, L., & Maizels, N. (2011). DNA nicks promote efficient and safe targeted gene correction. *PLoS One, 6*, e23981.
Deltcheva, E., Chylinski, K., Sharma, C. M., Gonzales, K., Chao, Y., Pirzada, Z. A., et al. (2011). CRISPR RNA maturation by trans-encoded small RNA and host factor RNase III. *Nature, 471*, 602–607.
Deng, D., Yan, C., Pan, X., Mahfouz, M., Wang, J., Zhu, J. K., et al. (2012). Structural basis for sequence-specific recognition of DNA by TAL effectors. *Science, 335*, 720–723.
Ding, Q., Lee, Y. K., Schaefer, E. A., Peters, D. T., Veres, A., Kim, K., et al. (2013). A TALEN genome-editing system for generating human stem cell-based disease models. *Cell Stem Cell, 12*, 238–251.
Ding, Q., Regan, S. N., Xia, Y., Oostrom, L. A., Cowan, C. A., & Musunuru, K. (2013). Enhanced efficiency of human pluripotent stem cell genome editing through replacing TALENs with CRISPRs. *Cell Stem Cell, 12*, 393–394.
Duncan, S. R., Scott, S., & Duncan, C. J. (2005). Reappraisal of the historical selective pressures for the CCR5-Delta32 mutation. *Journal of Medical Genetics, 42*, 205–208.
Ebert, A. D., Yu, J., Rose, F. F., Jr., Mattis, V. B., Lorson, C. L., Thomson, J. A., et al. (2009). Induced pluripotent stem cells from a spinal muscular atrophy patient. *Nature, 457*, 277–280.
Frank, S., Skryabin, B. V., & Greber, B. (2013). A modified TALEN-based system for robust generation of knock-out human pluripotent stem cell lines and disease models. *BMC Genomics, 14*, 773.
Gaj, T., Guo, J., Kato, Y., Sirk, S. J., & Barbas, C. F., 3rd. (2012). Targeted gene knockout by direct delivery of zinc-finger nuclease proteins. *Nature Methods, 9*, 805–807.
Guilinger, J. P., Thompson, D. B., & Liu, D. R. (2014). Fusion of catalytically inactive Cas9 to FokI nuclease improves the specificity of genome modification. *Nature Biotechnology, 32*(6), 577–582.
Hacein-Bey-Abina, S., Von Kalle, C., Schmidt, M., McCormack, M. P., Wulffraat, N., Leboulch, P., et al. (2003). LMO2-associated clonal T cell proliferation in two patients after gene therapy for SCID-X1. *Science, 302*, 415–419.
Handel, E. M., Gellhaus, K., Khan, K., Bednarski, C., Cornu, T. I., Muller-Lerch, F., et al. (2012). Versatile and efficient genome editing in human cells by combining zinc-finger nucleases with adeno-associated viral vectors. *Human Gene Therapy, 23*, 321–329.
Harrow, J., Nagy, A., Reymond, A., Alioto, T., Patthy, L., Antonarakis, S. E., et al. (2009). Identifying protein-coding genes in genomic sequences. *Genome Biology, 10*, 201.
Hemmi, H., Takeuchi, O., Kawai, T., Kaisho, T., Sato, S., Sanjo, H., et al. (2000). A toll-like receptor recognizes bacterial DNA. *Nature, 408*, 740–745.
Hockemeyer, D., Soldner, F., Beard, C., Gao, Q., Mitalipova, M., DeKelver, R. C., et al. (2009). Efficient targeting of expressed and silent genes in human ESCs and iPSCs using zinc-finger nucleases. *Nature Biotechnology, 27*, 851–857.
Hockemeyer, D., Wang, H., Kiani, S., Lai, C. S., Gao, Q., Cassady, J. P., et al. (2011). Genetic engineering of human pluripotent cells using TALE nucleases. *Nature Biotechnology, 29*, 731–734.
Holkers, M., Maggio, I., Liu, J., Janssen, J. M., Miselli, F., Mussolino, C., et al. (2013). Differential integrity of TALE nuclease genes following adenoviral and lentiviral vector gene transfer into human cells. *Nucleic Acids Research, 41*, e63.
Holt, N., Wang, J., Kim, K., Friedman, G., Wang, X., Taupin, V., et al. (2010). Human hematopoietic stem/progenitor cells modified by zinc-finger nucleases targeted to CCR5 control HIV-1 in vivo. *Nature Biotechnology, 28*, 839–847.
Hutter, G., Nowak, D., Mossner, M., Ganepola, S., Mussig, A., Allers, K., et al. (2009). Long-term control of HIV by CCR5 Delta32/Delta32 stem-cell transplantation. *New England Journal of Medicine, 360*, 692–698.

Hwang, W. Y., Fu, Y., Reyon, D., Maeder, M. L., Tsai, S. Q., Sander, J. D., et al. (2013). Efficient genome editing in zebrafish using a CRISPR-Cas system. *Nature Biotechnology, 31*, 227–229.

International Human Genome Sequencing, Consortium. (2004). Finishing the euchromatic sequence of the human genome. *Nature, 431*, 931–945.

Itzhaki, I., Maizels, L., Huber, I., Zwi-Dantsis, L., Caspi, O., Winterstern, A., et al. (2011). Modelling the long QT syndrome with induced pluripotent stem cells. *Nature, 471*, 225–229.

Jiang, W., Bikard, D., Cox, D., Zhang, F., & Marraffini, L. A. (2013). RNA-guided editing of bacterial genomes using CRISPR-Cas systems. *Nature Biotechnology, 31*, 233–239.

Jiang, J., Jing, Y., Cost, G. J., Chiang, J. C., Kolpa, H. J., Cotton, A. M., et al. (2013). Translating dosage compensation to trisomy 21. *Nature, 500*, 296–300.

Jinek, M., Chylinski, K., Fonfara, I., Hauer, M., Doudna, J. A., & Charpentier, E. (2012). A programmable dual-RNA-guided DNA endonuclease in adaptive bacterial immunity. *Science, 337*, 816–821.

Jinek, M., East, A., Cheng, A., Lin, S., Ma, E., & Doudna, J. (2013). RNA-programmed genome editing in human cells. *Elife, 2*, e00471.

Joung, J. K., & Sander, J. D. (2013). TALENs: A widely applicable technology for targeted genome editing. *Nature Reviews. Molecular Cell Biology, 14*, 49–55.

Kernstock, S., Davydova, E., Jakobsson, M., Moen, A., Pettersen, S., Maelandsmo, G. M., et al. (2012). Lysine methylation of VCP by a member of a novel human protein methyltransferase family. *Nature Communications, 3*, 1038.

Kim, H., & Kim, J. S. (2014). A guide to genome engineering with programmable nucleases. *Nature Reviews. Genetics, 15*, 321–334.

Kim, S., Kim, D., Cho, S. W., Kim, J., & Kim, J. S. (2014). Highly efficient RNA-guided genome editing in human cells via delivery of purified Cas9 ribonucleoproteins. *Genome Research, 24*(6), 1012–1019.

Kim, E., Kim, S., Kim, D. H., Choi, B. S., Choi, I. Y., & Kim, J. S. (2012). Precision genome engineering with programmable DNA-nicking enzymes. *Genome Research, 22*, 1327–1333.

Kim, H., Kim, M. S., Wee, G., Lee, C. I., Kim, H., & Kim, J. S. (2013). Magnetic separation and antibiotics selection enable enrichment of cells with ZFN/TALEN-induced mutations. *PLoS One, 8*, e56476.

Kim, Y., Kweon, J., Kim, A., Chon, J. K., Yoo, J. Y., Kim, H. J., et al. (2013). A library of TAL effector nucleases spanning the human genome. *Nature Biotechnology, 31*, 251–258.

Kim, H., Um, E., Cho, S. R., Jung, C., & Kim, J. S. (2011). Surrogate reporters for enrichment of cells with nuclease-induced mutations. *Nature Methods, 8*, 941–943.

Kim, Y. K., Wee, G., Park, J., Kim, J., Baek, D., Kim, J. S., et al. (2013). TALEN-based knockout library for human microRNAs. *Nature Structural & Molecular Biology, 20*, 1458–1464.

Lander, E. S., Linton, L. M., Birren, B., Nusbaum, C., Zody, M. C., Baldwin, J., et al. (2001). Initial sequencing and analysis of the human genome. *Nature, 409*, 860–921.

Lee, H. J., Kim, E., & Kim, J. S. (2010). Targeted chromosomal deletions in human cells using zinc finger nucleases. *Genome Research, 20*, 81–89.

Lee, H. J., Kweon, J., Kim, E., Kim, S., & Kim, J. S. (2012). Targeted chromosomal duplications and inversions in the human genome using zinc finger nucleases. *Genome Research, 22*, 539–548.

Lee, G., Papapetrou, E. P., Kim, H., Chambers, S. M., Tomishima, M. J., Fasano, C. A., et al. (2009). Modelling pathogenesis and treatment of familial dysautonomia using patient-specific iPSCs. *Nature, 461*, 402–406.

Li, H., Haurigot, V., Doyon, Y., Li, T., Wong, S. Y., Bhagwat, A. S., et al. (2011). In vivo genome editing restores haemostasis in a mouse model of haemophilia. *Nature, 475*, 217–221.

Li, L., Krymskaya, L., Wang, J., Henley, J., Rao, A., Cao, L. F., et al. (2013). Genomic editing of the HIV-1 coreceptor CCR5 in adult hematopoietic stem and progenitor cells using zinc finger nucleases. *Molecular Therapy*, *21*, 1259–1269.

Liu, J., Gaj, T., Patterson, J. T., Sirk, S. J., & Barbas, C. F., 3rd. (2014). Cell-penetrating peptide-mediated delivery of TALEN proteins via bioconjugation for genome engineering. *PLoS One*, *9*, e85755.

Lombardo, A., Cesana, D., Genovese, P., Di Stefano, B., Provasi, E., Colombo, D. F., et al. (2011). Site-specific integration and tailoring of cassette design for sustainable gene transfer. *Nature Methods*, *8*, 861–869.

Lombardo, A., Genovese, P., Beausejour, C. M., Colleoni, S., Lee, Y. L., Kim, K. A., et al. (2007). Gene editing in human stem cells using zinc finger nucleases and integrase-defective lentiviral vector delivery. *Nature Biotechnology*, *25*, 1298–1306.

Ma, N., Liao, B., Zhang, H., Wang, L., Shan, Y., Xue, Y., et al. (2013). Transcription activator-like effector nuclease (TALEN)-mediated gene correction in integration-free beta-thalassemia induced pluripotent stem cells. *Journal of Biological Chemistry*, *288*, 34671–34679.

Maier, D. A., Brennan, A. L., Jiang, S., Binder-Scholl, G. K., Lee, G., Plesa, G., et al. (2013). Efficient clinical scale gene modification via zinc finger nuclease-targeted disruption of the HIV co-receptor CCR5. *Human Gene Therapy*, *24*, 245–258.

Mak, A. N., Bradley, P., Cernadas, R. A., Bogdanove, A. J., & Stoddard, B. L. (2012). The crystal structure of TAL effector PthXo1 bound to its DNA target. *Science*, *335*, 716–719.

Makarova, K. S., Grishin, N. V., Shabalina, S. A., Wolf, Y. I., & Koonin, E. V. (2006). A putative RNA-interference-based immune system in prokaryotes: Computational analysis of the predicted enzymatic machinery, functional analogies with eukaryotic RNAi, and hypothetical mechanisms of action. *Biology Direct*, *1*, 7.

Mali, P., Esvelt, K. M., & Church, G. M. (2013). Cas9 as a versatile tool for engineering biology. *Nature Methods*, *10*, 957–963.

Mali, P., Yang, L., Esvelt, K. M., Aach, J., Guell, M., DiCarlo, J. E., et al. (2013). RNA-guided human genome engineering via Cas9. *Science*, *339*, 823–826.

Marchetto, M. C., Carromeu, C., Acab, A., Yu, D., Yeo, G. W., Mu, Y., et al. (2010). A model for neural development and treatment of Rett syndrome using human induced pluripotent stem cells. *Cell*, *143*, 527–539.

McConnell Smith, A., Takeuchi, R., Pellenz, S., Davis, L., Maizels, N., Monnat, R. J., Jr., et al. (2009). Generation of a nicking enzyme that stimulates site-specific gene conversion from the I-AniI LAGLIDADG homing endonuclease. *Proceedings of the National Academy of Sciences of the United States of America*, *106*, 5099–5104.

McCormack, M. P., & Rabbitts, T. H. (2004). Activation of the T-cell oncogene LMO2 after gene therapy for X-linked severe combined immunodeficiency. *New England Journal of Medicine*, *350*, 913–922.

Metzger, M. J., McConnell-Smith, A., Stoddard, B. L., & Miller, A. D. (2011). Single-strand nicks induce homologous recombination with less toxicity than double-strand breaks using an AAV vector template. *Nucleic Acids Research*, *39*, 926–935.

Miller, J. C., Tan, S., Qiao, G., Barlow, K. A., Wang, J., Xia, D. F., et al. (2011). A TALE nuclease architecture for efficient genome editing. *Nature Biotechnology*, *29*, 143–148.

Mojica, F. J., Diez-Villasenor, C., Garcia-Martinez, J., & Almendros, C. (2009). Short motif sequences determine the targets of the prokaryotic CRISPR defence system. *Microbiology*, *155*, 733–740.

Moscou, M. J., & Bogdanove, A. J. (2009). A simple cipher governs DNA recognition by TAL effectors. *Science*, *326*, 1501.

Nussbaum, R. L., McInnes, R. R., & Willard, H. F. (2007). *Thompson & Thompson Genetics in Medicine* (7th ed.). Philadelphia, PA: Elisevier.

Osborn, M. J., Starker, C. G., McElroy, A. N., Webber, B. R., Riddle, M. J., Xia, L., et al. (2013). TALEN-based gene correction for epidermolysis bullosa. *Molecular Therapy, 21*, 1151–1159.

Ousterout, D. G., Perez-Pinera, P., Thakore, P. I., Kabadi, A. M., Brown, M. T., Qin, X., et al. (2013). Reading frame correction by targeted genome editing restores dystrophin expression in cells from Duchenne muscular dystrophy patients. *Molecular Therapy, 21*, 1718–1726.

Pennisi, E. (2012). Genomics. ENCODE project writes eulogy for junk DNA. *Science, 337*, 1159–1161.

Perez, E. E., Wang, J., Miller, J. C., Jouvenot, Y., Kim, K. A., Liu, O., et al. (2008). Establishment of HIV-1 resistance in CD4+ T cells by genome editing using zinc-finger nucleases. *Nature Biotechnology, 26*, 808–816.

Porteus, M. H., & Baltimore, D. (2003). Chimeric nucleases stimulate gene targeting in human cells. *Science, 300*, 763.

Ramakrishna, S., Cho, S. W., Kim, S., Song, M., Gopalappa, R., Kim, J. S., et al. (2014). Surrogate reporter-based enrichment of cells containing RNA-guided Cas9 nuclease-induced mutations. *Nature Communications, 5*, 3378.

Ramakrishna, S., Kwaku Dad, A. B., Beloor, J., Gopalappa, R., Lee, S. K., & Kim, H. (2014). Gene disruption by cell-penetrating peptide-mediated delivery of Cas9 protein and guide RNA. *Genome Research, 24*(6), 1020–1027.

Ramirez, C. L., Certo, M. T., Mussolino, C., Goodwin, M. J., Cradick, T. J., McCaffrey, A. P., et al. (2012). Engineered zinc finger nickases induce homology-directed repair with reduced mutagenic effects. *Nucleic Acids Research, 40*, 5560–5568.

Ran, F. A., Hsu, P. D., Lin, C. Y., Gootenberg, J. S., Konermann, S., Trevino, A. E., et al. (2013). Double nicking by RNA-guided CRISPR Cas9 for enhanced genome editing specificity. *Cell, 154*, 1380–1389.

Rashid, S. T., Corbineau, S., Hannan, N., Marciniak, S. J., Miranda, E., Alexander, G., et al. (2010). Modeling inherited metabolic disorders of the liver using human induced pluripotent stem cells. *Journal of Clinical Investigation, 120*, 3127–3136.

Rouet, P., Smih, F., & Jasin, M. (1994). Introduction of double-strand breaks into the genome of mouse cells by expression of a rare-cutting endonuclease. *Molecular and Cellular Biology, 14*, 8096–8106.

Saha, K., & Jaenisch, R. (2009). Technical challenges in using human induced pluripotent stem cells to model disease. *Cell Stem Cell, 5*, 584–595.

Sander, J. D., & Joung, J. K. (2014). CRISPR-Cas systems for editing, regulating and targeting genomes. *Nature Biotechnology, 32*, 347–355.

Sapranauskas, R., Gasiunas, G., Fremaux, C., Barrangou, R., Horvath, P., & Siksnys, V. (2011). The Streptococcus thermophilus CRISPR/Cas system provides immunity in Escherichia coli. *Nucleic Acids Research, 39*, 9275–9282.

Schiffer, J. T., Aubert, M., Weber, N. D., Mintzer, E., Stone, D., & Jerome, K. R. (2012). Targeted DNA mutagenesis for the cure of chronic viral infections. *Journal of Virology, 86*, 8920–8936.

Schwank, G., Koo, B. K., Sasselli, V., Dekkers, J. F., Heo, I., Demircan, T., et al. (2013). Functional repair of CFTR by CRISPR/Cas9 in intestinal stem cell organoids of cystic fibrosis patients. *Cell Stem Cell, 13*, 653–658.

Sebastiano, V., Maeder, M. L., Angstman, J. F., Haddad, B., Khayter, C., Yeo, D. T., et al. (2011). In situ genetic correction of the sickle cell anemia mutation in human induced pluripotent stem cells using engineered zinc finger nucleases. *Stem Cells, 29*, 1717–1726.

Shalem, O., Sanjana, N. E., Hartenian, E., Shi, X., Scott, D. A., Mikkelsen, T. S., et al. (2014). Genome-scale CRISPR-Cas9 knockout screening in human cells. *Science, 343*, 84–87.

She, X., Jiang, Z., Clark, R. A., Liu, G., Cheng, Z., Tuzun, E., et al. (2004). Shotgun sequence assembly and recent segmental duplications within the human genome. *Nature, 431*, 927–930.

Smithies, O., Gregg, R. G., Boggs, S. S., Koralewski, M. A., & Kucherlapati, R. S. (1985). Insertion of DNA sequences into the human chromosomal beta-globin locus by homologous recombination. *Nature, 317*, 230–234.

Soldner, F., Laganiere, J., Cheng, A. W., Hockemeyer, D., Gao, Q., Alagappan, R., et al. (2011). Generation of isogenic pluripotent stem cells differing exclusively at two early onset Parkinson point mutations. *Cell, 146*, 318–331.

Song, K. Y., Schwartz, F., Maeda, N., Smithies, O., & Kucherlapati, R. (1987). Accurate modification of a chromosomal plasmid by homologous recombination in human cells. *Proceedings of the National Academy of Sciences of the United States of America, 84*, 6820–6824.

Steentoft, C., Vakhrushev, S. Y., Vester-Christensen, M. B., Schjoldager, K. T., Kong, Y., Bennett, E. P., et al. (2011). Mining the O-glycoproteome using zinc-finger nuclease-glycoengineered simple cell lines. *Nature Methods, 8*, 977–982.

Sun, N., Liang, J., Abil, Z., & Zhao, H. (2012). Optimized TAL effector nucleases (TALENs) for use in treatment of sickle cell disease. *Molecular Biosystems, 8*, 1255–1263.

Tebas, P., Stein, D., Tang, W. W., Frank, I., Wang, S. Q., Lee, G., et al. (2014). Gene editing of CCR5 in autologous CD4 T cells of persons infected with HIV. *New England Journal of Medicine, 370*, 901–910.

Thomas, C. E., Ehrhardt, A., & Kay, M. A. (2003). Progress and problems with the use of viral vectors for gene therapy. *Nature Reviews. Genetics, 4*, 346–358.

Tsai, S. Q., Wyvekens, N., Khayter, C., Foden, J. A., Thapar, V., Reyon, D., et al. (2014). Dimeric CRISPR RNA-guided FokI nucleases for highly specific genome editing. *Nature Biotechnology, 32*(6), 569–576.

Tupler, R., Perini, G., & Green, M. R. (2001). Expressing the human genome. *Nature, 409*, 832–833.

Urnov, F. D., Rebar, E. J., Holmes, M. C., Zhang, H. S., & Gregory, P. D. (2010). Genome editing with engineered zinc finger nucleases. *Nature Reviews. Genetics, 11*, 636–646.

Venter, J. C., Adams, M. D., Myers, E. W., Li, P. W., Mural, R. J., Sutton, G. G., et al. (2001). The sequence of the human genome. *Science, 291*, 1304–1351.

Voit, R. A., Hendel, A., Pruett-Miller, S. M., & Porteus, M. H. (2014). Nuclease-mediated gene editing by homologous recombination of the human globin locus. *Nucleic Acids Research, 42*, 1365–1378.

Wagner, H. (2001). Toll meets bacterial CpG-DNA. *Immunity, 14*, 499–502.

Wang, J., Friedman, G., Doyon, Y., Wang, N. S., Li, C. J., Miller, J. C., et al. (2012). Targeted gene addition to a predetermined site in the human genome using a ZFN-based nicking enzyme. *Genome Research, 22*, 1316–1326.

Wolfe, S. A., Nekludova, L., & Pabo, C. O. (2000). DNA recognition by Cys2His2 zinc finger proteins. *Annual Review of Biophysics and Biomolecular Structure, 29*, 183–212.

Wu, Z., Asokan, A., & Samulski, R. J. (2006). Adeno-associated virus serotypes: Vector toolkit for human gene therapy. *Molecular Therapy, 14*, 316–327.

Yin, H., Xue, W., Chen, S., Bogorad, R. L., Benedetti, E., Grompe, M., et al. (2014). Genome editing with Cas9 in adult mice corrects a disease mutation and phenotype. *Nature Biotechnology, 32*(6), 551–553.

Yusa, K., Rashid, S. T., Strick-Marchand, H., Varela, I., Liu, P. Q., Paschon, D. E., et al. (2011). Targeted gene correction of alpha1-antitrypsin deficiency in induced pluripotent stem cells. *Nature, 478*, 391–394.

Zou, J., Mali, P., Huang, X., Dowey, S. N., & Cheng, L. (2011). Site-specific gene correction of a point mutation in human iPS cells derived from an adult patient with sickle cell disease. *Blood, 118*, 4599–4608.

Zou, J., Sweeney, C. L., Chou, B. K., Choi, U., Pan, J., Wang, H., et al. (2011). Oxidase-deficient neutrophils from X-linked chronic granulomatous disease iPS cells: Functional correction by zinc finger nuclease-mediated safe harbor targeting. *Blood*, *117*, 5561–5572.

Zwaka, T. P., & Thomson, J. A. (2003). Homologous recombination in human embryonic stem cells. *Nature Biotechnology*, *21*, 319–321.

CHAPTER SIX

# Genome Editing in Human Stem Cells

**Susan M. Byrne[1], Prashant Mali, George M. Church**
Department of Genetics, Harvard Medical School, Boston, Massachusetts, USA
[1]Corresponding author: e-mail address: sbyrne@genetics.med.harvard.edu

## Contents



## Abstract

The use of custom-engineered sequence-specific nucleases (including CRISPR/Cas9, ZFN, and TALEN) allows genetic changes in human cells to be easily made with much greater efficiency and precision than before. Engineered double-stranded DNA breaks can efficiently disrupt genes, or, with the right donor vector, engineer point mutations and gene insertions. However, a number of design considerations should be taken into account to ensure maximum gene targeting efficiency and specificity. This is especially true when engineering human embryonic stem or induced pluripotent stem cells (iPSCs), which are more difficult to transfect and less resilient to DNA damage than immortalized tumor cell lines. Here, we describe a protocol for easily engineering genetic changes in human iPSCs, through which we typically achieve targeting efficiencies between 1% and 10% without any subsequent selection steps. Since this protocol

*Methods in Enzymology*, Volume 546
ISSN 0076-6879
http://dx.doi.org/10.1016/B978-0-12-801185-0.00006-4

only uses the simple transient transfection of plasmids and/or single-stranded oligonucleotides, most labs will easily be able to perform it. We also describe strategies for identifying, cloning, and genotyping successfully edited cells, and how to design the optimal sgRNA target sites and donor vectors. Finally, we discuss alternative methods for gene editing including viral delivery vectors, Cas9 nickases, and orthogonal Cas9 systems.

## 1. INTRODUCTION

The development of sequence-specific nucleases such as zinc finger nucleases, (ZFNs), transcription activator-like effector nucleases (TALENs), or CRISPR/Cas9 nucleases have enormously expanded our ability to engineer genetic changes in human cells (Joung & Sander, 2012; Mali, Yang, et al., 2013; Urnov, Rebar, Holmes, Zhang, & Gregory, 2010). These nucleases can be custom-engineered to create double-stranded DNA (dsDNA) breaks at a desired sequence in the genome. When these are repaired using the non-homologous end joining (NHEJ) pathway, small insertion and deletion mutations (indels) are produced and disrupt genes. Alternatively, the dsDNA break can be repaired by the homologous recombination pathway—specific base pair changes or gene insertions can be formed using a homologous donor targeting vector. Of these systems, Cas9 nucleases have been favored due to their easy construction and lower toxicity in human cells (Ding et al., 2013).

Human induced pluripotent stem cells (iPSCs) have been another great breakthrough for genetic studies in human cells. Their self-renewing capability allows them to be gene targeted, cloned, genotyped, and expanded. Successfully targeted iPSC clones can then be differentiated into a variety of other cell types to analyze the effects of the induced mutations. The ability to easily genetically modify human iPSCs also holds tremendous clinical promise for generating artificial organs and safer gene therapies. However, while immortalized human tumor cell lines have been edited with almost complete efficiency (Fu, Sander, Reyon, Cascio, & Joung, 2014), much lower success rates have been achieved in human iPSCs (Mali, Yang, et al., 2013; Yang et al., 2013). This difference may be due to the gross chromosomal abnormalities and an unusually robust response to DNA damage in tumor cell lines. In this chapter, we describe strategies to maximize the efficiency of genome editing in human iPSCs. Using these design considerations and the transient transfection protocol listed below, we typically achieve gene disruption frequencies of 1–25% and homologous gene targeting frequencies of 0.5–10% in human iPSCs without any subsequent selection steps.

## 2. GENE TARGETING STRATEGIES

For any gene targeting project, the structure of the gene must be considered and the nuclease targeting sites should be carefully chosen according to the experimental goals. For simple gene disruption, a single cut site can generate indel mutations using the NHEJ repair pathway. When within coding exons, such indels can cause frameshifts and disrupt protein expression. Targeting coding exons towards the beginning of the gene may be preferable, as mutations here may create more complete gene disruption and be less likely to accidentally generate truncated protein artifacts with residual biological activity. Areas possessing relatively unique genome sequences should be chosen, rather than a common domain shared by several homologous members of the same gene family (unless the goal is to target multiple members of the gene family).

Alternatively, one can design two nuclease sites to excise the intervening section of the genome. Regions from 100 bp to several kb can be excised with biallelic frequencies of over 10% (Cong et al., 2013). These junctions are often religated with perfect precision between the two dsDNA break sites, although indels are also sometimes found. This strategy allows nuclease sites within introns or outside genes to be used; this is particularly useful when no satisfactory nuclease sites can be found within an exon. Again, the organization of the gene must be carefully considered to avoid alternative exon splicing events or truncated products.

When specific mutations are desired, a donor targeting vector for homologous recombination is provided along with the nuclease elements. These donors can be single-stranded DNA oligonucleotides (ssODNs) or plasmids for engineering point mutations. Here, the nuclease site should be chosen as close to the intended mutation as possible, since homologous recombination targeting efficiencies drop precipitously as the dsDNA break becomes farther from the mutation. For ssODN donors, having the desired mutation in the center of the oligo showed the highest targeting efficiency. 90 bp ssODNs worked best, although lengths from 70 to 130 bp were able to produce targeting efficiencies >1%. The highest targeting frequencies occurred when the mutation was within 10 bp of the nuclease site; when the mutation was more than 40 bp away, gene targeting was barely detectable (Chen et al., 2011; Yang et al., 2013).

Alternatively, plasmid targeting vectors for homologous recombination can be used to generate desired point mutations, as well as larger "knock-in" gene insertions. Since the presence of a dsDNA break drastically increases the homologous recombination efficiency, shorter homology arms of 0.4–0.8 kb can be used (rather than the 2–14 kb arms used in conventional gene targeting vectors without nucleases), although increased homology may still improve targeting of difficult constructs (Beumer, Trautman, Mukherjee, & Carroll, 2013; Hendel et al., 2014; Hockemeyer et al., 2009; Orlando et al., 2010). Again, the dsDNA break must be positioned within 200 bp of the mutation, and gene targeting efficiency decreases with larger transgene insertions (Guye, Li, Wroblewska, Duportet, & Weiss, 2013; Moehle et al., 2007; Urnov et al., 2010).

## 3. CHOICE OF NUCLEASE TARGETING SITES

The *S. pyogenes* Cas9 nuclease (SpCas9) targets a 20 bp dsDNA sequence specified by the single guide RNA (sgRNA) next to a 3′-protospacer adjacent motif (PAM) of NGG, although PAM sequences of NAG can be targeted as well (Jinek et al., 2012; Mali, Aach, et al., 2013; Mali, Yang, et al., 2013). Upon binding to the sgRNA and complementary DNA targeting site, the Cas9 nuclease generates a blunt-ended, dsDNA break three base pairs upstream of the PAM. Cas9–sgRNA complexes can potentially tolerate 1–6 bp mismatches between the sgRNA and the target sequence, creating off-target cuts in genomic DNA. Although a "seed" sequence of the 8–13 nucleotides closest to the PAM appears to be more important for Cas9 nuclease specificity, mismatches can sometimes be tolerated here as well (Jinek et al., 2012; Mali, Aach, et al., 2013). Off-target Cas9 nuclease activity can also occur with small indel mismatches (Lin et al., 2014).

Several online tools and algorithms are available to identify specific nuclease targeting sites, including: the CRISPR Design Tool (crispr.mit.edu) (Hsu et al., 2013); ZiFiT targeter (zifit.partners.org/ZiFiT) (Fu et al., 2014); CasFinder (arep.med.harvard.edu/CasFinder/) (Aach, Mali, & Church, 2014); and E-Crisp (www.e-crisp.org/E-CRISP/) (Heigwer, Kerr, & Boutros, 2014). In addition, specific Cas9 sgRNA targets for disrupting human exons can be found from published sets of sgRNA screening libraries (Aach et al., 2014; Shalem et al., 2014; Wang, Wei, Sabatini, & Lander, 2014). These algorithms are constantly being refined to incorporate further discoveries about Cas9 targeting specificity.

The nuclease activity among different sgRNAs can also vary widely. Cas9 nuclease activity is positively correlated with areas of open chromatin (Kuscu, Arslan, Singh, Thorpe, & Adli, 2014; Yang et al., 2013); however, substantial variations in activity can still be found among neighboring sgRNAs in the same locus. Other characteristics associated with higher levels of sgRNA activity are: targeting sequences with between 20% and 80% GC content, sgRNAs targeting the nontranscribed strand, and purines in the last four bases of the spacer sequence (Wang et al., 2014). While these criteria were statistically significant, they still could not account for all of the observed variation in sgRNA activity.

Initial constructs used the human U6 polymerase III promoter to express the sgRNA due to its specific initiation and termination sites and its ubiquitous expression in human cells. Since the U6 promoter requires a G to initiate transcription, this led to the early restriction that only sequences fitting the form $GN_{20}GG$ could be targeted (Mali, Yang, et al., 2013). However, subsequent studies showed that up to 10 extra nucleotides could be added to the 5′-end of the sgRNA while retaining similar levels of nuclease activity and that these sgRNA extensions were being processed off (Mali, Aach, et al., 2013; Ran, Hsu, Lin, et al., 2013). Thus, any 20 bp sequence next to a PAM can be targeted, although an extra G is still required in the sgRNA expression construct to initiate transcription when the U6 promoter is used. Truncated sgRNAs with up to three base pairs missing from the 5′-end have been shown to increase specificity without much loss in activity, although truncations beyond 3 bp ablated activity (Fu et al., 2014). Appending up to 40 extra bp at the 3′-end of the sgRNA construct, after the hairpin backbone, resulted in slightly higher sgRNA activity, possibly due to increased half-life of the longer sgRNA (Mali, Aach, et al., 2013). Other promoters besides U6, such as H1 or pol-II, may also be used to express the sgRNA. The sgRNA constructs may also be transfected into cells as linear PCR products rather than plasmids (Ran, Hsu, Wright, et al., 2013).

Due to the ease of cloning sgRNAs, and the ongoing questions regarding sgRNA specificity and activity, we recommend that users select a few sgRNA target sites and test them empirically. While it is important to try to select sgRNAs that are as specific as possible, a perfectly unique sequence may not exist suitably close to your desired mutation. Alternative approaches are further discussed below.

## 4. EXPERIMENTAL PROCEDURES

With this protocol, we can consistently introduce plasmid DNA into human iPSCs with 60–70% transfection efficiency. While we have also had success using ZFNs and TALENs to edit iPSC genomes, the ease of cloning sgRNAs has made CRISPR/Cas the preferred method in our lab. Without any selection scheme, our overall gene disruption efficiencies using a single sgRNA in human iPSCs ranges from 1% to 25%, depending on the particular sgRNA used. Once the plasmids and cells are ready, the nucleofection process takes a few hours. After nucleofection, it takes 5–10 days of culture for the transient Cas9 transfection to subside and protein expression to turn over. Then, the potentially edited iPSCs can be cloned by single-cell FACS sorting. Eight days after sorting, individual iPSC have formed stable colonies, which can be further expanded and genotyped.

While this protocol focuses on human iPSC, it can be adapted for use in other cell types, using culture conditions and nucleofection protocols suitable for that cell type (although the amounts of plasmid/ssODN and promoters for Cas9 expression may need adjustment). Overall gene disruption efficiencies greater than 60% have thus been achieved in immortalized tumor cell lines.

## 4.1. Human iPSC culture and passaging

A number of different human iPSC lines are available from cell line resources such as Coriell (coriell.org), ATCC (atcc.org), and the Harvard Stem Cell Institute (hsci.harvard.edu), among many others. Furthermore, numerous academic and commercial facilities offer iPSC derivation services. Detailed protocols for culturing and passaging human ES and iPSC lines are available elsewhere (wicell.org, stembook.org). Here, we have used iPSC derived from open-consented participants in the Personal Genome Project (Lee et al., 2009), but this protocol is widely applicable to any human ES or iPS cell line. Cells used for gene targeting should be of a low passage number and free of karyotypic abnormalities. Cells should exhibit normal iPSC morphology and express pluripotency markers such as Tra-1/60 and SSEA4.

Human iPSCs for genome engineering are cultured under feeder-free conditions, in the defined mTesr-1 medium (StemCell Technologies) on Matrigel-coated tissue culture plates (BD). We have found lower transfection efficiencies (40–60%) when transfecting iPSCs growing on irradiated mouse embryonic fibroblasts (MEF), due to incomplete separation of the iPSCs from the MEFs immediately before nucleofection.

## 4.2. Preparation of plasmids for transient transfection

An increasingly wide selection of plasmids for ZFN, TALEN, and CRISPR/Cas9 genome editing, with instructions for cloning, are available from the Addgene plasmid repository (www.addgene.org/CRISPR/). This protocol was specifically developed with the plasmids to express human-codon optimized SpCas9 and sgRNAs from Mali, Yang, et al. (2013). However, an EF1α promoter was used to express Cas9 instead of the CMV promoter in iPSCs, as it produced a fivefold increase in gene disruption efficiency.

Plasmid donor vectors containing homology arms can be easily cloned using isothermal assembly or synthesized as gene fragments (Integrated DNA Technologies). Homology arm sequences should ideally be cloned from the cell line being targeted to obtain identical (isogenic) sequences. Any polymorphic differences between the targeting vector and the genomic locus can decrease gene targeting frequencies (Deyle, Li, Ren, & Russell, 2013). All plasmids for nucleofection into iPSCs should be endotoxin-free (Qiagen Endo-free Plasmid Maxi Kit) and at a concentration greater than 2 mg/ml, so as not to dilute the nucleofection buffer. Oligo donors (ssODN) should be HPLC-purified and resuspended in sterile distilled water.

## 4.3. Nucleofection protocol

This protocol uses the Amaxa 4D-Nucleofector X Unit (Lonza), but we have also gotten good transfection efficiencies in human iPSCs from the Neon Transfection system (Life Technologies). Traditional electroporation methods will produce lower transfection efficiencies in iPSCs, which will lower the overall gene targeting efficiency. The amounts listed below are for the 20-μl Nucleocuvette strips; if using the 100-μl single Nucleocuvettes, increase all quantities fivefold. A control reaction transfecting a fluorescent protein-expressing plasmid can be used to verify nucleofection efficiency.

Expand human iPSCs under feeder-free conditions in mTesr-1 medium on tissue culture plates coated with ES-qualified Matrigel (BD) according to the manufacturer's instructions. Each nucleofection reaction will need $0.5 \times 10^6$ cells, although a range of 0.2 to $2 \times 10^6$ iPSCs per reaction can be used. Depending on the number of reactions, 6-well plates or 10-cm dishes of cultured iPSC may be required.

Prepare Matrigel-coated 24-well tissue culture plates, one well per nucleofection reaction. Additional Matrigel-coated 96-well flat-bottom tissue culture plates may also be prepared to culture aliquots of transfected cells for analysis.

Pretreat human iPSC cultures with 10 $\mu M$ Rock inhibitor (Y-27632) (R&D Systems, EMD Millipore, or other source) in mTesr-1 for at least

30 min before nucleofection. Prepare additional mTesr with 10 μ*M* Rock inhibitor for use throughout the nucleofection procedure. Cells treated with Rock inhibitor should display the characteristic change in morphology of colonies with jagged edges.

Combine Nucleofector solution P3 with supplement according to manufacturer's instructions (Lonza). For each nucleofection reaction, dilute and combine the DNA mixtures in Nucleofector solution P3 (with supplement) to a final volume of 10 μl. Each nucleofection should contain 0.5 μg of Cas9-expressing plasmid and 1–1.5 μg of sgRNA-expressing plasmids. (When multiple sgRNA-expressing plasmids are used, mix them in equal amounts for a total of 1–1.5 μg plasmid.) If a plasmid targeting vector is being used, include 2 μg per nucleofection reaction. If an ssODN donor is being used, include up to 200 pmol per nucleofection reaction. DNA stock solutions must be concentrated enough such that the total volume of DNA does not exceed 10% of the nucleofection reaction (2 μl for a 20-μl Nucleocuvette). DNA amounts exceeding 4 μg per nucleofection may have an adverse effect on iPSC viability.

Remove the mTesr with Rock inhibitor media from the cells and incubate with Accutase dissociating enzyme (EMD Millipore, StemCell Technologies, or other source) for 5–10 min. Once iPSCs have detached, add an equal volume of mTesr with Rock inhibitor and pipet to achieve a single-cell suspension. Centrifuge the cells at 110 × *g* for 3 min at room temperature. Resuspend cell pellet in mTesr with Rock inhibitor and count live cells.

Centrifuge the required number of iPSCs at 110 × *g* for 3 min. Aspirate off the media. Resuspend cell pellet in 10-μl Nucleofector solution P3 (with supplement) for each reaction.

For each reaction, promptly combine 10 μl of DNA mixture with 10 μl of resuspended cells and transfer the whole 20 μl into a Nucleocuvette. Ensure that the sample is at the bottom of the cuvette.

Place Nucleocuvette into the Nucleofector device and run program CB-150.

Add 80 μl mTesr with Rock inhibitor medium into each Nucleocuvette well and pipet once or twice to resuspend cells. Transfer each reaction into one well of a Matrigel-coated 24-well plate containing 1 ml warm mTesr with Rock inhibitor medium. Alternatively, the nucleofected cells may also be distributed among one 24-well and one or two 96-wells, if analysis at intermediate time points is desired. (If Matrigel-coated 96-well plates are used, an optional centrifugation step (70 × *g*, 3 min, room temperature) can help plate the cells.) A high plating density post nucleofection is important for cell survival.

24 h post nucleofection, iPSCs transfected with a fluorescent protein-expressing plasmid may be examined to assess the transfection efficiency. Change the media to mTesr-1 without Rock inhibitor. Since the iPSCs were plated at a high density, they may appear confluent. As most of the Cas9-induced cell death occurs between 1 and 2 days post nucleofection, we advise waiting until 2 days post nucleofection to passage the iPSCs. Transfected iPSC can then be propagated using regular iPSC culture protocols. After 4 or 5 days post nucleofection, the transient transfection will have subsided, and the cell population can be assayed for gene editing efficiency.

## 4.4. Verification of successful cutting and gene targeting

As the isolation and genotyping of edited iPSC clones can be time consuming, laborious, and expensive, it is desirable to have intermediate ways to verify successful gene disruption and evaluate gene targeting efficiency. Examining a portion of the targeted cell population will help estimate how many clones should be genotyped and provide guidance for troubleshooting.

If the gene being disrupted or inserted is expressed by human iPSCs, the most straightforward assay is to check for expression of that protein by microscopy or flow cytometry. If the targeted gene is not expressed or lacks a convenient stain, then a control reaction using an sgRNA that does target an easily detectable expressed gene can be used to troubleshoot the overall protocol and vectors, although individual sgRNA activities may still vary widely.

If a gene segment is being inserted into the genome, a dilution PCR for the inserted segment can be done on genomic DNA from the edited cell population; however, care must be taken to ensure that the PCR reaction does not simply amplify residual amounts of the transfected donor fragment itself (De Semir & Aran, 2003). PCR primers designed to anneal to genomic DNA sequences outside of the targeted homology region may be used to ensure that only integrated segments are detected. Alternatively, a control targeting vector can be constructed with the same homology arms as the insertion targeting vector, except that a constitutively expressed fluorescent protein cassette is being inserted into the genome. This may provide a quick estimate of knock-in insertion frequencies at that locus using the same sgRNA and homology arms.

To directly measure the extent of gene disruption at a particular locus, a mismatch-specific endonuclease assay—either T7 endonuclease I (New

England Biolabs) or Cel-1 Surveyor nuclease (Transgenomic)—is commonly used (Kim, Lee, Kim, Cho, & Kim, 2009; Qiu et al., 2004). These assays involve PCR-amplifying a short region (roughly 500 bp) around the intended sgRNA targeting site from the genomic DNA of the population of potentially edited cells. These PCR products are melted and reannealed. Any mutations at the intended nuclease site will form a mismatch in the dsDNA, which will be recognized and cleaved by the mismatch-specific endonuclease. Cleaved PCR products can then be analyzed and quantitated by gel electrophoresis. If a restriction enzyme site is inserted or removed at the intended sgRNA targeting site, a restriction fragment length polymorphism assay may also assess Cas9 nuclease activity; here, PCR products around the intended sgRNA site are digested with the restriction enzyme to generate cleaved fragments (Chen et al., 2011).

While the endonuclease assays offer a rapid and cheap measure of gene disruption activity, the endonuclease digestion reaction can be sensitive to buffer and incubation conditions and the limit of detection is around 1–3% of sequences. We prefer a next-generation sequencing-based assay that has a much lower limit of detection (<0.1%) and provides additional sequence information about the edited sgRNA site (Yang et al., 2013). Here, a 100–200 bp region around the edited sgRNA targeting site is PCR amplified and sequenced on a MiSeq system (Illumina). The initial set of genome-specific PCR primers are designed with the requisite MiSeq adaptor sequences appended to the 5′-end. Then, a second round of nested PCR with standard index primers incorporates the barcodes (ScriptSeq from Epicentre or Nextera from Illumina). A detailed protocol with primer sequences has been published (Yang, Mali, Kim-Kiselak, & Church, 2014). While each MiSeq run (150 bp, paired-end) can be expensive, up to ~200 different samples can be barcoded, pooled, and sequenced in parallel to reduce costs (Yang et al., 2013). The resulting next-generation sequencing data can be analyzed by the online CRISPR Genome Analyzer platform, which accepts the sequencing reads, the genomic sequence being targeted, and a donor sequence for homologous recombination (if applicable), and calculates the rate of indels and successful homologous recombination (crispr-ga.net) (Guell, Yang, & Church, 2014).

## 4.5. Cloning by single cell FACS sorting

Several days post nucleofection, after the transiently transfected plasmids have been lost and the Cas9 nuclease activity has subsided, targeted iPSCs may be selected and cloned to generate a culture of successfully targeted cells. As was done for traditional gene targeting without nucleases, if a positive selection marker for antibiotic resistance has been integrated into the genome (such as those for neomycin, hygromycin, or puromycin), that antibiotic may be added to the culture to remove unrecombined antibiotic-sensitive cells. Emerging antibiotic-resistant stem cell clones can then be individually picked by hand and cultured.

Alternatively, human iPSCs may be cloned by FACS sorting individual cells into separate wells of a 96-well plate. To preserve the viability of the dissociated single iPSC, a cocktail of small molecule inhibitors (termed SMC4, from Biovision) is added to the culture (Valamehr et al., 2012). We find that the viability of isolated iPSCs is further enhanced by sorting the cells (previously cultured in feeder-free mTesr-1 media) onto a feeder layer of irradiated MEFs in human ES cell medium. Eight days after FACS sorting, colony formation should be apparent from the individually sorted iPSC, and the SMC4 inhibitors can be removed from the ES cell medium. Our detailed protocol for FACS sorting targeted human iPSCs has been published (Yang et al., 2014). We usually achieve 20–60% iPSC survival and colony formation post-sort. The gene targeting efficiency in the iPSC population (measured as described in Section 4.4) can be used to estimate the number of wells needed for sorting to obtain a successfully targeted viable clone. The iPSC colonies may then be cultured and expanded as usual on a MEF feeder layer for a few passages before being transitioned to feeder-free iPSC conditions. A portion of each potentially targeted iPSC clone may be taken for genomic DNA extraction and genotyping.

## 4.6. Genotyping of clones

Once potentially targeted iPSC clones have been expanded, they must then be genotyped to identify successful gene targeting. While the use of custom-engineered nucleases greatly increases the frequency of correctly targeted events, incorrect mutations still sometimes occur, including partial integrations, random integrations, homology arm duplications, and incorporation of plasmid backbone sequences. In addition, since the use of nucleases allows for potential targeting of both alleles, a genotyping scheme must be able to detect whether the targeted mutation is homozygous or heterozygous.

Typically, genomic DNA is purified from a portion of each expanded clone (while freezing or continuing to expand the remaining culture). For simple gene disruptions or small bp changes, PCR amplification and

Sanger sequencing of the targeted locus would suffice. Heterozygous base pair changes will be apparent as a double peak on the Sanger sequencing trace. Heterozygous indels can similarly be identified through programs that deconvolute a biallelic Sanger sequencing trace (Mutation Surveyor by Softgenetics). Alternatively, the biallelic PCR product can be subcloned into a plasmid vector (TOPO from Life Technologies) for each allele to be sequenced in a separate reaction. Any potential off-target nuclease sites may also be genotyped in this manner to check for mutations.

To genotype larger gene deletions, a PCR reaction with primers that span the two nuclease sites can be sequenced. A second PCR reaction with primers located within the two nuclease sites can identify any unexcised alleles and determine whether the gene deletion is homozygous or heterozygous.

For targeted knock-in gene insertions, one must not only ensure that the entire transgene has been incorporated into the genome, but also that both homology arms have been recombined to the correct site, without recombination into other areas or duplication of the homology arms. Southern blot screening has traditionally been used to determine this, using probes specific to the target locus outside of the homology arm regions. While non-radioactive Southern blot protocols now exist, this screening still requires a relatively large amount of genomic DNA, unique restriction enzyme patterns, and probes verified beforehand. A faster alternative is to use a series of PCR reactions to confirm complete and correct integration of the knock-in construct into the targeted locus. One set of PCR primers that spans the inserted gene can confirm complete insertion, while two other sets of primers that span each of the homology arms (with one primer annealing outside of the homology arm region) can confirm proper recombination on each end.

New screening techniques have been developed to genotype very long constructs or homology arms. Fluorescence *in situ* hybridization can measure the copy number of long homology arms to distinguish between correct targeting events (where copy number is maintained) and random gene integration (where an extra copy of the homology arm is added) (Yang & Seed, 2003). Single-molecule real-time DNA sequencing is capable of producing longer read lengths than Sanger sequencing, and has been used for genotyping nuclease-edited human cell lines with an average sequencing read length approaching 3 kb and ability to detect mutations down to 0.01% (Hendel et al., 2014).

## 4.7. Verify iPSC pluripotency and quality

Once successfully targeted iPSC clones have been identified through genotyping, they should be examined to confirm that they have not lost pluripotency or gained chromosomal abnormalities through the process. These checks are standard practice for any iPSC culture, even without nuclease-mediated gene targeting, as there is always a background level of differentiation and chromosomal rearrangement (Martí et al., 2013). However, these checks are particularly important when gene-targeted clones have been derived from a single iPS cell. Many academic stem cell core facilities and commercial suppliers offer these iPSC characterization services.

First, human iPSC can be stained for the expression of pluripotency markers both extracellular (Tra-1/60 or Tra-1/81, SSEA4) and intracellular (Oct4, Nanog) by either immunohistochemistry or flow cytometry. Gene-targeted iPSCs should also retain normal colony morphology. Second, cells should be karyotyped to ensure a normal chromosome number and lack of aberrant translocations. Finally, a teratoma assay is performed where iPSCs are injected into immunocompromised mice. The resulting iPSC-derived teratomas are histologically examined for generation of all three germ layers (mesoderm, ectoderm, and endoderm).

# 5. ALTERNATIVE APPROACHES

## 5.1. Low transfection

Using the above protocol, we typically achieve transfection efficiencies around 60–70% and gene targeting/disruption efficiencies around 1–25%. However, several strategies can be used to enrich for targeted clones in cases of low transfection or gene editing efficiencies. Positive selection markers and antibiotic selection schemes may still be used to select for rare gene insertions, although gene targeting frequencies around 1% are generally high enough for genotypic screening without the use of a positive or negative selection marker.

In case of low transfection efficiencies, the Cas9 nuclease and sgRNA construct can be expressed from the same plasmid, such that both components of the Crispr system are co-delivered into any transfected cells. Furthermore, several constructs have been developed to express multiple sgRNAs from the same plasmid rather than separate plasmids (Cong et al., 2013; Tsai et al., 2014).

Transiently transfected cells may be enriched using a fluorescent or antibiotic resistance selection marker that is either co-transfected or co-expressed with the Cas9 nuclease. Human iPSCs electroporated with a Cas9-T2A-GFP fusion protein can be FACS-sorted 24–48 h post-transfection (Ding et al., 2013). Cas9-T2A-Puromycin resistance constructs are also available, although the drug selection may need to be carefully optimized to match the period of transient Cas9 and resistance marker expression (Ran, Hsu, Wright, et al., 2013). Alternatively, a reporter construct plasmid could be co-transfected that possesses the sgRNA targeting site upstream of a fluorescent protein such that Cas9 nuclease editing brings the fluorescent protein in frame for expression. Cells with active Cas9–sgRNA complexes can then be enriched by flow cytometry (Ramakrishna et al., 2014).

## 5.2. Viral vectors

Viral vectors have also been used as alternatives to transient transfection. Lentiviral vectors are commonly used to introduce Cas9 and sgRNA components into a wide variety of cell types, both dividing and nondividing. As these retroviruses integrate the genetic construct into the chromosome, they are particularly useful when sustained nuclease activity is desired, including CRISPR library screens that require almost complete gene disruption efficiencies (Shalem et al., 2014; Wang et al., 2014). However, lentiviral vectors are limited by the size of insert that can practically be packaged into the capsid (roughly 7.5 kb) (Yacoub, Romanowska, Haritonova, & Foerster, 2007) and tend to recombine out repetitive DNA sequences (Holkers et al., 2012; Yang et al., 2013). Integrase-deficient lentiviral vectors (IDLV), that deliver gene constructs which do not integrate into the genome and are gradually lost through cell division, have also been used to deliver nucleases and homologous donor templates to edit human stem cells (Joglekar et al., 2013; Lombardo et al., 2007).

Adenoviral vectors have been a popular approach for administering gene therapies *in vivo* as well as gene editing human stem cells *in vitro* as they can transduce a wide variety of dividing and nondividing cells. The helper-dependent (or high-capacity) adenoviral vectors, where the viral genes have been removed, can deliver constructs up to 37 kb (Gonçalves & de Vries, 2006). Their linear dsDNA genome is generally not integrated into the chromosome, but maintained episomally until lost through cell division. Adenoviral vectors were better than lentiviral vectors at introducing constructs with repetitive elements such as TALENs into cells (Holkers et al., 2012). The higher packaging capacity of adenoviral vectors also allowed the 4.1 kb *S. pyogenes Cas9* nuclease gene to be efficiently delivered into human cells (Maggio et al., 2014).

Recombinant Adeno-Associated viruses have also been used for gene targeting a wide variety of cell types. Even without sequence-specific nucleases, they can induce higher rates of homologous recombination than a transfected plasmid (Russell & Hirata, 1998); however, like conventional plasmid targeting vectors, gene targeting with AAV vectors can be further enhanced by introducing a dsDNA break (Hirsch & Samulski, 2014). AAV vectors possess a single-stranded DNA genome with a packaging capacity limited to around 4.5 kb, although they may self-assemble or be directed to form concatamers, thereby producing longer constructs (Hirsch et al., 2013).

## 5.3. Off-targets

When targeting a particular human genomic locus, one can often not find a perfectly specific sgRNA site that possesses a unique 13 bp "seed" sequence not found alongside any other active PAM site. Even then, Cas9 nuclease activity may still occur at off-target sites containing one to two mismatches in the seed sequence. Of an initial set of 190,000 sgRNA sequences designed to target human exons, 99.96% were later computationally found to have off-target sites containing at least one mismatch in the seed sequence next to a NGG or NAG PAM (Mali, Aach, et al., 2013).

However, the frequency of off-target nuclease activity is a dose-dependent function of the on-target activity—one can titrate down the on-target Cas9 nuclease activity to remove the off-target activity (Fu et al., 2013). The initial reports studying off-target Cas9 nuclease activity were done on human immortalized tumor cell lines, which were able to disrupt genes with very high rates of on-target activity (40–80%) leading to substantial rates of off-target activity at certain sites (Fu et al., 2013; Hsu et al., 2013; Kuscu et al., 2014; Pattanayak et al., 2013). For cells with a lower rate of on-target nuclease activity, the off-target sites may be less of a concern. In mouse ES cells, a Cas9–sgRNA complex bound to hundreds of off-target sites, but nuclease activity was only found at one similar off-target site (Wu et al., 2014). With the above protocol transiently transfecting Cas9 nuclease and sgRNA into human iPSC, we typically find low off-target gene disruption frequencies around 0.1–0.2%, even with identical seed sequences. Robust tumor cell lines allow for very high on- and off-target gene disruption frequencies, whereas off-target mutations may be less of an issue for cells with a lower rate of on-target activity like human iPSCs.

Since neither the on-target nor off-target Cas9 nuclease activities can currently be completely predicted through computational analysis, we recommend that any close off-target sgRNA sites also be checked when assessing gene disruption frequency in the cell population and when genotyping successfully targeted clones. Off-target gene disruption frequencies can be lowered by titrating down Cas9 nuclease activity (by transfecting a smaller quantity of plasmid, or expressing the Cas9 nuclease under a weaker promoter), although this will also decrease the on-target nuclease activity.

## 5.4. Cas9 nickases

When further specificity is required, two Cas9 nickases may be used to generate the dsDNA break instead of a single Cas9 nuclease. In the nickase version of Cas9 (D10A), the RuvC endonuclease-like domain has been mutated, such that only a single-stranded DNA break is made on the complementary DNA strand (Jinek et al., 2012). (Gene targeting with an alternate Cas9 nickase where the HNH endonuclease-like domain has been mutated so only the noncomplementary strand is cleaved (H840A) has been less well characterized.) Two sgRNAs designed to target opposite strands at the same locus can be combined to generate an offset dsDNA break. Any off-target single-stranded DNA nicks will be unlikely to be repaired by NHEJ and result in very low indel rates (Cong et al., 2013; Mali, Aach, et al., 2013). A single-stranded DNA nick was sufficient to induce homologous recombination in a human tumor cell line, but not in a human ES cell line (Hsu et al., 2013). Offset dsDNA breaks generated by Cas9 nickases may be especially necessary for targeting genes that have many conserved family members, or for therapeutic applications that require more than a single accurately targeted cell clone.

For gene disruption using paired Cas9 nickases, the highest rate of indel formation was achieved using two offset sgRNAs where the double nicks resulted in a 5′-DNA overhang. Indel formation was greatest with a 20–50 bp 5′-overhang, although detectable up to 130 bp (Cho et al., 2014; Mali, Aach, et al., 2013; Ran, Hsu, Lin, et al., 2013). Genomic deletions could also be made with the Cas9 nickase and four sgRNAs that generated two offset dsDNA breaks. A 5′-DNA overhang produced by a double nick also showed a higher ratio of homologous recombination to NHEJ compared to a blunt dsDNA break, although the overall rate of homologous recombination was still higher with the Cas9 nuclease. Recently, further specificity has been achieved using catalytically-inactive nuclease-null Cas9 proteins fused to a FokI homodimer nuclease domain—a pair of sgRNAs can bring the attached FokI domains together at the target site to dimerize and generate a dsDNA break provided that they have the appropriate orientation (PAM sites facing outward) and spacing (14–25 bp apart depending on the fusion construct) (Guilinger, Thompson, & Liu, 2014; Tsai et al., 2014).

## 5.5. Orthogonal Cas9 systems

While the Cas9 nuclease from *S. pyogenes* has been most commonly used, Cas9 nucleases derived from other bacteria are also capable of editing human genomes. Cas9 nucleases from *S. thermophilus*, *N. meningitidis*, and *T. denticola* recognize different PAM sequences, thereby expanding the set of potentially targetable sgRNA sites (Aach et al., 2014; Esvelt et al., 2013; Hou et al., 2013). Specific unique sgRNA backbones have been developed for *S. pyogenes*, *S. thermophilus*, and *N. meningitidis*, which allow these three Cas9 systems to be used simultaneously in an orthogonal manner.

In addition to nuclease and nickase activity, the easily programmable DNA binding ability of Cas9 has been adapted for many other functions. A nuclease-null Cas9 can be used by itself to repress gene expression, or be fused to transcriptional activator domains, repressor domains, epigenetic regulators, or fluorescent proteins (Mali, Esvelt, & Church, 2013).

The technology for genetic engineering has progressed rapidly in the past few years and will certainly continue to improve. The ability to easily and efficiently edit human genomes using custom-engineered nucleases has already greatly expanded studies of gene function and holds great potential for constructing modified human iPSCs and safer gene therapies.

## REFERENCES

Aach, J., Mali, P., & Church, G. M. (2014). CasFinder: Flexible algorithm for identifying specific Cas9 targets in genomes. *BioRxiv*. http://dx.doi.org/10.1101/005074.

Beumer, K. J., Trautman, J. K., Mukherjee, K., & Carroll, D. (2013). Donor DNA utilization during gene targeting with zinc-finger nucleases. *G3: Genes Genomes Genetics*, *3*, 657–664.

Chen, F., Pruett-Miller, S. M., Huang, Y., Gjoka, M., Duda, K., Taunton, J., et al. (2011). High-frequency genome editing using ssDNA oligonucleotides with zinc-finger nucleases. *Nature Methods*, *8*, 753–755.

Cho, S. W., Kim, S., Kim, Y., Kweon, J., Kim, H. S., Bae, S., et al. (2014). Analysis of off-target effects of CRISPR/Cas-derived RNA-guided endonucleases and nickases. *Genome Research*, *24*, 132–141. http://dx.doi.org/10.1101/gr.162339.113.

Cong, L., Ran, F. A., Cox, D., Lin, S., Barretto, R., Habib, N., et al. (2013). Multiplex genome engineering using CRISPR/Cas systems. *Science*, *339*, 819–823.

De Semir, D., & Aran, J. M. (2003). Misleading gene conversion frequencies due to a PCR artifact using small fragment homologous replacement. *Oligonucleotides*, *13*, 261–269.

Deyle, D. R., Li, L. B., Ren, G., & Russell, D. W. (2013). The effects of polymorphisms on human gene targeting. *Nucleic Acids Research*, *42*, 3119–3124.

Ding, Q., Regan, S. N., Xia, Y., Oostrom, L. A., Cowan, C. A., & Musunuru, K. (2013). Enhanced efficiency of human pluripotent stem cell genome editing through replacing TALENs with CRISPRs. *Stem Cell*, *12*, 393–394.

Esvelt, K. M., Mali, P., Braff, J. L., Moosburner, M., Yaung, S. J., & Church, G. M. (2013). Orthogonal Cas9 proteins for RNA-guided gene regulation and editing. *Nature Methods*, *10*, 1116–1121.

Fu, Y., Foden, J. A., Khayter, C., Maeder, M. L., Reyon, D., Joung, J. K., et al. (2013). High-frequency off-target mutagenesis induced by CRISPR-Cas nucleases in human cells. *Nature Biotechnology*, *31*, 822–826.

Fu, Y., Sander, J. D., Reyon, D., Cascio, V. M., & Joung, J. K. (2014). Improving CRISPR-Cas nuclease specificity using truncated guide RNAs. *Nature Biotechnology*, *32*, 279–284.

Gonçalves, M. A. F. V., & de Vries, A. A. F. (2006). Adenovirus: From foe to friend. *Reviews in Medical Virology*, *16*, 167–186.

Guell, M., Yang, L., & Church, G. M. (2014). Genome editing assessment using CRISPR genome analyzer (CRISPR-GA). *Bioinformatics*. http://dx.doi.org/10.1093/bioinformatics/btu427.

Guilinger, J. P., Thompson, D. B., & Liu, D. R. (2014). Fusion of catalytically inactive Cas9 to FokI nuclease improves the specificity of genome modification. *Nature Biotechnology*, *32*, 577–582.

Guye, P., Li, Y., Wroblewska, L., Duportet, X., & Weiss, R. (2013). Rapid, modular and reliable construction of complex mammalian gene circuits. *Nucleic Acids Research*, *41*, e156.

Heigwer, F., Kerr, G., & Boutros, M. (2014). E-CRISP: Fast CRISPR target site identification. *Nature Methods*, *11*, 122–123.

Hendel, A., Kildebeck, E. J., Fine, E. J., Clark, J. T., Punjya, N., Sebastiano, V., et al. (2014). Quantifying genome-editing outcomes at endogenous loci with SMRT Sequencing. *Cell Reports*, *7*, 293–305.

Hirsch, M. L., Li, C., Bellon, I., Yin, C., Chavala, S., Pryadkina, M., et al. (2013). Oversized AAV transduction is mediated via a DNA-PKcs-independent, Rad51C-dependent repair pathway. *Molecular Therapy*, *21*, 2205–2216.

Hirsch, M. L., & Samulski, R. J. (2014). AAV-mediated gene editing via double-strand break repair. *Methods in Molecular Biology*, *1114*, 291–307.

Hockemeyer, D., Soldner, F., Beard, C., Gao, Q., Mitalipova, M., DeKelver, R. C., et al. (2009). Efficient targeting of expressed and silent genes in human ESCs and iPSCs using zinc-finger nucleases. *Nature Biotechnology*, *27*, 851–857.

Holkers, M., Maggio, I., Liu, J., Janssen, J. M., Miselli, F., Mussolino, C., et al. (2012). Differential integrity of TALE nuclease genes following adenoviral and lentiviral vector gene transfer into human cells. *Nucleic Acids Research*, *41*, e63.

Hou, Z., Zhang, Y., Propson, N. E., Howden, S. E., Chu, L.-F., Sontheimer, E. J., et al. (2013). Efficient genome engineering in human pluripotent stem cells using Cas9 from Neisseria meningitidis. *Proceedings of the National Academy of Sciences of the United States of America*, *110*, 15644–15649.

Hsu, P. D., Scott, D. A., Weinstein, J. A., Ran, F. A., Konermann, S., Agarwala, V., et al. (2013). DNA targeting specificity of RNA-guided Cas9 nucleases. *Nature Biotechnology*, *31*, 827–832.

Jinek, M., Chylinski, K., Fonfara, I., Hauer, M., Doudna, J. A., & Charpentier, E. (2012). A programmable dual-RNA-guided DNA endonuclease in adaptive bacterial immunity. *Science*, *337*, 816–821.

Joglekar, A. V., Hollis, R. P., Kuftinec, G., Senadheera, S., Chan, R., & Kohn, D. B. (2013). Integrase-defective lentiviral vectors as a delivery platform for targeted modification of adenosine deaminase locus. *Molecular Therapy*, *21*, 1705–1717.

Joung, J. K., & Sander, J. D. (2012). TALENs: A widely applicable technology for targeted genome editing. *Nature Reviews Molecular Cell Biology*, *14*, 49–55.

Kim, H. J., Lee, H. J., Kim, H., Cho, S. W., & Kim, J. S. (2009). Targeted genome editing in human cells with zinc finger nucleases constructed via modular assembly. *Genome Research*, *19*, 1279–1288.

Kuscu, C., Arslan, S., Singh, R., Thorpe, J., & Adli, M. (2014). Genome-wide analysis reveals characteristics of off-target sites bound by the Cas9 endonuclease. *Nature Biotechnology*, *32*, 677–683. http://dx.doi.org/10.1038/nbt.2916.

Lee, J.-H., Park, I.-H., Gao, Y., Li, J. B., Li, Z., Daley, G. Q., et al. (2009). A robust approach to identifying tissue-specific gene expression regulatory variants using personalized human induced pluripotent stem cells. *PLoS Genetics*, *5*, e1000718.

Lin, Y., Cradick, T. J., Brown, M. T., Deshmukh, H., Ranjan, P., Sarode, N., et al. (2014). CRISPR/Cas9 systems have off-target activity with insertions or deletions between target DNA and guide RNA sequences. *Nucleic Acids Research*, *42*(11), 7473–7485. http://dx.doi.org/10.1093/nar/gku402.

Lombardo, A., Genovese, P., Beausejour, C. M., Colleoni, S., Lee, Y.-L., Kim, K. A., et al. (2007). Gene editing in human stem cells using zinc finger nucleases and integrase-defective lentiviral vector delivery. *Nature Biotechnology*, *25*, 1298–1306.

Maggio, I., Holkers, M., Liu, J., Janssen, J. M., Chen, X., & Gonçalves, M. A. F. V. (2014). Adenoviral vector delivery of RNA-guided CRISPR/Cas9 nuclease complexes induces targeted mutagenesis in a diverse array of human cells. *Scientific Reports*, *4*, http://dx.doi.org/10.1038/srep05105.

Mali, P., Aach, J., Stranges, P. B., Esvelt, K. M., Moosburner, M., Kosuri, S., et al. (2013). CAS9 transcriptional activators for target specificity screening and paired nickases for cooperative genome engineering. *Nature Biotechnology*, *31*, 833–838.

Mali, P., Esvelt, K. M., & Church, G. M. (2013). Cas9 as a versatile tool for engineering biology. *Nature Methods*, *10*, 957–963.

Mali, P., Yang, L., Esvelt, K. M., Aach, J., Guell, M., DiCarlo, J. E., et al. (2013). RNA-guided human genome engineering via Cas9. *Science*, *339*, 823–826.

Martí, M., Mulero, L., Pardo, C., Morera, C., Carrió, M., Laricchia-Robbio, L., et al. (2013). Characterization of pluripotent stem cells. *Nature Protocols*, *8*, 223–253.

Moehle, E. A., Rock, J. M., Lee, Y.-L., Jouvenot, Y., DeKelver, R. C., Gregory, P. D., et al. (2007). Targeted gene addition into a specified location in the human genome using designed zinc finger nucleases. *Proceedings of the National Academy of Sciences of the United States of America*, *104*, 3055–3060.

Orlando, S. J., Santiago, Y., DeKelver, R. C., Freyvert, Y., Boydston, E. A., Moehle, E. A., et al. (2010). Zinc-finger nuclease-driven targeted integration into mammalian genomes using donors with limited chromosomal homology. *Nucleic Acids Research*, *38*, e152.

Pattanayak, V., Lin, S., Guilinger, J. P., Ma, E., Doudna, J. A., & Liu, D. R. (2013). High-throughput profiling of off-target DNA cleavage reveals rNA-programmed Cas9 nuclease specificity. *Nature Biotechnology*, *31*, 839–843.

Qiu, P., Shandilya, H., D'Alessio, J. M., O'Connor, K., Durocher, J., & Gerard, G. F. (2004). Mutation detection using Surveyor nuclease. *BioTechniques*, *36*, 702–707.

Ramakrishna, S., Cho, S. W., Kim, S., Song, M., Gopalappa, R., Kim, J.-S., et al. (2014). Surrogate reporter-based enrichment of cells containing RNA-guided Cas9 nuclease-induced mutations. *Nature Communications*, *5*, 3378.

Ran, F. A., Hsu, P. D., Lin, C.-Y., Gootenberg, J. S., Konermann, S., Trevino, A. E., et al. (2013). Double nicking by RNA-guided CRISPR Cas9 for enhanced genome editing specificity. *Cell, 154*, 1380–1389.

Ran, F. A., Hsu, P. D., Wright, J., Agarwala, V., Scott, D. A., & Zhang, F. (2013). Genome engineering using the CRISPR-Cas9 system. *Nature Protocols, 8*, 2281–2308.

Russell, D. W., & Hirata, R. K. (1998). Human gene targeting by viral vectors. *Nature Genetics, 18*, 325–330.

Shalem, O., Sanjana, N. E., Hartenian, E., Shi, X., Scott, D. A., Mikkelsen, T. S., et al. (2014). Genome-scale CRISPR-Cas9 knockout screening in human cells. *Science, 343*, 84–87.

Tsai, S. Q., Wyvekens, N., Khayter, C., Foden, J. A., Thapar, V., Reyon, D., et al. (2014). Dimeric CRISPR RNA-guided FokI nucleases for highly specific genome editing. *Nature Biotechnology, 32*, 569–576.

Urnov, F. D., Rebar, E. J., Holmes, M. C., Zhang, H. S., & Gregory, P. D. (2010). Genome editing with engineered zinc finger nucleases. *Nature Reviews Genetics, 11*, 636–646.

Valamehr, B., Abujarour, R., Robinson, M., Le, T., Robbins, D., Shoemaker, D., et al. (2012). A novel platform to enable the high-throughput derivation and characterization of feeder-free human iPSCs. *Scientific Reports, 2*, 213.

Wang, T., Wei, J. J., Sabatini, D. M., & Lander, E. S. (2014). Genetic screens in human cells using the CRISPR-Cas9 System. *Science, 343*, 80–84.

Wu, X., Scott, D. A., Kriz, A. J., Chiu, A. C., Hsu, P. D., Dadon, D. B., et al. (2014). Genome-wide binding of the CRISPR endonuclease Cas9 in mammalian cells. *Nature Biotechnology, 32*(7), 670–676. http://dx.doi.org/10.1038/nbt.2889.

Yacoub, N. A., Romanowska, M., Haritonova, N., & Foerster, J. (2007). Optimized production and concentration of lentiviral vectors containing large inserts. *The Journal of Gene Medicine, 9*, 579–584.

Yang, L., Guell, M., Byrne, S., Yang, J. L., De Los Angeles, A., Mali, P., et al. (2013). Optimization of scarless human stem cell genome editing. *Nucleic Acids Research, 41*, 9049–9061.

Yang, L., Mali, P., Kim-Kiselak, C., & Church, G. (2014). CRISPR-Cas-mediated targeted genome editing in human cells. *Methods in Molecular Biology, 1114*, 245–267.

Yang, Y., & Seed, B. (2003). Site-specific gene targeting in mouse embryonic stem cells with intact bacterial artificial chromosomes. *Nature Biotechnology, 21*, 447–451.

CHAPTER SEVEN

# Tagging Endogenous Loci for Live-Cell Fluorescence Imaging and Molecule Counting Using ZFNs, TALENs, and Cas9

**D. Dambournet[1], S.H. Hong, A. Grassart, D.G. Drubin**
Department of Molecular and Cell Biology, University of California, Berkeley, California, USA
[1]Corresponding author: e-mail address: dambournet@berkeley.edu

## Contents



## Abstract

The programmable ZFN, TALEN, and Cas9 nucleases allow genome editing of any cell line or organism. In this chapter, we describe methods to create gene fusions at endogenous loci in mammalian cells to express fluorescent fusions of proteins of interest at endogenous levels. The donor DNA, which includes the sequence encoding a fluorescent protein, is provided to the cell to repair a double-strand break induced by a nuclease. The engineered donor sequence is integrated by homology-directed repair into the genome in frame with the coding region of the gene of interest, resulting in expression of a fusion protein at physiological levels. We further describe techniques to study protein dynamics and numbers using the genome-edited cell lines. In contrast to cell lines stably overexpressing fusion proteins from modified cDNAs, genes encoding fluorescent proteins are targeted to the endogenous genetic locus, avoiding perturbation of alternative splicing and expression levels.

*Methods in Enzymology*, Volume 546
ISSN 0076-6879
http://dx.doi.org/10.1016/B978-0-12-801185-0.00007-6

# 1. INTRODUCTION

Zinc-finger nucleases (ZFN), transcription activator-like effector nucleases (TALEN), and clustered regulatory interspaced short palindromic repeat (CRISPR)/Cas9-based RNA-guided DNA endonucleases are powerful tools that have revolutionized our ability to edit the genomes of essentially any cell type (Gaj, Gersbach, & Barbas, 2013). The versatility and the efficiency of these nucleases allow precise genome editing in human and in other species with broad applications from personalized gene therapy to biotechnology (Li, Liu, Spalding, Weeks, & Yang, 2012; Perez et al., 2008; Santiago et al., 2008; Soldner et al., 2011).

These nucleases can be designed to bind to a specific target genomic DNA sequence. Each contains a nonspecific DNA nuclease domain that will generate DNA double-strand breaks at the specified site. The breaks are then repaired either by error prone nonhomologous end joining system (NHEJ) or by homology-directed repair (HDR) (Fig. 7.1). NHEJ generates a small

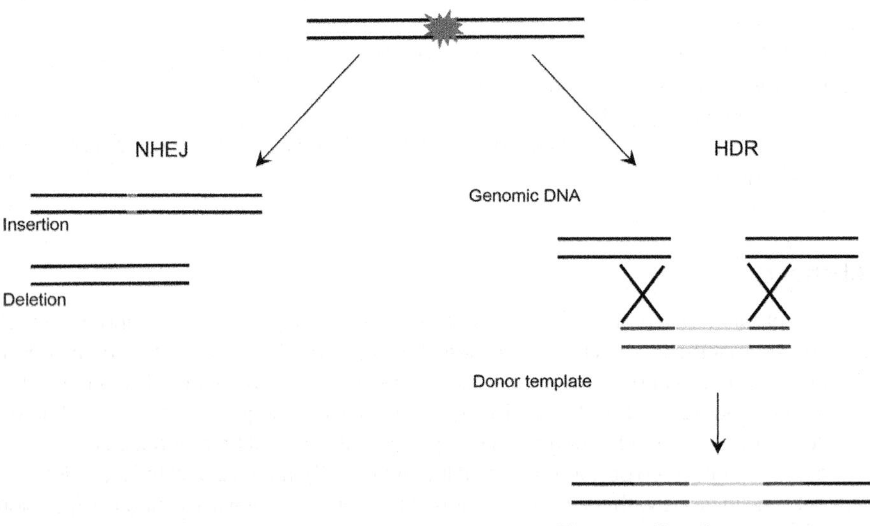


**Figure 7.1** Nuclease-induced genome editing. ZFN, TALEN, and Cas9 induce a single double-strand break in a gene locus. This break can be repaired either by the nonhomologous end joining (NHEJ, on the left) or by homology-directed repair (HDR, on the right). NHEJ repair generates deletion or insertion mutations of variable length. HDR uses homologous sequences on a donor template to repair the break while inserting a specific sequence or point mutation. (See the color plate.)

insertion or deletion, while HDR leads to sequence insertion or nucleotide correction directed by the donor DNA (Joung & Sander, 2013).

Spontaneous homologous recombination has been used to generate knock-out or knock-in mice (Hall, Limaye, & Kulkarni, 2009; Pucadyil & Schmid, 2008), but these events are rare, making more convenient and rapid approaches such as plasmid transfection appealing.

It is now possible by combining designer nucleases and donor DNAs to modify any genetic locus via sequence integration by HDR (Janke et al., 2004). Until very recently, such genomic gymnastics were largely restricted to model organisms such as yeast and mice.

Largely due to the difficulty associated with modifying genomes, investigators have often settled for overexpression of a protein of interest carrying a tag, such as green fluorescent protein (GFP). This approach has been used for decades to study the localization and the behavior of specific proteins in the cell, often under the control of a strong promoter. As a consequence, proteins are often expressed at levels that may be many fold greater than endogenous levels. Elevated levels of expression can lead to a host of problems, including protein misfolding, dominant effects, and disruption of the balance of protein–protein interactions (Gibson, Seiler, & Veitia, 2013). Overexpression is based on the transfection of a cDNA and is not compatible with mRNA splicing and the tissue specificity of expression of some isoforms. To minimize these effects, stable cell lines, in which a plasmid or a viral vector is integrated into the genome, are sometimes created. However, as the integration is not controlled, levels of the transgene may vary from one integration event to another, and expression of neighboring genes can also be altered by the presence of a strongly expressed transgene. Even more importantly, essential genes can be inactivated. Engineered ZFN, TALEN, and Cas9 nucleases enhance the efficiency of homologous recombination at a specific site, which may be so-called genomic safe harbors (Hockemeyer et al., 2011) or at the normal genomic locus (Doyon et al., 2011). Also ZFNs, TALENs, and Cas9 are able to generate double-strand breaks in diverse organisms including *Caenorhabditis elegans*, *Drosophila*, mouse, somatic human cells, and embryonic stem cells (Cui et al., 2011; Gratz et al., 2013; Lo et al., 2013).

In this chapter, we describe a protocol to create gene fusions at the endogenous genomic locus using one of three engineered nucleases: ZFNs, TALENs, or Cas9 in mammalian cells. The focus of this chapter is not on the design of the different types of nucleases, which is the subject of Chapters 2, 3, and 6. Instead, we describe how to prepare a donor template to improve efficient integration of the tag into human cancer cell lines and human-induced pluripotent stem cells (hiPSC), and how to isolate and characterize the cell lines. We also describe applications where the use of genome editing for expression of fluorescent fusion proteins has improved live-cell analysis of protein dynamics and has enabled protein counting.

## 2. METHODS

Genome editing of mammalian cells is based on the cell's ability to use an exogenous template, or a donor plasmid, to correct a double-strand break via the homology-directed repair pathway. Although it has been shown that homology of as little as 50 bp is sufficient to support homologous recombination in mammalian tissue culture cells (Orlando et al., 2010), we recommend a conservative approach and use about 500–700 bp of homology in each arm, centered around the nuclease cut site. Sequences up to 20 kb have been inserted in this manner (Jiang et al., 2013).

### 2.1. Donor plasmid design

In this section, we describe two methods to construct a donor plasmid, based on Gibson assembly and on conventional restriction enzyme-based cloning. The donor plasmid sequence contains two homology areas (HAs) flanking the XFP (green, red, or any fluorescent protein). The HAs are amplified from genomic DNA. We recommend the use of genomic DNA isolated from the cell line to be edited to preserve single nucleotide polymorphisms from the specific cell line. To ensure proper folding of both the protein of interest and the fluorescent protein, and to avoid interference with protein function, we use a linker of four to six amino acids between the protein and the XFP. The linker should be composed of neutral amino acids such as glycine and alanine, interspersed with serine residues to provide flexibility between the XFP and the protein of interest (Miyawaki, Sawano, & Kogure, 2003). As for construction of plasmids used for classical overexpression studies, some caution must be taken. For N-terminal tagging, the XFP followed by a linker should be inserted at the start codon (Fig. 7.2A, top). In such cases, it is advisable to remove the start codon of the protein of interest to avoid translation originating from the second ATG. For C-terminal tagging, the linker and XFP should be inserted directly before the stop codon of the protein of interest (Fig. 7.2A, middle). In this case, the start codon of the XFP is often removed. When generating an internally tagged protein, the XFP should be flanked by two linkers that maintain the open reading frame

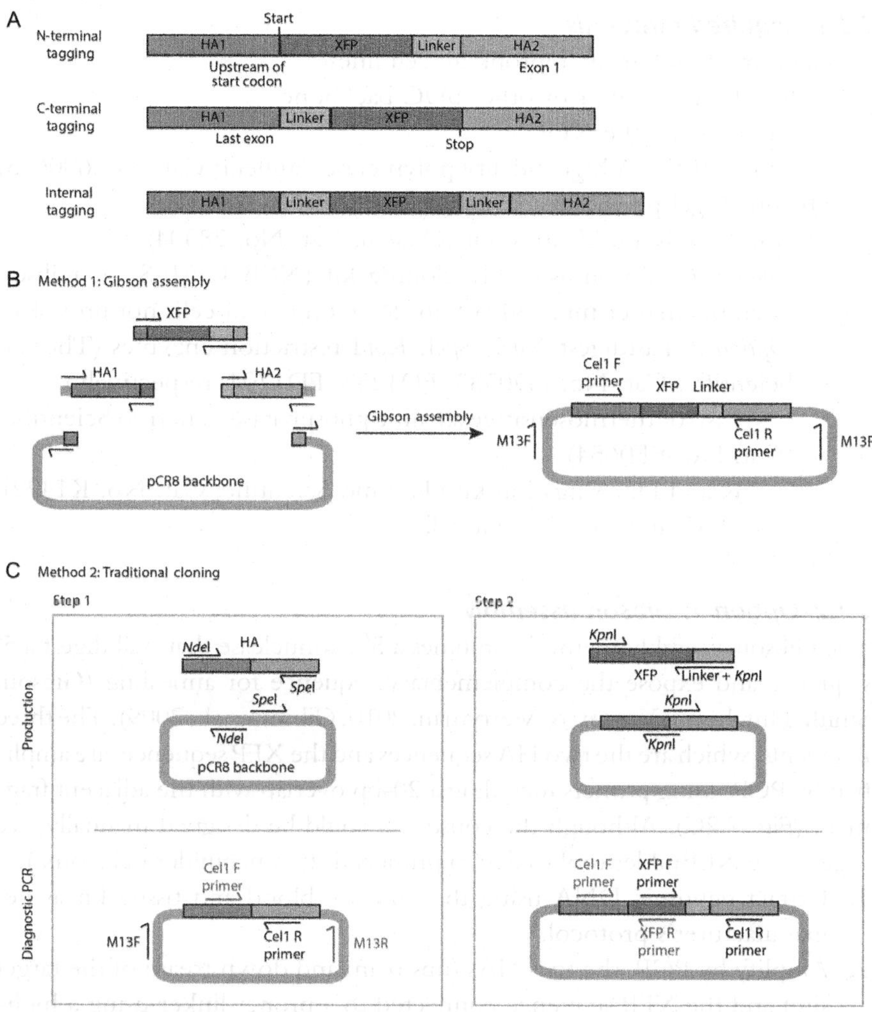


**Figure 7.2** Donor plasmid construction. (A) Possible organization of donor plasmids for N-terminal tagging (top), C-terminal tagging (middle), and internal tagging (bottom). (B) Gibson assembly of the donor plasmid. The schematic depicts a donor for N-terminal tagging. (C) Traditional cloning of the donor plasmid. (See the color plate.)

(Fig. 7.2A, bottom). To avoid mutations and to optimize the integration of the donor plasmid, the donor plasmid should be designed to introduce silent mutations at the ZFN/TALEN/Cas9 recognition site. Up to five mismatches are sufficient to prevent plasmid cutting. The backbone for the donor can be any vector that can be replicated in *Escherichia coli*. We introduce pCR8-TOPO as an example.

### 2.1.1 Required materials

Genomic DNA from appropriate cell lines
pCR8-TOPO vector or other pUC backbone
DNA encoding the XFP
Herculase II DNA high-fidelity polymerase (Agilent; Cat. No. 600675)
QIAquick gel purification kit (Qiagen; Cat. No. 28704)
MinElute PCR purification kit (Qiagen; Cat. No. 28004)

*Option 1*: Gibson assembly cloning kit (NEB E5510S) or Gibson assembly master mix (NEB E2611S) (competent cells not provided)

*Option 2*: Fastdigest *Nde*I, *Spe*I, *Kpn*I restriction enzymes (Thermo Scientific; Cat. No. FD0583, FD1253, FD1704, respectively)

FastAP thermosensitive alkaline phosphatase (Thermo Scientific; Cat. No. EF0654)

Rapid DNA ligation kit (Thermo Scientific; Cat. No. K1422)

DH5-alpha competent cells

### 2.1.2 Option 1: Gibson assembly

The Gibson assembly method combines a 5′ exonuclease that will digest a 5′ sequence and expose the complementary sequence for annealing (Gibson, Smith, Hutchison, Venter, & Merryman, 2010; Gibson et al., 2009). The three fragments, which are the two HA sequences and the XFP sequence, are amplified by PCR using primers including a 20-bp overlap with the adjacent fragment (Fig. 7.2C). Although the constructs could be designed manually, we suggest the NEBuilder tool to design primers (http://nebuilder.neb.com/).

1. Extract genomic DNA using the DNeasy blood and tissue kit as per manufacturer's protocol.
2. Amplify by PCR the two HAs (upstream and downstream of the target site) and the XFP sequence connected to a proper linker using a high-fidelity polymerase such as Herculase II.

| | Amount |
|---|---|
| Genomic DNA/backbone | 200 ng/50 ng |
| Herculase buffer 5 × | 10 μL |
| Primer forward (10 μ*M*) | 1.25 μL |
| Primer reverse (10 μ*M*) | 1.25 μL |
| dNTP (100 m*M*) | 0.5 μL |
| Herculase II | 0.5 μL |
| $ddH_2O$ | To 50 μL |

Place reaction on a PCR block preheated to 95 °C, and run the following protocol:

| | | |
|---|---|---|
| 95 °C | 2 min | |
| 95 °C | 10 s | 34 cycles |
| Ta°C | 20 s | |
| 72 °C | 1 min | |
| 72 °C | 3 min | |

*Note*: 3′ UTRs are rich in GC and are often difficult to amplify by PCR. The denaturation temperature can in such cases be increased to 98 °C.

3. Linearize the backbone either by PCR or by restriction enzyme.
4. Purify amplified HAs, backbone, and XFP sequence using QIAquick gel purification kit.
5. Use Gibson assembly cloning kit to assemble and transform the plasmid according to the manufacturer's instructions.
6. Pick four to eight colonies and identify positive clones by either colony PCR or restriction analysis.
7. Confirm by sequencing with appropriate primers to cover the whole sequence.

### *2.1.3 Option 2: Classical cloning method*

1. Design donor plasmids and select three restriction enzymes that do not have a restriction site within the HAs, the XFP sequence, or the backbone. In our example, *Nde*I, *Spe*I, and *Kpn*I are available.
2. Amplify the backbone and 1–1.5 kb of HA with primers designed to introduce the *Nde*I and *Spe*I restriction sites. Run the PCR product on an agarose gel.
3. Purify amplified HAs and backbone using QIAquick PCR purification kit. Elute with 20 μL EB with MinElute column.
4. Digest the backbone and the HA:

| | Amount |
|---|---|
| Backbone/HAs | 15 μL |
| *Nde*I Fastdigest | 1 μL |
| *Spe*I Fastdigest | 1 μL |
| Fastdigest buffer | 2 μL |
| $ddH_2O$ | 1 μL |

Add 1 μL FastAP alkaline phosphatase to the backbone digestion solution. Incubate 20 min at 37 °C

5. Purify digested HAs and backbone using QIAquick gel purification kit. Elute with 20 μL EB with MinElute column.
6. Ligate the digestion products. Use at least threefold molar excess of inserts.

| | Amount |
|---|---|
| Backbone | 50 ng |
| HAs | At least threefold molar excess |
| Quick T4 DNA ligase | 1 μL |
| Quick Ligation buffer 2× | 10 μL |
| $ddH_2O$ | To 20 μL |

Incubate for 15 min at room temperature.

7. Transform 2 μL ligation product into competent *E. coli*. Screen for the positives by PCR and sequence the plasmids. If the pCR8-TOPO backbone is used, M13FOR/Cel1REV or Cel1FOR/M13REV primer pairs can be used for diagnostic PCR (Fig. 7.2D).
8. Insert *Kpn*I restriction site by direct mutagenesis PCR. The *Kpn*I site should be located where the XFP and the linker sequences will be inserted.
9. Amplify the XFP and linker(s) sequences including *Kpn*I restriction sites on both sides by PCR.
10. Digest and ligate the pCR8-HA-*Kpn*I and XFP-linkers as described above.
11. Carry out diagnostic PCR on DNA from four to eight colonies using Cel1FOR/XFP-REV or XFP-FOR/Cel1REV primer pairs.
12. Sequence positive plasmids with M13FOR, M13REV, Cel1FOR, or Cel1REV. Make sure that XFP and the protein of interest are in frame.

## 2.2. Generation of genome-edited cell lines using CRISPR, TALENs, or ZFNs

### *2.2.1 Required materials*

MDA/SK-MEL2 cells lines and hiPSC can be obtained from various sources

hiPSC are cultivated in a feeder-free system

**Table 7.1** Nucleofactor solutions and Amaxa programs used for different cell lines

| Cell line | Nucleofactor solution | Amaxa program |
|---|---|---|
| U2OS | V | X-001 |
| SK-MEL2 | R | T-020 |
| HeLa | R | I-013 |
| MDA-MB-231 | V | X-013 |

Details for additional cell lines can be found on: http://bio.lonza.com/resources/product-instructions/protocols.

StemPro® Accutase® Cell Dissociation Agent (Gibco; Cat. No. A11105-01)
Matrigel hESC-qualified matrix (Corning; Cat. No. 354277)
mTESR™1 (STEMCELL Technologies; Cat. No. 05850)
ROCK inhibitor (Y27632) (Stemgent; Cat. No. 04-0012)
0.05% Trypsin–EDTA 1 × (Gibco; Cat. No. 25300-054)
DMEM/F12, Glutamax supplement (Gibco; Cat. No. 10565042)
DMEM/F12 no phenol red (Gibco; Cat. No. 11039047)
Fetal bovine serum, regular (Corning; Cat. No. 35-010-CV)
DPBS 1 × (Gibco; Cat. No. 14190-144)
Penicillin–streptomycin 100 × solution (10,000 U/mL) (P/S) (Gibco; Cat. No. 15140-122)
Nucleofector kit (see Table 7.1 for solution corresponding to the cell line or http://bio.lonza.com/resources/product-instructions/protocols/) and Supplement 1 (Lonza)
Amaxa™ certified 100 μL aluminum electrode cuvettes
Human stem cells Nucleofector kit 1 (Lonza)
Amaxa Nucleofector System (Lonza)
Tube, 5-mL with cell-strainer cap, round-bottom, polystyrene (Falcon; Cat. No. 352235)
Tube, 5-mL round-bottom, polypropylene (Falcon; Cat. No. 352063)
BD influx FACS sorter
Cloning disks
Tweezers

### *2.2.2 Preparation of cells*

**A.** Thawing and plating MDA cells

**1.** Quickly thaw the frozen MDA vials using a 37 °C water bath.

2. Transfer cells into 15-mL conical tube and add 5-mL warmed DMEM/F12 containing 10% FBS and 1 × P/S (hereafter referred to as "complete medium").
3. Spin down cells at 1000 rpm for 5 min.
4. Aspirate the medium, and resuspend the cell pellet in 10 mL complete medium.
5. Plate cells in a 10-cm culture dish.
6. Incubate cells at 37 °C, 5% $CO_2$ incubator.

**B.** Passaging and maintaining MDA cells

1. Cells should be passaged when they have reached 80–90% confluency. To ensure a good transfection rate, cells should not exceed passage 6.
2. Aspirate the medium and wash cells with 10 mL DPBS.
3. Add 2 mL Trypsin and incubate at 37 °C for 5 min.
4. Add 8 mL complete medium and pipette up and down to collect detached cells.
5. Transfer the detached cells to a 15-mL conical tube.
6. Spin down cells at 1000 rpm for 5 min.
7. Aspirate the medium, and resuspend cell pellets in 10 mL complete medium.
8. Plate cells onto a 10-cm culture dish (1:10 or 1:20 split).
9. Incubate cells at 37 °C, 5% $CO_2$ incubator.

**C.** Thawing and plating hiPSC cells

1. Prepare Matrigel-coated six-well plate: dilute 250 μL Matrigel into 25 mL DMEM/F12 without phenol red. Transfer 1 mL of diluted Matrigel into each well. Incubate for 1 h at room temperature. Store Matrigel-coated plates at 4 °C for up to 2 weeks.
2. Quickly thaw the frozen hiPSC vials using a 37 °C water bath.
3. Transfer cells into a 15-mL conical tube and add 5 mL warmed mTESR1.
4. Spin down cells at 1000 rpm for 3 min.
5. Aspirate the medium, and resuspend cell pellets in 2 mL mTESR1 supplemented with 10 μ*M* ROCK inhibitor (RI).
6. Plate cells in Matrigel-coated wells.
7. Incubate cells at 37 °C, 5% $CO_2$ incubator.
8. Change the medium every day.

**D.** Passaging and maintaining hiPSC cells

1. When cells reach 80–90% confluency, passage the cells.
2. Aspirate the medium and wash with 2 mL DPBS.
3. Add 1 mL Accutase and incubate at 37 °C for 2 min.

4. Add 5 mL mTESR1 and gently pipette up and down to collect detached cells.
5. Transfer the detached cells to a 15-mL conical tube.
6. Spin down cells at 1000 rpm for 3 min.
7. Aspirate the medium, and resuspend cell pellets with 2 mL mTESR1 supplemented with 10 μ*M* RI.
8. Plate cells in the Matrigel-coated wells (1:6 or 1:8 split).
9. Incubate cells at 37 °C, 5% $CO_2$ incubator.
10. Change the medium every day.

### *2.2.3 Electroporation*

Trypsinize cells 1 day before transfection and seed at around 70% confluency.

**A.** MDA electroporation

1. Gently wash cells with 10 mL room temperature DPBS.
2. Remove DPBS, add 2 mL Trypsin, and incubate at 37 °C for 5 min.
3. Resuspend cells with 10 mL complete medium.
4. Count cells to determine cell/mL (use Countess or a hemocytom eter) and calculate volume required for $2 \times 10^6$ cells (single transfection, scale as needed).
5. Place desired volume of cells into 15 mL centrifuge tube.
6. Spin at $200 \times g$ for 5 min at room temperature and remove supernatant.
7. Prepare the master mix:
   a. 82 μL Nucleofector solution recommended for the cell type
   b. 18 μL Supplement
8. Resuspend the cell pellet in the master mix and add 2.5 μg of each ZFN/TALEN/Cas9n or 5 μg Cas9 DNA and 15 μg of donor plasmid.
9. Transfer cells to a cuvette.
10. Place the cuvette into the Nucleofactor device.
11. Nucleofect cells using program recommended for the cells according to manufactures' specification or Table 7.1.
12. Quickly transfer cells into complete medium in a 10-cm dish.
13. Incubate cells at 37 °C, 5% $CO_2$ incubator for 3–7 days.
14. Harvest cells and use desired method to isolate edited cells.

**B.** hiPSC electroporation

For the hiPSC, we recommend recovery for 1 week after thawing, and supplementing the medium with 10 μ*M* RI the day before the electroporation.

1. Gently wash cells with 2 mL room temperature DPBS.

2. Aspirate DPBS, add 1 mL Accutase, and incubate at 37 °C for 5 min.
3. Resuspend cells in 5 mL mTESR1.
4. Count cells to determine cell/mL (use Countess or a hemocytometer) and calculate volume required for $2 \times 10^6$ cells (single transfection, scale as needed).
5. Place desired volume of cells into 15-mL centrifuge tube.
6. Spin at $200 \times g$ for 5 min at room temperature and aspirate supernatant.
7. Prepare the master mix:
   a. 82 μL Human stem cells Nucleofector kit 1 solution
   b. 18 μL Supplement 1
8. Incubate for 5 min at 37 °C.
9. Resuspend the cell pellet with the master mix and add 2.5 μg of each ZFN/TALEN/Cas9n or 5 μg Cas9 DNA and 15 μg of donor plasmid.
10. Transfer cells into a cuvette and incubate for 1–2 min on ice.
11. Place the cuvette in the Nucleofactor device.
12. Nucleofect cells using program T-020.
13. Quickly transfer cells into 4 mL mTESR1 + 10 μ*M* RI and plate the cells in two wells of a six-well plate coated with Matrigel.
14. Harvest cells 3–7 days after nucleofection and use desired method to isolate edited cells.

### 2.2.4 Isolation of genome-edited cells

**A.** FACS sorting

FACS parameters depend on the instrument model and the cell line used. We usually use a 100-μm nozzle with low pressure (20 psi). For the model that we used, BD influx FACS sorter, cells are sorted in "pure" mode, and the cytometer will sort one cell per drop. Appropriate nontransfected cells should be prepared in parallel for reference (Fig. 7.3A). As phenol red can interfere with fluorescence detection and can increase fluorescence background, complete medium should be replaced by DMEM/F12 phenol red free.

1. Detach the cells using Trypsin or Accutase.
2. Transfer the detached cells to a 15-mL conical tube.
3. Count cells.
4. Spin down cells at 1000 rpm for 5 min.
5. Resuspend the cells in 2 mL complete medium without phenol red/mTESR1 + RI. The concentration should not exceed $5\text{–}10 \times 10^6$ cells/mL.

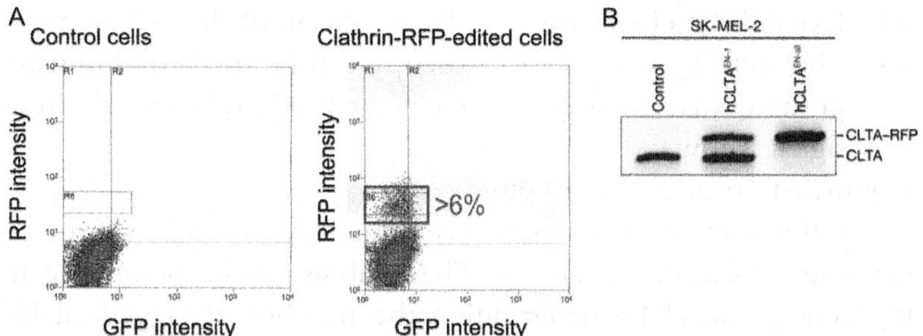


**Figure 7.3** Detection and collection of genome-edited cells. (A) FACS profiles of unstained control cells for reference on the left and CLTA–RFP edited cells on the right. The *x*-axis represents GFP fluorescence and the *y*-axis represents RFP fluorescence. (B) Genotyping PCR for control, single-tag- and all-tag-edited cells. The lower band represents amplification of the wild-type locus, while the upper band reflects amplification of locus containing XFP.

6. Put the cells through a 40 μm mesh to avoid cell aggregates.
7. Transfer the cells into a new 5-mL polypropylene tube.
8. Prepare 4–5 96-well plates with 100 μL complete medium and a 5 mL polypropylene tube containing 2 mL complete medium. For the hiPSC, we recommend sorting multiple cells (see step 9b, below) into each well of a six-well plate coated with Matrigel with 2 mL mTESR1 + RI per well. This strategy is used for cells that cannot survive as single cells.
9. Sorting RFP/GFP-positive cells (Fig. 7.3A):
   a. Sort single MDA cells into 96-well plates. Cells can also be sorted and collected in bulk in an additional tube and kept for freezing or further clone isolation.
   b. Sort 2000–4000 hiPSC per well.
10. Check single cells in the 96-well plates and change the medium every 2 days
11. Identify positive clones:
    a. For MDA cells: as the cells approach confluence, trypsinize the cells and plate them on a 24-well plate. Passage the cells onto a six-well plate and keep an aliquot of cells for genotyping.
    b. For hiPSC: using tweezers, soak a cloning paper disk with Accutase. Place it directly on a colony. To insure isolation of a clonal population, it is important to isolate only one colony. Quickly remove the cloning disk and transfer into a well of a 96-well plate. Remove the cloning disk from the well the next day. Change the medium 2 days later and then every day.

12. Expand the clones (see 2.2.2 Preparation of the cells section).
13. Live-cell fluorescence microscopy can be used to eliminate false positives that can be generated by the FACS when the fluorescence signal is low.

**B.** Confirmation of genome-edited cells

In this section, we describe several methods to confirm proper genome editing of the clones. This analysis can be performed first at the genome level by determining the number of tagged alleles and ensuring that the XFP is correctly integrated and that no mutations have been generated. Second, at the protein level, immune blots should be performed to confirm the expression and predicted size of the fusion protein. Finally, proper localization of the XFP fused protein can be assessed by immunostaining, and functional tests employed to insure that the fusion protein has not interfered with the process being investigated.

By sequencing

**Genomic DNA extraction**

1. Detach cells with Trypsin or Accutase.
2. Transfer the detached cells to a 15-mL conical tube.
3. Spin down cells at 1000 rpm for 5 min. Aspirate the medium completely. At this step, pelleted cells can be stored at −20 °C for at least 1 month.
4. Extract genomic DNA using the DNeasy blood and tissue kit as per manufacturer's protocol.

**PCR and sequencing**

PCR and sequencing can be performed using the primers used for the surveyor assay used to confirm cutting by the designed nucleases.

1. Prepare PCR mix as follows:

| | Amount |
|---|---|
| Genomic DNA | 200 ng |
| Herculase buffer 5 × | 10 μL |
| Primer forward (10 μ*M*) | 1.25 μL |
| Primer reverse (10 μ*M*) | 1.25 μL |
| dNTP (100 m*M*) | 0.5 μL |
| Herculase II | 0.5 μL |
| ddH2O | To 50 μL |

2. Place reaction on PCR block preheated to 95 °C and run the following protocol:

| | | |
|---|---|---|
| 95 °C | 2 min | |
| 95 °C | 10 s | 34 cycles |
| Ta°C | 20 s | |
| 72 °C | 30 s | |
| 72 °C | 3 min | |

3. Run PCR products on 1.0% TAE agarose gel and stain with ethidium bromide. When all alleles are not tagged, two bands should appear: the lower one corresponding to the wild-type and the upper one to the tagged allele (Fig. 7.3B).
4. Gel purify both the wild-type and edited bands using Qiagen MinElute gel extraction kit as per manufacturer's protocol. Elute in 10 μL $H_2O$.
5. Sequence wild-type and tagged alleles with appropriate primers (Note that it is crucial that the wild-type allele be sequenced to detect mutations introduced by NHEJ).

**By immune blot:**

1. For the preparation of lysates: cells were briefly washed in DPBS and incubated with 0.5 m*M* EDTA for 5 min.
2. Centrifuge 110 *g* for 2 min.
3. Remove supernatant and add warmed 2 × protein sample buffer (125 m*M* Tris–HCl at pH 6.8, 10% glycerol, 10% SDS, 130 m*M* dithiothreitol, 0.05% bromophenol blue and 12.5% β-mercaptoethanol).
4. Resuspend pellet by pipetting.
5. Heat samples to 95 °C, for 3 min.
6. Separate proteins by SDS-PAGE.
7. Transfer to Immobilon-FL PVDF transfer membrane (IPFL-00010, Millipore) or nitrocellulose membrane in transfer buffer (25 m*M* Trizma base, 200 m*M* glycine, 20% methanol, and 0.025% SDS) at 50 V and 4 °C for 1 h.
8. Rinse the membrane in TBS (Tris-buffered saline).
9. Incubate in Odyssey blocking buffer (diluted 1:1 in DPBS) at room temperature for 1 h.

10. Dilute appropriate primary antibodies in blocking buffer and incubate with the membrane at 4 °C overnight. The following antibodies can be used for XFP fusion proteins: anti-GFP (B2; 1:2000; Santa Cruz Biotechnology) or anti-tRFP (AB233; 1:500; Evrogen).
11. Blots were subsequently incubated in the dark for 1 h at room temperature with secondary IRDye 680 or 800CW antibodies (LI-COR Biosciences) diluted to 1:5000 in Odyssey blocking buffer/DPBS solution containing 0.1% Tween 20.
12. After washing with TBST (Tris-buffered saline Tween 20), the membrane was incubated in TBS and scanned using an Odyssey infrared imager (LI-COR Biosciences). Quantification of protein bands was performed using the Odyssey Infrared Imaging System (version 3.0).

**By immunofluorescence microscopy**

1. Grow cells on glass coverslips in 24-well plates overnight.
2. Fix cells in 4% paraformaldehyde at room temperature for 20 min.
3. After three washes with DPBS, quench the coverslips with 1 mg/mL NaBH4 for 15 min twice.
4. Permeabilize cells in 0.1% saponin/DPBS or 0.1% Triton X-100/DPBS. Wash briefly with DPBS.
5. Incubate coverslips with appropriate primary antibody overnight at 4 °C.
6. Wash with DPBS for 15 min.
7. Incubate coverslips with appropriate secondary antibody overnight for 2 h at room temperature.
8. Perform three washes with DPBS.
9. Mount the coverslips on glass slides using ProLong Gold Antifade Reagent (Invitrogen).

## 3. TAGGING/EDITING LIMITATIONS

Genome editing is a powerful tool that promises to facilitate more faithful fluorescence microscopy analysis of *in vivo* dynamics. This approach does, however, have limitations. Tagging a protein can impair the function of a protein even when it is expressed at the endogenous level. Some classes of proteins, such as Rab GTPases, can only be tagged at the N-terminus (Chavrier et al., 1991), whereas others, such as dynamin2, for instance, must be tagged on the C-terminus (Hinshaw & Schmid, 1995; Liu, Surka, Schroeter, Lukiyanchuk, & Schmid, 2008). Function of the μ subunit of the adaptor protein AP2 can be preserved by inserting the fluorescent protein into an internal loop (Nesterov, Carter, Sorkina, Gill, & Sorkin, 1999). We strongly recommend preliminary experiments to ensure that fusion protein function is preserved as much as possible. Since detection of function impairment can be done more sensitively in some simple organisms, like yeast, we have often modeled fusions for mammalian cells on fusions that preserve function in yeast.

The three nucleases also have limitations. The ZFNs and TALENs are highly sensitive to DNA methylation and thus are not optimal for GC-rich regions, which can be found in the UTR regions of genes (Kim et al., 2013). In the case of the Cas9/CRISPR system, off-target cutting is more likely to happen as Cas9 can tolerate a number of mismatches, although current efforts are addressing this issue (see below) (Cong et al., 2013; Fu et al., 2013; Mali et al., 2013). With the tolerance of mismatches, the short recognition site of CRISPR/Cas9—usually 20 nucleotides followed by the PAM sequence, NGG—is rarely unique in the genome. Multiple labs showed that the use of two pairs of guide RNAs can avoid the off-target cutting (Ran et al., 2013). They used a Cas9 nickase mutant that will generate a single-strand break. These breaks can be easily repaired by the high-fidelity base excision repair pathway without creating a deletion or an insertion (Dianov & Hübscher, 2013). This group increased the specificity for the target site by using a pair of offset sgRNAs complementary to opposite strands of the site.

In this chapter, we also provide a method to sort cells based on the fluorescence by FACS. Fluorescence levels correlate directly with the protein expression levels. Although FACS is highly sensitive, the expression level of the fusion protein might be barely higher than the fluorescence background. In such a case, positive cells will be indistinguishable from the negative cells. A resistance cassette could be added in the donor plasmid, followed/proceeded by a self-cleavage site, such as the 2A peptide (Hockemeyer et al., 2011; Kim et al., 2011). This sequence is thought to induce a "pause" and thus the release of the nascent peptide (Doronina et al., 2008). The resistance cassette will be expressed by the endogenous promoter. Another strategy is to integrate the resistance cassette into the locus of interest with its own promoter and flanked by the LoxP sequence. This strategy allows removal of the resistance cassette after selection by transient expression of the Cre recombinase. After selection and isolation of the clone, only the LoxP sequence will remain at the target site.

# 4. PERSPECTIVES

This section explores applications in which the effort needed to produce genome-edited cell lines was shown to be justified. We also explore some opportunities that this technique may provide

## 4.1. Efficiency of cellular processes: Example of clathrin-mediated endocytosis

We have compared the dynamics and function of endocytic proteins fused to fluorescent proteins in mammalian cells either overexpressing the proteins or genome-edited to express the proteins at endogenous levels. We showed that clathrin and dynamin have a more regular recruitment in the genome-edited cells and that endocytosis in those cells is more rapid and efficient (Fig. 7.4; Doyon et al., 2011). This example therefore highlights how protein overexpression can affect the dynamics and function of cell processes. We further note that genome editing can be used to express at endogenous levels proteins tagged with affinity tags, degron tags, etc., minimizing perturbation of their stoichiometries and their functions.

## 4.2. Quantification of protein stoichiometry in specific structures within genome-edited cells

Fluorescence live-cell imaging, which has transformed cell biology by revealing the spatiotemporal organization and molecular assembly during

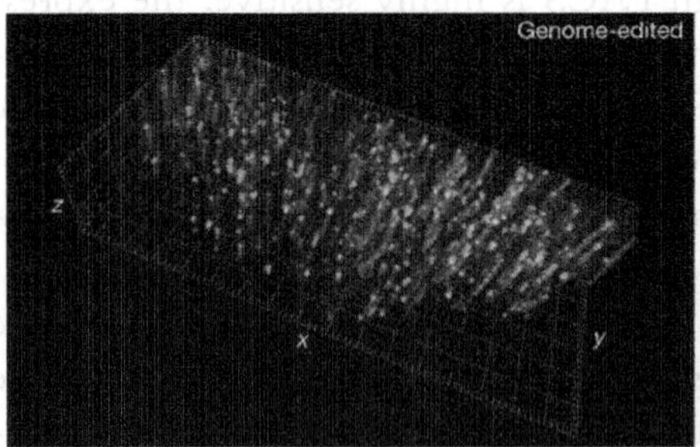


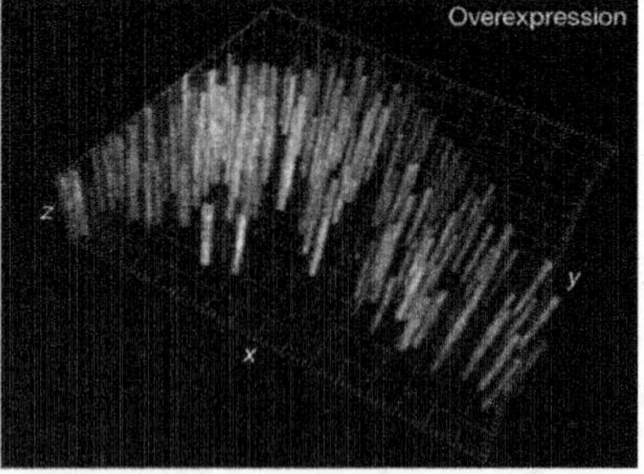


**Figure 7.4** Illustration of improved clathrin and dynamin dynamics in genome-edited cells. Three-dimensional kymographs from movies of a genome-edited cell line expressing clathrin-RFP and dynamin2-GFP on the left, or overexpressing clathrin-RFP and dynamin2-GFP on the right. *x–y* plane (2.5 μm$^2$ grid), cell plasma membrane; *z*-axis, time (240 s, 2 s per slice). *Originally published in Doyon et al. (2011).* (See the color plate.)

biological processes, is going through another revolution with single-molecule detection made possible by recent improvements in microscopy and camera technologies. The quantitative approach described here opens new avenues to count the number of molecules in specific structures. Such measurements allow more faithful *in vitro* reconstitutions and promise to be the basis for mathematical modeling. In yeast, fluorescence microscopy of cells expressing fusions of GFP to proteins of interest at the native gene locus, the number of molecules either globally or in particular structures of interest has been evaluated (Coffman & Wu, 2014; Joglekar, Bouck, Molk, Bloom, & Salmon, 2006; Wu & Pollard, 2005).

The use of ZFN, TALEN, and Cas9 nucleases to create fusions of endogenous and fluorescent proteins opened the door to the deployment of similar approaches in mammalian cells. In contrast to other methodologies, which require overexpression and complex statistical analysis (Cocucci, Aguet, Boulant, & Kirchhausen, 2012), the use of genome-edited cell lines, in which the protein of interest is expressed as GFP fusions at endogenous levels, allows rapid and simple quantification of protein recruitment with single-molecule sensitivity while preserving the stoichiometry of the proteins studied (Grassart et al., 2014; Cocucci, Gaudin, & Kirchhausen, 2014). Recently, our group successfully employed this approach to determine that one or multiple quanta of ~26 molecules of dynamin2 is recruited during endocytosis. Interestingly, our results validate previous estimations proposed by structural models and exemplify the benefits of such methodology. As genome editing can be readily implemented in any lab, it offers great promise to foster a quantitative understanding of the molecular biology of the cell.

## 4.3. Genome-edited stem cells: A new model for mammalian cell biology studies

When combined with genome editing, hiPSC represent another step forward in improvement of cell biological studies using fluorescent fusion proteins. Several labs have elucidated the signals that regulate various hiPSC differentiation pathways. As a consequence, it is now possible to differentiate hiPSC into a variety of cell types including cardiomyocytes, hepatocytes, and neurons (Chambers et al., 2009; Chambers, Mica, Studer, & Tomishima, 2011; Lian et al., 2013). Using genome editing to fluorescently label proteins that are specifically expressed in a certain tissue is also a good strategy to isolate a specific differentiated cell type derived from stem cells. Cells expressing the fluorescent protein can be FACS sorted. This strategy can then be used to study the fate of a cell population within a tissue during specific differentiation stages (Forster et al., 2014). hiPSC have been generated from patients and have been used as disease models (Chung et al., 2013; Li et al., 2013; Ryan et al., 2013). But differences in genetic background make comparisons difficult. We feel that hiPSC offer advantages for cell biology studies because they often are euploid, they are not cancer cells, they can be differentiated, and they can be edited to express fluorescent fusion proteins and to introduce human disease mutations. One promising avenue is to endogenously tag a wild-type or mutated protein with a fluorescent protein and study its behavior within different cell types to investigate how the mutation leads to disease phenotypes (Ding et al., 2013).

## ACKNOWLEDGMENTS

The authors thank Christa Cortesio and Jessica R. Marks for advice and comments on the manuscript. D. D. is supported by Human Frontier Science Program Fellowship and D.G.D. was supported by National Institutes of Health Grant R01 GM65462.

## REFERENCES

Chambers, S. M., Fasano, C. A., Papapetrou, E. P., Tomishima, M., Sadelain, M., & Studer, L. (2009). Highly efficient neural conversion of human ES and iPS cells by dual inhibition of SMAD signaling. *Nature Biotechnology*, *27*(3), 275–280.

Chambers, S. M., Mica, Y., Studer, L., & Tomishima, M. J. (2011). Converting human pluripotent stem cells to neural tissue and neurons to model neurodegeneration. *Methods in Molecular Biology (Clifton, N.J.)*, *793*, 87–97.

Chavrier, P., Gorvel, J. P., Stelzer, E., Simons, K., Gruenberg, J., & Zerial, M. (1991). Hypervariable C-terminal domain of rab proteins acts as a targeting signal. *Nature*, *353*(6346), 769–772.

Chung, C. Y., Khurana, V., Auluck, P. K., Tardiff, D. F., Mazzulli, J. R., Soldner, F., et al. (2013). Identification and rescue of α-synuclein toxicity in Parkinson patient-derived neurons. *Science (New York, N.Y.)*, *342*(6161), 983–987.

Cocucci, E., Aguet, F., Boulant, S., & Kirchhausen, T. (2012). The first five seconds in the life of a clathrin-coated pit. *Cell*, *150*(3), 495–507.

Cocucci, E., Gaudin, R., & Kirchhausen, T. (2014). Dynamin recruitment and membrane scission at the neck of a clathrin-coated pit. *Molecular Biology of the Cell*, (in press).

Coffman, V. C., & Wu, J.-Q. (2014). Every laboratory with a fluorescence microscope should consider counting molecules. *Molecular Biology of the Cell*, *25*(10), 1545–1548.

Cong, L., Ran, F. A., Cox, D., Lin, S., Barretto, R., Habib, N., et al. (2013). Multiplex genome engineering using CRISPR/Cas systems. *Science (New York, N.Y.)*, *339*(6121), 819–823.

Cui, X., Ji, D., Fisher, D. A., Wu, Y., Briner, D. M., & Weinstein, E. J. (2011). Targeted integration in rat and mouse embryos with zinc-finger nucleases. *Nature Biotechnology*, *29*(1), 64–67.

Dianov, G. L., & Hübscher, U. (2013). Mammalian base excision repair: The forgotten archangel. *Nucleic Acids Research*, *41*(6), 3483–3490.

Ding, Q., Regan, S. N., Xia, Y., Oostrom, L. A., Cowan, C. A., & Musunuru, K. (2013). Enhanced efficiency of human pluripotent stem cell genome editing through replacing TALENs with CRISPRs. *Cell Stem Cell, 12*(4), 393–394.

Doronina, V. A., Wu, C., de Felipe, P., Sachs, M. S., Ryan, M. D., & Brown, J. D. (2008). Site-specific release of nascent chains from ribosomes at a sense codon. *Molecular and Cellular Biology, 28*(13), 4227–4239.

Doyon, J. B., Zeitler, B., Cheng, J., Cheng, A. T., Cherone, J. M., Santiago, Y., et al. (2011). Rapid and efficient clathrin-mediated endocytosis revealed in genome-edited mammalian cells. *Nature Cell Biology, 13*(3), 331–337.

Forster, R., Chiba, K., Schaeffer, L., Regalado, S. G., Lai, C. S., Gao, Q., et al. (2014). Human intestinal tissue with adult stem cell properties derived from pluripotent stem cells. *Stem Cell Reports, 2*(6), 838–852.

Fu, Y., Foden, J. A., Khayter, C., Maeder, M. L., Reyon, D., Joung, J. K., et al. (2013). High-frequency off-target mutagenesis induced by CRISPR-Cas nucleases in human cells. *Nature Biotechnology, 31*(9), 822–826.

Gaj, T., Gersbach, C. A., & Barbas, C. F. (2013). ZFN, TALEN, and CRISPR/Cas-based methods for genome engineering. *Trends in Biotechnology, 31*(7), 397–405.

Gibson, T. J., Seiler, M., & Veitia, R. A. (2013). The transience of transient overexpression. *Nature Methods, 10*(8), 715–721.

Gibson, D. G., Smith, H. O., Hutchison, C. A., Venter, J. C., & Merryman, C. (2010). Chemical synthesis of the mouse mitochondrial genome. *Nature Methods,* 7(11), 901–903.

Gibson, D. G., Young, L., Chuang, R.-Y., Venter, J. C., Hutchison, C. A., & Smith, H. O. (2009). Enzymatic assembly of DNA molecules up to several hundred kilobases. *Nature Methods, 6*(5), 343–345.

Grassart, A., Cheng, A. T., Hong, S. H., Zhang, F., Zenzer, N., Feng, Y., et al. (2014). Actin and dynamin2 dynamics and interplay during clathrin-mediated endocytosis. *The Journal of Cell Biology, 205*(5), 721–735.

Gratz, S. J., Cummings, A. M., Nguyen, J. N., Hamm, D. C., Donohue, L. K., Harrison, M. M., et al. (2013). Genome engineering of Drosophila with the CRISPR RNA-guided Cas9 nuclease. *Genetics, 194*(4), 1029–1035.

Hall, B., Limaye, A., & Kulkarni, A. B. (2009). Overview: Generation of gene knockout mice. *Current Protocols in Cell Biology, 19*(12), 1–17.

Hinshaw, J. E., & Schmid, S. L. (1995). Dynamin self-assembles into rings suggesting a mechanism for coated vesicle budding. *Nature, 374*(6518), 190–192.

Hockemeyer, D., Wang, H., Kiani, S., Lai, C. S., Gao, Q., Cassady, J. P., et al. (2011). Genetic engineering of human pluripotent cells using TALE nucleases. *Nature Biotechnology, 29*(8), 731–734.

Janke, C., Magiera, M. M., Rathfelder, N., Taxis, C., Reber, S., Maekawa, H., et al. (2004). A versatile toolbox for PCR-based tagging of yeast genes: New fluorescent proteins, more markers and promoter substitution cassettes. *Yeast (Chichester, England), 21*(11), 947–962.

Jiang, J., Jing, Y., Cost, G. J., Chiang, J.-C., Kolpa, H. J., Cotton, A. M., et al. (2013). Translating dosage compensation to trisomy 21. *Nature, 500*(7462), 296–300.

Joglekar, A. P., Bouck, D. C., Molk, J. N., Bloom, K. S., & Salmon, E. D. (2006). Molecular architecture of a kinetochore-microtubule attachment site. *Nature Cell Biology, 8*(6), 581–585.

Joung, J. K., & Sander, J. D. (2013). TALENs: A widely applicable technology for targeted genome editing. *Nature Reviews Molecular Cell Biology, 14*(1), 49–55.

Kim, Y., Kweon, J., Kim, A., Chon, J. K., Yoo, J. Y., Kim, H. J., et al. (2013). A library of TAL effector nucleases spanning the human genome. *Nature Biotechnology, 31*(3), 251–258.

Kim, J. H., Lee, S.-R., Li, L.-H., Park, H.-J., Park, J.-H., Lee, K. Y., et al. (2011). High cleavage efficiency of a 2A peptide derived from porcine teschovirus-1 in human cell lines, zebrafish and mice. *PLoS One*, *6*(4), e18556.

Li, T., Liu, B., Spalding, M. H., Weeks, D. P., & Yang, B. (2012). High-efficiency TALEN-based gene editing produces disease-resistant rice. *Nature Biotechnology*, *30*(5), 390–392.

Li, Y., Wang, H., Muffat, J., Cheng, A. W., Orlando, D. A., Lovén, J., et al. (2013). Global transcriptional and translational repression in human-embryonic-stem-cell-derived Rett syndrome neurons. *Cell Stem Cell*, *13*(4), 446–458.

Lian, X., Zhang, J., Azarin, S. M., Zhu, K., Hazeltine, L. B., Bao, X., et al. (2013). Directed cardiomyocyte differentiation from human pluripotent stem cells by modulating Wnt/β-catenin signaling under fully defined conditions. *Nature Protocols*, *8*(1), 162–175.

Liu, Y.-W., Surka, M. C., Schroeter, T., Lukiyanchuk, V., & Schmid, S. L. (2008). Isoform and splice-variant specific functions of dynamin-2 revealed by analysis of conditional knock-out cells. *Molecular Biology of the Cell*, *19*(12), 5347–5359.

Lo, T.-W., Pickle, C. S., Lin, S., Ralston, E. J., Gurling, M., Schartner, C. M., et al. (2013). Precise and heritable genome editing in evolutionarily diverse nematodes using TALENs and CRISPR/Cas9 to engineer insertions and deletions. *Genetics*, *195*(2), 331–348.

Mali, P., Aach, J., Stranges, P. B., Esvelt, K. M., Moosburner, M., Kosuri, S., et al. (2013). CAS9 transcriptional activators for target specificity screening and paired nickases for cooperative genome engineering. *Nature Biotechnology*, *31*(9), 833–838.

Miyawaki, A., Sawano, A., & Kogure, T. (2003). Lighting up cells: Labelling proteins with fluorophores. *Nature Cell Biology*, (Suppl.), S1–S7.

Nesterov, A., Carter, R. E., Sorkina, T., Gill, G. N., & Sorkin, A. (1999). Inhibition of the receptor-binding function of clathrin adaptor protein AP-2 by dominant-negative mutant mu2 subunit and its effects on endocytosis. *The EMBO Journal*, *18*(9), 2489–2499.

Orlando, S. J., Santiago, Y., DeKelver, R. C., Freyvert, Y., Boydston, E. A., Moehle, E. A., et al. (2010). Zinc-finger nuclease-driven targeted integration into mammalian genomes using donors with limited chromosomal homology. *Nucleic Acids Research*, *38*(15), e152.

Perez, E. E., Wang, J., Miller, J. C., Jouvenot, Y., Kim, K. A., Liu, O., et al. (2008). Establishment of HIV-1 resistance in CD4+ T cells by genome editing using zinc-finger nucleases. *Nature Biotechnology*, *26*(7), 808–816.

Pucadyil, T. J., & Schmid, S. L. (2008). Real-time visualization of dynamin-catalyzed membrane fission and vesicle release. *Cell*, *135*(7), 1263–1275.

Ran, F. A., Hsu, P. D., Lin, C.-Y., Gootenberg, J. S., Konermann, S., Trevino, A. E., et al. (2013). Double nicking by RNA-guided CRISPR Cas9 for enhanced genome editing specificity. *Cell*, *154*(6), 1380–1389.

Ryan, S. D., Dolatabadi, N., Chan, S. F., Zhang, X., Akhtar, M. W., Parker, J., et al. (2013). Isogenic human iPSC Parkinson's model shows nitrosative stress-induced dysfunction in MEF2-PGC1α transcription. *Cell*, *155*(6), 1351–1364.

Santiago, Y., Chan, E., Liu, P.-Q., Orlando, S., Zhang, L., Urnov, F. D., et al. (2008). Targeted gene knockout in mammalian cells by using engineered zinc-finger nucleases. *Proceedings of the National Academy of Sciences of the United States of America*, *105*(15), 5809–5814.

Soldner, F., Laganière, J., Cheng, A. W., Hockemeyer, D., Gao, Q., Alagappan, R., et al. (2011). Generation of isogenic pluripotent stem cells differing exclusively at two early onset Parkinson point mutations. *Cell*, *146*(2), 318–331.

Wu, J.-Q., & Pollard, T. D. (2005). Counting cytokinesis proteins globally and locally in fission yeast. *Science (New York, N.Y.)*, *310*(5746), 310–314.

CHAPTER EIGHT

# Genome Editing Using Cas9 Nickases

**Alexandro E. Trevino**[*,†,‡,§], **Feng Zhang**[*,†,‡,§,1]
[*]Broad Institute of MIT and Harvard, 7 Cambridge Center, Cambridge, Massachusetts, USA
[†]McGovern Institute for Brain Research, Massachusetts Institute of Technology, Cambridge, Massachusetts, USA
[‡]Department of Brain and Cognitive Sciences, Massachusetts Institute of Technology, Cambridge, Massachusetts, USA
[§]Department of Biological Engineering, Massachusetts Institute of Technology, Cambridge, Massachusetts, USA
[1]Corresponding author: e-mail address: zhang@broadinstitute.org

## Contents



## Abstract

The RNA-guided, sequence-specific endonuclease Cas9 has been widely adopted as genome engineering tool due to its efficiency and ease of use. Derived from the microbial CRISPR (clustered regularly interspaced short palindromic repeat) type II adaptive immune system, Cas9 has now been successfully engineered for genome editing applications in a variety of animal and plant species. To reduce potential off-target mutagenesis by wild-type Cas9, homology- and structure-guided mutagenesis of *Streptococcus pyogenes* Cas9 catalytic domains has produced "nicking" enzymes (Cas9n) capable of inducing single-strand nicks rather than double-strand breaks. Since nicks are generally repaired with high fidelity in eukaryotic cells, Cas9n can be leveraged to mediate highly specific genome editing, either via nonhomologous end-joining or homology-directed repair. Here we describe the preparation, testing, and application of Cas9n reagents for precision mammalian genome engineering.

*Methods in Enzymology*, Volume 546
ISSN 0076-6879
http://dx.doi.org/10.1016/B978-0-12-801185-0.00008-8

## 1. INTRODUCTION

Targeted, rapid, and efficient genome editing using the RNA-guided Cas9 system is enabling the systematic interrogation of genetic elements in a variety of cells and organisms and holds enormous potential as a next-generation gene therapy (Hsu, Lander, & Zhang, 2014). In contrast to other DNA-targeting systems based on zinc-finger proteins (Klug, 2010) and transcription activator-like effectors (Boch & Bonas, 2010), which rely on protein domains to confer DNA-binding specificity, Cas9 forms a complex with a small-guide RNA that directs the enzyme to its DNA target via Watson–Crick base pairing. Consequently, the system is simple and fast to design and requires only the production of a short oligonucleotide to direct DNA binding to any locus.

The type II microbial CRISPR (clustered regularly interspaced short palindromic repeat) system (Chylinski, Makarova, Charpentier, & Koonin, 2014), which is the simplest among the three known CRISPR types (Barrangou & Marraffini, 2014; Gasiunas, Sinkunas, & Siksnys, 2014; Wiedenheft, Sternberg, & Doudna, 2012), consists of the CRISPR-associated (*Cas*) genes and a series of noncoding repetitive elements (direct repeats) interspaced by short variable sequences (spacers). These short ~30-bp spacers are often derived from foreign genetic elements such as phages and conjugating plasmids, and they constitute the basis for an adaptive immune memory of those invading elements (Barrangou et al., 2007). The corresponding sequences on the phage genomes and plasmids are called protospacers, and each protospacer is flanked by a short protospacer-adjacent motif (PAM), which plays a critical role in the target search and recognition mechanism of Cas9. The CRISPR array is transcribed and processed into short RNA molecules known as CRISPR RNAs (crRNA) that, together with a second short trans-activating RNA (tracrRNA) (Deltcheva et al., 2011), complex with Cas9 to facilitate target recognition and cleavage (Deltcheva et al., 2011; Garneau et al., 2010). Additionally, the crRNA and tracrRNA can be fused into a single guide RNA (sgRNA) to facilitate Cas9 targeting (Jinek et al., 2012).

The Cas9 enzyme from *Streptococcus pyogenes* (SpCas9), which requires a 5′-NGG PAM (Mojica, Diez-Villasenor, Garcia-Martinez, & Almendros, 2009), has been widely used for genome editing applications (Hsu et al., 2014). In order to target any desired genomic locus of interest that fulfills the PAM requirement, the enzyme can be "programmed" merely by altering the 20-bp guide sequence of the sgRNA. Additionally, the simplicity of targeting lends itself to easy multiplexing, such as simultaneous editing of several loci by including multiple sgRNAs (Cong et al., 2013; Wang et al., 2013).

Like other designer nucleases, Cas9 facilitates genome editing by inducing double-strand breaks (DSBs) at its target site, which in turn stimulates endogenous DNA damage repair pathways that lead to edited DNA: homology-directed repair (HDR), which requires a homologous template for recombination but repairs DSBs with high fidelity, and nonhomologous end-joining (NHEJ), which functions without a template and frequently produces insertions or deletions (indels) as a consequence of repair. Exogenous HDR templates can be designed and introduced along with Cas9 and sgRNA to promote exact sequence alteration at a target locus; however, this process typically occurs only in dividing cells and at low efficiency.

Certain applications—e.g., therapeutic genome editing in human stem cells—demand editing that is not only efficient but also highly specific. Nucleases with off-target DSB activity could induce undesirable mutations with potentially deleterious effects, an unacceptable outcome in most clinical settings. The remarkable ease of targeting Cas9 has enabled extensive off-target binding and mutagenesis studies employing deep sequencing (Fu et al., 2013; Hsu et al., 2013; Pattanayak et al., 2013) and chromatin immunoprecipitation in human cells (Kuscu, Arslan, Singh, Thorpe, & Adli, 2014; Wu et al., 2014). As a result, an increasingly complete picture of the off-target activity of the enzyme is emerging. Cas9 will tolerate some mismatches between its guide and a DNA substrate, a characteristic that depends strongly on the number, position (PAM proximal or distal), and identity of the mismatches. Off-target binding and cleavage may further depend on the organism being edited, the cell type, and epigenetic contexts.

These specificity studies, together with direct investigations of the catalytic mechanism of Cas9, have stimulated homology- and structure-guided engineering to improve its targeting specificity. The wild-type enzyme makes use of two conserved nuclease domains, HNH and RuvC, to cleave DNA by nicking the sgRNA-complementary and noncomplementary strands, respectively. A "nickase" mutant (Cas9n) can be generated by alanine substitution at key catalytic residues within these domains—SpCas9 D10A inactivates RuvC (Jinek et al., 2012), while N863A has been found to inactivate HNH (Nishimasu et al., 2014). Though an H840A mutation was also reported to convert Cas9 into a nicking enzyme, this mutant has

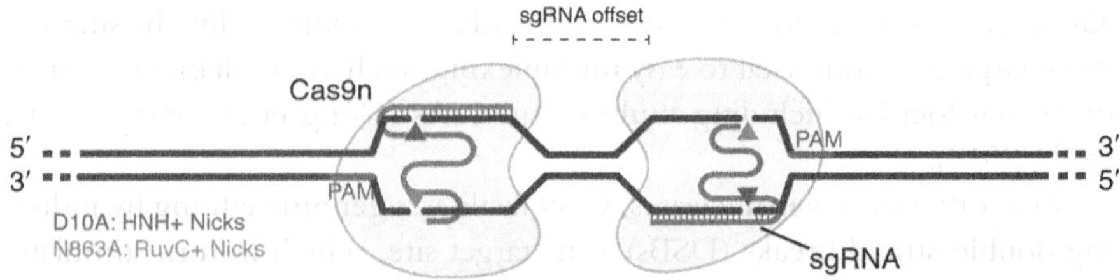


**Figure 8.1** Diagram of Cas9n enzymes in a double-nicking configuration. Offset nicking with the D10A mutant, which retains only the catalytic activity of the HNH nuclease domain, generates 5′ overhang products in the target genome by nicking the sgRNA-complementary DNA strand (nicks represented with red (dark gray in the print version) triangles). Alternatively, Cas9n N863A selectively nicks the noncomplementary strand (nicks represented with yellow (gray in the print version) triangles). sgRNA offset is defined as the distance between the 5′ (or PAM-distal) end of each sgRNA. The PAM sequences, represented in green (light gray in the print version), are present in the target genome but not the sgRNA.

reduced levels of activity in mammalian cells compared with N863A (Nishimasu et al., 2014) (Fig. 8.1).

Because single-stranded nicks are generally repaired via the non-mutagenic base-excision repair pathway (Dianov & Hubscher, 2013), Cas9n mutants can be leveraged to mediate highly specific genome engineering. A single Cas9n-induced nick can stimulate HDR at low efficiency in some cell types, while two nicking enzymes, appropriately spaced and oriented at the same locus, effectively generate DSBs, creating 3′ or 5′ overhangs along the target as opposed to a blunt DSB as in the wild-type case (Mali et al., 2013; Ran et al., 2013). The on-target modification efficiency of the double-nicking strategy is comparable to wild type, but indels at predicted off-target sites are reduced below the threshold of detection by Illumina deep sequencing (Ran et al., 2013).

The following protocol describes protocols and considerations for the design and testing of nickase reagents for high-precision mammalian genome editing, including target selection, sgRNA construction, transfection, detection of Cas9-induced indel mutations using the SURVEYOR nuclease assay, and design and quantification of homology-directed insertions.

## 2. TARGET SELECTION

SpCas9 targets can be any 20-bp DNA sequence followed at the 3′ end by 5′-NGG-3′. Our lab has developed an online tool that will accept a region of interest as input and output a list of all potential sgRNA target sites within that region. Each sgRNA target site is then associated with a list of predicted genomic off-targets (http://tools.genome-engineering.org).

The tool also generates double-nicking sgRNA pairs automatically. The most important consideration for double-nicking sgRNA design is the spacing between the two targets (Ran et al., 2013). If the "offset" between two guides is defined as the distance between the PAM-distal (5′) ends of an sgRNA pair, an offset of −4 to 20 bp is ideal, though offsets as large as 100 bp can induce DSB-mediated indels. sgRNA pairs for double nicking should target opposite DNA strands.

## 3. PLASMID sgRNA CONSTRUCTION

sgRNA expression vectors can be constructed by cloning 20-bp target sequences into a plasmid backbone encoding a human U6 promoter-driven sgRNA expression cassette and a CBh-driven Cas9-D10A (pSpCas9n(BB), Addgene #48873). The N863A nickase can be exchanged with D10A in all cases. It is recommended to prepare this plasmid as an endotoxin-free maxiprep. The generalized oligos needed to clone a new target into pSpCas9n(BB) are described in Table 8.1 and can be purchased from Integrated DNA Technologies (IDT). Note that the PAM sequence required for target recognition by Cas9 is never present as part of the sgRNA itself.

1. To clone a target sequence into an sgRNA backbone vector, first resuspend sgRNA-fwd and sgRNA-rev oligos to 100 μ*M*. Note that these oligos include an appended guanine (lowercase) not present in the target site in order to increase transcription from the U6 promoter.

**Table 8.1** General sgRNA cloning oligonucleotides

| **Primer** | **Sequence (5′ to 3′)** | **Description** |
|---|---|---|
| sgRNA-fwd | CACCGNNNNNNNNNNNNNNNNNNNN | Sticky overhang plus specific 20-bp genomic target to be cloned into sgRNA backbones |
| sgRNA-rev | AAACNNNNNNNNNNNNNNNNNNNNC | Complementary annealing oligo for cloning new target into sgRNA backbones |

2. Combine 1 μL of each oligo with 1 μL T4 ligation buffer, 10× (New England Biolabs (NEB) B0202S), 0.5 μL T4 PNK (NEB M0201S), and 6.5 μL $ddH_2O$ for a 10 μL reaction total. Treat with polynucleotide kinase to add 5′ phosphate and anneal the oligos in a thermocycler with the following protocol: 37 °C for 30 min, 95 °C for 5 min, ramp down to 25 °C at 5 °C/min.
3. Dilute the annealed oligos (10 μL reaction) by adding 90 μL $ddH_2O$.
4. Set up a Golden Gate digestion/ligation with pSpCas9n(BB) and the annealed oligos as a cloning insert. The plasmid contains twin *Bsm*BI restriction sites in place of the sgRNA target sequence such that digestion leaves overhangs complementary to the annealed oligo overhangs. In a 25 μL reaction, combine 25 ng pSpCas9n(BB), 1 μL diluted annealed oligos from step 3, 12.5 μL Rapid Ligation Buffer, 2× (Enzymatics L6020L), 1 μL FastDigest *Bsm*BI (ThermoScientific FD1014), 2.5 μL 10× BSA (NEB B9001), 0.125 μL T7 Ligase (Enzymatics L6020L), and 7 μL $ddH_2O$.
5. A negative control should be performed using the same conditions as above and substituting the insert oligos with $ddH_2O$.
6. Incubate the ligation in a thermocycler for six cycles of 37 °C for 5 min, 20 °C for 5 min. The ligation is stable for storage at −20 °C.
7. Transform 2 μL of the ligation reaction into a competent *E. coli* strain using the appropriate protocol—the Stbl3 strain is recommended—plate onto ampicillin selection plates (100 μg/mL ampicillin), and incubate overnight at 37 °C. Typically, transformation occurs at high efficiency; no colonies form on the negative control plate, and hundreds form when the sgRNA oligos have been successfully cloned into the backbone.
8. 14 h later, pick two or more colonies from the transformation with a sterile pipette tip and use the bacteria to inoculate 3 mL LB or TB broth with 100 μg/mL ampicillin. Shake the culture at 37 °C for 14 h.
9. Isolate plasmid DNA from the cultures using the Qiagen Spin Miniprep Kit (27104) and determine the DNA concentration by spectrophotometry. These constructs can be Sanger sequence-verified through the sgRNA scaffold to confirm correct insertion of the target sequence. For optimal transfection conditions downstream, endotoxin-free plasmid should be prepared.

## 4. VALIDATION OF sgRNAs IN CELL LINES

This protocol describes the functional validation of sgRNAs in HEK293FT cells; culture and transfection conditions may vary for other cell types.

1. Maintain HEK293FT cells (Life Technologies R700-07) in sterile D10 media (DMEM, high glucose (Life Technologies 10313-039) supplemented with 10% vol/vol fetal bovine serum (Seradigm 1500-500) and 10 m*M* HEPES (Life Technologies 15630-080)). For optimal health, cells should be passaged every day at a ratio of 1:2–2.5 and always kept under 80% confluence.
2. Plate cells for transfection. Seed 120,000 cells per well of a 24-well tissue-culture treated plate in a total volume of 500 μL. Cultures and transfections can be proportionally scaled up or down for different formats based on growth surface area. For many adherent cell types, poly-D-lysine coated plastic may improve adherence and viability.
3. Check the plates after 18 h to determine the confluence of the cells—generally 90% is ideal. Lipofectamine 2000 (Life Technologies 11668109) reagent can be used to transfect DNA according to the manufacturer's protocol. For a 24-well plate, we do not transfect more than 500 ng/well DNA total.
4. To deliver one nicking pSpCas9n(sgRNA) plasmid, transfect 500 ng; for multiple nicking constructs, e.g., delivering 2 sgRNAs for double nicking, mix different constructs up to 500 ng at equimolar ratios before transfection.
5. It is important to include transfection controls, such as untransfected wells and GFP plasmid, as well as experimental controls, such as Cas9n without guides or guides alone, in these experiments. Transfecting in technical triplicates will facilitate analysis.
6. Within 6 h of transfection, change the media to 2 mL of fresh, prewarmed D10 media per well. At 24 h, estimate transfection efficiency by examining GFP-transfected wells. >80% of cells should be GFP positive.
7. Harvest the cells for genomic DNA extraction and/or downstream analysis at 48–72 h. If harvesting a 72-h time point, change the media again at 48 h to maintain optimal cell health.

When working with different cell types, alternative transfection reagents should be compared for efficiency and toxicity. It may also be informative to titrate pSpCas9n(sgRNA) in order to find the optimal transfection concentration with highest efficacy.

## 5. CELL HARVEST AND DNA EXTRACTION

1. Harvest cells in 24-well plate format by aspirating the medium completely and adding 100 μL of TrypLE Express reagent (Life Technologies 12604013) to facilitate dissociation.

2. Collect the cell suspension in a 1.5-mL Eppendorf tube and spin for 5 min at 1500 × *g*, aspirate the supernatant completely, and resuspend the cell pellet in 200 μL DPBS (Life Technologies 14190-250) to wash.
3. Spin the cell suspension again for 5 min at 1500 × *g* and resuspend in 50 μL QuickExtract (Epicentre QE09050).
4. Transfer the QuickExtract suspension to a 0.2-mL PCR tube and extract genomic DNA according to the following thermocycler protocol adapted from the manufacturer's instructions: 65 °C for 15 min, 98 °C for 10 min.
5. Centrifuge the reaction product to pellet cell debris and transfer cleared supernatant into a fresh tube for further analysis.
6. Determine the DNA concentration of the extraction by spectrophotometry and normalize to 100–200 ng/μL with $ddH_2O$.

## 6. SURVEYOR INDEL ANALYSIS

The SURVEYOR assay (Transgenomic 706025) is a method for detecting polymorphisms and small indels. DNA samples are PCR-amplified, and the products are heated to denature and cooled slowly to form heteroduplexes. Mismatched duplexes are then cleaved by the SURVEYOR nuclease, and cleavage products are analyzed by gel electrophoresis.

1. Perform PCR on genomic DNA. The primers for SURVEYOR PCR should ideally produce a clean ~500 bp amplicon in untransfected cell samples. Genomic PCR primers may be designed using software such as Primer3. Set up a 50 μL reaction containing 1 μL of each 10 μ*M* SURVEYOR primer, 10 μL Herculase II Reaction Buffer 5 × (Agilent 600675), 0.5 μL of 100 m*M* dNTP, 0.5 μL Herculase II Fusion Polymerase, 2 μL of 25 m*M* $MgCl_2$, and 36 μL $ddH_2O$. Denature for 20 s at 95 °C, anneal for 20 s at 60 °C, and extend for 20 s at 72 °C.
2. Note that, since SURVEYOR was designed to detect mutations, it is crucial to use a high-fidelity polymerase to avoid false positives.
3. Run 2 μL of the PCR product on a 1% agarose gel to ensure that a single product of expected size has formed.
4. Purify the PCR product using the QIAquick PCR Purification Kit (28104) according to the instructions provided. Measure the DNA concentration of the eluate by spectrophotometry and normalize to 20 ng/μL using $ddH_2O$.

5. Mix 18 μL of normalized PCR product with 2 μL Taq PCR buffer, 10×, for a 20 μL reaction total. Melt and rehybridize the products gradually in a thermocycler: melt at 95 °C for 10 min, then ramp the temperature down to 85 °C at a rate of −0.3 °C/s. Hold at 85 °C for 1 min, then ramp to 75 °C at 0.3 °C/s. Hold at 75 °C for 1 min, then ramp to 65 °C and so on, until the temperature reaches 25 °C. From 25 °C, ramp down to 4 °C at 0.3 °C/s and hold.
6. Mix 2.5 μL of 0.15 *M* $MgCl_2$, 0.5 μL $ddH_2O$, 1 μL SURVEYOR nuclease S, and 1 μL SURVEYOR enhancer S with all of the annealed product from step (5) for a 25 μL total reaction volume. Perform the

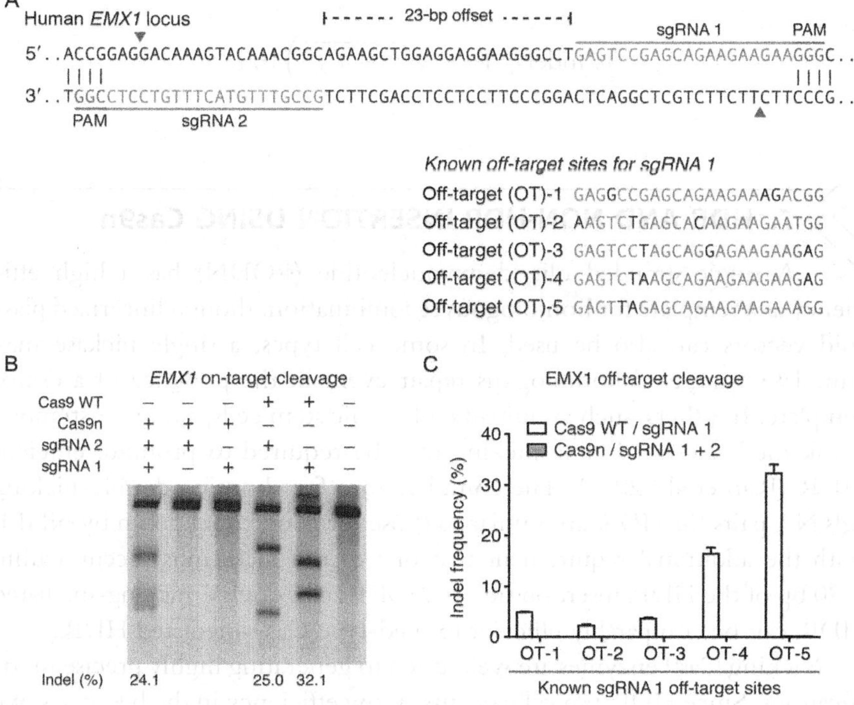


**Figure 8.2** Double nicking reduces off-target modification. (A) Diagram of a Cas9n D10A double-nicking sgRNA pair designed for the human EMX1 locus. Guide sequences are shown in blue, demonstrating a 23-bp offset. The PAM is shown in pink, and nicking sites are represented by red triangles. Five known genomic off-target sites (Hsu et al., 2013) for sgRNA 1 are listed. (B) Example SURVEYOR results showing modification of the EMX1 locus by Cas9 WT and Cas9n along with sgRNA 1 and/or 2. (C) Deep sequencing quantification of off-target modifications at five known off-target sites by Cas9 WT and sgRNA 1 or Cas9n with sgRNAs 1 and 2. *Adapted with permission from Ran et al. (2013).* (See the color plate.)

digestion by incubating the reaction at 42 °C for 30 min. Samples that have mutations within the rehybridized PCR amplicons will be cleaved by SURVEYOR.

7. The digestion products can be mixed with an appropriate loading dye and visualized by electrophoresis on a 4–20% polyacrylamide TBE gel (see example, Fig. 8.2B).
8. Genome modification rates can be estimated first by calculating the relative intensities of digestion products $a$ and $b$ and the undigested band $c$. The frequency of cutting $f_{cut}$ is then given by $(a+b)/(a+b+c)$. The following formula, based on the binomial probability distribution of duplex formation, estimates the percentage of indels in the sample.

$$\%\ \text{indel} = \left(1 - \sqrt{(1 - f_{cut})}\right)100$$

## 7. HDR AND NON-HDR INSERTION USING Cas9n

A single-stranded oligodeoxynucleotide (ssODN) has a high efficiency as a template for homologous recombination, though linearized plasmid vectors can also be used. In some cell types, a single nickase may stimulate a targeted homologous repair event in the presence of a donor template. In others, such as human embryonic stem cells, a double-stranded break mediated by double nicking may be required to promote efficient HDR (Ran et al., 2013). The considerations for choosing double-nicking sgRNA pairs for HDR are similar to those for gene knockdown by NHEJ, with the additional requirement that one of the nicks must occur within ~20 bp of the HDR insertion site. In 293FT cells, double-nicking-mediated HDR can be comparably efficient to wild-type Cas9-mediated HDR.

Nicking Cas9 enzymes are well suited to generating highly precise modifications. Since HDR typically occurs at low efficiency in the best cases, we also provide pSpCas9n plasmids encoding the polycistronic 2A linker followed by GFP and puromycin markers (Addgene #48140 and 48141) in order to facilitate enrichment of modified cells.

HDR in mammalian cells proceeds via the generation of 3′ overhangs followed by strand invasion of a homologous locus by the 3′ end. It is therefore possible that the generation of 3′ overhang products by N863A-mediated double nicking could increase HDR efficiency.

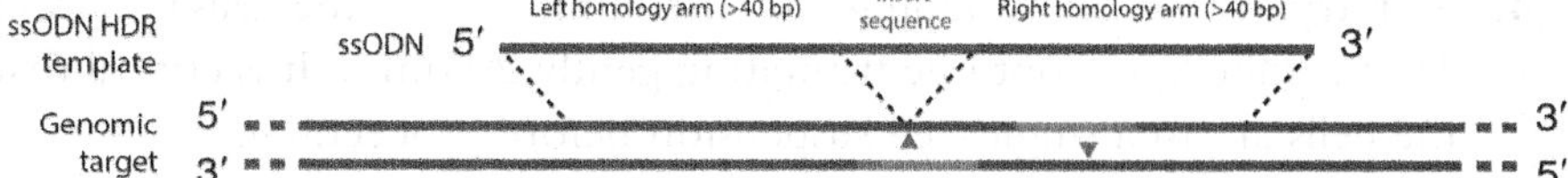


**Figure 8.3** General design of ssODN HDR templates. The ssODN consists of an insertion sequence (red (dark gray in the print version)) flanked by homology arms on the left and right sides (at least 40 bp each). The homology between the ssODN and its targeting region is indicated by black dashes. Double-nicking Cas9 target sites are shown in blue (light gray in the print version), and their corresponding PAM sequences are shown in pink (light gray in the print version). Nicking sites are represented by red (dark gray in the print version) triangles.

1. ssODN homology arms should be designed to be as long as possible, with at least 40 nucleotides of homology on either side of the sequence to be introduced. The Ultramer service provided by IDT allows the synthesis of oligos up to 200 bp in length. Homology templates should be diluted to 10 μ*M* and stored at −20 °C (see design example, Fig. 8.3).
2. Delivery by nucleofection is optimal for ssODNs. The 4D Nucleofector X Kit S (Lonza V4XC-2032) can be used for HEK293FT cells seeded in 6-well tissue-culture-treated plates. The manufacturer provides an optimal protocol for nucleofection of these and other cell types.
3. Mix 500 ng total pSpCas9n(sgRNA) plasmids with 1 μL of 10 μ*M* ssODN for nucleofection.

## 8. ANALYSIS OF HDR AND INSERTION EVENTS

HDR outcomes can be assessed and utilized in a variety of ways. Here, the FACS isolation of clonal pSpCas9n(sgRNA)-GFP 293FT cells is described. It is important to note that FACS procedures can vary between cell types.

1. Prepare FACS media (D10 without phenol red to facilitate fluorescence sorting): DMEM, high glucose, no phenol red (Life Technologies 31053-028) supplemented with 10% vol/vol fetal bovine serum and 10 m*M* HEPES supplemented with 1% penicillin–streptomycin (Life Technologies 15140122).
2. Prepare 96-well plates for clone sorting by adding 100 μL standard D10 media to each well.
3. 24 h after the transfection in Sections 7.2 and 7.3, aspirate the medium completely and dissociate the cells using sufficient TrypLE Express to cover the growth surface minimally.

4. Stop trypsinization by adding D10 medium, transfer the cells to a fresh 15-mL tube, and continue triturating gently 20 times. It is critical that the cells are in a single-cell suspension before proceeding.
5. Spin the cells for 5 min at 200 × *g*, aspirate the supernatant completely, and resuspend the pellet thoroughly and carefully in 200 μL FACS medium.
6. Filter the cells through a cell strainer (BD Falcon 352235) to filter out cell aggregates and place the cells on ice.
7. Sort single cells in the plates prepared in (2). The FACS machine can be gated on GFP+ cells in order to enrich for transfected cells. Wells can be visually inspected to check for the presence of one cell.
8. Incubate and expand the cells for 2–3 weeks, changing media to fresh D10 as necessary.
9. When cells exceed 60% confluence, clonal populations can be passaged into replica plates containing fresh D10 media. Dissociate cells, passage 20% of the cells into replica plates, and conserve 80% for DNA extraction as described in Section 5.
10. Genotyping can be performed by PCR amplification of the locus of interest, PCR purification, and Sanger sequencing of the products.

## 9. TROUBLESHOOTING

1. Colonies form on the negative control plate while cloning targets into pSpCas9n.
   a. The presence of negative colonies generally indicates an incomplete restriction digestion of the backbone plasmid. The Golden Gate reaction can be extended for 20–25 cycles in order to increase the efficiency of digestion. The amount of restriction enzyme used can be increased, though the volume of enzyme should not exceed 20% of the total reaction volume.
   b. Retransform the Cas9 backbone plasmid, isolate a new preparation of plasmid DNA, and sequence-verify the restriction site.
2. The transfection efficiency of Cas9 reagents is low.
   a. Low transfection efficiency may be the norm for some cell lines, and especially primary cells or stem cell lines. Cell populations can be enriched for transfected cells by using pSpCas9n(BB)-GFP or pSpCas9n(BB)-Puro plasmids to FACS on GFP fluorescence or perform antibiotic selection.

3. Double nicking does not produce indels.
   a. The individual double-nicking sgRNAs should be tested with the wild-type context to ensure that each of them functions separately as a valid Cas9 guide.
   b. Check the spacing of the sgRNA pair. Double nicking performs optimally when the guides are spaced 20 bp apart or less, and the guides should be oriented such that their respective 5′ PAM sequences face away from each other.
4. Efficiency of HDR is low.
   a. Silent mutations may be introduced within the target site on the ssODN to prevent cleavage of the successfully recombined genomic site.

## ACKNOWLEDGMENTS

This work is supported by the NIMH through a NIH Director's Pioneer Award (DP1-MH100706), the NINDS through a NIH Transformative R01 grant (R01-NS 07312401), NSF, the Keck, McKnight, Damon Runyon, Searle Scholars, Klingenstein, Vallee, Merkin, and Simons Foundations, and Bob Metcalfe. CRISPR reagents are available to the academic community through Addgene and associated protocols, support forum, and computational tools are available via the Zhang lab Web site (www.genome-engineering.org).

## REFERENCES

Barrangou, R., Fremaux, C., Deveau, H., Richards, M., Boyaval, P., Moineau, S., et al. (2007). CRISPR provides acquired resistance against viruses in prokaryotes. *Science*, *315*(5819), 1709–1712. http://dx.doi.org/10.1126/science.1138140.

Barrangou, R., & Marraffini, L. A. (2014). CRISPR-Cas systems: Prokaryotes upgrade to adaptive immunity. *Molecular Cell*, *54*(2), 234–244. http://dx.doi.org/10.1016/j.molcel.2014.03.011.

Boch, J., & Bonas, U. (2010). Xanthomonas AvrBs3 family-type III effectors: Discovery and function. *Annual Review of Phytopathology*, *48*, 419–436. http://dx.doi.org/10.1146/annurev-phyto-080508-081936.

Chylinski, K., Makarova, K. S., Charpentier, E., & Koonin, E. V. (2014). Classification and evolution of type II CRISPR-Cas systems. *Nucleic Acids Research*, *42*(10), 6091–6105. http://dx.doi.org/10.1093/nar/gku241.

Cong, L., Ran, F. A., Cox, D., Lin, S., Barretto, R., Habib, N., et al. (2013). Multiplex genome engineering using CRISPR/Cas systems. *Science*, *339*(6121), 819–823. http://dx.doi.org/10.1126/science.1231143.

Deltcheva, E., Chylinski, K., Sharma, C. M., Gonzales, K., Chao, Y., Pirzada, Z. A., et al. (2011). CRISPR RNA maturation by trans-encoded small RNA and host factor RNase III. *Nature*, *471*(7340), 602–607. http://dx.doi.org/10.1038/nature09886.

Dianov, G. L., & Hubscher, U. (2013). Mammalian base excision repair: The forgotten archangel. *Nucleic Acids Research*, *41*(6), 3483–3490. http://dx.doi.org/10.1093/nar/gkt076.

Fu, Y., Foden, J. A., Khayter, C., Maeder, M. L., Reyon, D., Joung, J. K., et al. (2013). High-frequency off-target mutagenesis induced by CRISPR-Cas nucleases in human cells. *Nature Biotechnology*, *31*(9), 822–826. http://dx.doi.org/10.1038/nbt.2623.

Garneau, J. E., Dupuis, M. E., Villion, M., Romero, D. A., Barrangou, R., Boyaval, P., et al. (2010). The CRISPR/Cas bacterial immune system cleaves bacteriophage and plasmid DNA. *Nature*, *468*(7320), 67–71. http://dx.doi.org/10.1038/nature09523.

Gasiunas, G., Sinkunas, T., & Siksnys, V. (2014). Molecular mechanisms of CRISPR-mediated microbial immunity. *Cellular and Molecular Life Sciences*, *71*(3), 449–465. http://dx.doi.org/10.1007/s00018-013-1438-6.

Hsu, P. D., Lander, E. S., & Zhang, F. (2014). Development and applications of CRISPR-Cas9 for genome engineering. *Cell*, *157*(6), 1262–1278. http://dx.doi.org/10.1016/j.cell.2014.05.010.

Hsu, P. D., Scott, D. A., Weinstein, J. A., Ran, F. A., Konermann, S., Agarwala, V., et al. (2013). DNA targeting specificity of RNA-guided Cas9 nucleases. *Nature Biotechnology*, *31*(9), 827–832. http://dx.doi.org/10.1038/nbt.2647.

Jinek, M., Chylinski, K., Fonfara, I., Hauer, M., Doudna, J. A., & Charpentier, E. (2012). A programmable dual-RNA-guided DNA endonuclease in adaptive bacterial immunity. *Science*, *337*(6096), 816–821. http://dx.doi.org/10.1126/science.1225829.

Klug, A. (2010). The discovery of zinc fingers and their applications in gene regulation and genome manipulation. *Annual Review of Biochemistry*, *79*, 213–231. http://dx.doi.org/10.1146/annurev-biochem-010909-095056.

Kuscu, C., Arslan, S., Singh, R., Thorpe, J., & Adli, M. (2014). Genome-wide analysis reveals characteristics of off-target sites bound by the Cas9 endonuclease. *Nature Biotechnology*, *32*(7), 677–683. http://dx.doi.org/10.1038/nbt.2916.

Mali, P., Aach, J., Stranges, P. B., Esvelt, K. M., Moosburner, M., Kosuri, S., et al. (2013). CAS9 transcriptional activators for target specificity screening and paired nickases for cooperative genome engineering. *Nature Biotechnology*, *31*(9), 833–838. http://dx.doi.org/10.1038/nbt.2675.

Mojica, F. J., Diez-Villasenor, C., Garcia-Martinez, J., & Almendros, C. (2009). Short motif sequences determine the targets of the prokaryotic CRISPR defence system. *Microbiology*, *155*(Pt. 3), 733–740. http://dx.doi.org/10.1099/mic.0.023960-0.

Nishimasu, H., Ran, F. A., Hsu, P. D., Konermann, S., Shehata, S. I., Dohmae, N., et al. (2014). Crystal structure of cas9 in complex with guide RNA and target DNA. *Cell*, *156*(5), 935–949. http://dx.doi.org/10.1016/j.cell.2014.02.001.

Pattanayak, V., Lin, S., Guilinger, J. P., Ma, E., Doudna, J. A., & Liu, D. R. (2013). High-throughput profiling of off-target DNA cleavage reveals RNA-programmed Cas9 nuclease specificity. *Nature Biotechnology*, *31*(9), 839–843. http://dx.doi.org/10.1038/nbt.2673.

Ran, F. A., Hsu, P. D., Lin, C. Y., Gootenberg, J. S., Konermann, S., Trevino, A. E., et al. (2013). Double nicking by RNA-guided CRISPR Cas9 for enhanced genome editing specificity. *Cell*, *154*(6), 1380–1389. http://dx.doi.org/10.1016/j.cell.2013.08.021.

Wang, H., Yang, H., Shivalila, C. S., Dawlaty, M. M., Cheng, A. W., Zhang, F., et al. (2013). One-step generation of mice carrying mutations in multiple genes by CRISPR/Cas-mediated genome engineering. *Cell*, *153*(4), 910–918. http://dx.doi.org/10.1016/j.cell.2013.04.025.

Wiedenheft, B., Sternberg, S. H., & Doudna, J. A. (2012). RNA-guided genetic silencing systems in bacteria and archaea. *Nature*, *482*(7385), 331–338. http://dx.doi.org/10.1038/nature10886.

Wu, X., Scott, D. A., Kriz, A. J., Chiu, A. C., Hsu, P. D., Dadon, D. B., et al. (2014). Genome-wide binding of the CRISPR endonuclease Cas9 in mammalian cells. *Nature Biotechnology*. http://dx.doi.org/10.1038/nbt.2889.

CHAPTER NINE

# Assaying Break and Nick-Induced Homologous Recombination in Mammalian Cells Using the DR-GFP Reporter and Cas9 Nucleases

**Lianne E.M. Vriend*,†, Maria Jasin*,1, Przemek M. Krawczyk*,†,1**

*Developmental Biology Program, Memorial Sloan-Kettering Cancer Center, New York, USA
†Department of Cell Biology and Histology, Academic Medical Center, University of Amsterdam, Amsterdam, The Netherlands
[1]Corresponding authors: e-mail address: m-jasin@mskcc.org; p.krawczyk@amc.uva.nl

## Contents



## Abstract

Thousands of DNA breaks occur daily in mammalian cells, including potentially tumorigenic double-strand breaks (DSBs) and less dangerous but vastly more abundant single-strand breaks (SSBs). The majority of SSBs are quickly repaired, but some can be converted to DSBs, posing a threat to the integrity of the genome. Although SSBs are usually repaired by dedicated pathways, they can also trigger homologous recombination (HR), an error-free pathway generally associated with DSB repair. While

*Methods in Enzymology*, Volume 546
ISSN 0076-6879
http://dx.doi.org/10.1016/B978-0-12-801185-0.00009-X

HR-mediated DSB repair has been extensively studied, the mechanisms of HR-mediated SSB repair are less clear. This chapter describes a protocol to investigate SSB-induced HR in mammalian cells employing the DR-GFP reporter, which has been widely used in DSB repair studies, together with an adapted bacterial CRISPR/Cas system.

## 1. INTRODUCTION

Mammalian cells endure continuous assault on the integrity of their genomes exerted by exogenous agents such as ionizing radiation, chemicals, and UV light. Additionally, by-products of endogenous metabolic activities (e.g., reactive-oxygen species and free radicals) and cellular processes that directly involve DNA (e.g., replication and transcription) cause DNA damage (Horton et al., 2008). Among the most dangerous, but least abundant (an estimated ten per cell per day), are DNA double-strand breaks (DSBs). By contrast, tens of thousands of less dangerous DNA single-strand breaks (SSBs) occur daily in mammalian cells (Caldecott, 2008). SSBs frequently arise as intermediates in excision repair of oxidatively damaged DNA bases (Hegde, Hazra, & Mitra, 2008), but may also form due to failed reactions of DNA maintenance enzymes, such as topoisomerase I (Pommier et al., 2003). Although SSBs, as such, do not pose a serious threat to the integrity of the genome, replication of nicked DNA can result in formation of a DSB (Haber, 1999; Saleh-Gohari et al., 2005).

Unless repaired in an appropriate manner, DSBs can cause chromosome loss or potentially cancer-causing chromosome rearrangements (Bunting & Nussenzweig, 2013; Weinstock, Richardson, Elliott, & Jasin, 2006). Molecular mechanisms of DSB repair have been investigated extensively in mammalian cells using rare-cutting endonucleases, primarily I-SceI, to introduce DSBs into the genome (Liang, Han, Romanienko, & Jasin, 1998; Rouet, Smih, & Jasin, 1994). These and other studies lead to the conclusion that cells have two robust DSB repair pathways, homologous recombination (HR) and nonhomologous end joining (NHEJ). NHEJ involves processing of the broken DNA ends followed by ligation to seal the break (Deriano & Roth, 2013; Lieber, 2010). Because end processing can lead to loss of DNA sequences, NHEJ is often considered to be error-prone, although it is certainly capable of precise DSB rejoining (Bétermier, Bertrand & Lopez, 2014).

HR involves a seemingly more complex set of enzymatic reactions that uses an intact DNA strand, usually the sister chromatid, to faithfully restore the original sequence at the break site (Jasin & Rothstein, 2013; San Filippo,

Sung, & Klein, 2008). Unlike NHEJ, which is functional throughout the cell cycle, HR activity is limited to the S and $G_2$ phases. HR is initiated by resection of the 5′ DNA ends to give rise to 3′ single-stranded (ss) DNA overhangs. In subsequent steps, the resected strand invades an intact, homologous DNA template, and forming heteroduplex DNA. The invading strand then acts as a primer for repair synthesis from the template, followed by the dissolution of the heteroduplex, reannealing of the newly synthesized strand to the second end of the DNA, and sealing of remaining gaps.

Like DSBs, endonuclease-induced SSBs also stimulate HR in mammalian cells (Davis & Maizels, 2014; McConnell Smith et al., 2009; Metzger, Stoddard, & Monnat, 2013). Interestingly, the mechanistic requirements of SSB-induced HR vary depending on whether the available template DNA is single or double stranded (Davis & Maizels, 2014). Although DSB and SSB repair pathways likely involve some mechanistically distinct but also overlapping steps, how nicks trigger HR is not well understood. For instance, it is unclear whether SSB-induced HR involves formation of a DSB intermediate or whether DNA replication influences the process.

One of the most common approaches to studying DSB-induced HR involves the use of a GFP reporter, DR-GFP, which allows flow cytometry-based detection of HR stimulated by I-SceI endonuclease-induced DSBs (Pierce & Jasin, 2005; Pierce, Johnson, Thompson, & Jasin, 1999; Fig. 9.1A). The recent adaptation of the bacterial adaptive immunity system, CRISPR/Cas (clustered regularly interspaced short palindromic repeats/CRISPR-associated), has enabled straightforward induction of DSBs and SSBs at desired loci by the RNA-guided Cas9 endonuclease (Gasiunas, Barrangou, Horvath, & Siksnys, 2012; Jinek et al., 2012; Mali, Yang, et al., 2013; Hsu, Lander, & Zhang, 2014). Here, we adapt the DR-GFP reporter for measuring both DSB- and SSB-induced HR, using wild-type Cas9 endonuclease and Cas9 nickases, respectively.

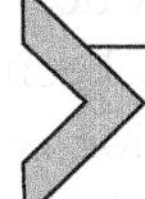

## 2. CLONING THE NICKASE AND CATALYTICALLY DEAD VARIANTS OF Cas9

### 2.1. The Cas9 endonuclease

Cas9 forms a nucleoprotein complex with a single guide RNA (sgRNA) containing a 19 nt sequence that determines binding specificity based on Watson–Crick base pairing (Cong et al., 2013; Jinek et al., 2012; Mali, Yang, et al., 2013; Fig. 9.1B). With the commonly used Cas9 protein from

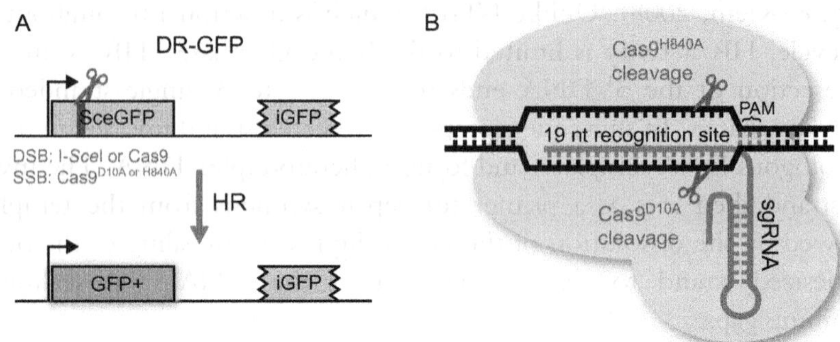


**Figure 9.1** Overview of the DR-GFP reporter and the RNA-guided Cas9 endonuclease. (A) The DR-GFP reporter consists of two copies of the GFP gene in tandem arrangement. The first copy (SceGFP) is inactive due to the presence of a stop codon within the I-SceI cleavage site (red bar (dark gray in the print version)), while the second copy (iGFP) is truncated at both ends. After cleavage within SceGFP by I-SceI or a Cas9 nuclease, HR uses iGFP as a template to restore the GFP open reading frame. (B) Cas9 nuclease has two catalytic domains, with each domain cleaving a single DNA strand guided by a short RNA (sgRNA, green (gray in the print version)) containing 19 nt complementary to the target site. Mutation of either catalytic domain (D10A or H840A) turns Cas9 into a nickase (as indicated), while mutation of both residues makes it catalytically dead (not shown). Cas9 requires a PAM motif (NGG) immediately downstream of the recognition site (red (gray in the print version)).

*Streptococcus pyogenes* (SpCas9), the only sequence requirement in the genomic target is an NGG (or, less optimally, an NAG) PAM motif (where N signifies any nucleotide) directly downstream from the binding sequence (Fig. 9.1B). Cas9 contains two catalytic domains, the modular RuvC-like domain and the C-terminal HNH-like domain. Each domain cleaves one of the DNA strands, resulting in a blunt-ended DSB or short overhang 3 bp upstream of the PAM motif (Fig. 9.1B). Mutation of the active site in either catalytic domain turns wild-type Cas9 ($Cas9^{WT}$) into a nicking enzyme (nCas9), while mutating both active sites renders it catalytically dead (dCas9), but still able to efficiently bind DNA, a feature that has been exploited for dCas9-mediated transcription regulation and visualization of DNA sequences in living cells (Chen et al., 2013; Mali, Aach, et al., 2013). The $Cas9^{D10A}$ variant with a mutation in the active site of the RuvC-like domain cleaves the DNA strand complementary to the sgRNA-binding sequence, while $Cas9^{H840A}$ with a mutation in the HNH-like domain cleaves the noncomplementary strand, and $Cas9^{D10A/H840A}$ is catalytically dead (Jinek et al., 2012).

There are two widely used sets of Cas9 expression vectors for mammalian cells available from Addgene (www.addgene.com), both of which include a codon-optimized Cas9 cDNA sequence. The set generated by the Zhang laboratory (www.addgene.org/crispr/zhang) is a bicistronic system wherein both Cas9 and sgRNA are expressed from a single plasmid. The advantage of this system is that only one plasmid needs to be generated and transfected if one type of Cas9 or sgRNA is to be used. However, this can also be disadvantageous if the experimental design requires using different Cas9 variants (e.g., for generating both DSBs and SSBs) or sgRNAs because multiple plasmids will need to be generated. The second commonly used set of Cas9 expression vectors, generated by the Church Laboratory (http://www.addgene.org/crispr/church), has separate plasmids for Cas9 and sgRNA expression. We routinely use the Church plasmid set, given our comparison of multiple Cas9 variants and sgRNAs, and the following protocol is based on these reagents. To obtain $Cas9^{H840A}$ and catalytically dead Cas9, we mutated the available wild-type and $Cas9^{D10A}$ variants, respectively, as described below.

## 2.2. Generating $Cas9^{H840A}$ and $Cas9^{D10A/H840A}$ expression vectors

**Step 1** Obtain the wild-type (ID 41815) and D10A (ID 41816) variants of Cas9 and empty sgRNA expression plasmid (ID 41824) from Addgene. Prepare plasmid stocks using midi- or maxi-prep kit of your choice. (We routinely use Life Technologies DNA purification kits.)

**Step 2** Order the following primers to sequence and generate mutant Cas9 nucleases:

| | |
|---|---|
| Cas9mutF | 5′-GACGTGGATGCTATCGTGCCCCAGTCTTTTCTCAA-3′ |
| Cas9mutR | 5′-AGCATCCACGTCGTAGTCGGAGAGCCGATTGATGTCCAG-3′ |
| Cas9seqF1 | 5′-GCTGTTTTGACCTCCATAGAAG-3′ |
| Cas9seqF2 | 5′-TGATAAGGCTGACTTGCGG-3′ |
| Cas9seqF3 | 5′-AGACGCCATTCTGCTGAGTG-3′ |
| Cas9seqF4 | 5′-CGCAAATCAGAAGAGACCATC-3′ |
| Cas9seqF5 | 5′-GAACGCTTGAAAACTTACGC-3′ |

| | |
|---|---|
| Cas9seqF6 | 5′-GCCCGAGAGAACCAAACTAC-3′ |
| Cas9seqF7 | 5′-GGCTTCTCCAAGGAAAGTATC-3′ |
| Cas9seqF8 | 5′-CGTGGAACAACACAAACACTAC-3′ |
| Cas9seqR1 | 5′-ACTGTAAGCGACTGTAGGAG-3′ |

Both mutagenesis primers (Cas9mutF and Cas9mutR) include 15 nt overlapping overhangs that are necessary for ligation-independent cloning using In-Fusion method (see Step 2 in Section 2.3).

## 2.3. Cloning and verifying the constructs

**Step 1** Setup the following two PCR reactions using a high-fidelity polymerase, such as in the iProof polymerase kit from Bio-Rad.

Cas9$^{WT}$ or Cas9$^{D10A}$ DNA (10 ng)
Cas9mutF primer (200 n*M*)
Cas9mutR primer (200 n*M*)
dNTPs (200 n*M*)
10 × reaction buffer (5 μL)
$H_2O$ (fill up to 25 μL)
iProof polymerase (0.2 μL)

Run PCRs in a heated-lid PCR thermocycler using the following program:

1. 98 °C for 30 s
2. 98 °C for 30 s
3. 55 °C for 30 s
4. 72 °C for 5 min
5. Repeat Steps 2–4 thirty times
6. 72 °C for 10 min
7. 12 °C ∞.

Electrophorese the PCR products on a 0.8% agarose gel and excise the 9.5-kb band. Purify the DNA using a gel extraction kit (e.g., from Life Technologies).

**Step 2** Circularize the purified PCR products using the In-Fusion HD cloning kit (Clontech), following the manufacturer's instructions. Transform competent bacteria using the obtained circularized DNA solution and streak on LB plates containing 100 μg/mL ampicillin. Typically, 10–100 colonies can be found on the plate after overnight incubation. Pick five colonies from each plate, inoculate 5 mL LB medium, incubate overnight at 37 °C, and isolate DNA using method of your choice. Verify the sequence of DNA isolated from two clones by sequencing using the primers shown in Section 2.2.

**Step 3** After verifying the correctly mutated Cas9$^{H840A}$ and Cas9$^{D10A/H840A}$ sequences, prepare plasmid stocks. DNA obtained by using Life Technologies midi- and maxi-prep kits is suitable for direct transfection without further purification steps.

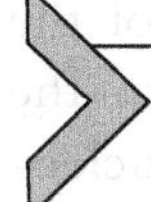

# 3. SELECTION OF THE TARGET SITE AND CLONING OF sgRNA CONSTRUCTS

## 3.1. Selecting suitable target sequences

Obtain the DR-GFP reporter from Addgene (ID 26475). To confirm proper functioning of the DR-GFP assay, we recommend also obtaining the I-SceI expression vector (pCBASceI; ID 26477). The following sequence in the SceGFP part of the DR-GFP reporter (Fig. 9.1A) contains the cleavage site for I-SceI, which gives rise to 4 bp overhangs (demarcated by gray arrowheads).

```
5′-...AGCGTGTCCGGCTAGGGATAACAGGGTAATACC...-3′
3′-...TCGCACAGGCCGATCCCTATTGTCCCATTATGG...-5′
```

Cas9 recognizes a 19 bp DNA sequence that binds the sgRNA followed by the NGG PAM motif in the following format: 3′-NNNNNNNNNNNNNNNNNN▼NNN-NCC-5′, where the arrowhead indicates the cleavage site. (See Jinek et al. (2012) for precise *in vitro* mapping of cleavage sites on both strands.) A PAM motif (underlined) is fortuitously present next to the I-SceI cleavage site, such that SceGFP can be targeted for cleavage by Cas9 using the 19 nt sequence (bold) adjacent to the PAM motif in the sgRNA. With this approach, Cas9 cleavage (open and filled arrowheads) results in a DSB at almost an identical position as I-SceI. Moreover, SceGFP is specifically cleaved because the iGFP repair template has five mismatches with the sgRNA and in addition the AGG PAM motif is not present. The Cas9$^{D10A}$ and Cas9$^{H840A}$ nick the DNA at the filled (black) and open arrowheads, respectively.

An alternative target sequence located on the opposite DNA strand is shown below. In this case, although the PAM motif (underlined) is present, iGFP has ten mismatches with the sgRNA sequence.

```
5′-...AGCGTGTCCGGCTAGGGATAACAGGGTAATACC...-3′
3′-...TCGCACAGGCCGATCCCTATTGTCCCATTATGG...-5′
```

In the following part of the protocol, we use the first target sequence.

## 3.2. Cloning the guide RNA constructs

The empty sgRNA expression vector generated by the Church Laboratory contains an U6 pol III promoter that drives the expression of the sgRNA. A specific protocol for cloning the target sequence into the sgRNA expression vector is provided using two specific 60 nt oligonucleotides (oligos) (http://www.addgene.org/static/cms/files/hCRISPR_gRNA_Synthesis.pdf) and we have successfully used this protocol. Below we detail an alternative protocol that requires one 57-nt oligo containing the specific target sequence and three short-universal oligos that can be reused for other target sequences. Cloning the new sgRNA expression construct involves running two separate PCR reactions using the empty sgRNA expression vector as a template. The first reaction with two universal primers produces a universal 2-kb fragment (which hence can be repetitively used). The second reaction with the specific forward primer and a universal reverse primer produces a 2-kb fragment containing the target 19-nt sequence. Both fragments, containing 15-nt overlaps, are then combined together using the seamless In-Fusion method, giving a circular final plasmid. (See also Zhang, Vanoli, LaRocque, Krawczyk, & Jasin, 2014.)

**Step 1** Order the following specific primer sgRNAF1 SceGFP containing the sequence for targeting SceGFP. Bold font indicates the specific target sequence, such that for each new sgRNA construct only this sequence needs to be changed.

5′-AAGGACGAAACACCG**GTGTCCGGCTAGGGATAAC**
GTTTTAGAGCTAGAAATAGCAAG-3′

**Step 2** Order the following universal primers:

| | |
|---|---|
| sgRNAF2 | 5′-CGTCAAGAAGGCGATAGAAG-3′ |
| sgRNAR1 | 5′-CGGTGTTTCGTCCTTTCCAC-3′ |
| sgRNAR2 | 5′-ATCGCCTTCTTGACGAGTTC-3′ |
| sgRNAseq | 5′-TGGACTATCATATGCTTACCGTAAC-3′ |

**Step 3** Setup the following two PCR reactions using a high-fidelity polymerase (e.g., from the iProof polymerase kit from Bio-Rad).

| *PCR 1 (specific)* | *PCR 2 (universal)* |
|---|---|
| empty sgRNA vector (10 ng) | empty sgRNA vector(10 ng) |
| sgRNAF1SceGFP (specific) primer (200 n*M*) | sgRNAF2 (universal) primer (200 n*M*) |
| sgRNAR2 (universal) primer (200 n*M*) | sgRNAR1 (universal) primer (200 n*M*) |
| dNTP (200 n*M*) | dNTP (200 n*M*) |
| 10 × reaction buffer (5 μL) | 10 × reaction buffer (5 μL) |
| $H_2O$ (fill up to 25 μL) | $H_2O$ (fill up to 25 μL) |
| iProof polymerase (0.2 μL) | iProof polymerase (0.2 μL) |

Run PCRs in a heated-lid PCR thermocycler using the following program:

1. 98 °C for 1 min
2. 98 °C for 30 s
3. 56 °C for 30 s
4. 72 °C for 1 min
5. Repeat Steps 2–4 thirty times
6. 72 °C for 5 min
7. 12 °C ∞.

Electrophorese the PCR products on a 1% agarose gel and excise the 2-kb bands. Purify the DNA using a gel extraction kit (e.g., from Life Technologies).

**Step 4** Combine the two purified PCR products using the In-Fusion HD cloning kit (Clontech), as described in Step 2 of Section 2.3. The sgRNA vector is kanamycin resistant, so use LB plates and media containing 50 μg/mL kanamycin. Verify the sequence of the sgRNA from two clones using the primer sgRNAseq.

*Tip:* The universal 2-kb fragment produced by PCR 2 can be reused for subsequent cloning reactions, such that for each new sgRNA vector only one (specific) PCR reaction needs to be run.

## 4. CELL TRANSFECTION AND FACS ANALYSIS

We describe the protocol for transfection of HEK293T cells using the Nucleofector-2b (Lonza). Either the commercial (Lonza) or home-made (Box 9.1) nucleofection solution can be used for transfections. In the case of HEK293T cells, we used program A-023 and the home-made nucleofection solution. Other cells might require different programs that can be found on the Lonza Website or in the program list of the nucleofector. Table 9.1 shows optimized nucleofection conditions for additional cell lines that we tested.

### BOX 9.1 Home-Made Nucleofection Solution

Prepare the following:

**Solution I**

2 g ATP-disodium salt

1.2 g $MgCl_2{\cdot}6H_2O$

10 mL $H_2O$

Sterilize the solution by passing it through a 0.22 μm filter and split into 80 μL aliquots. Store at −20 °C.

**Solution II**

6.0 g $KH_2PO_4$

0.6 g $NaHCO_3$

0.2 g glucose

300 mL $H_2O$

Adjust the pH to 7.4 with NaOH and add water to a final volume of 500 mL. Filter sterilize and split into aliquots of 4 mL. Store at −20 °C.

On the day of the experiment, thaw and mix one aliquot of Solution I with one aliquot of Solution II. The final solution can be stored at 4 °C for up to 2 weeks. Prewarm the final solution to 37 °C before transfection.

**Table 9.1** Optimized nucleofection conditions for tested cell lines

| Cell line | Nucleofector program | DNA quantity (μg) Cas9:sgRNA:DR-GFP |
|---|---|---|
| HEK293T | A-023 | 1:1:2 |
| U2OS | X-001 | 1:1:1 or 1:1:2 |
| AA8 (Chinese hamster cells) | D-023 | 5:2.5:5 |
| Mouse embryonic stem cells | A-023 | 4:4:4 (maximum cell survival) or 15:5:5 (maximum transfection efficiency) |

## 4.1. Transfection

See Fig. 9.2A for an overview of the procedure.

**Step 1** *Subculturing cells prior to transfection*: Plate 6–7 million cells into a 150-mm tissue culture plate 24 h before transfection such that cells are 70–80% confluent on the day of transfection. Subculturing cells 24 h before transfection significantly improves reproducibility of results.

**Step 2** *Preparation of tissue culture plates and media*: For each sample, prepare a 60-mm tissue culture plate containing 2.5 mL culture media at least 1 h before transfection. Incubate at 37 °C to warm up and to equilibrate the pH of the culture medium. In addition, warm up extra culture medium and nucleofection solution to 37 °C.

**Step 3** *Preparation of the plasmid mix*: In the case of HEK293T cells, Cas9 endonuclease (either WT, D10A, H840A or D10A/H840A), and sgRNA plasmids are used in a ratio of 1:1 (1 μg:1 μg). The amount and ratio of plasmids may need to be adjusted depending on the cell line used (e.g., Table 9.1). Unless cells harbor a genomically integrated DR-GFP copy, the DR-GFP plasmid is also cotransfected (2 μg). (*Note*: HEK293 DR-GFP cells have been developed, Nakanishi et al., 2005). As controls, we use either $Cas9^{WT}$ with an sgRNA expression vector containing no target sequence or $Cas9^{D10A/H840A}$ with the specific sgRNA of interest. For every sample, prepare a separate sterile eppendorf tube with the plasmid mix (Table 9.2), ideally starting with a mix of the common plasmids (e.g., in Samples 2–5, a mix of DR-GFP and the SceGFP sgRNA can be prepared and then aliquoted for each sample).

*Tip*: To confirm proper functioning of the DR-GFP assay, prepare an extra sample with the I-SceI endonuclease expression vector (pCBASceI, 1 μg). To determine overall transfection efficiency, prepare an extra sample with any GFP expression plasmid (e.g., NZE-GFP, 2 μg).

*Tip*: Do not exceed the total volume of ~10 μL of the plasmid mix. Larger volumes will significantly dilute the nucleofection solution which might lead to reduced or inconsistent transfection efficiencies.

*Tip*: Do not use very concentrated plasmid stocks to avoid pipetting errors. Dilute the plasmid stocks if necessary.

**Step 4** *Preparation of cells for transfection*: Trypsinize and count the cells. For each sample, dispense two million cells into in a sterile conical 15-mL tube and spin down (3 min at 1000 rpm). Carefully aspirate

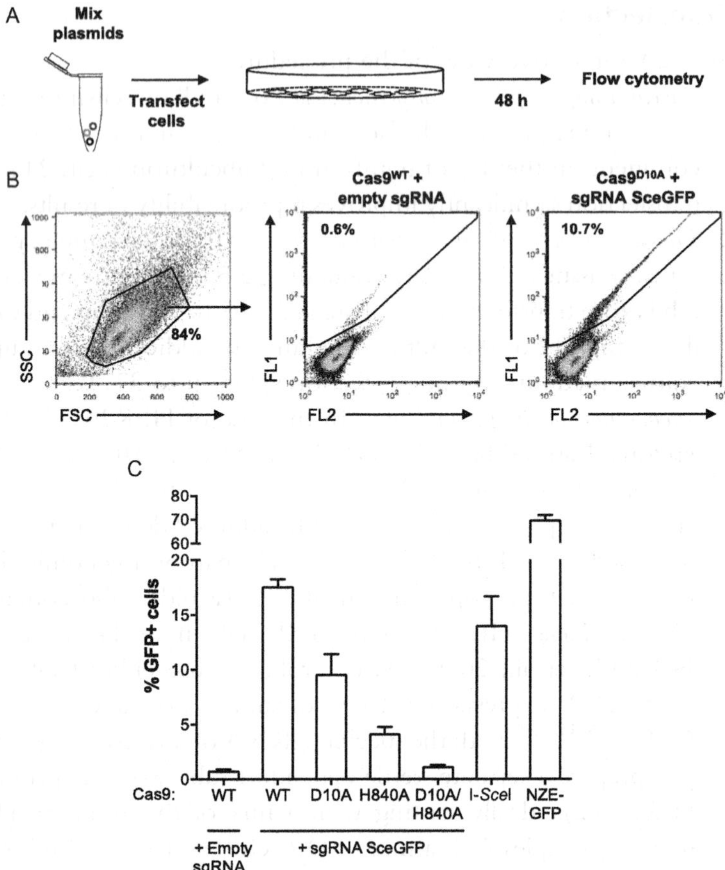


**Figure 9.2** Representative experiment testing different variants of Cas9 for HR in the DR-GFP reporter in HEK293T cells. (A) Schematic overview of the experiment. HEK293T cells were cotransfected with plasmid DNA (Table 9.2) using Nucleofector-2b (Lonza, program A-023). After 48 h, the percent GFP$^+$ cells, indicative of HR efficiency, was measured by flow cytometry. (B) FlowJo software was used to analyze the flow cytometry data. (Left) SSC versus FSC was used to set the gate for live cells. (Middle and Right) FL1 versus FL2 was used to determine the percent GFP$^+$ cells from a control sample (cells above diagonal line are GFP$^+$). (Middle) Note that there is a relatively high background when using the plasmid DR-GFP reporter. (Right) The percentage of GFP$^+$ cells is much higher when DNA damage is induced. (C) Results from three independent experiments. SSBs created by Cas9$^{D10A}$ and Cas9$^{H840A}$ are capable of inducing HR, albeit at reduced frequency, as compared to Cas9$^{WT}$. Error bars represent standard deviation from the mean. (See the color plate.)

**Table 9.2** Plasmid mixes for the experiment described in Fig. 9.2 (DNA quantity in microgram in parentheses)

| | Sample 1 | Sample 2 | Sample 3 | Sample 4 | Sample 5 | Sample 6 | Sample 7 |
|---|---|---|---|---|---|---|---|
| Cas9 | WT (1) | WT (1) | D10A (1) | H840A (1) | D10A/ H840A (1) | – | – |
| sgRNA | Empty (1) | SceGFP (1) | SceGFP (1) | SceGFP (1) | SceGFP (1) | – | – |
| DR-GFP | (2) | (2) | (2) | (2) | (2) | (2) | – |
| Other | | | | | | pCBASceI (1) +pCAGGS (1) | NZE-GFP (2) +pCAGGS (2) |

To equalize the total amount of DNA in each sample, an empty plasmid (pCAGGS) is added to Samples 6 and 7.

the supernatant, resuspend cells in 2 mL sterile PBS by vortexing at low speed. Spin down cells and carefully aspirate PBS, leaving the pellet as dry as possible. Residual PBS dilutes the transfection solution and might influence the transfection efficiency.

**Step 5** *Nucleofection*: Select the appropriate program on the nucleofector. Resuspend the prepared plasmid mix in 100 μL nucleofection solution and transfer to the 15-mL tube containing the cell pellet. Resuspend the cells gently by pipetting up and down and transfer to a 2-mm Gene pulser cuvette. Make sure no bubbles are formed during the transfer as this may reduce transfection efficiency. Place the cuvette in the nucleofector and press the start button. Remove the cuvette, gently add 1 mL prewarmed media into the cuvette, and transfer the cell suspension into the 60-mm culture dish prepared in Step 1. Repeat this step for all experimental samples and controls. Incubate the cells at 37 °C for 48 h.

**Step 6** *Flow cytometry*: After 48 h measure the frequency of $GFP^+$ cells using a flow cytometer. Any instrument able to excite and detect GFP fluorescence is suitable. Trypsinize cells and resuspend them in 0.5 mL culture media in a flow cytometer tube. Analyze the samples using cell line-specific cytometer settings. Forward (FSC) versus side (SSC) scatter plots are used to select the live cells (Fig. 9.2B). Typically, 30,000 live cells per sample are analyzed.

## 4.2. Analysis and interpretation of the results

Data collected with the flow cytometer are analyzed with FlowJo software (Tree Star). Typical results obtained with HEK293T cells are shown in Fig. 9.2B. Gate the live cells based on the FSC and SSC (Fig. 9.2B, left panel). Use the negative control Sample 1 to set the gate for GFP analysis (Fig. 9.2B, middle panel) and apply the same gate to experimental samples (Fig. 9.2B, right panel). DSBs induce relatively high levels of HR, either created by $Cas9^{WT}$ (~17.5%) or I-SceI (~14%). Nicks on the transcribed ($Cas9^{D10A}$) or nontranscribed ($Cas9^{H840A}$) DNA strand are also able to induce HR in ~9.5% and ~4% of cells, respectively. These results were obtained using transient transfection of the DR-GFP reporter, together with Cas9 and sgRNA vectors, and so cells likely contain multiple copies of the DR-GFP reporter. Therefore, lower HR frequencies are to be expected when using cells harboring a single, genomically integrated copy of the reporter. However, an advantage of cells with an integrated reporter is that the $GFP^+$ background in the absence of nuclease expression is very low. It should be noted that SSBs are usually repaired within minutes (Caldecott, 2008), but nCas9s have the potential to nick the DNA again, in a repetitive breakage-repair cycle, until HR destroys the Cas9 recognition site while restoring the GFP open reading frame. Consequently, the induction of HR by SSBs measured using the DR-GFP reporter may be overestimated relative to physiological SSBs, which may be rapidly restored by SSB-specific repair pathways without the intervention of HR.

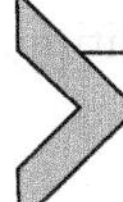

## 5. MATERIALS

### 5.1. Cloning

Primers
Plasmids (Addgene)
Mini and midi-/maxi-prep DNA extraction kit
iProof polymerase kit (Bio-Rad)
Gel extraction kit
In-Fusion HD cloning kit (Clontech)
LB plates and media containing 100 μg/mL ampicillin or 50 μg/mL kanamycin
Competent bacteria
PCR thermocycler
Incubator for bacteria (37 °C)

### 5.2. Cell culture, transfections, data collection, and analysis

Plasmids
Sterile eppendorf tubes
Cell type specific culture media
Sterile Trypsin, 0.2%
Sterile PBS
Sterile 15-mL conical tubes
Cell culture dishes, 60- and 150-mm diameter
Commercial (Lonza) or home-made (Box 9.1) nucleofection solution
Gene pulser cuvettes, 2 mm (Bio-Rad)
FACS tubes (if necessary with cell strainer cap; BD falcon)
Nucleofector-2b (Lonza)
FACS analyzer
FlowJo software (Tree Star)

## 6. SUMMARY

SSBs can induce HR but the underlying mechanisms are not well understood. The DR-GFP reporter has been used widely to study factors involved in DSB-induced HR. In this chapter, we have presented a straightforward protocol to assay SSB-induced HR using the Cas9 nicking endonuclease and DR-GFP. This approach can be used to investigate mechanisms of SSB-induced HR and may be also adaptable to explore other applications requiring targeted induction of SSBs, such as genome editing.

## REFERENCES

Bétermier, M., Bertrand, P., & Lopez, B. S. (2014). Is non-homologous end-joining really an inherently error-prone process? *PLoS Genetics*, *10*, e1004086.

Bunting, S. F., & Nussenzweig, A. (2013). End-joining, translocations and cancer. *Nature Reviews. Cancer*, *13*, 443–454.

Caldecott, K. W. (2008). Single-strand break repair and genetic disease. *Nature Reviews. Genetics*, *9*, 619–631.

Chen, B., Gilbert, L. A., Cimini, B. A., Schnitzbauer, J., Zhang, W., Li, G.-W., et al. (2013). Dynamic imaging of genomic loci in living human cells by an optimized CRISPR/Cas system. *Cell*, *155*, 1479–1491.

Cong, L., Ran, F. A., Cox, D., Lin, S., Barretto, R., Habib, N., et al. (2013). Multiplex genome engineering using CRISPR/Cas systems. *Science*, *339*, 819–823.

Davis, L., & Maizels, N. (2014). Homology-directed repair of DNA nicks via pathways distinct from canonical double-strand break repair. *Proceedings of the National Academy of Sciences of the United States of America*, *111*, E924–E932.

Deriano, L., & Roth, D. B. (2013). Modernizing the nonhomologous end-joining repertoire: Alternative and classical NHEJ share the stage. *Annual Review of Genetics*, *47*, 433–455.

Gasiunas, G., Barrangou, R., Horvath, P., & Siksnys, V. (2012). Cas9-crRNA ribonucleoprotein complex mediates specific DNA cleavage for adaptive immunity in bacteria. *Proceedings of the National Academy of Sciences of the United States of America*, *109*, E2579–E2586.

Haber, J. E. (1999). DNA recombination: The replication connection. *Trends in Biochemical Sciences*, *24*, 271–275.

Hegde, M. L., Hazra, T. K., & Mitra, S. (2008). Early steps in the DNA base excision/single-strand interruption repair pathway in mammalian cells. *Cell Research*, *18*, 27–47.

Horton, J. K., Watson, M., Stefanick, D. F., Shaughnessy, D. T., Taylor, J. A., & Wilson, S. H. (2008). XRCC1 and DNA polymerase beta in cellular protection against cytotoxic DNA single-strand breaks. *Cell Research*, *18*, 48–63.

Hsu, P. D., Lander, E. S., & Zhang, F. (2014). Development and applications of CRISPR-Cas9 for genome engineering. *Cell*, *157*, 1262–1278.

Jasin, M., & Rothstein, R. (2013). Repair of strand breaks by homologous recombination. *Cold Spring Harbor Perspectives in Biology*, *5*, a012740.

Jinek, M., Chylinski, K., Fonfara, I., Hauer, M., Doudna, J. A., & Charpentier, E. (2012). A programmable dual-RNA-guided DNA endonuclease in adaptive bacterial immunity. *Science*, *337*, 816–821.

Liang, F., Han, M., Romanienko, P. J., & Jasin, M. (1998). Homology-directed repair is a major double-strand break repair pathway in mammalian cells. *Proceedings of the National Academy of Sciences of the United States of America, 95*, 5172–5177.

Lieber, M. R. (2010). The mechanism of double-strand DNA break repair by the non-homologous DNA end-joining pathway. *Annual Review of Biochemistry, 79*, 181–211.

Mali, P., Aach, J., Stranges, P. B., Esvelt, K. M., Moosburner, M., Kosuri, S., et al. (2013). CAS9 transcriptional activators for target specificity screening and paired nickases for cooperative genome engineering. *Nature Biotechnology, 31*, 833–838.

Mali, P., Yang, L., Esvelt, K. M., Aach, J., Guell, M., DiCarlo, J. E., et al. (2013). RNA-guided human genome engineering via Cas9. *Science, 339*, 823–826.

McConnell Smith, A., Takeuchi, R., Pellenz, S., Davis, L., Maizels, N., Monnat, R. J., Jr., et al. (2009). Generation of a nicking enzyme that stimulates site-specific gene conversion from the I-AniI LAGLIDADG homing endonuclease. *Proceedings of the National Academy of Sciences of the United States of America, 106*, 5099–5104.

Metzger, M. J., Stoddard, B. L., & Monnat, R. J., Jr. (2013). PARP-mediated repair, homologous recombination, and back-up non-homologous end joining-like repair of single-strand nicks. *DNA Repair, 12*, 529–534.

Nakanishi, K., Yang, Y.-G., Pierce, A. J., Taniguchi, T., Digweed, M., D'Andrea, A. D., et al. (2005). Human Fanconi anemia monoubiquitination pathway promotes homologous DNA repair. *Proceedings of the National Academy of Sciences of the United States of America, 102*, 1110–1115.

Pierce, A. J., & Jasin, M. (2005). Measuring recombination proficiency in mouse embryonic stem cells. *Methods in Molecular Biology, 291*, 373–384.

Pierce, A. J., Johnson, R. D., Thompson, L. H., & Jasin, M. (1999). XRCC3 promotes homology-directed repair of DNA damage in mammalian cells. *Genes & Development, 13*, 2633–2638.

Pommier, Y., Redon, C., Rao, V. A., Seiler, J. A., Sordet, O., Takemura, H., et al. (2003). Repair of and checkpoint response to topoisomerase I-mediated DNA damage. *Mutation Research, 532*, 173–203.

Rouet, P., Smih, F., & Jasin, M. (1994). Expression of a site-specific endonuclease stimulates homologous recombination in mammalian cells. *Proceedings of the National Academy of Sciences of the United States of America, 91*, 6064–6068.

Saleh-Gohari, N., Bryant, H. E., Schultz, N., Parker, K. M., Cassel, T. N., & Helleday, T. (2005). Spontaneous homologous recombination is induced by collapsed replication forks that are caused by endogenous DNA single-strand breaks. *Molecular and Cellular Biology, 25*, 7158–7169.

San Filippo, J., Sung, P., & Klein, H. (2008). Mechanism of eukaryotic homologous recombination. *Annual Review of Biochemistry*, 77, 229–257.

Weinstock, D. M., Richardson, C. A., Elliott, B., & Jasin, M. (2006). Modeling oncogenic translocations: Distinct roles for double-strand break repair pathways in translocation formation in mammalian cells. *DNA Repair, 5*, 1065–1074.

Zhang, Y., Vanoli, F., LaRocque, J. R., Krawczyk, P. M., & Jasin, M. (2014). Biallelic targeting of expressed genes in mouse embryonic stem cells using the Cas9 system. *Methods, 69*, 171–178.

CHAPTER TEN

# Adapting CRISPR/Cas9 for Functional Genomics Screens

**Abba Malina*, Alexandra Katigbak*, Regina Cencic*, Rayelle Itoua Maïga*, Francis Robert*, Hisashi Miura*, Jerry Pelletier*,†,‡,1**

*Department of Biochemistry, McGill University, Montreal, Quebec, Canada
†Department of Oncology, McGill University, Montreal, Quebec, Canada
‡The Rosalind and Morris Goodman Cancer Research Center, McGill University, Montreal, Quebec, Canada
[1]Corresponding author: e-mail address: jerry.pelletier@mcgill.ca

## Contents



## Abstract

The use of CRISPR/Cas9 (clustered regularly interspaced short palindromic repeats/CRISPR-associated protein) for targeted genome editing has been widely adopted and is considered a "game changing" technology. The ease and rapidity by which this approach can be used to modify endogenous loci in a wide spectrum of cell types and organisms makes it a powerful tool for customizable genetic modifications as well as for large-scale functional genomics. The development of retrovirus-based expression platforms to simultaneously deliver the Cas9 nuclease and single guide (sg) RNAs provides unique opportunities by which to ensure stable and reproducible expression of the editing tools and a broad cell targeting spectrum, while remaining compatible with *in vivo* genetic screens. Here, we describe methods and highlight considerations for designing and generating sgRNA libraries in all-in-one retroviral vectors for such applications.

*Methods in Enzymology*, Volume 546
ISSN 0076-6879
http://dx.doi.org/10.1016/B978-0-12-801185-0.00010-6

# 1. INTRODUCTION

The rise of functional genetic screens in mammalian cells and animal models owes a considerable debt to RNA interference (RNAi) technology. RNAi allows for broad, systemic, and unbiased inquiry into complex biological systems in a wide variety of contexts due in large part to development of genome-wide multiplexed pooled short-hairpin RNA (shRNA) library-based screening methods. Yet despite its proven track record, state-of-the-art RNAi-based screens have their drawbacks: (1) targets are limited to the exome; (2) a substantial portion of shRNAs often yield incomplete and unpredictable knockdown efficiencies, which can be insufficient to elicit the desired phenotype of interest; (3) many shRNAs have "off-target" effects that increase the number of spurious hits and lead to erroneous interpretations; and (4) although this can be partially mitigated by increasing the diversity and gene coverage of shRNA targets, it comes at the expense of increased library-pool size and assay complexity.

Applying modern genome editing tools to genetic screens aims to solve many of these problems. While modular transcription factor-based genome editing technologies such as zinc-finger and transcription activator-like effector-based nucleases (ZFNs and TALENs, respectively) have been demonstrated to be reliable and powerful on a one-by-one gene targeting basis, they are all but impractical to implement at a genome-wide scale due to their inherent bulky, pair-wise, and iterative design parameters. In contrast, CRISPR/Cas9 (clustered regularly interspaced short palindromic repeats/CRISPR-associated protein)-based genome editing has shown tremendous promise as a versatile and practical gene targeting technique that would be amenable to genetic screening approaches. Based on a bacterial adaptive immune response that targets invading foreign viral and plasmid DNA, the type II CRISPR system uses an RNA-guided DNA endonuclease (Cas9) to cleave DNA in a sequence-specific manner through a ~20 nt RNA–DNA base match (Jinek et al., 2012). Thus, Cas9 can be readily programmed to introduce double-stranded breaks in virtually any genomic locus through simple alteration of a ~20-bp cognate single guide RNA (sgRNA) when coexpressed in a cell (Cong et al., 2013; Jinek et al., 2013; Mali, Yang, et al., 2013). This inherent flexibility and design simplicity makes CRISPR/Cas9 genome editing easily adaptable for mammalian whole-genome screens and indeed we are beginning to see its application in such settings (Koike-Yusa, Li, Tan, Velasco-Herrera, & Yusa, 2013;

Shalem et al., 2014; Wang, Wei, Sabatini, & Lander, 2014; Zhou et al., 2014). Here, we present methodology and discuss issues pertaining to the use of CRISPR/Cas9 for positive-selection screens.

## 2. ALTERING THE VECTOR DESIGN FOR HIGH-THROUGHPUT SCREENS

The first step in any successful CRISPR/Cas9-based genetic screen is choosing the appropriate method of expression of the two key editing components, Cas9 and its cognate sgRNA. Two approaches dominate the literature: either expressing Cas9 and sgRNA from separate vectors or expressing both in an "all-in-one" vector design. While there are some advantages to independently expressing each part (e.g. the ability to use two different selection markers), we opt for simultaneous delivery given the convenience of a linked single-vector format and, importantly, a more consistent level of expression of either Cas9 and sgRNA not only in terms of selectability but also in terms of stoichiometry, the latter of which has been shown to be important for mitigating off-target cleavage events (Hsu et al., 2013; Pattanayak et al., 2013). Retroviral plasmids provide a convenient way for achieving this given their broad tropism, adjustable levels of infections and expression, and the ability to enrich for permanent and successful integration with a selectable marker (either fluorescence or drug resistance). While we have previously reported the construction and characterization of "all-in-one" retroviral-based vectors coexpressing sgRNAs and Cas9 (from the murine U6 small nucleolar RNA promoter and from the SV40 or Spleen focus-forming virus (SFFV) promoters, respectively) (Malina et al., 2013), we have since modified their design to better suit high-throughput screening purposes by engineering unique restriction sites 17 nucleotides upstream of the U6 transcription start site and at the junction of the crRNA/tracrRNA fusion, in order to facilitate the insertion of oligonucleotides harboring guide sequences, streamlining the process for the generation of sgRNA-based libraries (Fig. 10.1A and B). To distinguish these vectors from our first generation pQCiG and pLC series, we refer to them as pQCiG2 and pLCiG2. These vectors, like their predecessors, express human codon-optimized Cas9 from *Streptococcus pyogenes* (SpCas9) with a 3xFlag epitope tag and two NLS (nuclear localization signal) tags at the N-terminus. Cas9 expression can be explicitly monitored given that its transcription is linked to GFP via an EMCV IRES (Fig. 10.1A and B). We made certain that these subtle changes in sequence would not interfere with

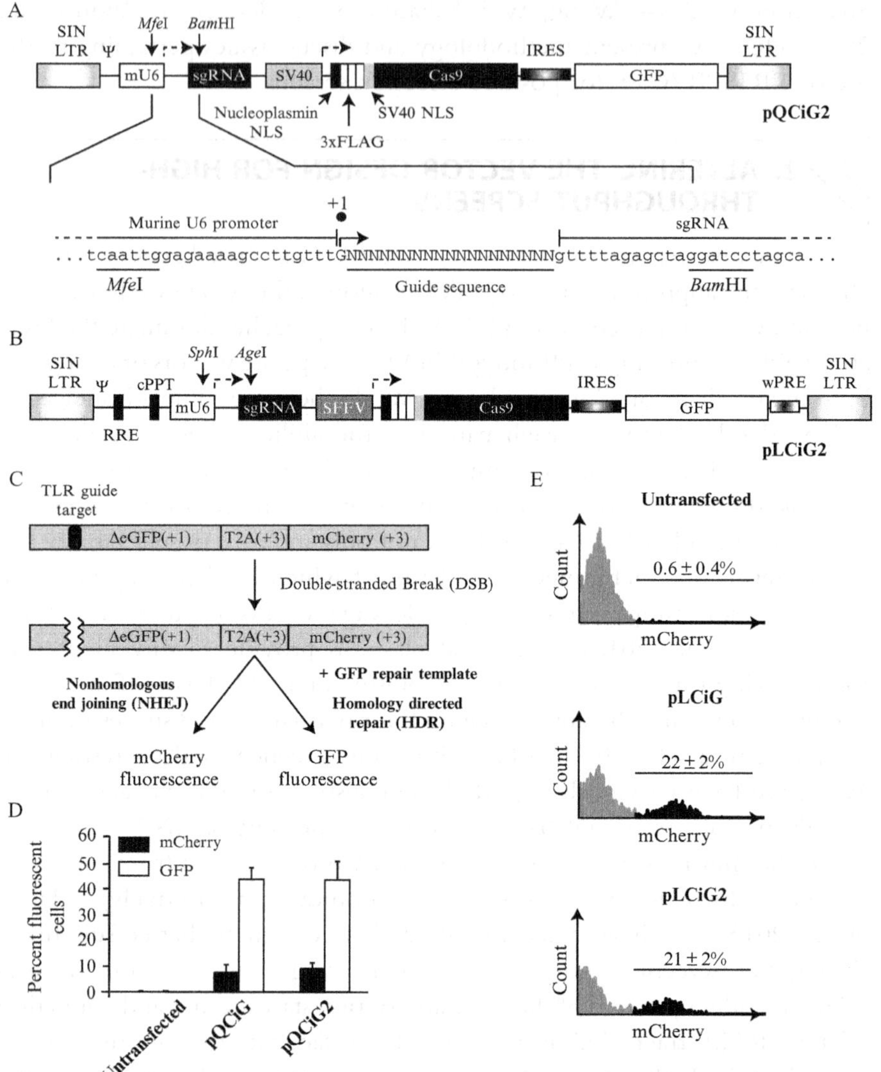


**Figure 10.1** Design of retrovirus vectors for codelivery of Cas9 and sgRNAs that are compatible with large-scale guide library generation. (A) Schematic diagram of pQCiG2-based vector driving expression of Cas9, GFP, and sgRNAs. The unique *Mfe*I and *Bam*HI sites are indicated and are present within the murine U6 promoter and sgRNA, respectively. Right-angled arrows denote the site of transcription initiation. The expanded view illustrates the nucleotide sequence spanning a portion of the mU6 promoter, the start of transcription (G at +1), the 19 nt guide sequence, and the 5′ end of the sgRNA. (B) Schematic diagram of pLCiG2-based lentivirus vector driving expression of Cas9, GFP, and sgRNAs. The unique *Sph*I and *Age*I sites are indicated

Cas9-driven genome editing by assaying the relative cleavage efficiencies of the new version compared to the old version using the "traffic light reporter" (TLR) system, an assay that simultaneously measures the frequency of NHEJ and HDR following a Cas9-induced DSB (Certo et al., 2011). Both versions of the Cas9/sgRNA retroviral vectors stimulated NHEJ to similar extents in 293T cells, indicating that the introduced changes had not impaired editing activity (Figs. 10.1D and E).

Our original sgRNA design incorporated elements from published work by Church and coworkers (Mali, Yang, et al., 2013; Fig. 10.2A, top design), but more recent publications have used a significantly altered sgRNA layout. Two new versions are notable: (1) a version termed sgRNA.2.1, which extends the crRNA:tracrRNA scaffold by four nucleotides, has been reported to improve cleavage efficiency but with concomitant decrease in on-target versus off-target specificity (Pattanayak et al., 2013) and (2) a version incorporating the aforementioned extension and which also mutates a U-rich stretch immediately downstream of the guide sequence, which has been suggested to function as an RNA Pol III transcription termination signal (Fig. 10.2A, sgRNA.3). This has been reported to reduce nucleolar localization of Cas9 (Chen et al., 2013). We evaluated whether these

and cleave within the murine U6 promoter and sgRNA, respectively. (C) A schematic of the traffic light reporter (TLR) assay (Certo et al., 2011). The sgRNA guide target sequence is engineered in the GFP open reading frame (ORF) and shifts the reading frame leading to premature translation termination. The GFP ORF (+1 frame) is fused out of frame to the T2A ribosome "skipping" sequence (Szymczak-Workman, Vignali, & Vignali, 2012) and the mCherry ORF (+3 frame). Induction of a DSB at the guide target sequence will result in mutagenic repair by NHEJ which, in one of three cases, will place the disabled GFP ORF in-frame with mCherry, yielding mCherry$^{+}$ cells. Exogenously supplying a truncated GFP donor plasmid *in trans* will result in GFP fluorescence as a result of HDR of the TLR GFP ORF. Since our vectors express GFP as a reporter, we could not score for HDR activity but rather used the percentage of GFP$^{+}$ cells as an assessment of transfection efficiency and mCherry fluorescence as a gauge of relative NHEJ repair efficiency. Both vectors harbored a previously described guide sequence (TLR: $^{5'}$GAGCAGCGTCTTCGAGAGTG$^{3'}$) that targets a unique site embedded within the GFP ORF of the TLR (C) (Malina et al., 2013). (D) Assessment of pQCiG- and pQCiG2-mediated NHEJ in a stably integrated TLR reporter 293T cell line with a stably integrated TLR reporter locus. Cells were transfected with pQCiG or pQCiG2 (1.5–3 μg) and analyzed by flow cytometry 6 days later. GFP fluorescence measures transfection efficiency whereas mCherry fluorescence scores for NHEJ repair events. $n=3$, error bars represent SEM. (E) Assessment of pLCiG- and pLCiG2-mediated NHEJ in a stably integrated TLR reporter 293T cell line. Cells were transfected with pLCiG or pLCiG2 and analyzed by flow cytometry 6 days later. Shown is a representative histogram illustrating the percent of mCherry$^{+}$ cells and the mean percent fluorescence. $n=4$, error is SEM.

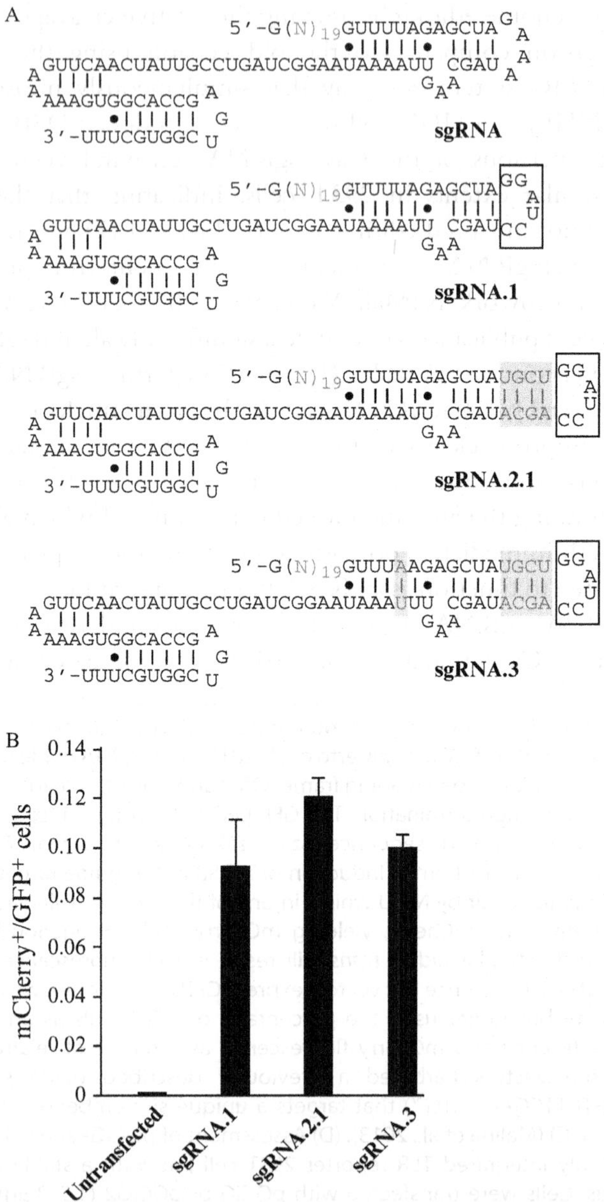


**Figure 10.2** Assessment of NHEJ repair efficiency mediated by different sgRNA variants. (A) Predicted secondary structure (http://rna.tbi.univie.ac.at/cgi-bin/RNAfold.cgi) of chimeric RNAs showing the first guanine arising from the transcription initiation site followed by the guide region $(N)_{19}$, for four different sgRNAs. The open box denotes the crRNA/tracrRNA junction where a *Bam*HI site was inserted to generate sgRNA.1 (*Age*I

changes would produce any significant functional differences, but could not detect any differences in gene editing efficiencies among the three sgRNAs (Fig. 10.2B). In light of this and given our preference to maintain a high ratio of on-target to off-target specificity, our retroviral vectors retain the original sgRNA.1 configuration.

# 3. CONSTRUCTION OF sgRNA LIBRARIES

## 3.1. Guide sequence prediction

Prediction of guide sequences can be accomplished by manually inspecting annotated gene sequences (when only a small number of guides is required) or by using one of several design tools available at the time of this writing (Table 10.1). In either case, one first locates a sequence of interest bearing a protospacer-adjacent motif (PAM), which is essential for recognition by Cas9. Our vectors use a humanized version of Cas9 protein that originates from *S. pyogenes* and is the one most frequently used in the literature, owing to its short PAM target sequence ($^{5'}$NGG$^{3'}$) and thus high prevalence in the genome (Jiang, Bikard, Cox, Zhang, & Marraffini, 2013; Jinek et al., 2012). Although it has been reported that $^{5'}$NAG$^{3'}$ can also be used as a PAM by *S. pyogenes* Cas9, it is much less efficiently recognized (Jiang et al., 2013), and probably very rarely so at limiting Cas9 cellular concentrations (Wu et al., 2014), thus we generally do not consider it when designing guide sequences. After locating a PAM sequence, the adjacent 20 upstream nucleotides to the PAM are chosen as the guide sequence. Should the 20th nucleotide not end with guanosine, we forcibly terminate the sequence with a 5′ guanosine, which is a necessary requirement for U6 transcription initiation but has little effect on the rate of target cleavage even when unmatched (Fu, Sander, Reyon, Cascio, & Joung, 2014; see also Fig 10.1A). Mismatches between the PAM proximal region of the target and the sgRNA are known to more adversely affect Cas9 endonuclease activity (Fu et al., 2013; Hsu et al., 2013; Jinek et al., 2012; Mali, Aach, et al., 2013;

in the case of pLCiG2). The grey shaded areas denote sequence differences between sgRNA.1, sgRNA2.1, and sgRNA.3. Note that our sgRNA.2.1 and sgRNA.3 designs differ from the originals in harboring a *Bam*HI site at the crRNA/tracrRNA junction. (B) Assessment of normalized NHEJ repair efficiency in a stably integrated TLR reporter 293T cell line. Cells were transfected with pQCiG2 (1.5–3 μg) expressing the indicated sgRNAs and analyzed by flow cytometry 6 days later. GFP fluorescence was used to track transfection efficiency, whereas mCherry was used to monitor NHEJ. GFP values (transfection efficiency) ranged from 32% to 51%. $n=3$, error bars represent SEM.

**Table 10.1** CRISPR/Cas9 guide design tools

| Tool name | Web interface | Target genomes[a] | URL | Off-target analysis[b] |
|---|---|---|---|---|
| CRISPR design | Yes | 15 | http://crispr.mit.edu http://www.broadinstitute.org/mpg/crispr_design/ | Yes |
| E-CRISP | Yes | 18 | http://www.e-crisp.org/E-CRISP/designcrispr.html | Yes |
| Cas9 design | Yes | 7 | http://cas9.cbi.pku.edu.cn/index.jsp | No |
| CasOT | No | Any | http://eendb.zfgenetics.org/casot/index.php | Yes |
| CRISPR sgRNA design tool | Yes | 3 | https://www.dna20.com/eCommerce/cas9/input | No |
| CasFinder | No | Any | http://arep.med.harvard.edu/CasFinder/ | Yes |
| flyCRISPR | Yes | Fly | http://flycrispr.molbio.wisc.edu/tools | Yes |
| DRSC CRISPR finder | Yes | Fly | http://www.flyrnai.org/crispr/ | Yes |
| ZiFiT Targeter | Yes | Any | http://zifit.partners.org/ZiFiT/ChoiceMenu.aspx | No |
| CRISPy | Yes | CHO | http://staff.biosustain.dtu.dk/laeb/crispy/ | Yes[c] |
| GT-Scan | Yes | 32 | http://gt-scan.braembl.org.au/gt-scan/submit | Yes |
| CHOPCHOP | Yes | 9 | https://chopchop.rc.fas.harvard.edu/ | Yes[d,e] |

[a]Refers to the number or nature of species that the software allows one to analyze.
[b]Refers to whether the software is capable of predicting off-target sites based on sequence similarity and location adjacent to a PAM.
[c]Can only scan for sites matching 13 nucleotides + NGG.
[d]Can scan alternate Cas9 PAM motifs.
[e]Can also output flanking primer sequences to, and identify restriction sites within, the target site.

Pattanayak et al., 2013), which lends support to the notion that an 8–12 nts "seed" sequence upstream of the PAM drives Cas9-mediated cleavage efficiencies (Jinek et al., 2012; Semenova et al., 2011). In order to minimize potential off-target cleavage sites, and given the more stringent requirement for homology between the "seed" region and the sgRNA, we typically heuristically align only the first 12 nucleotides of the chosen sequence plus all four iterations of the PAM to annotated online genome databases, with sequences that result in the least number of perfect matches being preferred. More recent genome-wide ChIP-seq-based analyses have suggested a far greater tolerance for mismatches driving Cas9 DNA binding, with enriched genomic regions being frequently characterized by "seed" sequences that can be as short as five nucleotides, although most of these sites were only rarely altered when sequenced directly (Kuscu, Arslan, Singh, Thorpe, & Adli, 2014; Wu et al., 2014). Nevertheless, as a precaution, we recommend designing at least three sgRNAs for each locus to control for potential off-target effects. Further points in sgRNA design that should also be considered:

1. When targeting genes encoding mRNAs, sgRNAs targeting the last coding exon have been reported to be less effective than those targeting earlier exons (Wang et al., 2014). As well, we would recommend that users avoid targeting the region that harbors the first AUG codon since genes may have in-frame downstream AUG (and even non-AUG) initiation codons that can be used and give rise to functional truncated products (Ellison & Bishop, 1996). Rather it is probably a safer bet to target somewhere in the middle of a gene when disruption of function is desired.
2. It has been reported that sgRNAs that target the transcribed strand are less effective than those targeting the nontranscribed strand (Wang et al., 2014).
3. Be aware that the sgRNAs are transcribed by RNA Polymerase III, whose termination signal is a stretch of four or more sequential Us (Nielsen, Yuzenkova, & Zenkin, 2013; Orioli et al., 2011) and guide sequences that are U-rich have been shown to decrease sgRNA abundance (Wu et al., 2014). Therefore, avoid guides that have stretches of three or more Us.
4. Recent crystal structure data of sgRNA-bound Cas9 have revealed protein:RNA interactions between residues Arg71-G18 and Arg447-U16, which correspond to the 3rd and 5th residue upstream from the crRNA: tracrRNA scaffold (Nishimasu et al., 2014). This is in line with other recently reported data that high-performing sgRNAs displayed a preference for four purines adjacent to the PAM (Wang et al., 2014), so it might be worthwhile to prioritize guides with a G residue three nucleotides upstream of the PAM, if one has that option when choosing.

5. sgRNAs with very high or very low-GC content should be avoided (Wang et al., 2014).
6. Finally, ensure that your guide sequences are absent of restriction sites used for cloning (*Mfe*I/*Bam*HI or *Sph*I/*Age*I).

## 3.2. Cloning of guide templates

The construction of guide libraries uses either pools of oligonucleotides derived from small-scale synthesis or from highly parallel approaches (Fig. 10.3).

### 3.2.1 Layout of the guide template

The template for cloning into pQCiG2 is: $^{5'}$CAATTG-GAGAAAAGCCTTGTTTG$(N)_{19}$ GTTTTAGAGCTAGGATCCTAGC$^{3'}$ (where the *Mfe*I and *BamHI* sites are underlined and 19 nucleotide guide region represented by N). For pLCiG2, the template is: $^{5'}$GCATGC-GAGAAAAGCCTTGTTTG$(N)_{19}$ GTTTTAGAGCTAACCGGTTAGC$^{3'}$ (where the *Sph*I and *Age*I sites are underlined).

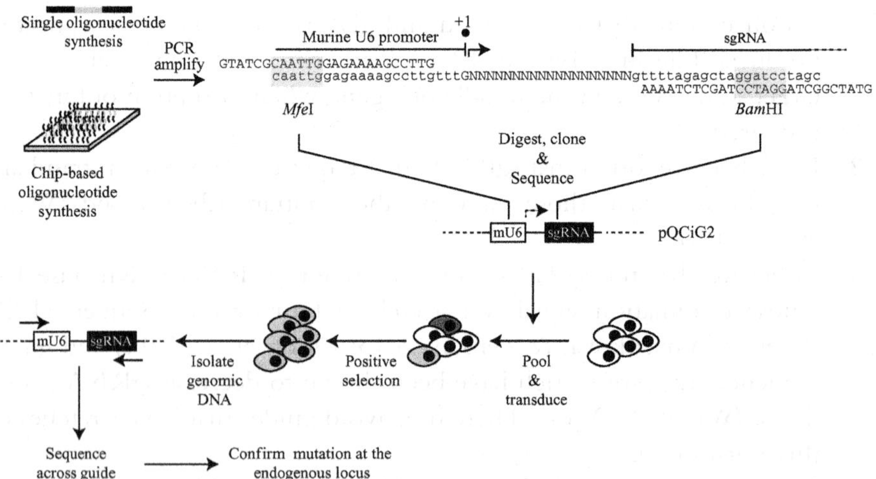


**Figure 10.3** Schematic representation of sgRNA library generation and pooled screening strategy. Oligonucleotides are individually synthesized or *en masse* on a microarray chip. These are then PCR amplified to incorporate vector compatible restriction sites. The sequence of the primers and template shown are compatible with cloning into pQCiG2 (see text for details for cloning into pLCiG2). Library pools of guides are then used for screening purposes. Following isolation of genomic DNA from positively selected cells, amplification by PCR across the guide region is performed and the guide identified by sequencing. Modification at the expected locus is then confirmed using the T7 endonuclease I assay, SURVEYOR assay, or sequencing of PCR products.

### 3.2.2 Initial guide library preparation

Depending on the size and complexity of the required sgRNA library, one pools oligos ordered either individually prealiquoted in 96- or 384-well dishes (e.g., IDT, Coralville, IA), or synthesized *en masse* on a chip as an array and liberated following acid hydrolysis (e.g., Oligomix from LC Sciences Inc., Houston, TX). In our experience, we get much lower rates of mutant clones when derived from pools of individually synthesized oligonucleotides than those from arrays (~80–90% vs. 30–50% produce error-free clones, respectively).

### 3.2.3 PCR amplification of pooled oligonucleotide templates

If ordered on an individual basis, oligonucleotides should first be pooled at an equimolar ratio and then amplified using Forward ($^{5'}$GTATCGCAATTGGAGAAAAGCCTTG$^{3'}$ for pQCiG2 and $^{5'}$GTATCGGCATGCGAGAAAAGCCTTG$^{3'}$ for pLCiG2) and Reverse ($^{5'}$GTATCGGCTAGGATCCAGCTCTAAAA$^{3'}$ for pQCiG2 and $^{5'}$GTATCGGCTAACCGGTTAGCTCTAAAA$^{3'}$ for pLCiG2) primers (Fig. 10.3). PCR conditions are as follows:

Reagent amounts

- 5 μl 10 × ThermoPol buffer with $MgCl_2$
- 2.5 μl Forward Primer (10 μ*M*)
- 2.5 μl Reverse Primer (10 μ*M*)
- 1 μl of Oligonucleotide Template (100 ng/μl)
- 1 μl dNTPs (10 m*M*)
- 0.25 μl Vent DNA Polymerase (NEB) (2 U/μl)
- 37.75 μl $dH_2O$

Thermocycler reaction conditions

- 94 °C for 3 min (Initial denaturation)
- 30 cycles of 94 °C for 30 s, 52 °C for 30 s, and 72 °C for 1 min
- 72 °C for 10 min (Final extension)

The PCR conditions and reagents are slightly different if amplifying oligos from arrays, and are as follows:

Reagent amounts

- 10 μl 5 × Phusion buffer
- 1 μl Forward Primer (20 μ*M*)
- 1 μl Reverse Primer (20 μ*M*)
- 1 μl Oligo Template (0.5 ng/μl)
- 1 μl dNTPs (10 m*M*)
- 0.5 μl Phusion High-Fidelity DNA polymerase (NEB) (2 U/μl)

1 μl 30% DMSO
34.5 μl $dH_2O$

Thermocycler reaction conditions
98 °C for 30 s (Initial denaturation)
30 cycles of 98 °C for 10 s, 54 °C for 30 s, and 72 °C for 25 s
72 °C for 5 min (Final extension)

Confirm amplification of the desired PCR products by analyzing an aliquot (5 μl) on a 2% agarose gel to verify the presence of a single band (76 bp).

### 3.2.4 Digestion and ligation of the guides into vector backbone

The PCR product is purified using a PCR Purification Kit (e.g., QIAquick kits [Qiagen] or EZ-10 Spin Column PCR Products Purification Kit [Bio Basic Inc.]) following the manufacturer's recommendations and the eluent digested with *Mfe*I/BamHI-HF or *Sph*I/*Age*I (NEB) depending on the desired target vector. Ligations into the appropriate vector are performed according to standard techniques (Green & Sambrook, 2012). Make sure to include a "vector-only" ligation control. Following ligations, 2 μg of glycogen is added to each ligation and the volume is increased to 100 μl with $ddH_2O$, followed by two consecutive ethanol precipitations and 70% ethanol washes. The precipitate is resuspended in 20 μl $ddH_2O$ and is ready for transformation by electroporation.

### 3.2.5 Assessing ligation efficiency

An aliquot (1 μl) of the ligation is used for a test chemical transformation and the ratio of colonies from the "vector + insert" ligation reaction to "vector-only" reaction is determined. We proceed with large-scale transformation if we obtain at least a 10:1 ratio of colonies in the "vector + insert" plate relative to the "vector-only" plate. We also process at least 24 minipreps and have them sequenced to assess the quality of the clones and representation of the library.

### 3.2.6 Large-scale transformation of the guide library

To generate the guide RNA-expressing retroviral plasmid library as bacterial clones, we use Electromax-competent DH10B cells for pQCiG2 vectors or Electromax-competent Stbl4 for pLCiG2 (Life Technologies) and a Bio-Rad Gene Pulser using the following conditions of 2.0 kV, 200 Ω, and 25 μF. We follow the manufacturer's recommendation using 1 μl of ligation/100 μl of competent cells. From 1 ml of culture following transformation, aliquots of 1, 2, and 10 μl are taken and plated onto LB + 100 μg/ml carbenicillin plates to assess the efficiency of transformation. The remaining culture is kept at 4 °C overnight. Once transformation efficiency has been determined, the reserved culture material is plated onto large square LB plates (245 mm × 245 mm) containing 100 μg/ml carbenicillin to obtain 1000–2000 (if colonies are to be individually picked) or 10,000 (if colonies are to be pooled) colonies/plate. We prefer the use of carbenicillin over ampicillin since it is more stable than the latter and results in fewer satellite colonies during library generation.

### 3.2.7 Checking the quality of the guide library

Ninety-six colonies are seeded into deep-well 96-well plates (VWR® 96-Well Deep Well Plates; cat. no.: 82006–448) containing 1.5 ml of Terrific broth (TB) + 100 μg/ml carbenicillin. The plates are sealed with an air-pore sheet (Qiagen cat. no.: 19571). Following growth in a 37 °C shaker for 24 h, we isolate plasmid DNA using a QIAprep 96 Turbo Miniprep Kit (QIAgen), which is then submitted for sequencing.

### 3.2.8 Bulk harvesting of bacterial-transformed guide library

For some applications, it may be sufficient to harvest the plated colonies in bulk and use the resulting pool directly in a screen. This is achieved by pipetting 50 ml of TB + 100 μg/ml carbenicillin directly onto each plate and using a flat rubber policeman to gently scrape the colonies off the plate into a sterile 2-l flask. The nature of the screen will determine the desired library complexity, but we use 500 ml of TB + 100 μg/ml carbenicillin per pool aiming for complexities of 10,000–20,000 clones/pool. After growth at 37 °C for ~6 h, the plasmid DNA can be isolated using standard procedures (Green & Sambrook, 2012) or a commercial maxiprep kit (e.g., Plasmid Maxi Kit; Qiagen).

### 3.2.9 Arraying individual bacterial guide library clones

Although significantly more expensive and labor intensive than a nonarrayed library, our preference is to generate arrayed, sequence-verified libraries since these are renewable resources with greater flexibility. Here, individual colonies are picked into deep-well 96-well plates containing 1.5 ml of TB + 100 μg/ml carbenicillin and covered with an air-pore sheet (Qiagen cat. no.: 19571). Following growth for 24 h at 37 °C, 50 μl aliquots are transferred to two 96-well plates (Falcon cat. no.: 353910) containing 50 μl TB + 100 μg/ml carbenicillin + 50% glycerol, sealed and stored at −70 °C as Master plates. The remainder of the culture is processed to prepare miniprep DNA that is then used for sequencing across the sgRNA insert. Alternatively, colonies can be picked into 384-well plates containing 65 μl of TB + 100 μg/ml carbenicillin. Following growth overnight at 37 °C, 2 μl of a 1/10th dilution of the culture is directly used in a PCR as template for amplification across the guide sequence. The PCR product is purified using Agencourt AMPure XP—PCR Purification (Beckman-Coulter) and used directly for sequencing. The position and identity of each clone in the Master Plates is recorded.

Once arrayed, the library pools are made by identifying the coordinates of the clones of interest and then thawing the plates at room temperature. The plates are then briefly centrifuged, a small hole is made by piercing through the aluminium foil cover and 1 μl of the desired bacterial culture corresponding to the clone of interest is removed. The puncture hole is sealed using a small aluminium foil patch. This method avoids potential well cross-contamination due to aerosol generation that could arise if the entire cover was removed. The 1 μl aliquot is used to seed 1 ml of TB + 100 μg/ml carbenicillin and grown at 37 °C to saturation (~24 h). The following day, the individual bacterial cultures are pooled into a 2-l flask containing 500 ml TB + 100 μg/ml carbenicillin, grown for 6 h and processed for plasmid DNA isolation.

## 4. RETROVIRAL TRANSDUCTION OF THE GUIDE LIBRARY

The resulting library is then used for virus preparation using standard techniques (Barde, Salmon, & Trono, 2001; Swift, Lorens, Achacoso, & Nolan, 2001). Depending on the viral vector backbone, either helper-free stable virus producing Phoenix cell line is used (in the case of pQCXiG2-based libraries), or 293T/17 (ATCC) cells are cotransfected with packaging and VSV-G envelope vectors (in the case of pLCiG2-based libraries). One advantage of using pseudotyped lentivirus is the ability to generate large quantities of library-pool transducing viral supernatant preps which can then be concentrated, titered, aliquoted, and frozen for later use (for further details, see Kutner, Zhang, & Reiser, 2009). The viral MOI (multiplicity of infection) is determined by serial dilution of the preparation on 293T cells and through measurement of the fraction of GFP expressing cells as determined by FACS (we aim to get to at least 5–10% $GFP^+$ cells, which is in the linear range of viral transduction). The amount of cells plated for viral preparation will depend on the desired library complexity: generally speaking, we want to make enough virus to infect cells at an MOI of ~0.2–0.1 (which ensures that only one sgRNA is expressed per cell for the vast majority of the population) with also at least 1000 infected cells/construct (which maintains the library complexity—see notes below).

## 5. NOTES ON SCREENING DESIGN PARAMETERS

Each genetic screen will entail designs unique to the respective experiment. Rather than present specifics about one particular screen, it is more practical to consider general attributes that will impact on most screens:

1. *The nature of the phenotype and the strength of the selective pressure.* One of the most important determinants of the success of a screen is how well the desired phenotype can be distinguished from baseline. It should be robust and with little variation. Positive-selection screens looking for cooperating tumor suppressors or lesions that impart drug resistance embody these features. The time to phenotype onset will dictate the duration of the experiment and the strength of the selective pressure. Greater selective pressure enhances the phenotypic shifts in sgRNA representation under shorter periods of time, but can also lead to increased variability among replicates and can lead to a loss of representation of sgRNA species (especially those at the lower end of abundance) due to a sudden population bottleneck. This can be partially mitigated by increasing the number of infected cells per construct. While the optimal amount of selective pressure will have to be determined empirically in pilot screens (ideally with help of positive controls), we generally strive for ~25% loss of cell population following a given toxic treatment, which balances good reproducibility, selective pressure, and maintenance of sgRNA construct abundance.
2. *Maintaining library complexity during propagation of cells.* Many factors will determine the appropriate pool size and consequent sgRNA library representation in a cell population over the course of the screening process (e.g., cell line infectability, number of replicates, rate of allele modification, etc.). Following a successful screen, we generally will infer the representation of sgRNAs through the use of next-generation sequencing. The number of spurious reads (or baseline noise level) that arise from a massive parallel sequencer is typically in the range of 50–100 counts, and therefore as a rule of thumb we typically try to ensure that at least 1000 cells/construct are infected at the onset (which should result on average in roughly a ~10–20-fold increase in sgRNA read counts above baseline noise). Moreover, if over the course of a screen the cells need to be split,

it is critical to ensure that at each split the full library representation that was initially used at the start of the experiment will be maintained. If too many cells are removed during propagation, the representation of the library becomes skewed.

3. *The availability of positive and negative controls.* Although one does not always have access to positive controls when undertaking novel screens, their availability will significantly facilitate assay development and optimization. Pilot screens testing a series of serial dilutions of the positive control can be used to tease out the limits of detection for a given sgRNA and inform on the required library complexity. As well, we make sure to include multiple negative controls, both "scrambled" sgRNAs that do not match to any region in the genome as well sgRNAs that are known to cleave genes or loci that when disrupted are neutral for most phenotypes (e.g., AAVS1 for human cells or the ROSA26 locus for mouse). These are vital to score for the relative increase or decrease in sgRNA output following a successful screen.
4. *Tracking each step.* Our vectors harbor a GFP marker (which is neutral in most settings) allowing us to document infection efficiencies throughout the experiment.
5. Is monoallelic, biallelic, or multiallelic (in the case of pseudodiploid cells) modification required for the phenotype of interest? The efficiency of locus modification by CRISPR/Cas9 in high-throughput screens has been reported to range from 13% to >90% (Koike-Yusa et al., 2013; Shalem et al., 2014; Wang et al., 2014; Zhou et al., 2014). Although the reasons for this variation are unclear, it could relate to differences in guide targeting efficiency, MOI, cell line, the ratio of Cas9:sgRNA cellular levels, and the methods of library delivery. Given these potential issues, it is important to try to understand the phenotype(s) that is expected and whether all alleles of the target need to be inactivated and how the delivery system chosen for the screen will impact on this.
6. Different guides to the same target should yield the same phenotype. If this is not the case, we recommend generating additional sgRNAs to resolve the discrepancy. A recent publication has indicated that guide sequences with 17 or 18 nucleotides complementarity (called "tru-gRNAs") show reduced mutagenesis at off-target sites without sacrificing on-target editing efficiencies (Fu et al., 2014) and this feature could easily be incorporated into guide library design.
7. Be aware that loss of a particular sgRNA can occur during virus generation, which can happen due to a given sgRNA-affecting viral

replication and/or packaging, or may simply be due to the inactivation of an essential host gene in the packaging cell line. Deep sequencing of the library pool before and after virus production will shed information on this and is recommended.

**8.** To date, four large-scale screens have been published using CRISPR/Cas9 and nonarrayed sgRNA libraries (Koike-Yusa et al., 2013; Shalem et al., 2014; Wang et al., 2014; Zhou et al., 2014) and there are several lessons to be learnt from these:

**A.** Two screens engineered their cell lines to constitutively express Cas9 (Koike-Yusa et al., 2013; Zhou et al., 2014), whereas a third engineered a doxycycline-inducible Cas9 in the line of interest (Wang et al., 2014). Zhang and colleagues performed negative and positive-selection screens with a delivery system similar to the one described above by us (Shalem et al., 2014). Developing cell lines that express Cas9 is more labor intensive and requires prescreening of cell clones to identify the ones with highest editing efficiency since this can vary between clones and may be a consequence of variations in Cas9 expression levels (Zhou et al., 2014). Furthermore, the clonal nature of the cell might influence the phenotypic outcome of particular screen rendering it less widely applicable.

**B.** Wei and colleagues (Zhou et al., 2014) also ectopically expressed OCT1, a transcription factor shown to boost U6 promoter activity (Lin & Natarajan, 2012) in their line of interest. This added feature may increase sgRNA expression and should be piloted to assess whether the gain in sgRNA levels obtained with higher OCT1 levels translates into higher mutation efficiency, which could also influence the measured phenotype.

**C.** RNAi was *not* universally successful in validating the sgRNAs identified from the screens. The ability to phenocopy the results obtained with sgRNAs tended to correlate with knockdown efficiency (Koike-Yusa et al., 2013; Shalem et al., 2014).

**D.** In one screen, complementation with cDNAs was successful at reverting the phenotype (Koike-Yusa et al., 2013) and may be a better approach at validating "hits" than using shRNAs, assuming that the mutant allele is not functioning in a dominant-negative or gain-of-function manner.

**9.** The recent description of CRISPR/Cas9 gene editing in mice both *ex vivo* (where CRISPR/Cas9 was expressed in cultured primary lymphoma cells via retroviral transduction and later reimplanted) and *in vivo* (in which the CRISPR/Cas9 system was delivered via hydrodynamic injections to directly modify hepatocytes *in situ*), raises the exciting possibility of performing CRISPR-based sgRNA screens in a live mammalian model organism (Malina et al., 2013; Yin et al., 2014).

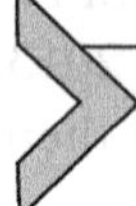

## 6. DECODING "HITS" FROM POSITIVE SELECTION SCREENS INVOLVING sgRNA LIBRARY POOLS

Once cells are obtained following a positive selection screen, we identify the guide sequence responsible for the phenotype by amplifying across the guide of the integrated retroviral-derived construct in the cells of interest. Genomic DNA from the clone(s) of interest is isolated using standard techniques (Green & Sambrook, 2012) and the guide region amplified by PCR. In our experience, the guide region can be amplified quite specifically.

Reagent amounts

- 5 μl 5 × Phusion Buffer
- 1 μl Primer Mix (10 μ*M* each; Trigger ID F: $^{5'}$AGCCCTTTGTACACCCTAAGCCTC$^{3'}$ Trigger ID R: $^{5'}$CTAACTGACACACATTCCACAGGG$^{3'}$)
- 0.5 μl dNTPs (10 m*M*)
- 1 μl Genomic DNA from pQCiG2 infected cells (100 ng/μl)
- 0.15 μl of Phusion High-Fidelity DNA polymerase (NEB) (2 U/μl)
- 17.35 μl dd$H_2O$

Thermocycler reaction conditions

- 98 °C for 30 s (Initial denaturation)
- 25 cycles of 98 °C for 10 s, 57 °C for 30 s, and 72 °C for 30 s
- 72 °C for 10 s (final extension)

The PCR product is then purified using a PCR Purification Kit (e.g., Qiaquick kits [Qiagen] or EZ-10 Spin Column PCR Products Purification Kit [Bio Basic Inc.]) following the manufacturer's recommendations and directly sequenced using the sequencing primer Psi: $^{5'}$AGCCCTTTGTACACCCTAAGC$^{3'}$. Once the guide sequence has been successfully identified as a potential "hit," we then confirm that the endogenous locus has been mutated using the original genomic preps and perform either a T7 endonuclease I assay (Reyon et al., 2012) or SURVEYOR assay (Transgenomic), or, if a more thorough examination of the kinds of sequence alterations is desired, through sequencing on an Ion Torrent personal genome machine (Malina et al., 2013).

## 7. CONCLUSION

CRISPR/Cas9 has much to offer in complementing RNAi-based screens. The larger targeting range of CRISPR/Cas9 relative to RNAi extends to the whole genome and offers the opportunity to probe structure/function relationships beyond the transcriptome. As well, the potential exists for Cas9-driven cleavage events to yield not only loss-of-function but also gain-of-function and dominant-negative, alleles—thus extending the mutational "depth" beyond the straight suppression possible with RNAi. Whereas somatic cell genetics provided stunning insights into gene organization and regulation in the 1970s and 1980s (Caskey, Robbins, North Atlantic Treaty Organization, & Scientific Affairs Division, 1982), the remarkable progress that has been made in applying CRISPR/Cas9 to genome engineering since 2013 and the potential it holds for genetic analysis of almost any cell type at an unprecedented scale would suggest an up-and-coming rebirth of this discipline. It will be exciting to participate in this new adventure as CRISPR/Cas9 is used to uncover novel genome functionalities.

## REFERENCES


Barde, I., Salmon, P., & Trono, D. (2001). Production and titration of lentiviral vectors. *Current protocols in neuroscience*. New York, NY: John Wiley & Sons, Inc.

Caskey, C. T., Robbins, D. C., North Atlantic Treaty Organization, & Scientific Affairs Division. (1982). *Somatic cell genetics*. New York: Plenum Press, published in cooperation with NATO Scientific Affairs Division.

Certo, M. T., Ryu, B. Y., Annis, J. E., Garibov, M., Jarjour, J., Rawlings, D. J., et al. (2011). Tracking genome engineering outcome at individual DNA breakpoints. *Nature Methods*, *8*(8), 671–676. http://dx.doi.org/10.1038/nmeth.1648.

Chen, B., Gilbert, L. A., Cimini, B. A., Schnitzbauer, J., Zhang, W., Li, G. W., et al. (2013). Dynamic imaging of genomic loci in living human cells by an optimized CRISPR/Cas system. *Cell*, *155*(7), 1479–1491. http://dx.doi.org/10.1016/j.cell.2013.12.001.

Cong, L., Ran, F. A., Cox, D., Lin, S., Barretto, R., Habib, N., et al. (2013). Multiplex genome engineering using CRISPR/Cas systems. *Science*, *339*(6121), 819–823. http://dx.doi.org/10.1126/science.1231143, science.1231143 [pii].

Ellison, A. R., & Bishop, J. O. (1996). Initiation of herpes simplex virus thymidine kinase polypeptides. *Nucleic Acids Research*, *24*(11), 2073–2079.

Fu, Y., Foden, J. A., Khayter, C., Maeder, M. L., Reyon, D., Joung, J. K., et al. (2013). High-frequency off-target mutagenesis induced by CRISPR-Cas nucleases in human cells. *Nature Biotechnology*, *31*(9), 822–826. http://dx.doi.org/10.1038/nbt.2623, nbt.2623 [pii].

Fu, Y., Sander, J. D., Reyon, D., Cascio, V. M., & Joung, J. K. (2014). Improving CRISPR-Cas nuclease specificity using truncated guide RNAs. *Nature Biotechnology*, *32*(3), 279–284. http://dx.doi.org/10.1038/nbt.2808, nbt.2808 [pii].

Green, M. R., & Sambrook, J. (2012). *Molecular cloning: A laboratory manual* (4th ed.). Cold Spring Harbor, NY: Cold Spring Harbor Laboratory Press.

Hsu, P. D., Scott, D. A., Weinstein, J. A., Ran, F. A., Konermann, S., Agarwala, V., et al. (2013). DNA targeting specificity of RNA-guided Cas9 nucleases. *Nature Biotechnology*, *31*(9), 827–832. http://dx.doi.org/10.1038/nbt.2647, nbt.2647 [pii].

Jiang, W., Bikard, D., Cox, D., Zhang, F., & Marraffini, L. A. (2013). RNA-guided editing of bacterial genomes using CRISPR-Cas systems. *Nature Biotechnology*, *31*(3), 233–239. http://dx.doi.org/10.1038/nbt.2508, nbt.2508 [pii].

Jinek, M., Chylinski, K., Fonfara, I., Hauer, M., Doudna, J. A., & Charpentier, E. (2012). A programmable dual-RNA-guided DNA endonuclease in adaptive bacterial immunity. *Science*, *337*(6096), 816–821. http://dx.doi.org/10.1126/science.1225829, science.1225829 [pii].

Jinek, M., East, A., Cheng, A., Lin, S., Ma, E., & Doudna, J. (2013). RNA-programmed genome editing in human cells. *eLife*, *2*, e00471. http://dx.doi.org/10.7554/eLife.00471 00471 [pii].

Koike-Yusa, H., Li, Y., Tan, E. P., Velasco-Herrera, M. D., & Yusa, K. (2013). Genome-wide recessive genetic screening in mammalian cells with a lentiviral CRISPR-guide RNA library. *Nature Biotechnology*, *32*, 267–273. http://dx.doi.org/10.1038/nbt.2800.

Kuscu, C., Arslan, S., Singh, R., Thorpe, J., & Adli, M. (2014). Genome-wide analysis reveals characteristics of off-target sites bound by the Cas9 endonuclease. *Nature Biotechnology*, *32*, 677–683. http://dx.doi.org/10.1038/nbt.2916, nbt.2916 [pii].

Kutner, R. H., Zhang, X.-Y., & Reiser, J. (2009). Production, concentration and titration of pseudotyped HIV-1-based lentiviral vectors. *Nature Protocols*, *4*(4), 495–505. http://dx.doi.org/10.1038/nprot.2009.22.

Lin, B. R., & Natarajan, V. (2012). Negative regulation of human U6 snRNA promoter by p38 kinase through Oct-1. *Gene*, *497*(2), 200–207. http://dx.doi.org/10.1016/j.gene.2012.01.041.

Mali, P., Aach, J., Stranges, P. B., Esvelt, K. M., Moosburner, M., Kosuri, S., et al. (2013). CAS9 transcriptional activators for target specificity screening and paired nickases for cooperative genome engineering. *Nature Biotechnology*, *31*(9), 833–838. http://dx.doi.org/10.1038/nbt.2675, nbt.2675 [pii].

Mali, P., Yang, L., Esvelt, K. M., Aach, J., Guell, M., DiCarlo, J. E., et al. (2013). RNA-guided human genome engineering via Cas9. *Science*, *339*(6121), 823–826. http://dx.doi.org/10.1126/science.1232033 science.1232033 [pii].

Malina, A., Mills, J. R., Cencic, R., Yan, Y., Fraser, J., Schippers, L. M., et al. (2013). Repurposing CRISPR/Cas9 for in situ functional assays. *Genes & Development*, *27*(23), 2602–2614. http://dx.doi.org/10.1101/gad.227132.113.

Nielsen, S., Yuzenkova, Y., & Zenkin, N. (2013). Mechanism of eukaryotic RNA polymerase III transcription termination. *Science*, *340*(6140), 1577–1580. http://dx.doi.org/10.1126/science.1237934.

Nishimasu, H., Ran, F. A., Hsu, P. D., Konermann, S., Shehata, S. I., Dohmae, N., et al. (2014). Crystal structure of cas9 in complex with guide RNA and target DNA. *Cell*, *156*(5), 935–949. http://dx.doi.org/10.1016/j.cell.2014.02.001.

Orioli, A., Pascali, C., Quartararo, J., Diebel, K. W., Praz, V., Romascano, D., et al. (2011). Widespread occurrence of non-canonical transcription termination by human RNA polymerase III. *Nucleic Acids Research*, *39*(13), 5499–5512. http://dx.doi.org/10.1093/nar/gkr074.

Pattanayak, V., Lin, S., Guilinger, J. P., Ma, E., Doudna, J. A., & Liu, D. R. (2013). High-throughput profiling of off-target DNA cleavage reveals RNA-programmed Cas9 nuclease specificity. *Nature Biotechnology*, *31*(9), 839–843. http://dx.doi.org/10.1038/nbt.2673, nbt.2673 [pii].

Reyon, D., Tsai, S. Q., Khayter, C., Foden, J. A., Sander, J. D., & Joung, J. K. (2012). FLASH assembly of TALENs for high-throughput genome editing. *Nature Biotechnology*, *30*(5), 460–465. http://dx.doi.org/10.1038/nbt.2170.

Semenova, E., Jore, M. M., Datsenko, K. A., Semenova, A., Westra, E. R., Wanner, B., et al. (2011). Interference by clustered regularly interspaced short palindromic repeat (CRISPR) RNA is governed by a seed sequence. *Proceedings of the National Academy of Sciences of the United States of America*, *108*(25), 10098–10103. http://dx.doi.org/10.1073/pnas.1104144108, 1104144108 [pii].

Shalem, O., Sanjana, N. E., Hartenian, E., Shi, X., Scott, D. A., Mikkelsen, T. S., et al. (2014). Genome-scale CRISPR-Cas9 knockout screening in human cells. *Science*, *343*(6166), 84–87. http://dx.doi.org/10.1126/science.1247005.

Swift, S., Lorens, J., Achacoso, P., & Nolan, G. P. (2001). Rapid production of retroviruses for efficient gene delivery to mammalian cells using 293T cell-based systems. *Current protocols in immunology*. New York, NY: John Wiley & Sons, Inc.

Szymczak-Workman, A. L., Vignali, K. M., & Vignali, D. A. (2012). Design and construction of 2A peptide-linked multicistronic vectors. *Cold Spring Harbor Protocols*, *2012*(2), 199–204. http://dx.doi.org/10.1101/pdb.ip067876.

Wang, T., Wei, J. J., Sabatini, D. M., & Lander, E. S. (2014). Genetic screens in human cells using the CRISPR-Cas9 system. *Science*, *343*(6166), 80–84. http://dx.doi.org/10.1126/science.1246981.

Wu, X., Scott, D. A., Kriz, A. J., Chiu, A. C., Hsu, P. D., Dadon, D. B., et al. (2014). Genome-wide binding of the CRISPR endonuclease Cas9 in mammalian cells. *Nature Biotechnology*, *32*, 670–676. http://dx.doi.org/10.1038/nbt.2889, nbt.2889 [pii].

Yin, H., Xue, W., Chen, S., Bogorad, R. L., Benedetti, E., Grompe, M., et al. (2014). Genome editing with Cas9 in adult mice corrects a disease mutation and phenotype. *Nature Biotechnology*, *32*, 551–553. http://dx.doi.org/10.1038/nbt.2884, nbt.2884 [pii].

Zhou, Y., Zhu, S., Cai, C., Yuan, P., Li, C., Huang, Y., et al. (2014). High-throughput screening of a CRISPR/Cas9 library for functional genomics in human cells. *Nature*, *509*(7501), 487–491. http://dx.doi.org/10.1038/nature13166, nature13166 [pii].

CHAPTER ELEVEN

# The iCRISPR Platform for Rapid Genome Editing in Human Pluripotent Stem Cells

**Zengrong Zhu[1], Federico González[1], Danwei Huangfu[2]**
Developmental Biology Program, Sloan-Kettering Institute, New York, USA
[1]These authors contributed equally to this work.
[2]Corresponding author: e-mail address: huangfud@mskcc.org

## Contents



*Methods in Enzymology*, Volume 546
ISSN 0076-6879
http://dx.doi.org/10.1016/B978-0-12-801185-0.00011-8

## Abstract

Human pluripotent stem cells (hPSCs) have the potential to generate all adult cell types, including rare or inaccessible human cell populations, thus providing a unique platform for disease studies. To realize this promise, it is essential to develop methods for efficient genetic manipulations in hPSCs. Established using TALEN (transcription activator-like effector nuclease) and CRISPR (clustered regularly interspaced short palindromic repeats)/Cas (CRISPR-associated) systems, the iCRISPR platform supports a variety of genome-engineering approaches with high efficiencies. Here, we first describe the establishment of the iCRISPR platform through TALEN-mediated targeting of inducible Cas9 expression cassettes into the *AAVS1* locus. Next, we provide a series of technical procedures for using iCRISPR to achieve one-step knockout of one or multiple gene(s), "scarless" introduction of precise nucleotide alterations, as well as inducible knockout during hPSC differentiation. We present an optimized workflow, as well as guidelines for the selection of CRISPR targeting sequences and the design of single-stranded DNA (ssDNA) homology-directed DNA repair templates for the introduction of specific nucleotide alterations. We have successfully used these protocols in four different hPSC lines, including human embryonic stem cells and induced pluripotent stem cells. Once the iCRISPR platform is established, clonal lines with desired genetic modifications can be established in as little as 1 month. The methods described here enable a wide range of genome-engineering applications in hPSCs, thus providing a valuable resource for the creation of diverse hPSC-based disease models with superior speed and ease.

## 1. INTRODUCTION

Functional analysis of sequence variants affecting diverse human traits, including disease susceptibility, is a key to understanding human biology and disease mechanisms. With their unlimited self-renewal capacity and the potential to generate all adult cell types, human pluripotent stem cells (hPSCs), including human embryonic stem cells (hESCs) and human induced pluripotent stem cells (hiPSCs), offer an ideal platform for biological and disease studies (Zhu & Huangfu, 2013). To meet this goal, it is mandatory to develop efficient methods for genome engineering in hPSCs.

The development of programmable site-specific nucleases has significantly facilitated targeted-genome editing in a wide range of organisms and cultured cell types (Joung & Sander, 2013; Ran et al., 2013; Urnov, Rebar, Holmes, Zhang, & Gregory, 2010). These customized nucleases induce DNA double-strand breaks (DSBs) at desired genomic loci, triggering the endogenous DNA repair machinery through two competing pathways: error-prone nonhomologous end-joining (NHEJ), leading to insertion/deletion mutations (Indels), or homology-directed repair (HDR), which can be co-opted to introduce precise nucleotide alterations using a homologous DNA template (Jasin, 1996; Rouet, Smih, & Jasin, 1994). Among various customized nuclease systems developed so far, the transcription activator-like effector (TALE) nuclease (TALEN) and the clustered, regularly interspaced, short palindromic repeat (CRISPR) technologies have emerged as powerful and versatile tools for genome editing in hPSCs.

The DNA target specificity of TALENs is guided by the TALE DNA-binding domain. Originally discovered in the plant pathogenic bacteria *Xanthomonas*, TALEs can bind to the promoter of various genes of the plant host, hijacking the transcriptional machinery to promote bacterial infection (Rossier, Wengelnik, Hahn, & Bonas, 1999; Szurek, Marois, Bonas, & Van den Ackerveken, 2001). The DNA-binding domain of TALE is composed of ~34 amino acids repeats (TALE repeats) arranged in tandem. Each repeat contains two variable adjacent amino acids called the "repeat variable diresidue," which determine the single base-recognition specificity. Thus, each TALE repeat independently specifies one target base (Boch et al., 2009; Moscou & Bogdanove, 2009). To introduce DSBs, TALENs are designed as pairs, recognizing the genomic sequences flanking the target site. Each TALEN consists of a programmable, sequence-specific TALE DNA-binding domain fused to the cleavage domain of the bacterial endonuclease *Fok*I. The binding of a TALEN pair to DNA allows *Fok*I dimerization and DNA cleavage (Cermak et al., 2011; Miller et al., 2011).

Recently, the CRISPR technology has been developed for genome engineering in mammalian systems (Cho, Kim, Kim, & Kim, 2013; Cong et al., 2013; Jinek et al., 2013; Mali, Yang, et al., 2013; Wang et al., 2013). The CRISPR/Cas system is derived from *Streptococcus pyogenes* where it functions as part of an immune system to provide acquired resistance against invading viruses (van der Oost, Westra, Jackson, & Wiedenheft, 2014). CRISPR/Cas-mediated genome engineering requires two components: the constant RNA-guided DNA endonuclease Cas9 protein required for DNA cleavage and a variable CRISPR RNA (crRNA) and *trans*-activating crRNA (tracrRNA) duplex that specifies DNA target recognition (Jinek et al., 2012). Most applications now replace the crRNA/tracrRNA duplex with a single chimeric guide RNA (sgRNA), which works more efficiently than the original duplex design (Hsu et al., 2013; Jinek et al., 2012).

sgRNA directs Cas9 to its target genomic locus by recognizing a 20 nucleotide (nt) sequence (protospacer) followed by an NGG motif (protospacer-associated motif or PAM, where N can be A, T, G, or C), and DNA cleavage occurs 3 bp upstream of the PAM sequence. In our experience with hPSCs, the CRISPR/Cas system tends to outperform TALENs, which has also been observed by others (Ding et al., 2013). Compared to TALENs, the CRISPR/Cas system is easier to engineer and simplifies multiplexing. However, there have also been concerns regarding its off-target effects (Cho et al., 2014; Fu et al., 2013; Hsu et al., 2013; Mali, Aach, et al., 2013; Pattanayak et al., 2013), which will be discussed further in Section 6.

A number of studies have now used CRISPR/Cas to establish modified hPSC lines with variable efficiencies. Several studies use HDR-mediated editing to target a selectable marker into the locus of interest, which allows enrichment of correctly targeted cells after selection (An et al., 2014; Hou et al., 2013; Ye et al., 2014). Although efficient, the construction of the targeting construct could be time consuming, and it is often desirable to remove the selectable marker to allow more precise modeling of the disease conditions. Alternatively, the CRISPR/Cas system also supports efficient NHEJ or HDR-mediated genome editing without the need for drug selection (Ding et al., 2013; Gonzalez et al., 2014; Horii, Tamura, Morita, Kimura, & Hatada, 2013; Wang et al., 2014).

To further improve the efficiency, and to also achieve multiplexable and inducible genome editing in hPSCs, we have developed a genome-engineering platform called iCRISPR (Gonzalez et al., 2014). Through TALEN-mediated gene targeting, hPSC lines are engineered for doxycycline-inducible expression of Cas9 (referred to as iCas9 hPSCs). Upon doxycycline treatment, these lines can then be transfected with (a) a single or multiple sgRNA(s) to generate biallelic knockout hPSC lines for individual or multiple genes; (b) a sgRNA together with a HDR template to generate knockin alleles; and (c) a sgRNA at specific stages of hPSC differentiation to achieve inducible gene knockout.

Below we describe an optimized protocol for the establishment of the iCRISPR platform through TALEN-mediated targeting of inducible Cas9 expression cassettes into the *AAVS1* locus of hPSCs (Fig. 11.1). We have successfully used this protocol on four different hPSC lines and obtained similar results: ~50% of the lines are correctly targeted with no additional random integrations. Next, we provide detailed protocols for using iCRISPR to achieve one-step knockout of one or multiple gene(s),

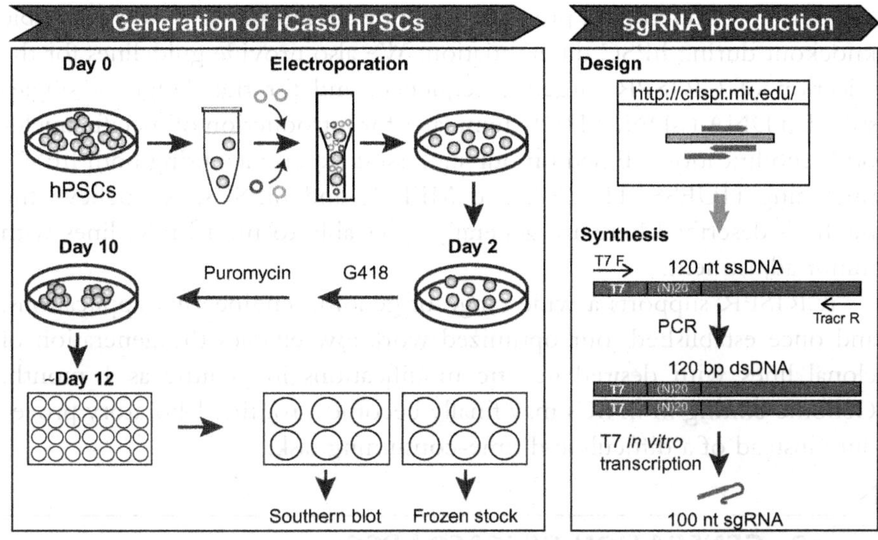


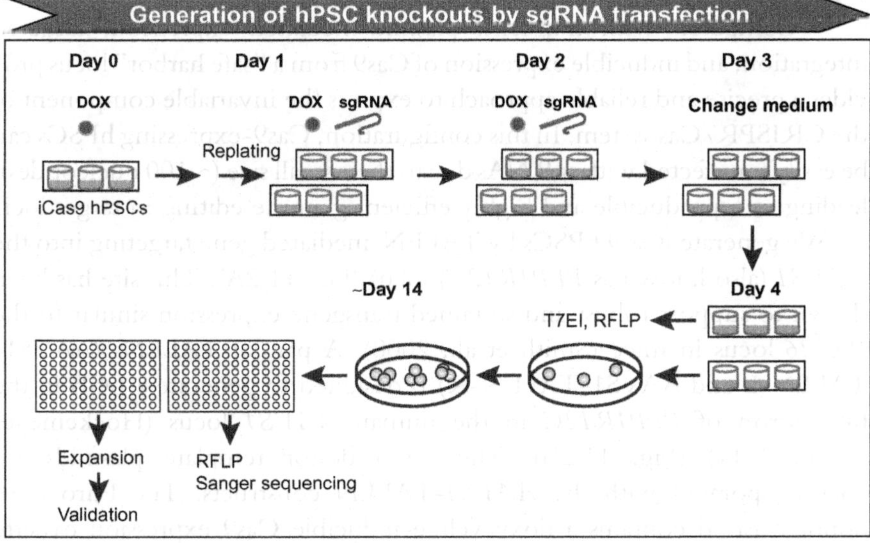


**Figure 11.1** The iCRISPR platform for rapid genome editing in hPSCs. Instead of transient expression of Cas9 and sgRNAs from electroporated plasmids in hPSCs, targeted integration and inducible expression of Cas9 into the *AAVS1* locus provides a precise and reliable approach to express the invariable component of the CRISPR/Cas system (Section 2). Then, sgRNAs targeting specific loci are designed and synthesized by PCR amplification of *in vitro* transcription DNA templates (Sections 3.1 and 3.2). Due to their small size (~100 nucleotides), transfection of iCas9 hPSCs with sgRNAs is highly efficient, leading to reproducible and highly efficient gene knockout in hPSCs (Section 3.3). (See the color plate.)

"scarless" introduction of precise nucleotide alterations, as well as inducible knockout during hPSC differentiation. We also provide guidelines for the selection of CRISPR targeting sequence, and for the design of single-stranded DNA (ssDNA) HDR templates for introduction of specific nucleotide modifications. Based on our successful experience using both hESCs (including HUES8, HUES9, and MEL-1) and hiPSCs, we believe the methods described here are generally applicable to most hPSC lines with minor adjustment.

iCRISPR supports a wide range of genome-engineering applications, and once established, our optimized workflow enables the generation of clonal lines with desired genetic modifications in as little as 1 month. Genome editing in hPSCs may finally become a routine laboratory procedure instead of a difficult and time-consuming task.

## 2. GENERATION OF iCAS9 hPSCs

Compared with transient, plasmid-mediated expression, targeted integration, and inducible expression of Cas9 from a "safe harbor" locus provides a precise and reliable approach to express the invariable component of the CRISPR/Cas system. In this configuration, Cas9-expressing hPSCs can be easily transfected with sgRNAs due to their small size (~100 nucleotides), leading to reproducible and highly efficient genome editing in target loci.

We generate iCas9 hPSCs by TALEN-mediated gene targeting into the *AAVS1* (also known as *PPP1R12C*) locus (Fig. 11.2A). This site has been shown to support robust and sustained transgene expression similar to the *Rosa26* locus in mice (Smith et al., 2008). A pair of TALENs (AAVS1-TALEN-L and AAVS1-TALEN-R) is designed to generate a DSB in the first intron of *PPP1R12C* in the human *AAVS1* locus (Hockemeyer et al., 2011) (Fig. 11.2B). Then two donor template plasmids are coelectroporated with the *AAVS1*-TALEN constructs. The Puro-Cas9 donor plasmid contains a doxycycline-inducible Cas9 expression cassette selectable with puromycin, and the Neo-M2rtTA donor carries a constitutive reverse tetracycline transactivator (M2rtTA) expression cassette selectable with G418 (Geneticin) (DeKelver et al., 2010) (Fig. 11.2C). HDR of the DSBs allows simultaneous introduction of both Puro-Cas9 and Neo-M2rtTA cassettes into both *AAVS1* alleles *in trans*. Cas9 expression is then induced trough doxycycline treatment in established clonal iCas9 lines (Fig. 11.2A).

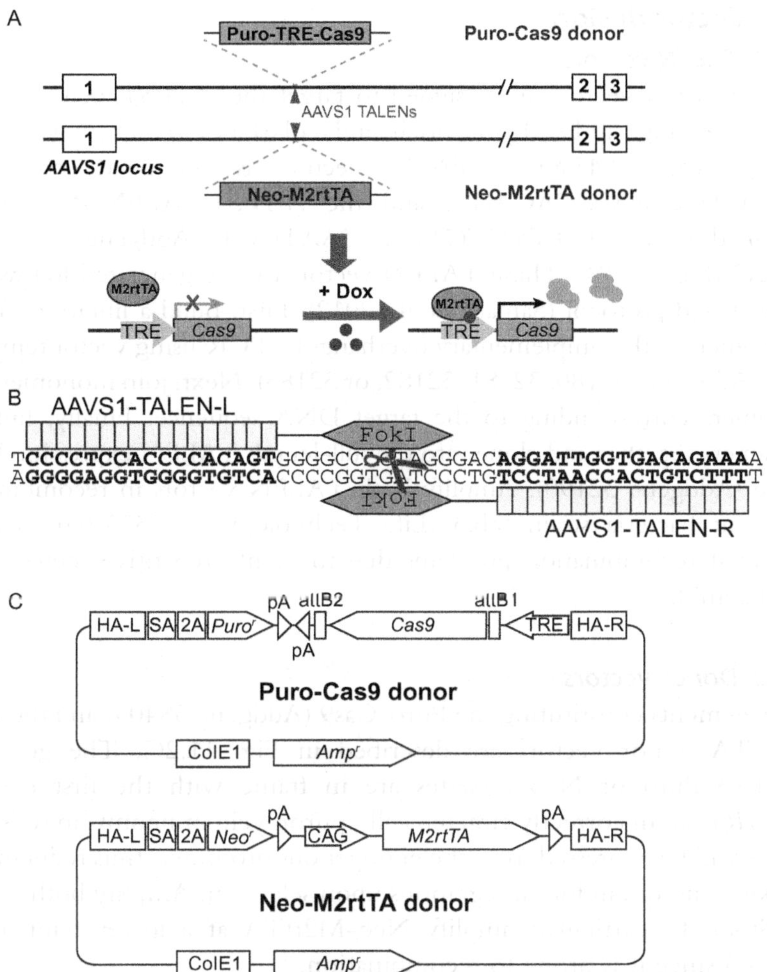


**Figure 11.2** Generation of iCas9 hPSC through TALEN-mediated gene targeting into the *AAVS1* locus. (A) TALEN-mediated gene targeting of Puro-Cas9 and Neo-M2rtTA into the *AAVS1* locus. Homology-directed repair (HDR) of the DSBs introduced by a pair of TALENs allows simultaneous introduction of both Puro-Cas9 and Neo-M2rtTA cassettes into both *AAVS1* alleles *in trans*. Cas9 expression is then induced through doxycycline treatment in established clonal iCas9 lines. (B) The AAVS1-TALEN-L vector is designed to target CCCCTCCACCCCACAGT, and the AAVS1-TALEN-R vector is designed to target TTTCTGTCACCAATCCT. Dimerization of the *Fok*I domain induces DSB in the spacer between the TALEN binding sites. (C) Vector maps of Puro-Cas9 and Neo-M2rtTA vectors. HA-L and HA-R, left and right homology arm; SA, splicing acceptor; 2A, self-cleaving 2A peptide; pA, polyadenylation signal sequence; attB1 and attB2, GATEWAY attB sequence; TRE, tetracycline response element; Amp$^r$, Ampicillin resistance gene; ColE1, replication origin.

## 2.1. Vector design

### 2.1.1 TALEN vectors

Two TALEN vectors are designed to target the *AAVS1* locus based on gene targeting studies by Jaenisch and colleagues (Hockemeyer et al., 2011): the *AAVS1*-TALEN-L vector is designed to target CCCCTCCACCCCACAGT, and the *AAVS1*-TALEN-R vector is designed to target TTTCTGTCACCAATCCT (Addgene 59025 and 59026) (Fig. 11.2B). These TALEN vectors can be generated following a PCR-based protocol (Sanjana et al., 2012). First, build a library of TALE monomers with complementary overhangs by PCR using vector templates from Addgene (32180, 32181, 32182, or 32183). Next, join monomers into hexamers corresponding to the target DNA sequence. Finally, link the hexamers together and clone into the full-length TALEN expression backbone (Addgene 32190). Amplify both TALEN vectors in recombination deficient bacteria strain Stbl3 (Life Technologies, C7373-03) to avoid potential recombination problems due to highly repetitive sequences in the plasmid.

### 2.1.2 Donor vectors

The elements constituting the Puro-Cas9 (Addgene 58409) and the Neo-M2rtTA donor vectors are described in Fig. 11.2C. The gene-trap SA-P2A-Puro or Neo cassettes are in frame with the first exon of *PPP1R12C*. In correctly targeted cells, puromycin or neomycin resistance genes will be expressed from the endogenous promoter, thus reducing the background of random integrations upon selection. Amplify both vectors in Stbl3. In particular, amplify Neo-M2rtTA at a lower temperature (30 °C) since it is prone to recombination.

## 2.2. hPSC electroporation

The electroporation protocol is optimized for hPSCs cultured on feeders. We routinely culture hPSCs on irradiated mouse embryonic fibroblasts (iMEFs, ~0.3 M cells per well of a 6-well plate) in DMEM/F12 medium (Life Technologies, 11320-082) supplemented with 20% KnockOut Serum Replacement (Life Technologies, 10828-028), 1 × nonessential amino acids (Life Technologies, 11140076), 1 × GlutaMAX (Life Technologies, 35050079), 100 U/ml Penicillin/100 μg/ml Streptomycin (Life Technologies, 15070063), 0.055 m*M* 2-mercaptoethanol (Life Technologies,

21985023), and 10 ng/ml recombinant human basic FGF (Life Technologies, PHG0263). Cultures are generally passaged at 1:6 to 1:12 split ratios every 4–6 days using TrypLE Select enzyme (Life Technologies, 12563-011). Add 5 μ*M* of Rho-associated protein kinase (ROCK) inhibitor Y-27632 (R&D, 1254) into the culture medium when passaging or thawing cells. It may be necessary to adjust the following protocol for hPSCs cultured under different conditions such as the feeder-free TeSR and Essential 8 culture conditions (Chen, Gulbranson, et al., 2011; Ludwig et al., 2006).

1. *Day - 1*: On the day before electroporation, change to hPSC medium with 5 μ*M* ROCK inhibitor. Typically, there are ~$1–2 \times 10^7$ hPSCs in one 10-cm dish when it is confluent on Day 0, which is sufficient for one targeting experiment (need ~$1 \times 10^7$ hPSCs).

   Seed irradiated DR4-MEFs (ATCC, SCRC-1045) (DR4-iMEFs) on gelatinized 10-cm culture dishes ($2 \times 10^6$ cells/10-cm dish). The number of dishes needed depends on the survival rate after electroporation, which is line dependent. We generally prepare three 10-cm dishes. DR4-MEFs are resistant to neomycin, hygromycin, puromycin, and 6-thioguanine and thus can be used as feeders to support the growth of hPSCs with multiple drug selections.
2. *Day 0*: On the day of electroporation, dissociate hPSCs using TrypLE Select enzyme. Stop the reaction by adding double amount of hPSC medium. Pipet gently to break cell aggregates into single-cell suspension and then pass cells through a 40 μm cell strainer to filter out cell clumps.
3. Pellet cells at $200 \times g \times 5$ min at room temperature (RT).
4. Gently resuspend cells in cold (4 °C) PBS and adjust cell density to $12.5 \times 10^6$ cells/ml.
5. Add the following plasmid mix into 800 μl cell suspension and mix well but gently.

| Plasmid | Amount (μg) |
|---|---|
| AAVS1-TALEN-L | 5 |
| AAVS1-TALEN-R | 5 |
| Puro-Cas9 | 40 |
| Neo-M2rtTA | 40 |
| hPSC ($12.5 \times 10^6$ cells/ml) | 800 μl |

We recommend including a negative control (without the Puro-Cas9 and Neo-M2rtTA donor vectors) when performing this experiment for the first time.

6. Transfer the DNA/cell mixture into a 0.4-cm electroporation cuvette and keep on ice for 5 min.
7. Electroporate cells using Gene Pulser Xcell Electroporation System (Bio-Rad, 165-2660) at 250 V and 500 μF (Costa et al., 2007). The time constant observed after electroporation is typically ~9–13.
8. After electroporation, transfer cells into a 15-ml conical tube with 5 ml prewarmed hPSC medium. Handle cells gently when transfer, resuspend, and plate cells after electroporation.
9. Pellet cells at $200 \times g \times 5$ min RT. Make sure to remove all floating dead cells and debris, which may impair the viability of the surviving hPSCs.
10. Resuspend cells in 10 ml hPSC medium with ROCK inhibitor gently and plate 5, 2.5, and $1 \times 10^6$ cells onto each of the three 10-cm dishes preseeded with DR4-iMEFs. This ensures that at least one of the plates would have sufficient colonies at clonal density for colony picking.
11. *Day 1*: Change hPSC medium. It is normal to observe significant cell death.

## 2.3. Selection and expansion of clonal lines

1. *Days 2–5*: Perform G418 (Life Technologies, 10131-035) selection. Change media daily with freshly made G418 (50 μg/ml) selection media for 4 days. We expect to observe ~200 colonies emerging from $2.5 \times 10^6$ hPSCs plated on Day 0, whereas no colonies should be present in the negative control dish. The optimal concentration and duration of G418 treatment may need to be adjusted depending on the G418 batch and the hPSC line.
2. *Day 6*: Four days after G418 selection, change to hPSC medium without G418. Supplement with additional DR4-iMEFs ($1 \times 10^6$ cells/10-cm dish), as iMEFs typically support hPSC growth for up to 7 days. Rock the plates gently to spread the feeder cells.
3. *Days 7–9*: Perform puromycin (Sigma, P8833) selection and change media daily with freshly made puromycin (0.5 μg/ml) selection media for 3 days. One may need to adjust the concentration and duration of puromycin treatment depending on the puromycin batch and the hPSC line.
4. *Day 10*: Change to regular hPSC medium without puromycin from Day 10 onward. hPSC colonies can be picked around Day 11–14 when they

reach ~2 mm in diameter. We typically observe ~50 colonies in the 10-cm dish with $2.5 \times 10^6$ hPSCs plated on Day 0.

5. Because *AAVS1* targeting using this approach is highly efficient, we typically pick only 12–24 colonies from each targeting experiment. Using a stereomicroscope and a 23G needle (a 200 μl pipette tip is also fine), mechanically disaggregate hPSC colonies into small pieces (~10 pieces per colony) and transfer cells directly into a 24-well plate preseeded with iMEFs. Pipet gently in the 24-well plate to further break hPSC colonies. Generally, it takes ~30 min or less to pick 24 colonies.
6. Change medium daily until cells become confluent. Passage cells in each well of 24-well plates into two wells of a 6-well plate.
7. When cells become confluent in 6-well plates, use one well for a frozen stock and the other well for genomic DNA extraction.

   *Freezing medium*: 10% DMSO, 40% FBS, and 50% hPSC medium.

## 2.4. Genotyping by Southern blot

From our experience with four hPSC lines, the efficiencies of biallelic integration of the Puro-Cas9 and Neo-M2rtTA transgenes into the *AAVS1* locus are usually close to 100%. It is therefore unnecessary to perform prescreening using PCR. We recommend amplifying the clones directly to obtain enough DNA for Southern blot, which would not only identify clones with correct biallelic transgene integrations in the *AAVS1* locus but also discriminate clones carrying random integrations (representing approximately 50% of the clones).

1. *Digoxigenin (DIG)-labeled probe synthesis*: We routinely perform Southern blot using nonradioactive probes. Two DIG-labeled probes are used for *AAVS1* Southern blot genotyping. A 3′-external probe and a 5′-internal probe (Fig. 11.3, EXT and INT). They are synthesized using the PCR DIG Probe Synthesis Kit (Roche, 11636090910).

   For 3′-external probe, PCR amplify the probe template from human genomic DNA with high-fidelity Herculase II Fusion DNA Polymerase (Agilent, 600679) using 3′F and 3′R primers. Clone the PCR product using Zero Blunt TOPO PCR Cloning Kit (Life Technologies, 450245) and use a sequence verified clone as template for amplifying the DIG-labeled probe.

   For 5′ external probe, directly amplify the DIG-labeled probe using Puro-Cas9 donor as the template using 5′F and 5′R primers.

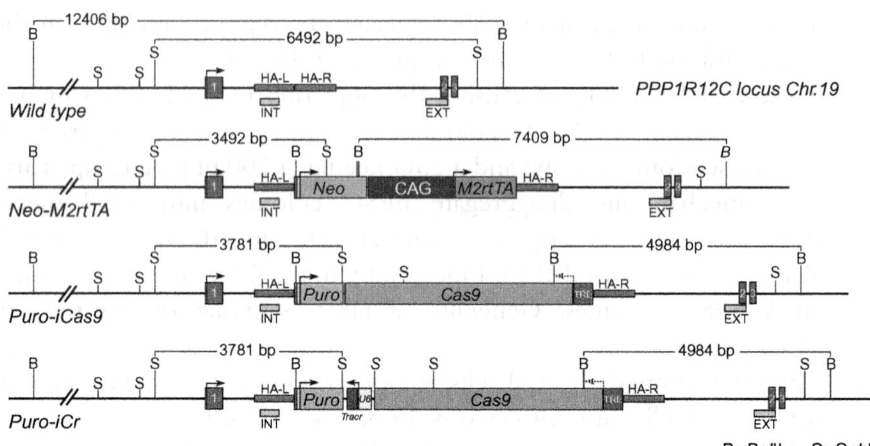


**Figure 11.3** Southern blot genotyping of *AAVS1*-targeted hPSC iCas9 lines. DNA is digested with either *Bgl*II (B), for 3′ external probe hybridization, or *Sph*I (S), for 5′ internal probe hybridization. *Bgl*II digestion and 3′ external probe hybridization yield bands of 12,406, 7409, and 4984 bp for wild type, Neo-M2rtTA, Puro-iCas9 (or Puro-iCr); whereas *Sph*I digestion and 5′ internal probe hybridization yields bands of 6492, 3492, and 3781 bp, respectively. HA-L, Left homology arm; HA-R, Right homology arm; INT, 5′ internal probe; EXT, 3′ external probe.

| Primer | Sequence |
|---|---|
| 3′F | ACAGGTACCATGTGGGGTTC |
| 3′R | CTTGCCTCACCTGGCGATAT |
| 5′F | AGGTTCCGTCTTCCTCCACT |
| 5′R | GTCCAGGCAAAGAAAGCAAG |

PCR DIG-labeled probe synthesis mix

| | Full-labeled | Half-labeled | Unlabeled |
|---|---|---|---|
| $ddH_2O$ | 33.25 | 33.25 | 33.25 |
| 10 × PCR buffer | 5 | 5 | 5 |
| PCR DIG synthesis mix | 5 | 2.5 | 0 |
| dNTP stock | 0 | 2.5 | 5 |
| Primer mix (10 μ*M*) | 5 | 5 | 5 |
| Enzyme mix | 0.75 | 0.75 | 0.75 |
| Plasmid DNA (50 pg/μl) | 1 | 1 | 1 |
| Total (μl) | 50 | 50 | 50 |

PCR DIG-labeled probe cycling condition

| Cycle number | Denature | Anneal | Extend |
|---|---|---|---|
| 1 | 95 °C, 2 min | | |
| 2–31 | 95 °C, 30 s | 60 °C, 30 s | 72 °C, 1 min |
| 32 | | | 72 °C, 7 min |

Examine the PCR product on 2% agarose gel. *Note*: DIG-labeled DNA probe migrates slower than nonlabeled DNA.

2. Extract genomic DNA from expanded iCas9 lines as well as non-targeted control wild-type hPSCs using DNeasy Blood & Tissue Kit (Qiagen, 69504). Typically, confluent hPSCs in one well of a 6-well plate yield 25–50 μg genomic DNA.
3. Digest 10 μg genomic DNA in a 20 μl reaction at 37 °C over night using 20 U of *Bgl*II (for 3′ external probe hybridization) or 20 U of *Sph*I (for 5′ internal probe hybridization) (Fig. 11.3).
4. Run digestion samples on a 1% TAE agarose gel (including a lane with a DNA ladder). Migrate at 80 V for 8 h. Image the gel in parallel with a ruler for scaling the image and determining band size upon film development.
5. Incubate the gel in denaturation buffer (1.5 *M* NaCl and 0.5 *M* NaOH) for 30 min at RT to denature DNA. If your loading buffer contains bromophenol blue, it should turn green during this step.
6. Remove denaturation buffer, wash for 1 min with $ddH_2O$ before incubating the gel in neutralization buffer (1.5 *M* NaCl, 0.5 *M* Tris, and 0.001 *M* EDTA; pH 6.9) for 30 min at RT, and incubate for another 15 min with fresh neutralization buffer.
7. *Transfer*
   (1) Preincubate the Hybond-N membrane (GE healthcare, RPN203N) in 10× SSC buffer for 5 min.
   (2) Mount the gel as described before for capillarity transfer (Maniatis, Fritsch, & Sambrook, 1982). Make sure there is no bubble between the gel and membrane.
   (3) Transfer overnight with 10× SSC buffer.
8. After transfer, wash the membrane in NaPi buffer (1 *M* $Na_2HPO_4$, pH 7.2) for 5 min.
9. Fix DNA on the membrane by baking at 80 °C for 2 h.
10. Place the membrane in a hybridization tube (Fisher, K736500-3515). Add 20 ml NaPi buffer. Incubate for 5 min in a hybridization oven at RT.

**11.** Prehybridize the membrane with 20 ml hybridization buffer for 1 h in a hybridization oven at 65 °C.

Hybridization buffer (500 ml)

| Component | Amount (ml) |
|---|---|
| NaPi 1 *M*, pH 7.2 | 250 |
| SDS 20% | 175 |
| EDTA 0.5 *M*, pH 8 | 1 |
| ddH$_2$O | 74 |

**12.** *Probe preparation*: Denature 5 μl of the probe from Step 1 in 100 μl hybridization buffer in boiling water for 10 min. Then dilute the 100 μl probe in 20 ml hybridization buffer (prewarmed at 65 °C).

*Note*: After use, the probe can be collected and stored at −20 °C and reused for at least five additional times. For subsequent uses, thaw the diluted probe and denature in boiling water for 10 min.

**13.** Remove prehybridization buffer and add 20 ml of the diluted probe to the hybridization tube. Hybridize the membrane in a hybridization oven at 65 °C overnight.

**14.** Wash the membrane twice with washing buffer at 65 °C for 5 min.

Washing buffer (2000 ml)

| Component | Amount (ml) |
|---|---|
| NaPi 1 *M*, pH 7.2 | 80 |
| SDS 20% | 100 |
| ddH$_2$O | 1920 |

**15.** Wash the membrane in DIG buffer I (0.1 M Maleic Acid, 0.15 M NaCl; PH 7.5) with 0.3% Tween at RT under gentle agitation for 5 min.

**16.** Block the membrane in 20 ml DIG buffer I with 1% blocking reagent (Roche, 11096176001) at RT for 30 min in a sealed plastic bag.

**17.** Centrifuge the anti-DIG-AP antibody (Roche, 11093274910) at 4 °C for 5 min before use. Dilute 1 μl of the antibody in 20 ml DIG buffer I with 1% blocking reagent.

**18.** Incubate the membrane with 20 ml of the diluted antibody at RT for 30 min in a sealed plastic bag.

19. Wash the membrane twice in DIG buffer I with 0.3% Tween in a plastic container for 30 min.
20. Wash the membrane with DIG buffer III (0.1 *M* NaCl, 0.1 *M* Tris, pH 9.5) for 5 min.
21. *Chemiluminescent detection*: Incubate the membrane with 12 μl CDP-star (Roche, 11759051001) diluted in 2 ml DIG buffer III for 5 min.
22. Remove excess liquid without drying. Transfer the membrane to a clean sealed bag. Expose with Amersham Hyperfilm (Fisher, 45-001-507) in Amersham Hypercassette (Fisher, 45-000-758). Exposure time varies with probe concentration and quality but is typically 30 min.

## 2.5. Validation

Correctly targeted iCas9 lines are further validated through qRT-PCR for inducible expression of Cas9, immunohistochemistry to assess proper expression of the pluripotency markers (e.g., OCT4, SOX2, and NANOG) and teratoma assay for functional assessment of pluripotency. We also recommend verifying that the newly established iCas9 cells have normal karyotypes.

### *2.5.1 RT-PCR analysis*

1. Treat iCas9 cells with or without doxycycline (2 μg/ml) (Fisher, BP26535) for 2 days.
2. Isolate total RNA using RNeasy Plus Mini Kit (Qiagen, 74134).
3. Synthesize cDNA using High Capacity cDNA Reverse Transcription Kit (Life Technologies, 4368814).
4. Perform quantitative PCR using SYBR Green low ROX mix (Fisher, AB4322B) and 7500 Real-Time PCR System (Life Technologies, 4351104). The following primers are used. GAPDH is used as the internal control. Doxycycline treatment typically induces ~1000-fold increase in Cas9 mRNA level.

| Primer | Sequence |
|---|---|
| Cas9-F | CCGAAGAGGTCGTGAAGAAG |
| Cas9-R | GCCTTATCCAGTTCGCTCAG |
| GAPDH-F | GGAGCCAAACGGGTCATCATCTC |
| GAPDH-R | GAGGGGCCATCCACAGTCTTCT |

### 2.5.2 Immunohistochemical analysis of pluripotency marker expression

1. Rinse cells with PBS once and fix cells with 4% PFA (Fisher, 50980495) in PBS for 10 min directly in the culture plate.
2. Permeabilize and wash cells three times with PBST (0.1% Triton X-100 in PBS) for 5 min.
3. Block cells with blocking buffer (5% serum in PBST) for 5 min.
   *Note*: Serum should be from species different from primary antibody.
4. Incubate with primary antibody diluted in blocking solution at RT for 1 h (or at 4 °C overnight).

| Antibody | Company | Dilution ratio |
|---|---|---|
| Goat anti-OCT4 | Santa Cruz, sc-8628 | 1:100 |
| Rabbit anti-NANOG | Cosmobio Japan, REC-RCAB0004P-F | 1:100 |
| Goat anti-SOX2 | Santa Cruz, sc-17320 | 1:100 |

5. Remove primary antibody and wash three times with PBST for 5 min.
6. Incubate with fluorescence conjugated secondary antibody diluted in blocking solution at RT for 1 h (also add DAPI if nuclear staining is desired).
7. Remove secondary antibody and wash cells three times with PBST for 5 min.

### 2.5.3 Teratoma assay

1. Expand iCas9 lines to 80–100% confluency on 10-cm dishes.
2. Add 7 ml collagenase type IV (Life Technologies, 17104-019) working solution (1 mg/ml in DMEM) to each 10-cm dish and incubate at 37 °C for 10 min.
3. Aspirate collagenase solution.
4. Add 5 ml hPSC medium.
5. Scrape cells with cell scraper (Fisher, 087711A) and collect cell suspension into 15-ml conical tube.
6. Pellet cells at 200 × *g* for 5 min.
7. Resuspend cell pellet from each 10-cm dish in 400 μl PBS.
8. Inject subcutaneously, 100 μl cell suspension into the right hind leg of an anesthetized immunocompromised mice. We routinely use severe combined immunodeficient mice for this purpose.

# 3. GENERATION OF KNOCKOUT hPSCs USING iCRISPR

## 3.1. sgRNA design

NHEJ-mediated repair of DNA DSBs leads to random Indel mutations, which is useful for generating loss-of-function mutations or knockouts. Thus, the design of the gene targeting strategy aims at generating premature stop codons through the creation of frameshift Indel mutations, and the target sequence need to be strategically chosen to maximize the possibility of disrupting the function of the corresponding protein. It is important to identify all possible splice variants of the gene of interest, and design sgRNAs to target a functional region that is at least present in the isoform relevant for the study. For well-annotated genes, we recommend targeting sequences upstream of an essential functional domain of the corresponding protein. Alternatively, one may target a region adjacent to and downstream of the start codon of the major isoform. The second strategy may sometimes create a hypomorphic allele instead of a null allele if an alternative start site is used to produce a partially active protein.

Proceed to designing specific sgRNAs to target the genomic region of interest after a targeting strategy is decided. We routinely use the CRISPR design tool developed by the Feng Zhang group at MIT (http://crispr.mit.edu/). The software not only identifies all possible CRISPR targets in an input DNA sequence but also uncovers potential off-target sites, thus predicting sgRNAs with the highest targeting specificity. For each gene of interest, we routinely design three sgRNAs and in most cases all of them efficiently induce Indel mutations (>20% by T7EI assay) in target loci. Nevertheless, it is still a good idea to at least verify efficient mutagenesis before investing the time picking colonies.

## 3.2. sgRNA production

### *3.2.1 PCR amplification of* in vitro *transcription (IVT) DNA templates*

Different from most commonly used approaches that start with the construction of a sgRNA-expressing plasmid, we have found that it is faster and more cost-effective to use an oligo template. This method also allows easy scaling up for higher throughput knockout studies. Design a 120 nucleotide oligo including the T7 promoter sequence (blue and underlined) and the variable 20 nt crRNA recognition sequence $(N)_{20}$ followed by a constant chimeric sgRNA sequence (Fig. 11.4A). PCR amplify the oligo using the T7 F and Tracr R universal primers.

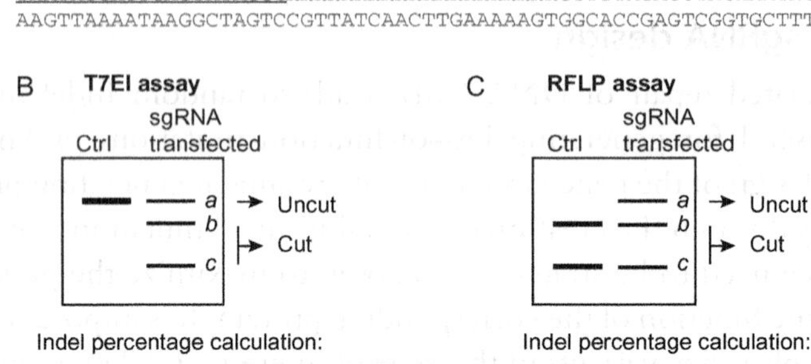


Indel percentage calculation:
$100 \times (1 - (1 - (b + c)/(a + b + c))^{1/2})$

Indel percentage calculation:
$100 \times a/(a+b+c)$

**Figure 11.4** sgRNA production, T7EI and RFLP assay. (A) The 120 nt ssDNA template for sgRNA production. T7, T7 promoter sequence; (N)20, 20 nucleotide sgRNA target sequence; (B) T7EI assay. Ctrl, none-transfection control; (C) RFLP assay.

| Primer | Sequence |
|---|---|
| T7 F | TAATACGACTCACTATAGGG |
| Tracr R | AAAAGCACCGACTCGGTGCC |

PCR reaction mix (50 μl)

| Component | Amount (μl) |
|---|---|
| $ddH_2O$ | 35.5 |
| 5 × Herculase II reaction buffer | 10 |
| dNTP mix (25 m*M*) | 0.5 |
| T7 F (10 μ*M*) | 1.25 |
| Tracr R (10 μ*M*) | 1.25 |
| T7-sgRNA IVT template (25 n*M*) | 1 |
| Herculase II Fusion DNA Polymerase | 0.5 |

PCR cycling conditions

| Cycle number | Denature | Anneal | Extend |
|---|---|---|---|
| 1 | 94 °C, 2 min | | |
| 2–31 | 94 °C, 20 s | 60 °C, 20 s | 72 °C, 1 min |
| 32 | | | 72 °C, 2 min |

### 3.2.2 *In vitro transcription and purification of sgRNAs*

Use the MEGAshortscript T7 Transcription kit (Life Technologies, AM1354) to generate sgRNAs.

*In vitro* transcription mix (20 μl)

| Component | Amount (μl) |
|---|---|
| T7 ATP | 2 |
| T7 CTP | 2 |
| T7 GTP | 2 |
| T7 UTP | 2 |
| T7 10 × buffer | 2 |
| T7 enzyme mix | 2 |
| PCR-amplified T7-sgRNA IVT template | 8 |

Incubate for 4 h to overnight at 37 °C.

Add 1 μl TURBO DNase and incubate for 15 min at 37 °C.

Proceed to RNA purification using the MEGAclear Transcription Clean-Up Kit (Life Technologies, AM1908) following manufacturer's instructions and elute sgRNAs (typically ~50–100 μg) in 100 μl RNase-free water. When possible adjust concentration to 320 ng/μl (10 μ*M*) and store at −80 °C until use.

## 3.3. Single or multiplex sgRNA transfection in hPSCs

We routinely transfect hPSCs cultured in 24-well dishes in duplicating wells: one well is used for T7EI and/or restriction fragment length polymorphism (RFLP) assays to confirm efficient mutagenesis, and the other well is used for replating and colony picking for establishment of clonal knockout lines. We also recommend using three different cell densities for transfection to achieve an optimal balance between Indel efficiency and cell survival.

*Day 0*: Plate iMEFs on a gelatin-coated 24-well plate and treat 60% confluent iCas9 hPSC with hPSC medium containing 2 μg/ml doxycycline.

*Day 1*: Dissociate hPSCs using TrypLE and resuspend at 0.2, 0.4, and $1 \times 10^6$ cells/ml in hPSC media with 5 μ*M* ROCK inhibitor and 2 μg/ml doxycycline. Replate on a 24-well plate with 0.5 ml cells per well. For each dilution, plate an additional well to serve as nontransfected control for T7EI and RFLP assays.

*1st transfection*: For each 0.5 ml cell suspension, prepare separately:
Mix 1: 25 μl Opti-MEM + 0.5 μl sgRNA (160 ng).
Mix 2: 25 μl Opti-MEM + 1.5 μl Lipofectamine RNAiMAX (Life Technologies, 13778-150).
Mix 1 + 2, incubate for 5 min at RT and add dropwise into dissociated hPSCs.
The final concentration of the sgRNA is 10 n*M*.

*Day 2*: Change hPSC medium with 2 μg/ml doxycycline.

*2nd transfection (optional)*: Same conditions as 1st transfection.

*Days 3–4*: Change media: 0.5 ml/well of hPSC medium.

*For multiplexed sgRNA transfection, use the same strategy but mix and transfect equal amounts of each sgRNA keeping the sum equal to 160 ng/well.*

## 3.4. Assessment of Indel frequency

Two or three days after the last sgRNA transfection, extract genomic DNA from transfected and nontransfected control cells using DNeasy Blood & Tissue Kit. Adjust final concentration to 50 ng/μl.

### *3.4.1 PCR amplification of the CRISPR target region*

PCR amplify a stretch of genomic region (~400 to 600 bp) flanking the CRISPR target site using the high-fidelity Herculase II Fusion DNA Polymerase to minimize potential PCR-induced mutations.

PCR reaction mix (20 μl)

| Component | Amount (μl) |
|---|---|
| dd$H_2O$ | 13.6 |
| 5 × Herculase II reaction buffer | 4 |
| dNTP mix (25 m*M*) | 0.2 |
| Primer-F (10 μ*M*) | 0.5 |
| Primer-R (10 μ*M*) | 0.5 |
| Genomic DNA (50 ng/μl) | 1 |
| Herculase II Fusion DNA Polymerase | 0.2 |

PCR cycling conditions

| Cycle number | Denature | Anneal | Extend |
|---|---|---|---|
| 1 | 94 °C, 2 min | | |
| 2–36 | 94 °C, 20 s | N* °C, 20 s | 72 °C, 30 s |
| 37 | | | 72 °C, 2 min |

*Annealing temperature of the T7EI primer pair, ideally between 55 and 65 °C.

### 3.4.2 Quantification of Indels through T7EI assay

For this step, we use the T7 Endonuclease I (New England Biolabs, M0302S)

#### 3.4.2.1 Hybridization

Mix (16 μl)

| Component | Amount (μl) |
|---|---|
| PCR product (do not purify) | 8 |
| NEB Buffer 2 10 × | 1.6 |
| $ddH_2O$ | 6.4 |

Conditions (use thermocycler).
95 °C, 5 min, 95–85 °C at −2 °C/s, 85–25 °C at −0.1 °C/s.
Hold at 4 °C.

#### 3.4.2.2 Digestion

Mix (20 μl)

| Component | Amount (μl) |
|---|---|
| Hybridized PCR product | 16 |
| NEB Buffer 2 10 × | 0.4 |
| T7EI (10 U/μl) | 0.2 |
| $ddH_2O$ | 3.4 |

Incubate at 37 °C for 30 min, immediately proceed to Section 3.4.2.3 or store at −20 °C to inhibit the reaction.

#### 3.4.2.3 Quantification

Resolve samples by electrophoresis through a 2.5% ethidium bromide stained agarose gel. Use Xylene Orange loading dye to avoid overlap with the expected bands. Run the gel at 5 V/cm enough time to separate the digested bands. The background decreases with migration time and bands resolve better.

UV illuminate the gel and image (avoid saturation) with a Gel Doc gel imaging system (Bio-Rad). Indel frequency is calculated based on relative band intensities quantified using ImageJ software (National Institutes of Health, Bethesda, MD) (Fig. 11.3B).

Indel percentage is determined by the formula: $100\times(1-(1-(b+c)/(a+b+c))^{1/2})$, where $a$ is the integrated intensity of the undigested PCR product, and $b$ and $c$ are the integrated intensities of each cleavage product (Hsu et al., 2013) (Fig. 11.3B).

### *3.4.3 Quantification of Indels through RFLP assay*

One may also design CRISPRs so that the Cas9 cleavage site (3 bp upstream of the PAM sequence) is in close proximity (~5 bp) to a restriction site, thus a RFLP assay can be performed to quantify Indel frequency. The assay requires the same PCR amplification of the CRISPR cutting region (see Section 3.4.1). Since this assay directly measures the loss of a restriction site, it is more sensitive at detecting Indels compared to the T7EI assay as long as the restriction site is located within ~5 bp of the Cas9 cleavage site.

#### 3.4.3.1 Digestion

Mix (20 μl)

| Component | Amount (μl) |
|---|---|
| PCR product | 8 |
| Buffer 10× | 2 |
| Appropriate restriction enzyme (10 U/μl) | 0.5 |
| $ddH_2O$ | 9.5 |

Incubate for at least 1 h, and proceed to Section 3.4.3.2 or store at −20 °C.

#### 3.4.3.2 Quantification

Follow the same approach described in Section 3.4.2.3.

Indel percentage is determined by the formula: $100\times a/(a+b+c)$ (Fig. 11.3C).

## 3.5. Clonal expansion of knockout lines

After performing T7EI and/or RFLP assay using cells from one of the duplicating wells, identify the best transfection conditions (with high Indel efficiency and good cell survival) and use the corresponding well for clonal expansion of mutant lines. We generally proceed to clonal expansion when Indel frequencies >20% are observed, which is usually achieved with at least one out of three sgRNAs tested.

### *3.5.1 Replating and colony picking*

Two to three days after the last sgRNA transfection, hPSCs are dissociated into single cells and replated at ~2000 cells per 10-cm dish pre-seeded with iMEFs. Cells are allowed to grow until colonies reach ~2 mm in diameter (~10 days).

Single colonies are then picked and plated in an uncoated 96-well plate containing 100 μl of hPSC medium with ROCK inhibitor. Each clone is mechanically disaggregated by gently pipetting up and down five times and replated in duplicated 96-well plates (50 μl each) preseeded with iMEFs and containing 100 μl of hPSC medium with ROCK inhibitor. Alternatively, colonies could first be cultured in one 96-well plate and then split into two 96-well plates when becoming confluent (in ~5 days).

Depending on the frequencies of Indel observed by T7EI and RFLP assays, pick between 48 (>20% targeting efficiency for most genes, applicable to making single-gene knockouts) and 288 clones (1–10% targeting efficiency, applicable to making double- or triple-gene knockouts). Media change and passaging using multichannel aspirator and dispenser pipettes greatly facilitates this step.

### *3.5.2 Colony screening*

Grow duplicated 96-well plates with daily change of fresh hPSC media. Use one plate for analysis in 3 days and the other plate for maintenance. One may use RFLP to identify clonal mutant lines when feasible. We routinely screen for mutant clones through direct sequencing of PCR products using a simplified procedure. This procedure involves first using a simple protocol (without phenol/chloroform extraction) to extract genomic DNA, then PCR amplifying the region of interest, and finally sequencing the unpurified PCR product. Since the entire procedure can be accomplished using multichannel pipettes, it enables the screening of a large number of clones rapidly. In our experience, this procedure works well as long as the primers are pretested before the targeting experiment, and it is feasible for a single person to process up to 288 samples (three 96-well plates) at a time.

#### 3.5.2.1 Lysis

Remove media from the wells, wash once with PBS, and add 50 μl of lysis buffer.

Lysis buffer (50 μl/well)

| Component | Amount (μl) |
|---|---|
| Proteinase K (10 mg/ml) | 1.5 |
| JumpStart PCR buffer 10 × (Sigma) | 5 |
| $ddH_2O$ | 43.5 |

To avoid excessive evaporation, seal the plate using a qPCR film sticker and incubate overnight at 55 °C.

The next day, mix and transfer the cell lysates into a 96-well PCR plate and incubate for 10 min at 96 °C in a thermocycler to inactivate Proteinase K.

#### 3.5.2.2 PCR and sequencing

PCR amplify using the same primers as for T7EI or RFLP assay and 1 μl of cell lysate as template. Use 1 μl of the PCR product and perform Sanger sequencing using an internal primer to minimize sequencing issues caused by non-specific PCR product. To generate knockout lines, amplifying clones with frameshift Indel mutations for frozen stocks. We also recommend amplifying and freezing a couple of wild-type and heterozygous clones from the same targeting experiment to serve as control lines.

In the case of heterozygous and compound heterozygous mutants, the traces can be complicated to interpret due to overlapping peaks. To further characterize the mutations present in each allele, we amplify the sgRNA target region of these clones with high-fidelity Herculase II Fusion DNA Polymerase using T7EI PCR primers and clone the PCR product using Zero Blunt TOPO PCR Cloning Kit (Life Technologies, 450245). Sanger sequencing of 10 individual bacterial colonies allows monoallelic characterization of the mutations.

If the sequencing turnaround time is relatively short (e.g., within a day), it is possible to directly amplify the desired mutant clones from the maintenance plate before they become confluent. Alternatively, if the sequencing turnaround time is longer, we recommend freezing down cells in the entire maintenance plate. Dissociate cells with 25 μl TrypLE Select enzyme in each well of 96-well plates, stop the reaction by adding 50 μl hPSC medium per well. Then add 75 μl of 2 × freezing medium directly into the well and mix gently.

Wrap the plate with plastic film, put into a Styrofoam box and transfer to −80 °C. The frozen plate can be stored for at least 3 months at −80 °C.

### 3.5.3 Validation

#### 3.5.3.1 Validation of the mutant alleles

It is important to realize that CRISPR-mediated Indel mutations may create complete loss-of-function, partial loss-of-function, or occasionally gain-of-function (e.g., dominant-negative or neomorphic) alleles. Therefore, it is important to perform additional analysis to determine the exact nature of individual mutant alleles. For instance, Western blot may be performed to verify the absence of the wild-type protein in a mutant clone, as long as a good antibody is available for the protein of interest. One may also validate the absence of a functional protein based on its known biological function. For instance, based on the requirement of TET proteins for catalyzing the conversion of 5-methylcytosine to 5-hydroxymethylcytosine (5hmC) (Ito et al., 2010; Tahiliani et al., 2009), quantification of the 5hmC levels can be used to verify the TET1/2/3 triple-knockout hESC lines (Gonzalez et al., 2014).

#### 3.5.3.2 Off-target analysis

Due to mismatch tolerance of CRISPR sgRNA paring (Hsu et al., 2013), there are concerns about the off-target mutagenic effects of the CRISPR/Cas system. Recently, whole-genome sequencing of CRISPR targeted hPSC clonal lines has detected very rare off-target mutations attributable to CRISPR (Veres et al., 2014). Thus off-target mutations may not be a significant concern for disease modeling and biological studies using hPSCs.

One may identify potential ectopic sgRNAs targets using the CRISPR design tool (http://crispr.mit.edu/). The most likely off-targets falling in gene coding sequences (4–5 sites per sgRNA-mediated targeting experiment) can be analyzed through sequencing using the same approach described for analyzing mutations at sgRNA target sites. PCR amplify a stretch of genomic region (~500 bp) flanking a putative CRISPR off-target site using Herculase II Fusion DNA Polymerase and analyze by Sanger sequencing using a primer binding inside the PCR product.

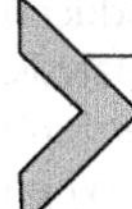

## 4. GENERATION OF PRECISE NUCLEOTIDE ALTERATIONS USING iCRISPR

Without a DNA repair template, NHEJ repair of DSBs introduces random Indels. However, in the presence of a DNA repair template, HDR of DSBs could lead to precise nucleotide alterations of the hPSC genome, which is important for either creating disease-specific variants in wild-type cells or correcting disease-associated mutations in patient cells. Compared with double-stranded DNAs, synthetic ssDNAs (~80–200 nt) are easy to produce and can be used as DNA repair template without selection (Chen, Pruett-Miller, et al., 2011).

## 4.1. Design of ssDNA as HDR templates

To introduce specific nucleotide alterations, we cotransfect sgRNA with ssDNA as HDR template.

1. Design 2–3 sgRNAs in close proximity to the genetic alteration locus.
2. Design ssDNA templates containing the genetic alternation flanked by ~40–80 nt of homology on each side of the space between the genetic alternation site and the CRISPR cutting site (Fig. 11.5A). Because of the close proximity of the sgRNAs designed, only one ssDNA template is necessary in most cases.
3. We highly recommend introducing a silent mutation to eliminate the PAM sequence in the ssDNA template. This would prevent further

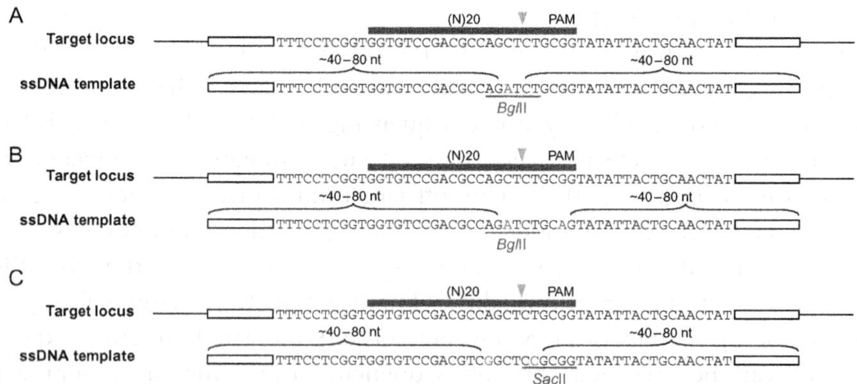


**Figure 11.5** Examples illustrating the design of homology-directed repair (HDR) ssDNA templates. (A) sgRNAs are designed close to the genetic alteration locus. Arrow head points to the Cas9 cutting site 3 nt upstream of the PAM sequence. The ssDNA HDR template contains the required genetic alternation (C > A, labeled in red (dark gray in the print version)) flanked by ~40–80 nt of homology on each side of the space between the genetic alternation site and the CRISPR cutting site. A new restriction site (*Bgl*II) introduced by the genetic alteration allows assessment of HR efficiency through RFLP analysis. (B) Introducing a silent mutation (G > A, labeled in green (dark gray in the print version)) in the PAM sequence would prevent further editing of correctly modified alleles. (C) When the required genetic alternation does not generate a restriction site, introducing a silent mutation (T > C, labeled in green (dark gray in the print version)) nearby that creates a new restriction site (*Sac*II) enables assessment of HR efficiency through RFLP.

editing of correctly modified alleles (Fig. 11.5B, G>A in the PAM sequence).

4. If the genetic alternation introduces or disrupts a restriction site, HR efficiency can be assessed through RFLP analysis (Fig. 11.5A, C>A introduces a *Bgl*II site).
5. If the genetic alternation does not introduce or disrupt a restriction site (Fig. 11.5C, A>G) we recommend introducing a silent mutation that creates a new restriction site (Fig. 11.5C, T>C generates a *Sac*II site) to enable assessment of HR efficiency through RFLP.

## 4.2. ssDNA/sgRNA cotransfection in hPSCs

1. *Day 0*: Pretreatment with doxycycline. Treat iCas9 cells with doxycycline for 24 h before transfection when hPSCs are ~60% confluent so that on the day of transfection, Cas9 will already be expressed at a high level. Seed iMEFs on a gelatin-coated 24-well plate.
2. *Day 1*: *1st transfection*. On the day of transfection, dissociate and replate cells as described in Section 3.3.

   Perform cotransfection of sgRNA and ssDNA in one well of a 24-well plate as follows:

| Component | Amount (μl) |
|---|---|
| Opti-MEM | 50 |
| RNAiMAX reagent | 1.5 |
| 320 ng/μl sgRNA | 0.5 |
| 300 ng/μl ssDNA | 2.5 |

3. *Day 2*: *2nd transfection (optional)*. Twenty-four hours after the 1st transfection, change hPSC medium with doxycycline and repeat the transfection as Day 1.
4. *Days 3–4*: Change hPSC medium.

## 4.3. Establishment of clonal lines

HDR frequency will be assessed through RFLP assays as described in Section 3.4.3, followed by procedures to establish clonal lines as described in Section 3.5.

## 5. INDUCIBLE GENE KNOCKOUT IN hPSCs USING iCRISPR

Inducible gene knockout during differentiation of hPSCs into specific cell types is of great importance for studying genes with pleiotropic effects. With the iCRISPR platform, inducible gene knockout could be achieved by inducible Cas9 expression and temporally regulated delivery of sgRNA due to the low toxicity of lipid-mediated sgRNA transfection (Fig. 11.6A). Alternatively, inducible gene knockout could also be achieved through generation of iCr lines including a constitutive sgRNA expression module in addition to the doxycycline-inducible Cas9 expression cassette (Puro-iCr)

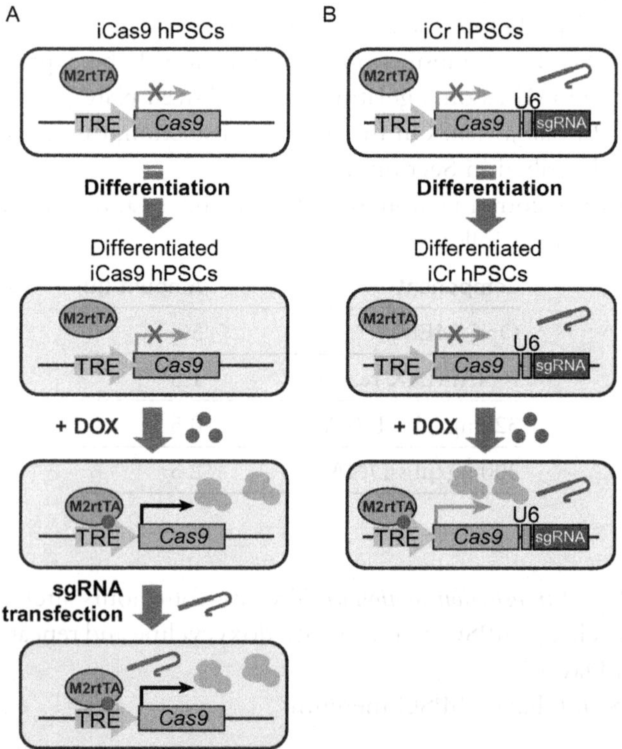


**Figure 11.6** Inducible gene knockout in hPSCs using iCRISPR. (A) iCas9 hPSCs are first differentiated into required cell types. Induced expression of Cas9 with doxycycline treatment and sgRNA transfection in differentiated cells result in inducible gene knockout. (B) In iCr hPSCs, a constitutive sgRNA expression module is inserted 3′ of the Cas9 expression cassette. Treatment of differentiated iCr hPSCs with doxycycline induces Cas9 expression thus inducing gene knockout.

targeted into the *AAVS1* locus (Fig. 11.6B). Compared with temporally regulated delivery of sgRNA, generation of iCr lines allows inducible gene targeting in all the cells upon doxycycline treatment.

## 5.1. Inducible gene knockout through sgRNA transfection

1. Differentiate iCas9 into target cell types following established protocol.
2. Treat cells with doxycycline 24 h before transfection.
3. FACS sort target cells when cell surface marker(s) is available and replate cells.
4. Perform sgRNA transfection as described in Section 3.3.

## 5.2. Inducible gene knockout through using iCr hPSC lines

To generate iCr hPSC lines for inducible gene knockout, we use the same cloning approach developed by Cong et al. (2013) to insert the CRISPR target sequence ((N)19) into a single *Bbs*I site found in the piCRg Entry plasmid (Addgene 58904) (Fig. 11.7A and B).

Then through LR reaction using piCRg Entry and Puro-iDEST plasmids, the sgRNA expression module and Cas9 expression cassette are

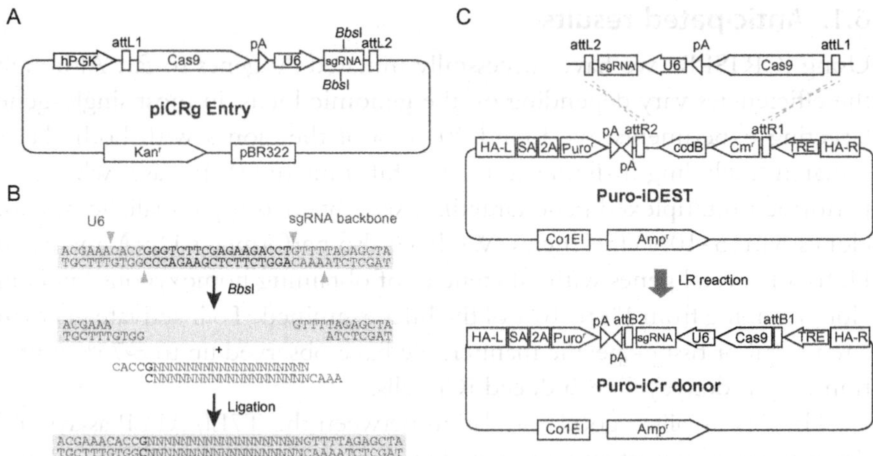


**Figure 11.7** Generation of Puro-iCr donor plasmid. (A) The pEntr-Cr vector contains the chimeric sgRNA expression module and Cas9 expression cassette. (B) Cloning of the target DNA sequence into the pEntr-Cr vector via digestion with *BbsI* and ligation. *Note*: Due to the requirement of U6 promoter to have a "G" base at the transcription start site, one base "G" followed by 19 Ns is cloned into the pEntr-Cr backbone. (C) Through GATEWAY LR reaction, the chimeric sgRNA expression module and Cas9 expression cassette in the pEntr-cr is transferred to Puro-iDEST to generate Puro-iCr donor plasmid.

transferred from piCRg Entry: to Puro-iDEST, generating the Puro-iCr donor (Fig. 11.7C).

1. piCRg Entry: piCRg Entry plasmid was constructed by introducing the chimeric sgRNA expression module and Cas9 coding sequence, PCR amplified from pX330 (Cong et al., 2013), into a pEntr plasmid (Fig. 11.7A).
2. Puro-iDEST: Puro-iDEST plasmid was constructed by replacing the Cas9 coding sequence in the Puro-Cas9 donor with a gateway destination cassette (Life Technologies, Gateway system) (Fig. 11.7C).
3. Puro-iCr: Donor CRISPR/Cas9 expression vectors targeting specific genomic loci are generated through LR reaction between pEntr-Cr and Puro-iDEST (Invitrogen, Gateway Technology) (Fig. 11.7C).
4. Generate iCr hPSCs as described in Section 2. Prescreen after Southern blot to make sure there's no Indel in the gene of interest.

# 6. CONCLUSIONS AND FUTURE DIRECTIONS

Below we discuss expected results, common issues and considerations, and potential additional use and extension of the iCRISPR platform.

## 6.1. Anticipated results

Using iCRISPR, we have successfully mutated 20 genes so far. Although the efficiencies vary depending on the genomic locus, in most single-gene targeting experiments, we found 20–60% of the clones with both alleles mutated (including in-frame and frameshift mutations). In cases where we performed multiplexed gene targeting, we obtained triple biallelic mutant clones with 5–10% efficiencies. We have also performed ssDNA-mediated HDR of several genes with efficiencies of obtaining homozygous knockin clones ranging from 1% to 10% of the lines examined. To inactivate genes in a temporal or tissue-specific manner, we have observed up to ~75% mutation rate in doxycycline-induced iCr cells.

There is usually a good correlation between the T7EI/RFLP assays and the number of mutant lines recovered upon clonal expansion. This highlights the importance of performing these assays in parallel with clonal expansion. One may also test several sgRNAs through performing T7EI/RFLP assays beforehand, and pick the most efficient one(s) for establishing mutant lines.

## 6.2. In-frame mutations

Occasionally, we have encountered situations where the majority of mutant alleles carry the same in-frame mutations. This appears to be caused by microhomology-mediated repair that utilizes short stretches of repeated sequences flanking the DSB site. This issue can be overcome simply by designing sgRNAs that target different sequences. One may also attempt to predict the possibility of microhomology-associated Indels at CRISPR target sites, and potentially make use of microhomology-mediated DNA repair to maximize the frequency of frameshift mutations (Bae, Kweon, Kim, & Kim, 2014).

Of note, a failure to recover frameshift mutant clones may also reflect a requirement of the gene for the survival or self-renewal of hPSCs. One should consider this possibility when multiple sgRNAs with high Indel efficiencies have produced only in-frame biallelic mutations, and there is no apparent association with microhomology-mediated DNA repair. In this situation one may perform inducible knockout (see Section 5) or knockdown by RNA interference.

## 6.3. Cross contamination

While we have been able to routinely generate mutant lines that are homogeneous with respect to the mutated allele(s), it is occasionally possible that a given mutant line has arisen from more than one cell. Therefore, care should be taken to pick only well-spaced colonies for line establishment, and to avoid cross contamination when handling multiple clonal lines. After a clonal line is established, we recommend confirming the presence of mutant allele(s) through TA cloning. One may further ensure the establishment of clonal lines through subcloning. Although not routinely used in our laboratory, single-cell deposition flow cytometry can be used to assist this procedure (Davis et al., 2008).

## 6.4. Time and throughput considerations

Based on the generally high targeting efficiencies obtained with iCRISPR, the analysis of ~24–48 colonies should be sufficient to establish multiple mono- and biallelic mutant lines for each gene, and the entire process takes about 1–2 months. We have found that it is feasible for a trained individual to pick around 288 colonies at a time, which takes approximately 3 h. With this throughput, an experienced individual could also mutant several individual genes in parallel.

## 6.5. Off-target considerations

iCRISPR offers an efficient platform to generate isogenic wild-type and mutant hPSC clones allowing rapid functional studies of single and multiple genes. However, studies performed in pools of immortalized cell lines suggest that both TALEN and CRISPR/Cas systems can induce unwanted mutations at loci sharing high sequence similarity with the target site (Cho et al., 2014; Fu et al., 2013; Hsu et al., 2013; Mali, Aach, et al., 2013; Pattanayak et al., 2013). These studies raise concerns that off-target effects could make it difficult to interpret findings from studies using TALEN or CRISPR/Cas systems. Recent studies have identified only low incidence of off-target mutations from whole-genome sequencing analysis in individual hPSC clones obtained through genome manipulation with TALENs or CRISPR/Cas (Smith et al., 2014; Suzuki et al., 2014; Veres et al., 2014). Therefore, although off-target mutations are still a risk for therapeutic applications, they are not a significant concern for biological studies and disease modeling.

Nevertheless, unintended off-target mutations could still obscure functional studies in hPSCs. To minimize confounding phenotypes introduced by off-target effects, we recommend generating independent mutant lines using at least two different sgRNAs targeting different sequences in the same gene. Wild-type clones identified from the same targeting experiment can be used as control lines. One should also consider complementing the loss-of-function approaches with rescue experiments when feasible. To further demonstrate the generality of any experimental finding, it may be desirable to generate mutant lines using more than one hPSC iCas9 line.

## 6.6. Additional use and extension of the iCRISPR platform

The iCRISPR platform is likely to facilitate other types of genome editing in hPSCs. Also, by replacing Cas9 with Cas9 variants such as dCas9, dCas9-KRAB, or dCas9-VP16 (Gilbert et al., 2013; Qi et al., 2013), one may repurpose iCRISPR for gene regulation. Among the many possible applications of CRISPR/Cas, we would like to highlight two areas that we feel have not yet been widely discussed. First, CRISPR/Cas may be used to create deletions in noncoding RNAs or gene regulatory regions such as promoters and enhancers. To efficiently generate regulatory mutants using iCRISPR, one may design sgRNAs to disrupt the binding site (usually a short sequence) of a DNA-binding protein including but not restricted to the basal transcription machinery or a tissue-specific transcription factor. Alternatively, specific protein binding sites may be mutated through ssDNA-mediated HDR.

iCRISPR may also facilitate the generation of more complex genomic modifications, such as the creation of reporter alleles through HDR-mediated gene targeting using long donor DNA templates encoding protein tags or fluorescent proteins. Conventional approaches to generate knockin reporters require a drug-resistance cassette due to low targeting efficiencies. An additional step is therefore often necessary to remove the drug-resistance cassette, the presence of which may affect reporter gene expression. Although similar targeting experiments have been performed with CRISPR/Cas zygote injection in mice and rats (Ma et al., 2014; Yang et al., 2013), this has not yet been achieved in mouse or human pluripotent stem cells. With sufficiently high targeting efficiencies, it might be possible to perform this type of targeting in hPSCs without drug selection, which would be a significant technical advance in our opinion.

## ACKNOWLEDGMENTS

We thank Feng Zhang and Rudolf Jaenisch for providing vectors through Addgene, and Dirk Hockemeyer for the Neo-M2rtTA donor. This study was funded in part by NIH (R01DK096239), NYSTEM (C029156), and March of Dimes Foundation (Basil O'Connor Starter Scholar Research Award Grant 5-FY12-82). F.G. and Z.Z. were supported by the New York State Stem Cell Science fellowship from the Center for Stem Cell Biology of the Sloan-Kettering Institute

## REFERENCES

An, M. C., O'Brien, R. N., Zhang, N., Patra, B. N., De La Cruz, M., Ray, A., et al. (2014). Polyglutamine disease modeling: Epitope based screen for homologous recombination using CRISPR/Cas9 system. *PLoS Currents*, *6*, In Press.

Bae, S., Kweon, J., Kim, H. S., & Kim, J. S. (2014). Microhomology-based choice of Cas9 nuclease target sites. *Nature Methods*, *11*, 705–706.

Boch, J., Scholze, H., Schornack, S., Landgraf, A., Hahn, S., Kay, S., et al. (2009). Breaking the code of DNA binding specificity of TAL-type III effectors. *Science*, *326*, 1509–1512.

Cermak, T., Doyle, E. L., Christian, M., Wang, L., Zhang, Y., Schmidt, C., et al. (2011). Efficient design and assembly of custom TALEN and other TAL effector-based constructs for DNA targeting. *Nucleic Acids Research*, *39*, e82.

Chen, G., Gulbranson, D. R., Hou, Z., Bolin, J. M., Ruotti, V., Probasco, M. D., et al. (2011). Chemically defined conditions for human iPSC derivation and culture. *Nature Methods*, *8*, 424–429.

Chen, F., Pruett-Miller, S. M., Huang, Y., Gjoka, M., Duda, K., Taunton, J., et al. (2011). High-frequency genome editing using ssDNA oligonucleotides with zinc-finger nucleases. *Nature Methods*, *8*, 753–755.

Cho, S. W., Kim, S., Kim, J. M., & Kim, J. S. (2013). Targeted genome engineering in human cells with the Cas9 RNA-guided endonuclease. *Nature Biotechnology*, *31*, 230–232.

Cho, S. W., Kim, S., Kim, Y., Kweon, J., Kim, H. S., Bae, S., et al. (2014). Analysis of off-target effects of CRISPR/Cas-derived RNA-guided endonucleases and nickases. *Genome Research*, *24*, 132–141.

Cong, L., Ran, F. A., Cox, D., Lin, S., Barretto, R., Habib, N., et al. (2013). Multiplex genome engineering using CRISPR/Cas systems. *Science*, *339*, 819–823.

Costa, M., Dottori, M., Sourris, K., Jamshidi, P., Hatzistavrou, T., Davis, R., et al. (2007). A method for genetic modification of human embryonic stem cells using electroporation. *Nature Protocols*, *2*, 792–796.

Davis, R. P., Costa, M., Grandela, C., Holland, A. M., Hatzistavrou, T., Micallef, S. J., et al. (2008). A protocol for removal of antibiotic resistance cassettes from human embryonic stem cells genetically modified by homologous recombination or transgenesis. *Nature Protocols*, *3*, 1550–1558.

DeKelver, R. C., Choi, V. M., Moehle, E. A., Paschon, D. E., Hockemeyer, D., Meijsing, S. H., et al. (2010). Functional genomics, proteomics, and regulatory DNA analysis in isogenic settings using zinc finger nuclease-driven transgenesis into a safe harbor locus in the human genome. *Genome Research*, *20*, 1133–1142.

Ding, Q., Regan, S. N., Xia, Y., Oostrom, L. A., Cowan, C. A., & Musunuru, K. (2013). Enhanced efficiency of human pluripotent stem cell genome editing through replacing TALENs with CRISPRs. *Cell Stem Cell*, *12*, 393–394.

Fu, Y., Foden, J. A., Khayter, C., Maeder, M. L., Reyon, D., Joung, J. K., et al. (2013). High-frequency off-target mutagenesis induced by CRISPR-Cas nucleases in human cells. *Nature Biotechnology*, *31*, 822–826.

Gilbert, L. A., Larson, M. H., Morsut, L., Liu, Z., Brar, G. A., Torres, S. E., et al. (2013). CRISPR-mediated modular RNA-guided regulation of transcription in eukaryotes. *Cell*, *154*, 442–451.

Gonzalez, F., Zhu, Z., Shi, Z. D., Lelli, K., Verma, N., Li, Q. V., et al. (2014). An iCRISPR platform for rapid, multiplexable, and inducible genome editing in human pluripotent stem cells. *Cell Stem Cell*, *15*, 215–226.

Hockemeyer, D., Wang, H., Kiani, S., Lai, C. S., Gao, Q., Cassady, J. P., et al. (2011). Genetic engineering of human pluripotent cells using TALE nucleases. *Nature Biotechnology*, *29*, 731–734.

Horii, T., Tamura, D., Morita, S., Kimura, M., & Hatada, I. (2013). Generation of an ICF syndrome model by efficient genome editing of human induced pluripotent stem cells using the CRISPR system. *International Journal of Molecular Sciences*, *14*, 19774–19781.

Hou, Z., Zhang, Y., Propson, N. E., Howden, S. E., Chu, L. F., Sontheimer, E. J., et al. (2013). Efficient genome engineering in human pluripotent stem cells using Cas9 from *Neisseria meningitidis*. *Proceedings of the National Academy of Sciences of the United States of America*, *110*, 15644–15649.

Hsu, P. D., Scott, D. A., Weinstein, J. A., Ran, F. A., Konermann, S., Agarwala, V., et al. (2013). DNA targeting specificity of RNA-guided Cas9 nucleases. *Nature Biotechnology*, *31*, 827–832.

Ito, S., D'Alessio, A. C., Taranova, O. V., Hong, K., Sowers, L. C., & Zhang, Y. (2010). Role of Tet proteins in 5mC to 5hmC conversion, ES-cell self-renewal and inner cell mass specification. *Nature*, *466*, 1129–1133.

Jasin, M. (1996). Genetic manipulation of genomes with rare-cutting endonucleases. *Trends in Genetics*, *12*, 224–228.

Jinek, M., Chylinski, K., Fonfara, I., Hauer, M., Doudna, J. A., & Charpentier, E. (2012). A programmable dual-RNA-guided DNA endonuclease in adaptive bacterial immunity. *Science*, *337*, 816–821.

Jinek, M., East, A., Cheng, A., Lin, S., Ma, E., & Doudna, J. (2013). RNA-programmed genome editing in human cells. *eLife*, *2*, e00471.

Joung, J. K., & Sander, J. D. (2013). TALENs: A widely applicable technology for targeted genome editing. *Nature Reviews Molecular Cell Biology*, *14*, 49–55.

Ludwig, T. E., Bergendahl, V., Levenstein, M. E., Yu, J., Probasco, M. D., & Thomson, J. A. (2006). Feeder-independent culture of human embryonic stem cells. *Nature Methods*, *3*, 637–646.

Ma, Y., Ma, J., Zhang, X., Chen, W., Yu, L., Lu, Y., et al. (2014). Generation of eGFP and Cre knockin rats by CRISPR/Cas9. *The FEBS Journal*, *281*, 3779–3790.

Mali, P., Aach, J., Stranges, P. B., Esvelt, K. M., Moosburner, M., Kosuri, S., et al. (2013). CAS9 transcriptional activators for target specificity screening and paired nickases for cooperative genome engineering. *Nature Biotechnology*, *31*, 833–838.

Mali, P., Yang, L., Esvelt, K. M., Aach, J., Guell, M., DiCarlo, J. E., et al. (2013). RNA-guided human genome engineering via Cas9. *Science*, *339*, 823–826.

Maniatis, T., Fritsch, E. F., & Sambrook, J. (1982). *Molecular cloning: A laboratory manual*. Cold Spring Harbor, N.Y.: Cold Spring Harbor Laboratory.

Miller, J. C., Tan, S., Qiao, G., Barlow, K. A., Wang, J., Xia, D. F., et al. (2011). A TALE nuclease architecture for efficient genome editing. *Nature Biotechnology*, *29*, 143–148.

Moscou, M. J., & Bogdanove, A. J. (2009). A simple cipher governs DNA recognition by TAL effectors. *Science*, *326*, 1501.

Pattanayak, V., Lin, S., Guilinger, J. P., Ma, E., Doudna, J. A., & Liu, D. R. (2013). High-throughput profiling of off-target DNA cleavage reveals RNA-programmed Cas9 nuclease specificity. *Nature Biotechnology*, *31*, 839–843.

Qi, L. S., Larson, M. H., Gilbert, L. A., Doudna, J. A., Weissman, J. S., Arkin, A. P., et al. (2013). Repurposing CRISPR as an RNA-guided platform for sequence-specific control of gene expression. *Cell*, *152*, 1173–1183.

Ran, F. A., Hsu, P. D., Wright, J., Agarwala, V., Scott, D. A., & Zhang, F. (2013). Genome engineering using the CRISPR-Cas9 system. *Nature Protocols*, *8*, 2281–2308.

Rossier, O., Wengelnik, K., Hahn, K., & Bonas, U. (1999). The Xanthomonas Hrp type III system secretes proteins from plant and mammalian bacterial pathogens. *Proceedings of the National Academy of Sciences of the United States of America*, *96*, 9368–9373.

Rouet, P., Smih, F., & Jasin, M. (1994). Introduction of double-strand breaks into the genome of mouse cells by expression of a rare-cutting endonuclease. *Molecular and Cellular Biology*, *14*, 8096–8106.

Sanjana, N. E., Cong, L., Zhou, Y., Cunniff, M. M., Feng, G., & Zhang, F. (2012). A transcription activator-like effector toolbox for genome engineering. *Nature Protocols*, *7*, 171–192.

Smith, C., Gore, A., Yan, W., Abalde-Atristain, L., Li, Z., He, C., et al. (2014). Whole-genome sequencing analysis reveals high specificity of CRISPR/Cas9 and TALEN-based genome editing in human iPSCs. *Cell Stem Cell*, *15*, 12–13.

Smith, J. R., Maguire, S., Davis, L. A., Alexander, M., Yang, F., Chandran, S., et al. (2008). Robust, persistent transgene expression in human embryonic stem cells is achieved with AAVS1-targeted integration. *Stem Cells*, *26*, 496–504.

Suzuki, K., Yu, C., Qu, J., Li, M., Yao, X., Yuan, T., et al. (2014). Targeted gene correction minimally impacts whole-genome mutational load in human-disease-specific induced pluripotent stem cell clones. *Cell Stem Cell*, *15*, 31–36.

Szurek, B., Marois, E., Bonas, U., & Van den Ackerveken, G. (2001). Eukaryotic features of the Xanthomonas type III effector AvrBs3: Protein domains involved in transcriptional activation and the interaction with nuclear import receptors from pepper. *The Plant Journal*, *26*, 523–534.

Tahiliani, M., Koh, K. P., Shen, Y., Pastor, W. A., Bandukwala, H., Brudno, Y., et al. (2009). Conversion of 5-methylcytosine to 5-hydroxymethylcytosine in mammalian DNA by MLL partner TET1. *Science*, *324*, 930–935.

Urnov, F. D., Rebar, E. J., Holmes, M. C., Zhang, H. S., & Gregory, P. D. (2010). Genome editing with engineered zinc finger nucleases. *Nature Reviews Genetics*, *11*, 636–646.

van der Oost, J., Westra, E. R., Jackson, R. N., & Wiedenheft, B. (2014). Unravelling the structural and mechanistic basis of CRISPR-Cas systems. *Nature Reviews Microbiology*, *12*, 479–492.

Veres, A., Gosis, B. S., Ding, Q., Collins, R., Ragavendran, A., Brand, H., et al. (2014). Low incidence of off-target mutations in individual CRISPR-Cas9 and TALEN targeted human stem cell clones detected by whole-genome sequencing. *Cell Stem Cell, 15*, 27–30.

Wang, G., McCain, M. L., Yang, L., He, A., Pasqualini, F. S., Agarwal, A., et al. (2014). Modeling the mitochondrial cardiomyopathy of Barth syndrome with induced pluripotent stem cell and heart-on-chip technologies. *Nature Medicine, 20*, 616–623.

Wang, H., Yang, H., Shivalila, C. S., Dawlaty, M. M., Cheng, A. W., Zhang, F., et al. (2013). One-step generation of mice carrying mutations in multiple genes by CRISPR/Cas-mediated genome engineering. *Cell, 153*, 910–918.

Yang, H., Wang, H., Shivalila, C. S., Cheng, A. W., Shi, L., & Jaenisch, R. (2013). One-step generation of mice carrying reporter and conditional alleles by CRISPR/Cas-mediated genome engineering. *Cell, 154*, 1370–1379.

Ye, L., Wang, J., Beyer, A. I., Teque, F., Cradick, T. J., Qi, Z., et al. (2014). Seamless modification of wild-type induced pluripotent stem cells to the natural CCR5Delta32 mutation confers resistance to HIV infection. *Proceedings of the National Academy of Sciences of the United States of America, 111*, 9591–9596.

Zhu, Z., & Huangfu, D. (2013). Human pluripotent stem cells: An emerging model in developmental biology. *Development, 140*, 705–717.

CHAPTER TWELVE

# Creating Cancer Translocations in Human Cells Using Cas9 DSBs and nCas9 Paired Nicks

**Benjamin Renouf*, Marion Piganeau*, Hind Ghezraoui*, Maria Jasin[†,1], Erika Brunet[*,1]**
*Museum National d'Histoire Naturelle, INSERM U1154, CNRS 7196, Paris, France
[†]Developmental Biology Program, Memorial Sloan-Kettering Cancer Center, New York, USA
[1]Corresponding authors: e-mail address: ebrunet@mnhn.fr; m-jasin@ski.mskcc.org

## Contents



## Abstract

Recurrent chromosomal translocations are found in numerous tumor types, often leading to the formation and expression of fusion genes with oncogenic potential. Creating chromosomal translocations at the relevant endogenous loci, rather than ectopically expressing the fusion genes, opens new possibilities for better characterizing molecular mechanisms driving tumor formation. In this chapter, we describe methods to create cancer translocations in human cells. DSBs or paired nicks generated by either wild-type

*Methods in Enzymology*, Volume 546
ISSN 0076-6879
http://dx.doi.org/10.1016/B978-0-12-801185-0.00012-X

Cas9 or the Cas9 nickase, respectively, are used to induce translocations at the relevant loci. Using different PCR-based methods, we also explain how to quantify translocation frequency and to analyze breakpoint junctions in the cells of interest. In addition, PCR detection of translocations is used as a very sensitive method to detect off-target effects, which has general utility.

## 1. INTRODUCTION

The discovery that a chromosomal translocation was associated with oncogenesis was a watershed event in tumor biology research (Chandra et al., 2011; Rowley, 1973). Hundreds of recurrent reciprocal translocations have now been found in a variety of human cancers, including hematological malignancies, sarcomas, and epithelial tumors (Mani & Chinnaiyan, 2010; Mitelman, Johansson, & Mertens, 2007). These translocations are considered primary causes of many cancers and have been important for the development of targeted therapies. A typical consequence of a chromosomal translocation is the formation of a gene fusion that leads to expression of a novel protein with oncogenic potential; alternatively, translocations can lead to a gene coming under the control of a strong promoter, such that overexpression confers an oncogenic property. The cellular consequences of fusion gene expression have been widely investigated using ectopic expression in cells without the translocation or gene silencing in cells carrying the translocation. However, these studies are not optimal for several reasons. Ectopic expression does not recapitulate the genetics of the disease and may lead to nonphysiological levels of fusion gene expression, and silencing the fusion gene in tumor cells does not take into account the numerous other mutations that the tumor cells have acquired. Thus, having methods to induce translocations at will in a variety of cell types would provide a significant advantage to cancer researchers.

Translocations involve chromosome breakage and aberrant joining of DNA ends. Two concurrent DNA double-strand breaks (DSBs), one on each chromosome, have been shown to induce translocations (Richardson & Jasin, 2000). Joining of the DNA ends typically involves some type of nonhomologous end-joining (NHEJ) repair (Weinstock, Elliott, & Jasin, 2006). The simplest system for inducing concurrent DSBs is the expression of a nuclease which cleaves specific sites. Initial studies performed in model systems used the I-SceI endonuclease, a homing endonuclease from yeast, in which I-SceI sites were introduced at specific loci.

However, the development of designer nucleases (ZFNs and TALENs) has allowed DSBs to be introduced into genomes without prior modification (Gaj, Sirk, & Barbas, 2014; Urnov, Rebar, Holmes, Zhang, & Gregory, 2010). As a result, chromosomal translocations could be readily induced in human and mouse cells at endogenous loci (Brunet et al., 2009; Piganeau et al., 2013; Simsek et al., 2011), including translocations associated with human tumors. EWS–FLI1 translocations associated with Ewing sarcoma were induced by ZFNs directed to the EWS and FLI1 loci in mesenchymal cells (Piganeau et al., 2013). NPM–ALK translocations associated with anaplastic large cell lymphoma (ALCL) were induced by TALENs directed to the NPM and ALK loci in Jurkat cells (Piganeau et al., 2013). Of note, the ALCL translocation could also be reversed in patient cell lines carrying the translocation, by TALENs directed to the NPM–ALK and ALK–NPM loci. The versatility of TALENs suggests that designer nucleases could be used to induce translocations involving any loci.

The most recently developed designer nucleases are RNA-guided, which as such are easiest to design. Currently, the most commonly used nuclease is Cas9 from *S. pyogenes* (Cong et al., 2013; Hsu, Lander, & Zhang, 2014; Mali, Yang, et al., 2013). The guide RNA (sgRNA) has ~20 nucleotides of sequence complementary to a target site, followed by a Protospacer Adjacent Motif (PAM) sequence (NGG) which is critical for binding to Cas9. When both Cas9 and the sgRNA are expressed in cells, the target site is cleaved on both strands a few nucleotides away from the PAM, creating a DSB (Jinek et al., 2012). Because Cas9 has two active sites, each cleaving a defined strand, Cas9 can be converted to a nickase by mutation of one active site (nickase Cas9 or nCas9). For example, Cas9 D10A only cleaves the DNA strand complementary to the sgRNA. When two sgRNAs are provided which bind opposite DNA strands in close proximity, however, paired nicks can be introduced, also creating a DSB but with a long overhang (Mali, Yang, et al., 2013; Ran et al., 2013). Paired nicks are considered to have fewer potential off-target sites, since two distinct sgRNAs are required for double-strand cleavage. Following the principles established with TALENs and ZFNs (Piganeau et al., 2013), Cas9 has also been used to induce NPM-ALK and other tumor-relevant translocations (Choi & Meyerson, 2014; Ghezraoui et al., 2014; Torres et al., 2014). More recently, nCas9-induced paired nicks have also been used to induce translocations (Ghezraoui et al., 2014). In this chapter, we detail methods for the induction of translocations by Cas9 and nCas9, using a PCR screen for translocation junctions (Brunet et al., 2009).

# 2. MATERIALS

## 2.1. Cas9, nCas9, and sgRNA expression plasmid preparation

1. Expression plasmids can be obtained from Addgene (https://www.addgene.org/CRISPR/). We use pCas9_GFP (Addgene plasmid 44719) to induce DSBs and pCas9D10A_GFP to induce paired nicks (nCAS9) (Addgene plasmid 44720), together with sgRNA expression plasmids derived from MLM3636 (Addgene plasmid 43860).
2. Competent bacteria, e.g., DH5α.
3. LB agar plates with antibiotic (ampicillin for the plasmids above).
4. LB medium.
5. PureLink® HiPure Plasmid Filter *Maxiprep* Kit (Invitrogen).
6. NanoDrop 2000c (Thermo Scientific).

## 2.2. Cell culture and transfection

1. RPE1 (hTert-RPE1) cells or mesenchymal stem cells (MSC) are used in this chapter, although the approach is applicable to any other cell type that can be transfected.
2. Dulbecco's Modified Eagle Medium: Nutrient Mixture F-12 (DMEM/F-12, Life Technologies) with 10% Fetal Bovine Serum (FBS) for RPE1 cells; alpha-Minimum Essential Eagle Medium (αMEM, Life Technologies), supplemented with 10% FBS and 2 ng/mL bFGF (Recombinant Human FGF basic (146 aa) 233-FB-025 R&D systems) for MSC.
3. T150 flasks, 150 $cm^2$.
4. Cas9, nCas9, and sgRNA expression plasmids, at concentrations $\geq 2$ μg/μL.
5. 6-well plates.
6. 48-well plates.
7. 96-well plates.
8. 0.05% Trypsin–EDTA (1 ×).
9. Dulbecco's Phosphate-Buffered Saline (DPBS, Life Technologies).
10. Nucleofector II device (Lonza).
11. Cell Line Nucleofector Kit V with cuvettes (Lonza).

## 2.3. T7 endonuclease I assay

1. QIAamp DNA Mini Kit (Qiagen).
2. Primers: Primers may be designed using a variety of programs such as Primer3Plus (http://www.bioinformatics.nl/cgi-bin/primer3plus/primer3plus.cgi/). Settings are chosen to yield 22 bp primers with melting temperatures ~62 °C.
3. Phusion High-Fidelity DNA Polymerase (Thermo Scientific).
4. T7 endonuclease I (New England Biolabs).
5. NEBuffer 2.1 (New England Biolabs).
6. 2× T7 loading buffer: 50% sucrose, bromophenol blue, 260 μg/mL proteinase K.
7. 2.4% agarose gel with Ethidium Bromide (EtBr).
8. 0.5× TBE running buffer (Life Technologies).
9. UV station.

## 2.4. PCR detection of translocations

1. Primers for nested PCR (two sets of primers). Settings are chosen to yield 20 bp primers with melting temperatures ~60 °C.
2. FastStart Taq DNA polymerase (Roche).
3. 1% agarose gel with EtBr.
4. 0.5× TBE running buffer.
5. UV station.

## 2.5. PCR quantification of translocations

1. 10× Lysis buffer: 100 m*M* Tris–HCl (pH 8), 4.5% NP40, 4.5% Tween20.
2. Proteinase K (New England Biolabs).
3. 2× Master Mix 1:1× GC-RICH solution (Roche FastStart Taq), 2× PCR Buffer with 20 m*M* $MgCl_2$ (Roche FastStart Taq), 400 μ*M* dNTP mix (Roche FastStart Taq), 4% DMSO, 0.01% Tween20, 0.01% NP40. Store at 4 °C for up to 2 weeks.
4. 2× Master Mix 2:1× GC-RICH solution (Roche FastStart Taq), 2× PCR Buffer with 20 m*M* $MgCl_2$ (Roche FastStart Taq), 400 μ*M* dNTP mix (Roche FastStart Taq), 4% DMSO, 0.01% Tween20, 0.01% NP40, 0.25 μL of SYBR® Green I (10,000× in DMSO, Sigma–Aldrich) diluted in 2 mL of 20% DMSO and Rox (concentrations depend on

the real-time PCR machine; 30 n*M* of Rox is used with MX3005P from Agilent). Store at 4 °C for up to 2 weeks.

5. Primers for nested PCR.
6. FastStart Taq DNA polymerase (Roche).
7. Real-time quantitative PCR machine (MX3005P, Agilent).

# 3. METHODS TO INDUCE AND DETECT CANCER TRANSLOCATIONS IN HUMAN CELLS

Both Cas9 and nCas9 can be used to generate translocations. We have found that Cas9 is somewhat more efficient (Ghezraoui et al., 2014), but nCas9 has fewer off-target concerns. Two translocations are used to illustrate the methods described in this chapter: NPM–ALK found in ALCL (Elmberger, Lozano, Weisenburger, Sanger, & Chan, 1995; Kuefer et al., 1997; Morris et al., 1994) (Fig. 12.1A) and EWS–FLI1 found in Ewing sarcoma (May et al., 1993) (Fig. 12.2A).

## 3.1. sgRNA design and expression plasmid construction

1. For the translocation of interest, a good starting point for designing sgRNA target sequences is close to reported breakpoint junctions in patients, if available. If the translocation generates a fusion gene, target sites within introns involved in the translocation are requisite. When using wild-type Cas9, design two sgRNAs, one to each chromosome; when using nCas9 to induce paired nicks, design two pairs of sgRNAs to each chromosome. The ~20 bp target sequence is located upstream of the requisite NGG PAM sequence. For a 20 bp target sequence, follow the simple rule $N_{20}$NGG for Cas9 and CC$N_{20}$-spacer-$N_{20}$NGG for nCas9. The spacer can be a few bp or longer (Mali, Aach, et al., 2013; Ran et al., 2013). Specific Websites can be used to minimize the number of off-target sites of the sgRNAs (e.g., http://crispr.mit.edu/, http://zifit.partners.org/ZiFiT/Disclaimer.aspx, https://chopchop.rc.fas.harvard.edu/; see also Montague, Cruz, Gagnon, Church, & Valen, 2014; Xie, Shen, Zhang, Huang, & Zhang, 2014). The U6 promoter, which prefers G as the transcription start site, drives sgRNA expression in MLM3636; thus, either the target sequence should begin with G or a G should be inserted before the target sequence.

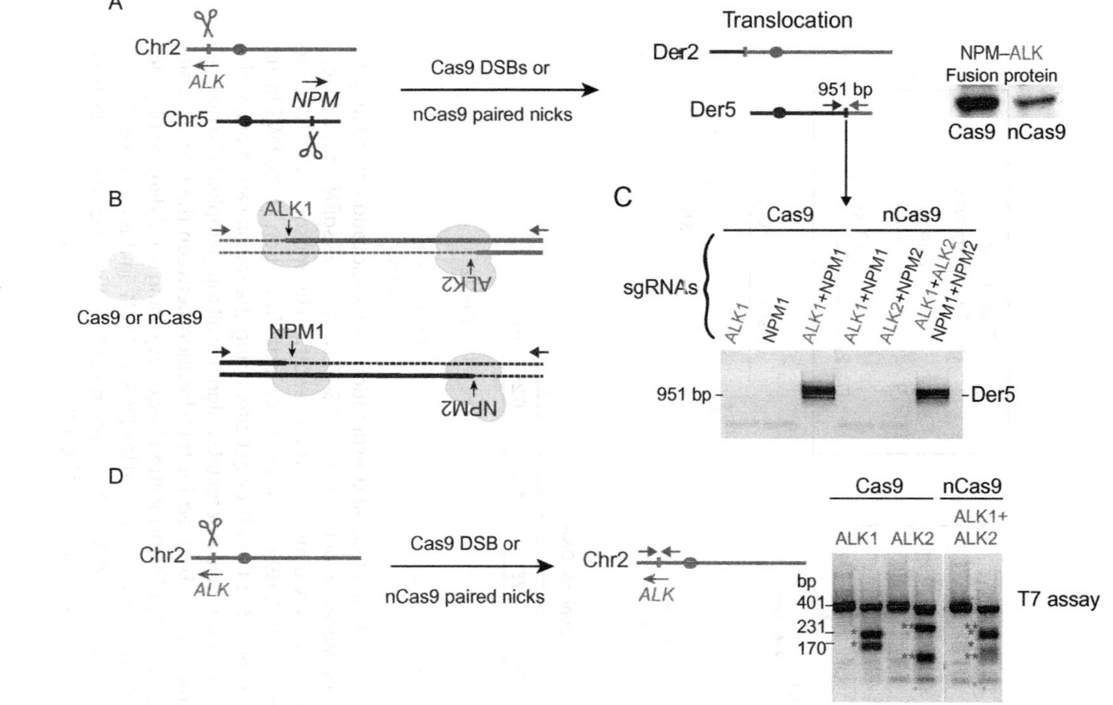


**Figure 12.1** Induction of the NPM–ALK translocation by Cas9 DSBs or nCas9 paired nicks. (A) Induction of DSBs or paired nicks at the *NPM* and *ALK* loci can lead to chromosomal translocations and expression of the NPM–ALK fusion protein from Der5. (B) Position of the sgRNAs. With wild-type Cas9, DSBs are formed using sgRNAs NPM1 and ALK1. With nCas9, paired nicks are formed using NPM1 + NPM2 (37 bp apart) and ALK1 + ALK2 (41 bp apart), resulting in 5′-overhangs if the strands are separated. (C) Translocations are only detected when two DSBs or two paired nicks are introduced. PCR is performed with primers indicated in (A). (D) Estimate of cleavage efficiency at the ALK locus using wild-type Cas9 and either the ALK1 or ALK2 sgRNA and nCas9 with both the ALK1 and ALK2 sgRNAs. The T7 assay detects about 60% indel formation in all three instances.

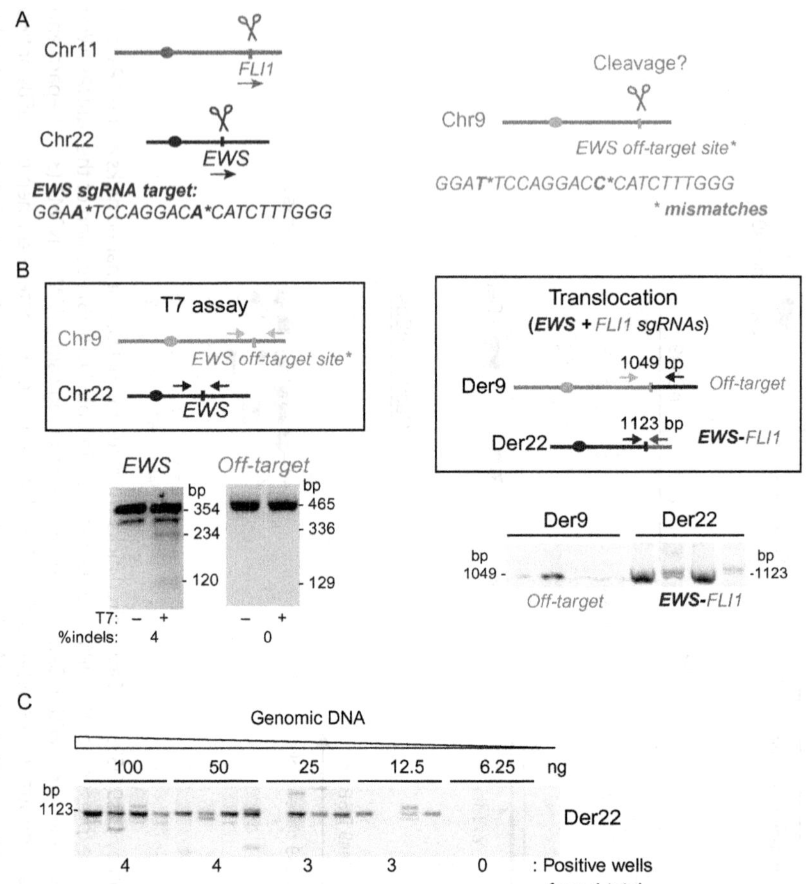


**Figure 12.2** sgRNA off-target analysis and estimation of translocation frequency by serial dilution. (A) The translocation of interest is EWS–FLI1. The EWS sgRNA has a potential off-target site that contains two mismatches on another chromosome. (B) Using the T7 assay, the EWS sgRNA with Cas9 gives detectable indel formation at the EWS locus, while no cleavage is observed at the off-target site. Using the translocation assay, the EWS and FLI1 sgRNAs with Cas9 give a detectable signal in all four samples (150 ng DNA each); however, a signal is also observed for translocation between the off-target site and the EWS locus. (C) Estimate of translocation frequency using serial dilution of genomic DNA from RPE1 cells using Cas9 and sgRNAs EWS and FLI1. In this example, the frequency of Der22 formation is $\geq 0.375 \times 10^{-3}$, determined for three positive wells from $(((4 \text{ total wells} \times 12.5 \text{ ng DNA})/6.25 \text{ ng}) \times 10^3 \text{ cells})$.

**NPM–ALK**

Cas9-generated DSBs with sgRNAs NPM1 and ALK1; nCas9-generated paired nicks with sgRNAs NPM1 + NPM2 and ALK1 + ALK2 (Fig. 12.1B).

NPM genomic sequence followed by sgRNA sequences:

5′-**CCT**CGAACTGCTACTGGGTTCACCTCAGCCTCTGGAA-TAGCTAGAACTAC**AGG**-3′

sgRNA NPM1: 5′-GTGAACCCAGTAGCAGTTCG-3′

sgRNA NPM2: 5′-GCCTCTGGAATAGCTAGAACTAC-3′

ALK genomic sequence followed by sgRNA sequences:

5′-**CCT**CAGGTAACCCTAATCTGATCACGGTCGGTCCATT-GCATAGAGG**AGG**-3′

sgRNA ALK1: 5′-GATCAGATTAGGGTTACCTG-3′

sgRNA ALK2: 5′-GTCGGTCCATTGCATAGAGG-3′

2. For each sgRNA, order the synthesis of two DNA oligonucleotides (sense and antisense) that would anneal to BsmBI-linearized plasmid MLM3636. (http://zifit.partners.org/ZiFiT/CSquare9GetOligos.aspx)

   Resuspend at 100 μ*M* in annealing buffer (10 m*M* Tris (pH 7.5), 1 m*M* EDTA, 50 m*M* NaCl).

sgRNA NPM1 oligonucleotides:

5′-ACACCGTGAACCCAGTAGCAGTTCGG-3′

5′-AAAACCGAACTGCTACTGGGTTCACG-3′

sgRNA ALK1 oligonucleotides:

5′-ACACCGATCAGATTAGGGTTACCTGG-3′

5′-AAAACCAGGTAACCCTAATCTGATCG-3′

3. Anneal the two oligonucleotides to generate a duplex. Add 10 μL each of the sense and antisense oligonucleotides to 80 μL of annealing buffer. Place tube in a standard heating block at 95 °C for 5 min. Remove the heating block from the apparatus and allow to cool to room temperature on the workbench. Slow cooling to room temperature should take 45–60 min. Store on ice or at 4 °C until ready to use.
4. Ligate the duplex into BsmBI-linearized sgRNA plasmid MLM3636 vector.
5. Transform into competent bacteria. Spread on a LB agar plate and incubate overnight at 37 °C.
6. Proceed to plasmid isolation with PureLink® Quick Plasmid *Miniprep* Kit (Invitrogen). Because the BsmBI overhangs are distinct (GTGT

and TTTT), a single duplex oligonucleotide should ligate in the correct orientation. Sequencing is recommended to confirm.

7. For a positive clone, proceed to plasmid isolation with PureLink® HiPure Plasmid Filter *Maxiprep* Kit (Invitrogen). Measure the DNA concentration (e.g., using a NanoDrop 2000c spectrophotometer). The final DNA plasmid concentration should be >2 μg/μL.
8. Note: If the translocation rate is very low, sgRNAs can be cloned directly in the same plasmid as Cas9 (pX330-U6-Chimeric_BB-CBh-hSpCas9, Addgene plasmid 42230) or nCas9 (pX335-U6-Chimeric_BB-CBh-hSpCas9n(D10A), Addgene plasmid 42335). Other sgRNA plasmids contain markers such as GFP (Addgene plasmid PX458) for FACS sorting or the puromycin resistance gene (Addgene plasmid PX459) for selection to increase the recovery of transfected cells.

## 3.2. Cell transfections with sgRNA and Cas9 or nCas9 expression plasmids

The protocol has been optimized for RPE1 and MSC cells, but any other cell line of interest can be used.

Transfection can induce high cell lethality, which can be explained by (i) inappropriate cell culture conditions, (ii) inappropriate transfection program, (iii) lack of sgRNA specificity, and (iv) induction of a lethal translocation. Check these parameters to reduce mortality.

1. RPE1 and MSC cells are cultured in a T150 flask in 20 mL of DMEM/F-12 medium and supplemented αMEM medium, respectively, at 37 °C and 5% $CO_2$. Passage every 2–3 days at a split ratio of 1:5 to 1:10, keeping them at ≤80% confluency.
2. The day before transfection, pass the cells without antibiotics (to decrease lethality during electroporation) to reach ~70–80% confluency on the day of transfection.
3. On the day of transfection, prepare Eppendorf tubes containing 3.5 μg of each sgRNA plasmid and 3.5 μg of the Cas9 or nCas9 expression plasmid. As controls, transfect one of the two sgRNA expression plasmids for Cas9 or two of the four sgRNA expression plasmids for nCas9 and the same quantity of the sgRNA expression plasmid in which no sgRNA was cloned. For nCas9, you can transfect one sgRNA each targeting the two chromosomes to exclude nick-induced translocations (Fig. 12.1C). The total volume of transfected DNA should not exceed 10 μL.
4. Prefill 2–3 wells of a 6-well plate with 1 mL medium and prewarm to 37 °C.

5. Prefill three 96-well plates with 50 μL medium per well and prewarm to 37 °C.
6. Trypsinize the RPE1 or MSC cells and resuspend in prewarm supplemented medium DMEM/F-12 or αMEM, respectively (typically 10 mL). Do not use 4 °C media at any time.
7. Count the cells and for each transfection put $7.5 \times 10^5$ cells in a 15-mL Falcon tube. The program and the number of cells must be adjusted for each cell line (refer to Lonza Nucleofector Protocols). Conditions described here are optimized for MSC and RPE1 cells.
8. Centrifuge 10 min at 90 × *g* at room temperature.
9. Carefully remove the medium without aspirating the cells. FBS from the medium may inhibit transfection. A supplemental wash with DPBS may increase the transfection efficiency.
10. Carefully resuspend cells in 100 μL of Cell Line Nucleofector Kit V (Lonza) solution. To prevent cell lethality, avoid all unnecessary cell agitation.
11. Transfer the cells into the tube containing DNA and then transfer the cell/DNA mix into the Amaxa DNA cuvettes.
12. Electroporate using the Nucleofector II system (Lonza) on program B-016 for MSCs and X-001 for RPE1 cells.
13. Transfer transfected cells to 5 mL of prewarmed medium.
14. Dilute the cells as follows to a final volume of 6 mL: 1.2 mL (1/5), 600 μL (1/10), and 300 μL (1/20). Plate 50 μL of each cell dilution into each well of a prewarmed 96-well plate. In addition, transfer 1 mL of the 5 mL of cells to a well of a 48-well plate for cell counts (see step 3.6.1).
15. Transfer the rest of the cells into wells of a 6-well plate (step 3.2.4) for further DNA, protein, or chromosome analysis. If the translocation frequency is high enough, translocations can be directly detected by Western blotting using antibodies directed against the fusion protein (frequency $\geq 10^{-3}$) or by microscopy using FISH probes (frequency $\geq 10^{-2}$) (Piganeau et al., 2013). Prepare supplementary wells during the transfection if you plan these additional analyses.

---

Induction of the *NPM–ALK* translocation is frequent enough that expression of the NPM–ALK fusion protein can be detected in the bulk population of transfected cells (Fig. 12.1A).

---

16. Incubate cells at 37 °C, 5% $CO_2$.

17. Note: To estimate transfection efficiency, our experiments utilize pCas9_GFP and pCasD10A9_GFP in which GFP is expressed as a 2A fusion. The percent GFP-positive cells is determined by flow cytometry 48 h after transfection. Poor transfection can result in low translocation efficiency. Test several programs to optimize the transfection efficiency for each cell line.

## 3.3. T7 endonuclease I assay to estimate cleavage efficiency

This assay estimates the efficiency of sgRNAs to direct cleavage by quantifying insertions and deletions (indels) resulting from DSB repair via NHEJ (Guschin et al., 2010). For nCas9 paired nicks, the estimate is done using each of the sgRNAs separately with wild-type Cas9 as well as with the two sgRNAs together with nCas9 (Fig. 12.1D).

Several sgRNAs can be designed to target the same locus, such that sgRNAs that are found to efficiently induce indels can be used to generate translocations.

1. Five days after transfection, trypsinize cells in the 6-well plate (from step 3.2.15).
2. Centrifuge at 200 × *g* at 4 °C for 5 min.
3. Remove supernatant and wash the pellet with 1 mL of DPBS.
4. Repeat centrifugation.
5. Remove supernatant. Cell pellets can be stored at −80 °C.
6. Extract genomic DNA with the QIAamp DNA Mini Kit (Qiagen). Genomic DNA can be stored at −20 °C.
7. Design a set of primers amplifying a 300–500 bp region encompassing the target sites on nontranslocated chromosomes. Cleavage sites should not be located in the middle of the amplicon, in order to obtain two distinct bands after T7 endonuclease I cleavage.

---

ALK locus primers generate a 401-bp fragment from the wild-type locus and smaller fragments from the modified locus (Fig. 12.1D):

ALK-F 5′-AGATGGGCAGAGGCTTGAAAAG-3′

ALK-R 5′-TGAGGATGTTCTGGAAGGCAAA-3′

---

8. Perform a 35-cycle PCR on 50 ng of genomic DNA in a total of 25 μL to amplify regions encompassing the target sites. We typically use Phusion High-Fidelity DNA Polymerase (Thermo Scientific).
9. Verify the quality of the amplification by running 5 μL of the PCR reaction on a 1% agarose gel with EtBr in 0.5 × TBE. Only one band should be amplified.

10. Mix 10 μL PCR reaction with 10 μL 2 × NEBuffer 2.1 in two different PCR tubes.
11. Melt and reanneal amplicons by incubating as follows: 95 °C for 5 min, 95 °C to 25 °C at −0.5 °C/s, and 15 min at 4 °C. This step converts fragments with and without indels into mismatched heteroduplex DNA. Add 1.5 U of T7 endonuclease I in one of the two tubes. Add the same volume of buffer in the second one as a control.
12. Incubate at 37 °C for 20 min. This step allows T7 endonuclease I to cleave mismatched DNA duplexes.
13. Add 10 μL of 2 × T7 loading buffer containing Proteinase K to 10 μL DNA. Incubate at room temperature for 5 min. In this step, T7 endonuclease I is degraded by Proteinase K.
14. Load the PCR product on a 2.4% agarose gel with EtBr and run at 100 V for 20 min in 0.5 × TBE buffer.
15. Capture the gel image with a UV imaging station (Fig. 12.1D). The amount of cleaved DNA estimates the Cas9-induced mutation rate. Each sgRNA should induce detectable levels of indels when tested with Cas9. If the T7 endonuclease I assay reveals a low indel efficiency for a chosen sgRNA, increase the amount of the sgRNA expression vector or redesign the sgRNA. If using the nCas9 approach, the four sgRNAs must induce indels separately.

### 3.4. PCR-based translocation detection

1. With a low translocation frequency, a nested PCR approach is typically used. Design two sets of primers on each side of the translocation junction, the second set located within the first product (see Section 2.4). For a frequency $>10^{-4}$, however, translocations can be detected with a single set of PCR primers. The amplified product should be 600–1000 bp to recover potentially long deletions from the DNA ends during translocation formation.

---

Der5, corresponding to the NPM–ALK fusion:
External primers:
Der5-F 5′-CAGTTGCTTGGTTCCCAGTT-3′
Der5-R 5′-AGGAATTGGCCTGCCTTAGT-3′
Internal primers:
Der5-NF 5′-GGGGAGAGGAAATCTTGCTG-3′
Der5-NR 5′-GCAGCTTCAGTGCAATCACA-3′

---

2. Perform a 23-cycle PCR on 100–150 ng genomic DNA (extracted in step 3.3.6) with the external set of primers. We classically use FastStart Taq DNA polymerase (Roche).
3. Perform a 40-cycle PCR on 0.5–1 μL of the first PCR reaction with the internal set of primers. Nested PCR is a highly sensitive method, presenting contamination risks. Work carefully in a dedicated place, wear gloves, take care not to contaminate a tube with already amplified PCR products, and clean up your bench and your material after each PCR.
4. Load the PCR product on a 1% agarose gel with EtBr in 0.5 × TBE buffer.
5. Capture the gel image with a UV imaging station (Fig. 12.1C).
6. Note: Translocations require DSBs or paired nicks on both chromosomes (Fig. 12.1C). PCR amplification of translocation junctions may lead to larger or smaller products than expected due to indels generated during translocations formation (see below Fig. 12.2C), which can be confirmed by sequencing.

## 3.5. Quantification of potential off-target cleavage

When using Cas9, potential off-target sites should be determined during sgRNA design, because mismatches to the ~20 nt target sequence do not necessary abrogate cleavage. Sequences containing up to five mismatches compared to the target sequence, especially in the PAM-distal region (Hsu et al., 2013), should be considered as potential sites for cleavage by Cas9. Potential off-target effects from nCas9 and single sgRNAs are significantly reduced since nicks are generated. However, off-target paired nicks are still possible with nCas9 if two sgRNAs bind by chance relatively close to each other. Thus, all combinations of sgRNAs should be considered for potential off-target binding.

---

**EWS–FLI1**

EWS genomic sequence followed by sgRNA sequence:

5′-GGAATCCAGGACACATCTTT**AGG**-3′

sgRNA EWS: 5′-GGAATCCAGGACACATCTTT

Potential off-target site for the sgRNA EWS on Chr9 (two mismatches) (Fig. 12.2A):

5′-GGA*T**TCCAGGAC*C**CATCTTT**GGG**-3′

FLI1 genomic sequence followed by sgRNA sequence:

5′-**CCT**CCCATGGCTGTCTCTAGACC-3′

sgRNA FLI1: 5′-GGTCTAGAGACAGCCATGGG

---

1. After determining potential sgRNA off-target sites, design a set of primers amplifying a 300–500 bp region encompassing the potential off-target site. Then perform the T7 endonuclease I assay (Section 3.3).

---

EWS locus primers generate a 354-bp fragment from the wild-type locus and smaller fragments from the modified locus (Fig. 12.2B):
 EWS-F 5′-CCTCAGCCACCCAGAGTGTT-3′
 EWS-R 5′-TAGCTGCCTCCCCACTTTACAT-3′
EWS off-target locus primers generate a 465-bp fragment from the wild-type locus and smaller fragments from the modified locus
 EWS-OFF-F 5′-ACTCACCTGGTTGGGTTGTCTT-3′
 EWS-OFF-R 5′-GTCCGTACTATGAAGGGGTCGT-3′

---

2. Design primers for detection of translocations between one of the target chromosomes and the potential off-target site. Then proceed to nested PCR to detect translocations (Section 3.4)

---

Der22, corresponding to the EWS–FLI1 fusion (Fig. 12.2B):
 External primers:
 Der22-F 5′-ATCCTACAGCCAAGCTCCAA-3′
 Der22-R 5′-GGCCTCATTGTTTCTGGCTA-3′
 Internal primers:
 Der22-NF 5′-CTACGGGCAGCAGAGTGAGT-3′
 Der22-NR 5′-TTCCTCAAGGCTCTGGAAAA-3′
Der9, corresponding to the off-target-FLI1 fusion (Fig. 12.2B):
 External primers:
 Der9-F 5′-TAGTGGGGGAAGGAGAGACA-3′
 Der9-B 5′-GCCAGGTTTCTTAGGGCTTT-3′
 Internal primers:
 Der9-NF 5′-GAAAAGGCTCCATTTCATGC-3′
 Der9-NB 5′-GGGCTGAGCTCCATAAATCA-3′

---

A positive signal in one of these two assays (presence of indels after T7 assay or translocation detection by PCR) will reveal off-target cleavage of the tested sgRNA. However, the T7 assay is less sensitive than the translocation assay. The T7 assay sensitivity is ~1–2%, while translocations can be easily detected at a frequency $\geq 10^{-5}$. For example, in Fig. 12.2B a translocation between the target site and the off-target site can be amplified when no signal can be detect in the T7 assay.

Thus, PCR detection of translocation formation is a highly sensitive assay to detect off-target cleavage and can be used as a general method in other applications to evaluate off-target sites.

### 3.6. Quantification of translocation frequency using a 96-well plate screen

Translocation frequencies can be accurately determined using a 96-well plate format (Brunet et al., 2009; Piganeau et al., 2013). The basic strategy is that small pools of cells are placed in each well, such that most wells on a plate contain <1 translocation. The translocation frequency can be determined from the number of negative and positive wells, with correction for the number of wells with two or more translocations. Having a single event in most positive wells also means that unique translocation junctions can be scored.

1. 48-well plate (step 3.2.14): Trypsinize cells in the well ≤24 h after transfection, before cells have had a chance to divide. Use a small volume of trypsin (typically 100 μL) and count the cells contained in this well. This number represents 1/5th the transfected cells surviving transfection in each of the 96-well plates. Do not exceed this time frame to ensure an accurate cell number to calculate translocation frequency.
2. 96-well plates (step 3.2.14): Remove all of the culture medium 48 h to 5 days after transfection, depending on the growth of the cells. Plates can be stored wrapped in Parafilm at −80 °C and thawed for later use.
3. Add Proteinase K to 2.5 mL of 1 × lysis solution at final concentration of 100 μg/mL. Place 25 μL Proteinase K solution into each well of the thawed 96-well plate.
4. Incubate at 55 °C for 120 min in a humid chamber, such as a closed plastic container with moist towels at the bottom.
5. Transfer the lysates into 96 tubes on PCR tube strips.
6. Incubate at 95 °C for 10 min to inactivate Proteinase K.
7. Add 200 n*M* of the external primers (same primers used in Section 3.4) to 5 mL of 1 × PCR Master Mix 1, and add 25 μL of FastStart Taq polymerase. Perform the first round of PCR on 4–7 μL cell lysate in a total volume of 50 μL.
8. Add 200 n*M* of the internal primers (same primers used in Section 3.4) to 2 mL of 1 × PCR Master Mix 2 (containing the SYBR Green), and add 10 μL of FastStart Taq polymerase. Perform the second round of PCR on 0.5–1 μL of the first PCR in a total volume of 20 μL. The PCR program has to contain a denaturation curve cycle.
9. Count the number of wells with a positive denaturation curve. Since nested PCR fragments corresponding to translocation junctions are

designed to amplify typically 600–900 bp, all wells with a $T_m > 85$ °C are considered to be positive, corresponding to fragments >100–150 bp.

**10.** ≤12–14 positive wells per 96-well plate: Each well is estimated to contain no more than one translocation and the translocation frequency is ($p$), the number positive wells divided by the number of plated cells (determined in step 3.6.1).

**11.** >12–14 positive wells per 96-well plate: The translocation frequency is determined using a correction following a beta cumulative distribution function $k(x,a,b)$, where $k$ = number of translocations per well, $p$ = total number of positive wells per number of plated cells, $n$ = number of cells per well, and $x = 1 - p$, $a = n - k$, $b = k + 1$. If the number of wells with two or more translocations is high to interfere with precise frequency measurements, calculate from a plate with fewer cells per well (step 3.2.14).

PCR products from positive wells can be sent for sequencing to analyze breakpoint junctions (e.g., Der5, Fig. 12.3 for Cas9 and Fig. 12.4 for nCas9).

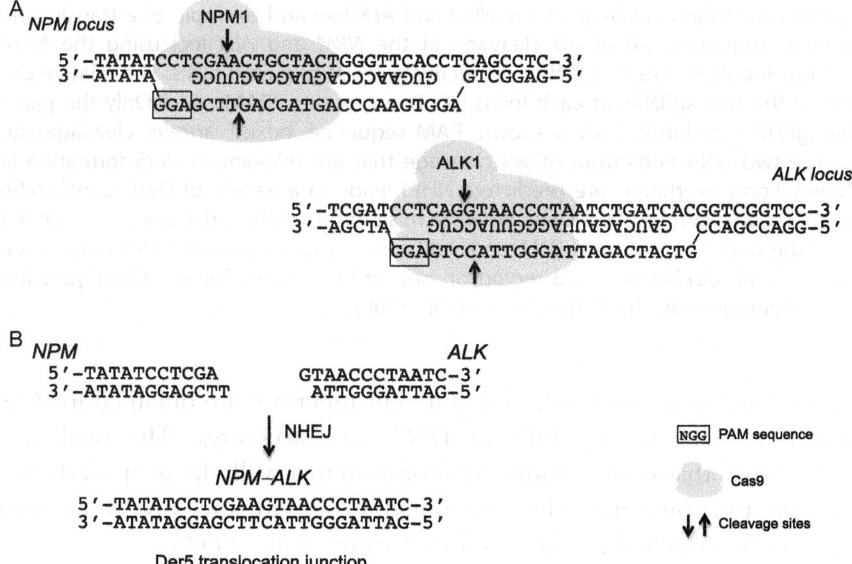


**Figure 12.3** Cas9 cleavage at the *NPM* and *ALK* loci and example of a translocation junction sequence. (A) Cas9 cleavage at the *NPM* and *ALK* loci using the NPM1 and ALK1 sgRNAs, respectively, leads to DSBs. Only the part of the sgRNA that binds DNA is shown. PAM sequence, boxed; arrows, cleavage sites. (B) DNA ends from Cas9 cleavage that are relevant to Der5 formation are shown. NHEJ leads to a variety of Der5 translocation junctions, but a common junction (shown) is direct ligation of the DNA ends, presumably involving fill-in of the 1-base 5′ overhang. For additional junctions, see Ghezraoui et al. (2014). (See the color plate.)

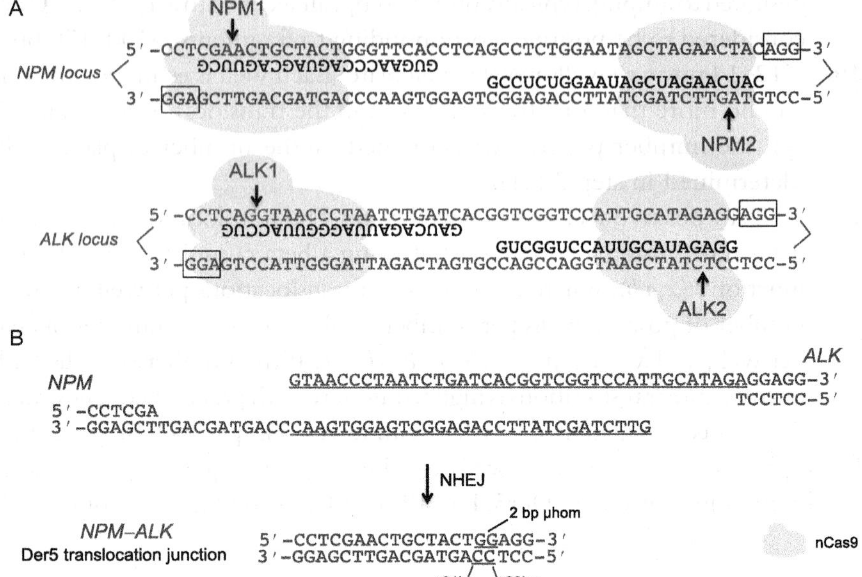


**Figure 12.4** nCas9 cleavage at the *NPM* and *ALK* loci and example of a translocation junction sequence. (A) nCas9 cleavage at the *NPM* and *ALK* loci using the NPM1 + NPM2 and ALK1 + ALK2 sgRNAs, respectively, leads to paired nicks. The relative position of the two sgRNAs at each locus has been termed "PAMs out." Only the part of the sgRNA that binds DNA is shown. PAM sequence, boxed; arrows, cleavage sites. (B) The two DNA ends from nCas9 cleavage that are relevant to Der5 formation are shown. Long overhangs are predicted. NHEJ leads to a variety of Der5 translocation junctions, an example of which is shown. In this typical junction, deletions occur primarily in the overhangs at both DNA ends, although in other junctions deletions extend well into the double-stranded region on one or both sides. For additional junctions, see Ghezraoui et al. (2014) (μhom = microhomology).

At the local sequence level, very different junctions are obtained for Cas9 and nCas9 due to the different DNA end structures. The small pool PCR means that with a unique translocation in a well, the reciprocal translocation junction can also be determined. Cells with a reciprocal translocation can be enriched by sib selection (Brunet et al., 2009).

## 3.7. Translocation frequency determination by serial dilution

The protocol described in Section 3.6 requires removal of medium from the 96-well plate. Thus, with nonadherent cells, cell loss is a potential problem. Here, we suggest an alternative method for translocation frequency assessment when frequencies are $>10^{-4}$, which is not atypical in our experience.

1. Perform serial dilutions on genomic DNA in quadruplicates starting from 100 to 1.56 ng DNA (extraction in step 3.3.6).
2. Perform PCR on each dilution as in step 3.4.3.
3. Load on a 1% agarose gel with EtBr.
4. Capture the gel image with a UV imaging station. Considering that one human diploid cell contains 6 pg of DNA, 6.25 ng represent about $10^3$ cells. Consider all the quadruplicates to determine the translocation frequency (Fig. 12.2C for Der22).

## 4. CONCLUSIONS

In this chapter, we described a method to induce cancer translocations using Cas9 DSBs or nCas9 paired nicks in human cells. The ease at which potential target sites can be found at any locus of interest makes this approach readily adaptable for expanding the repertoire of possible cancer translocations that can be induced in any cell type of interest. Concerns about off-target effects are diminished by the use of nCas9. In addition, nCas9 will better recapitulate some types of breakpoints junctions found in patient cells (Ghezraoui et al., 2014). Cells carrying the chosen translocations provide a more accurate model to understand mechanisms of tumor initiation than standard ectopic expression of a fusion protein. Results from such studies may assist the design of targeted therapies, for example, to avoid therapy-induced secondary tumors with specific translocations (Cowell et al., 2012). Our approach can also be used to decipher the repair mechanisms that are involved in translocation formation, because PCR screening of translocation junctions allows accurate frequency determination in parallel with the precise determination of breakpoint junctions (Ghezraoui et al., 2014). Finally, translocation assays provide a highly sensitive method to detect off-target cleavage for any application, and so has general utility.

## ACKNOWLEDGMENTS

We thank Carine Giovannangeli, Anne De Cian and Jean-Paul Concordet (MNHN) and Matt Krawczyk (MSKCC) for their helpful discussions. This work was supported by ANR-12-JSV6-0005 (B.R.), Le Canceropole IDF (M.P.), La Ligue Nationale contre le Cancer (H.G.), and NIH grant GM054668 (M.J.).

## REFERENCES

Brunet, E., Simsek, D., Tomishima, M., DeKelver, R., Choi, V. M., Gregory, P., et al. (2009). Chromosomal translocations induced at specified loci in human stem cells. *Proceedings of the National Academy of Sciences of the United States of America*, *106*, 10620–10625.

Chandra, H. S., Heisterkamp, N. C., Hungerford, A., Morrissette, J. J., Nowell, P. C., Rowley, J. D., et al. (2011). Philadelphia chromosome symposium: Commemoration of the 50th anniversary of the discovery of the Ph chromosome. *Cancer Genetics*, *204*, 171–179.

Choi, P. S., & Meyerson, M. (2014). Targeted genomic rearrangements using CRISPR/Cas technology. *Nature Communications*, *5*, 3728.

Cong, L., Ran, F. A., Cox, D., Lin, S., Barretto, R., Habib, N., et al. (2013). Multiplex genome engineering using CRISPR/Cas systems. *Science*, *339*, 819–823.

Cowell, I. G., Sondka, Z., Smith, K., Lee, K. C., Manville, C. M., Sidorczuk-Lesthuruge, M., et al. (2012). Model for MLL translocations in therapy-related leukemia involving topoisomerase IIbeta-mediated DNA strand breaks and gene proximity. *Proceedings of the National Academy of Sciences of the United States of America*, *109*, 8989–8994.

Elmberger, P. G., Lozano, M. D., Weisenburger, D. D., Sanger, W., & Chan, W. C. (1995). Transcripts of the npm-alk fusion gene in anaplastic large cell lymphoma, Hodgkin's disease, and reactive lymphoid lesions. *Blood*, *86*, 3517–3521.

Gaj, T., Sirk, S. J., & Barbas, C. F., 3rd. (2014). Expanding the scope of site-specific recombinases for genetic and metabolic engineering. *Biotechnology and Bioengineering*, *111*, 1–15.

Ghezraoui, H., Piganeau, M., Renouf, B., Renaud, J.-B., Aallmyr, A., Ruis, B., et al. (2014) Chromosomal translocations in human cells are generated by canonical nonhomologous end-joining. *Molecular Cell*, *55*, 829–842.

Guschin, D. Y., Waite, A. J., Katibah, G. E., Miller, J. C., Holmes, M. C., & Rebar, E. J. (2010). A rapid and general assay for monitoring endogenous gene modification. *Methods in Molecular Biology*, *649*, 247–256.

Hsu, P. D., Lander, E. S., & Zhang, F. (2014). Development and applications of CRISPR-Cas9 for genome engineering. *Cell*, *157*, 1262–1278.

Hsu, P. D., Scott, D. A., Weinstein, J. A., Ran, F. A., Konermann, S., Agarwala, V., et al. (2013). DNA targeting specificity of RNA-guided Cas9 nucleases. *Nature Biotechnology*, *31*, 827–832.

Jinek, M., Chylinski, K., Fonfara, I., Hauer, M., Doudna, J. A., & Charpentier, E. (2012). A programmable dual-RNA-guided DNA endonuclease in adaptive bacterial immunity. *Science*, *337*, 816–821.

Kuefer, M. U., Look, A. T., Pulford, K., Behm, F. G., Pattengale, P. K., Mason, D. Y., et al. (1997). Retrovirus-mediated gene transfer of NPM-ALK causes lymphoid malignancy in mice. *Blood*, *90*, 2901–2910.

Mali, P., Aach, J., Stranges, P. B., Esvelt, K. M., Moosburner, M., Kosuri, S., et al. (2013). CAS9 transcriptional activators for target specificity screening and paired nickases for cooperative genome engineering. *Nature Biotechnology*, *31*, 833–838.

Mali, P., Yang, L., Esvelt, K. M., Aach, J., Guell, M., DiCarlo, J. E., et al. (2013). RNA-guided human genome engineering via Cas9. *Science*, *339*, 823–826.

Mani, R. S., & Chinnaiyan, A. M. (2010). Triggers for genomic rearrangements: insights into genomic, cellular and environmental influences. *Nature Reviews Genetics*, *11*, 819–829.

May, W. A., Gishizky, M. L., Lessnick, S. L., Lunsford, L. B., Lewis, B. C., Delattre, O., et al. (1993). Ewing sarcoma 11;22 translocation produces a chimeric transcription factor that requires the DNA-binding domain encoded by FLI1 for transformation. *Proceedings of the National Academy of Sciences of the United States of America*, *90*, 5752–5756.

Mitelman, F., Johansson, B., & Mertens, F. (2007). The impact of translocations and gene fusions on cancer causation. *Nature Reviews Cancer*, *7*, 233–245.

Montague, T. G., Cruz, J. M., Gagnon, J. A., Church, G. M., & Valen, E. (2014). CHOP-CHOP: A CRISPR/Cas9 and TALEN web tool for genome editing. *Nucleic Acids Research*, *42*, W401–W407.

Morris, S. W., Kirstein, M. N., Valentine, M. B., Dittmer, K. G., Shapiro, D. N., Saltman, D. L., et al. (1994). Fusion of a kinase gene, ALK, to a nucleolar protein gene, NPM, in non-Hodgkin's lymphoma. *Science, 263*, 1281–1284.

Piganeau, M., Ghezraoui, H., De Cian, A., Guittat, L., Tomishima, M., Perrouault, L., et al. (2013). Cancer translocations in human cells induced by zinc finger and TALE nucleases. *Genome Research, 23*, 1182–1193.

Ran, F. A., Hsu, P. D., Lin, C. Y., Gootenberg, J. S., Konermann, S., Trevino, A. E., et al. (2013). Double nicking by RNA-guided CRISPR Cas9 for enhanced genome editing specificity. *Cell, 154*, 1380–1389.

Richardson, C., & Jasin, M. (2000). Frequent chromosomal translocations induced by DNA double-strand breaks. *Nature, 405*, 697–700.

Rowley, J. D. (1973). Letter: A new consistent chromosomal abnormality in chronic myelogenous leukaemia identified by quinacrine fluorescence and Giemsa staining. *Nature, 243*, 290–293.

Simsek, D., Brunet, E., Wong, S. Y., Katyal, S., Gao, Y., McKinnon, P. J., et al. (2011). DNA ligase III promotes alternative nonhomologous end-joining during chromosomal translocation formation. *PLoS Genetics, 7*, e1002080.

Torres, R., Martin, M. C., Garcia, A., Cigudosa, J. C., Ramirez, J. C., & Rodriguez-Perales, S. (2014). Engineering human tumour-associated chromosomal translocations with the RNA-guided CRISPR-Cas9 system. *Nature Communications, 5*, 3964.

Urnov, F. D., Rebar, E. J., Holmes, M. C., Zhang, H. S., & Gregory, P. D. (2010). Genome editing with engineered zinc finger nucleases. *Nature Reviews Genetics, 11*, 636–646.

Weinstock, D. M., Elliott, B., & Jasin, M. (2006). A model of oncogenic rearrangements: Differences between chromosomal translocation mechanisms and simple double-strand break repair. *Blood, 107*, 777–780.

Xie, S., Shen, B., Zhang, C., Huang, X., & Zhang, Y. (2014). sgRNAcas9: A software package for designing CRISPR sgRNA and evaluating potential off-target cleavage sites. *PLoS One, 9*, e100448.

CHAPTER THIRTEEN

# Genome Editing for Human Gene Therapy

**Torsten B. Meissner*[,1], Pankaj K. Mandal*[,†,1], Leonardo M.R. Ferreira*, Derrick J. Rossi*[,†,‡,§], Chad A. Cowan*[,§,¶,2]**

*Department of Stem Cell and Regenerative Biology, Harvard University, Cambridge, Massachusetts, USA
†Program in Cellular and Molecular Medicine, Division of Hematology/Oncology, Boston Children's Hospital, Boston, Massachusetts, USA
‡Department of Pediatrics, Harvard Medical School, Boston, Massachusetts, USA
§Harvard Stem Cell Institute, Sherman Fairchild Biochemistry, Cambridge, Massachusetts, USA
¶Center for Regenerative Medicine, Massachusetts General Hospital, Boston, Massachusetts, USA
[1]These authors contributed equally to this work
[2]Corresponding author: e-mail address: chad_cowan@harvard.edu

## Contents



## Abstract

The rapid advancement of genome-editing techniques holds much promise for the field of human gene therapy. From bacteria to model organisms and human cells, genome editing tools such as zinc-finger nucleases (ZNFs), TALENs, and CRISPR/Cas9 have been successfully used to manipulate the respective genomes with unprecedented precision. With regard to human gene therapy, it is of great interest to test the feasibility of genome editing in primary human hematopoietic cells that could potentially be used to treat a variety of human genetic disorders such as hemoglobinopathies, primary immunodeficiencies, and cancer. In this chapter, we explore the use of the CRISPR/Cas9 system for the efficient ablation of genes in two clinically relevant primary human cell types, CD4+ T cells and CD34+ hematopoietic stem and progenitor cells. By using two guide RNAs directed at a single locus, we achieve highly efficient and predictable deletions that ablate gene function. The use of a Cas9-2A-GFP fusion

*Methods in Enzymology*, Volume 546
ISSN 0076-6879
http://dx.doi.org/10.1016/B978-0-12-801185-0.00013-1

protein allows FACS-based enrichment of the transfected cells. The ease of designing, constructing, and testing guide RNAs makes this dual guide strategy an attractive approach for the efficient deletion of clinically relevant genes in primary human hematopoietic stem and effector cells and enables the use of CRISPR/Cas9 for gene therapy.

## 1. INTRODUCTION

The goal of gene therapy is the correction of a mutation or genetic defect using recombinant DNA technologies. Historically, the manipulation of genomes relied on homologous recombination (HR)-based strategies, where a donor template with significant homology to the targeted region was used to introduce the desired changes (Capecchi, 1989). While HR-mediated gene targeting has proven extremely successful in certain animal models, it has only been met with moderate success in human cell lines, mainly because of the low efficiency of HR in human cells.

The development of genome-editing tools such as zinc-finger nucleases (ZFNs), transcription activator-like effector nucleases (TALENs), and clustered, regularly interspaced, short palindromic repeat (CRISPR)/CRISPR-associated (Cas) nucleases has changed this scenario dramatically (Hsu, Lander, & Zhang, 2014). These programmable nucleases generate double-strand breaks (DSBs) at high efficiency that are usually repaired by nonhomologous end joining (NHEJ), an error-prone process, which frequently results in the disruption of the targeted gene (Sander & Joung, 2014). Alternatively, genome editing tools can also be used for gene repair by increasing the incorporation of HR templates, such as single-stranded oligonucleotides or homology-based targeting vectors (Sander & Joung, 2014).

While there has been some success in gene repair using ZFNs and TALENs in cells of hematopoietic origin and human pluripotent stem cells (Ding, Lee, et al., 2013; Genovese et al., 2014; Kiskinis et al., 2014; Mali et al., 2013), the more recently developed CRISPR/Cas9 system has quickly become the tool of choice by virtue of its efficacy and ease of use (Hsu et al., 2014). Within less than 2 years from its first application in human cell lines (Cong et al., 2013; Jinek et al., 2013; Mali et al., 2013), CRISPR/Cas9-mediated genome editing has been successfully employed in a variety of model organisms, human embryonic and induced pluripotent stem cells, as well as in human adult stem cells (Ding, Regan, et al., 2013; Sander & Joung, 2014; Schwank et al., 2013).

Recent proof-of-principle studies in small animal models suggest that gene correction using CRISPR/Cas9 may even be possible in patients (Ding et al., 2014; Yin et al., 2014). Changes introduced by genome editing are permanent, which in certain cases can be superior over short-lived biologics or small-molecule drug treatments. Together, these studies demonstrate that genome editing has the potential for correction of human genetic diseases and might translate into new clinical therapies.

In this section, we focus on the transient delivery of CRISPR/Cas9 to human hematopoietic cells, which can readily be isolated from peripheral blood, expanded *ex vivo*, and subsequently reinfused into the patient (Bryder, Rossi, & Weissman, 2006). The *ex vivo* culture/expansion step opens up a window of opportunity for the manipulation of the desired cell type, such as human T cells or hematopoietic progenitor cells, by genome editing.

The first protocol in this section describes the targeting of the beta-2-microglobulin (*B2M*) gene in primary human CD4+ T cells. B2M-deficient cells are devoid of major histocompatibility complex-I (MHC-I) surface expression (Gussow et al., 1987), which normally represents the major barrier in donor-derived (allogeneic) transplantation.

Antigen-specific T cells are already used in adoptive cell transfer (ACT)-based therapies, and provide a promising treatment for a variety of malignancies, including melanoma, and acute and chronic lymphoma (June, Rosenberg, Sadelain, & Weber, 2012). More recently, the design of chimeric antigen receptors has allowed the redirection of effector T cells to fight a variety of hematological malignancies (Maus, Grupp, Porter, & June, 2014). A major drawback of adoptive T cell therapies, however, is their current limitation to an autologous setting, where tumor-specific T cells are isolated from a patient, expanded and given back to the same patient. Moreover, the derivation of antigen-specific T cells for ACT and their expansion can take up to several weeks. It is thus desirable to generate universally transferable donor T cells that are readily available and can be used in an allogeneic transplantation setting and thus be administered to multiple recipients with disparate MHC expression.

The second protocol describes a strategy to delete chemokine (C-C motif) receptor 5 (*CCR5*) in CD34+ HSPCs using the CRISPR/Cas9 system. As CCR5 is an essential co-receptor for entry of the human immunodeficiency virus-1 (HIV-1) (Trkola et al., 1996), targeting *CCR5* in HSPCs represents an attractive way of creating an HIV-1-resistant immune system. Upon transplantation, CCR5-deficient HSPCs can engraft, expand and ultimately give rise to CCR5-deficient CD4+ T cells that are resistant to HIV-1 virus infection (Holt et al., 2010). Further studies will be required to evaluate the safety and extent of off-target effects, particularly *in vivo*, and strategies to reduce off-target effects, such as Cas9 nickase (Ran, Hsu, Lin, et al., 2013) or the recently developed Cas9-FokI fusion protein (Tsai et al., 2014), will help to translate these strategies more safely into the clinic. Ultimately, combining both approaches described in this section, deletion of *B2M* and *CCR5*, may extend adoptive cell therapy for HIV-1 infection beyond the autologous setting and make this form of treatment more accessible to a larger number of patients.

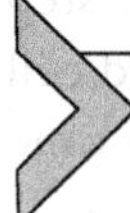

## 2. GENOME EDITING OF *B2M* IN PRIMARY HUMAN CD4+ T CELLS

Overcoming the MHC barrier in transplantation is one of the major goals in regenerative medicine, and several strategies have been envisioned to reduce the immunogenicity of transplanted cells by targeting either HLA molecules directly or the accessory chain B2M (Lu et al., 2013; Riolobos et al., 2013; Torikai et al., 2013). B2M is constitutively expressed on virtually all nucleated cells and required for the proper surface trafficking of MHC class I molecules (Gussow et al., 1987). Deletion of the *B2M* gene results in loss of MHC-I surface expression (Koller, Marrack, Kappler, & Smithies, 1990; Zijlstra et al., 1990), and is a common strategy of tumors to evade immune rejection (Challa-Malladi et al., 2011; D'Urso et al., 1991; Rosa et al., 1983).

In this protocol, we focus on the delivery of *B2M*-specific CRISPR/Cas9 to primary human CD4+ T cells obtained from peripheral blood. CD4+ T cells are isolated by negative selection and are subsequently transiently transfected by nucleofection. The use of a Cas9-2A-GFP construct allows the identification and sorting of the transfected cells based on the GFP fluorescence. In the case of B2M, which is expressed at the cell surface, successful targeting can be confirmed by monitoring loss of B2M expression by means of a simple surface staining with an anti-B2M antibody, followed by fluorescence-activated cell sorting (FACS) analysis.

In this study, we chose to employ a dual guide strategy, which uses two CRISPR guides directed against the *B2M* locus (see Fig. 13.1). We previously observed improved cutting efficiency for certain guide combinations in primary human hematopoietic cells (P. Mandal & L. Ferreira, unpublished) using this dual guide approach. Similar results have been

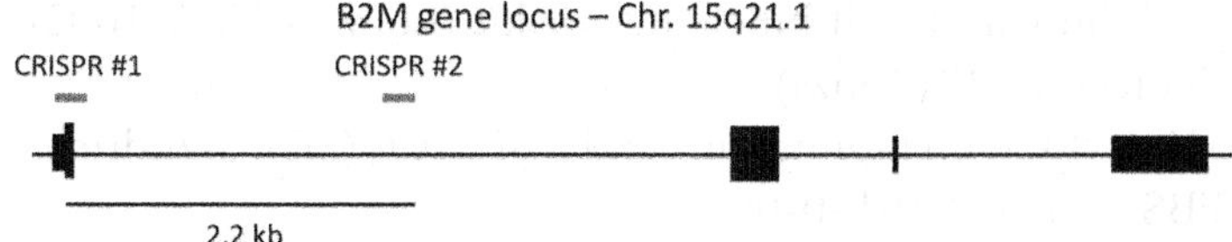


**Figure 13.1** Dual guide strategy. Schematic representation of the *B2M* locus on chromosome 15q21.1 indicating the binding sites of the CRISPRs used in this study (red (light gray in the print version) bars). CRISPR#1 targets the first exon, while CRISPR#2 will introduce a double-strand break 2.2 kb downstream in the first intron of the *B2M* gene. Cutting of the individual CRISPRs had been previously established in 293T cells.

observed in mice and deletion of the *B2M* gene was recently accomplished using CRISPR/Cas9 multiplexing in mice and rats (Cong et al., 2013; Ma et al., 2014; Zhou et al., 2014). Another advantage of the dual guide strategy lies in the fact that a defined region is excised rather than the unpredictable indels produced as a result of NHEJ.

A limitation of the proposed method is the high degree of cell death that results from electroporation. Ultimately, the use of modified RNA or non-integrating lentiviral vectors may reduce cell mortality and allow for more efficacious CRISPR/Cas9 delivery (Banasik & McCray, 2010; Warren et al., 2010). Whether the dual guide approach will result in an increase in off-target effects will have to be determined empirically on a case-by-case basis. Nevertheless, this strategy should be applicable to a wide variety of target loci and allows the rapid assessment of the amenability of a locus to genome editing in a variety of human primary cells that are most valuable for adoptive cell therapy.

## 2.1. Required materials

- Citrate anticoagulated blood obtained from leukophoresis (leukopac) from a healthy donor (MGH Blood bank, Boston, MA, USA)
- Rosette Sep™ human CD4+ T Cell Enrichment Cocktail (StemCell™ Technologies #15022)
- Hemocytometer or equivalent cell counter
- Trypan blue solution (Sigma #T8154)
- Ficoll-Paque™ (GE Healthcare #17-1440-03)
- Plastic transfer pipettes (BD Falcon #357575)
- Dulbecco's Phosphate buffered saline (Corning #21-031-CV)
- Freshly isolated CD4+ T cells (see 2.2)
- Cas9-2A-GFP plasmid (Addgene #44719)
- Guide plasmid (Addgene #41824)

- Amaxa® human T cell Nucleofector kit (Lonza #VPA-1002)
- Nucleofector® II (Lonza)
- RPMI-1640, with L-glutamine and $NaHCO_3$ (Sigma-Aldrich #R8758)
- HI FBS (Gibco #16140-063)
- Penicillin/Streptomycin Solution (VWR #45000-652)
- HEPES (Invitrogen #15630)
- GlutaMax™ (Gibco #35050061)
- Human IL-2, animal free (Peprotech #AF-200-02)
- FACS blocking buffer (4% FBS in PBS)
- FACS staining buffer (1% FBS in PBS)
- FACS tubes with cell-strainer cap (BD Falcon #352235)
- 7-AAD viability staining solution (BioLegend #420404)
- Zombie Aqua™ fixable viability kit (BioLegend #423101)
- Anti-CD4-PE, clone RPA-T4, IgG1, κ (BD Pharmingen #555347)
- Anti-B2M-APC, clone 2M2, IgG1, κ (BioLegend #316302)
- Anti-human HLA-A,B,C-Alexa647, clone W6/32, IgG2a, κ (BioLegend #311416)
- FACS Calibur™ or LSR II (BD Bioscience) for cell analysis
- FACS Aria™ (BD Bioscience) for cell sorting
- Genomic DNA isolation kit (Qiagen #69506)
- Phusion Green Hot Start II High-Fidelity DNA Polymerase (Thermo Scientific F-537S)
- Gene-specific primers
- dNTPs (Thermo Scientific #R0192)
- Agarose (GeneMate #E-3120-500)
- TBE buffer (Thermo Scientific #B52)
- Gel running station (BioRad)

## 2.2. Isolation of CD4+ T cells from peripheral blood

A variety of kits are available to isolate primary human leukocytes. In this protocol, we chose to carry out negative selection using Rosette Sep™ (Stem Cell Technologies), which allows the isolation of untouched CD4+ T lymphocytes. Purity can be assessed by flow cytometry after staining with a fluorophore-conjugated anti-CD4 antibody, possibly in combination with other T cell-specific markers if desired (see Fig. 13.2).

1. Pour blood from a leukopac into a 50-ml Falcon tube (see Note 1).
2. Add Rosette Sep™ human CD4+ T Cell Enrichment Cocktail at 50 μl/ml of blood.

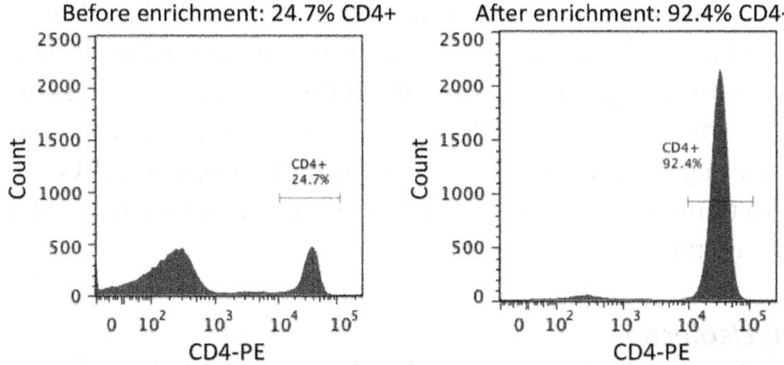


**Figure 13.2** Enrichment of human CD4+ T cells using Rosette Sep™. Human CD4+ T cells can be enriched to a purity >90% using the human CD4+ T Cell Enrichment Cocktail. CD4+ T cells were isolated from leukopacs and purity was assessed before (left panel) and after enrichment (right panel) using a phycoerythrin-labeled anti-CD4 antibody.

3. Incubate mixture for 20 min at room temperature (RT).
4. Dilute sample with an equal volume of PBS and mix gently by inverting.
5. Layer the diluted sample over an adequate volume of Ficoll-Paque™ (see Note 2).
6. Centrifuge for 25 min at 1800 rpm without break at RT.
7. Collect cells at interface with a plastic transfer pipette and transfer to a fresh tube.
8. Dilute cells with 10 ml PBS and centrifuge for 5 min at 1800 rpm at RT.
9. Wash cells again with 10 ml PBS.
10. Resuspend cell pellet in 5 ml PBS and count cells using a hemocytometer.
11. Keep at 4 °C until use.

**Notes**

1. We recommend the use of a leukocyte-enriched source for the isolation of CD4+ T cells such as leukopacs or buffy coats. From a leukopac, the yield is typically $10–20 \times 10^6$ CD4+ T cells/ml, which is about 10 times higher than from freshly isolated peripheral blood ($1–2 \times 10^6$ CD4+ T cells/ml).
2. Alternatively, layer the density medium underneath the diluted sample.

## 2.3. Delivery of CRISPR/Cas9 by nucleofection

In this protocol, both the Cas9 and the CRISPR guide plasmids are transiently introduced into CD4+ T cells by nucleofection. Cas9 expression is driven by the *CAG* promoter, followed by a 2A-linked GFP, which allows the FACS-based enrichment of transfected cells based on the GFP signal (see Fig. 13.3). There are several protocols and online tools available that assist in the design and cloning of CRISPR single guide RNAs (Peters, Cowan, & Musunuru, 2013; Ran, Hsu, Wright, et al., 2013). We recommend verifying the cutting efficiency of the individual CRISPRs in cells that are easy to transfect, e.g., 293T cells, before switching to a more difficult-to-transfect cell type.

### 2.3.1 Nucleofection

1. Supplement the Nucleofector® solution according to the manufacturer's instructions (Lonza) and keep at RT until use.
2. Count CD4+ T cells and determine live cell count (see Note 1).
3. Prepare tubes with the required amount of cells (5–10 × $10^6$ cells per sample).
4. Centrifuge tubes at 1800 rpm, aspirate supernatants and keep cell pellets on ice (see Note 2).
5. Aliquot DNAs according to the table below (see Note 3).
6. Mix the CD4+ T cells in 100 μl of Nucleofector® solution and subsequently mix the cell suspension with the DNA aliquot. Immediately proceed to the next step (see Note 4).

   One Nucleofector® sample contains

   5–10 × $10^6$ CD4+ T cells
   5 μg Cas9-2A-GFP plasmid
   2.5 μg Guide plasmid#1
   2.5 μg Guide plasmid#2
   100 μl Human CD4+ T cell nucleofector solution
7. Transfer cell/DNA suspension to cuvettes provided by the kit and insert cuvette into nucleofector.
8. Use program U-014 for nucleofection (see Note 5).
9. Immediately transfer cells to a 15-ml Falcon tube filled with 7 ml prewarmed RPMI-10 using the provided plastic pipettes (see Note 6).
10. Pellet cells by centrifugation at 1800 rpm for 5 min.
11. Aspirate medium and resuspend in fresh RPMI-10.
12. Replate in 0.5 ml RPMI-10 in an appropriately sized well.

    RPMI-10 medium

    RPMI
    10% FBS
    10 m*M* HEPES
    1× GlutaMax
    1× Pen/Strep

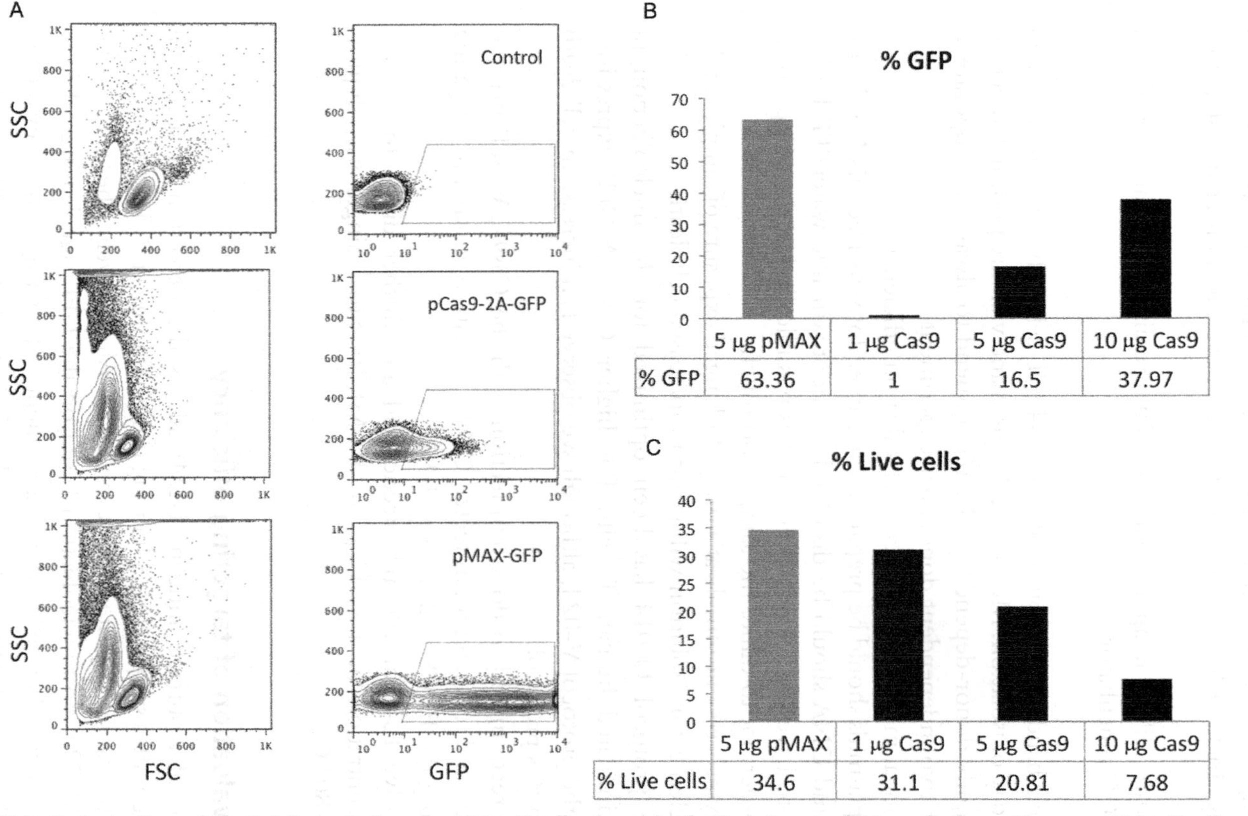


**Figure 13.3** Optimization of Cas9 delivery to human CD4+ T cells via nucleofection. Successful transfection of human CD4+ T cells can be analyzed based on the green fluorescent protein (GFP) signal when compared to an untransfected or a mock electroporated sample. pMax-GFP was used as a positive control (A). Increasing the amount of the Cas9-2A-GFP plasmid per nucleofection results in a higher percentage of GFP+ cells (B). However, it also induces an increase in cell death, thus reducing the overall number of live GFP+ cells (C).

### *2.3.2 Postnucleofection*

1. Change medium within the first 6–12 h posttransfection to improve viability. Include 50 IU/ml human IL-2 in the medium at this step (see Note 7).
2. Culture cells in an appropriate sized culture dish until analysis at 37 °C, 5% $CO_2$ incubator.

**Notes**

1. In our hands, the viability of the CD4+ T cells is best when processed the day of isolation. Moreover, the yield and viability postnucleofection is highly donor-dependent. We recommend to always electroporate at least two independent donors per experiment.
2. Residual medium can affect nucleofection efficiency.
3. All plasmids should be prepared using an endotoxin-free DNA isolation kit and DNA should be dissolved in endotoxin-free water/TE buffer. Try to have DNAs as concentrated as possible (in less than 10 μl total) in order not to dilute the sample too much.
4. Only process one sample at a time. Prolonged exposure of samples in the Nucleofector® solution will result in increased cell death.
5. The protocol U-014 has been optimized for the nucleofection of unstimulated human T cells. For higher Cas9-2A-GFP expression, use the protocol V-024, although we observed an increase in cell death with this protocol.
6. We recommend to do a transfection with the Cas9-2A-GFP plasmid alone, and to keep untransfected CD4+ T cells to ensure proper gating during the FACS analysis (see 2.4.1).
7. Dissolve IL-2 in 100 m*M* acetic acid and further dilute to a stock concentration of 100 IU/μl in RPMI-10. Store single use aliquots at −80 °C.

## 2.4. Evaluation of targeting efficiency

The use of a surface antigen (B2M) allows the detection of successful targeting with a simple FACS staining using a fluorescently labeled antibody against the targeted protein (Section 2.4.1). Alternatively, the cells can be sorted 48–72 h postnucleofection based on the GFP signal using a FACS Aria or comparable cell sorter. Subsequently, targeting can be evaluated by PCR on genomic DNA isolated from the pooled population (Section 2.4.2).

### 2.4.1 FACS-based analysis

FACS analysis is performed 48–72 h postnucleofection. Cells are harvested, stained with the desired fluorophore-conjugated antibodies and analyzed using a BD LSR II (BD Bioscience) or equivalent cell sorter. Because of the high number of dead cells following electroporation, we strongly recommend to include a dead cell marker, such as 7-AAD or Zombie Aqua™, in the staining protocol. The gating strategy used to detect successfully targeted cells is depicted in Fig. 13.4.

1. Spin down cells at 1800 rpm for 5 min.
2. Resuspend pellets in 750 μl PBS and repeat centrifugation.
3. Resuspend pellet in 200 μl PBS including 2 μl Zombie Aqua™ (see Note 1).
4. Incubate at RT for 15–20 min. Protect from light! (see Note 2).
5. Pellet cells and wash once with staining buffer (1% FBS in PBS).

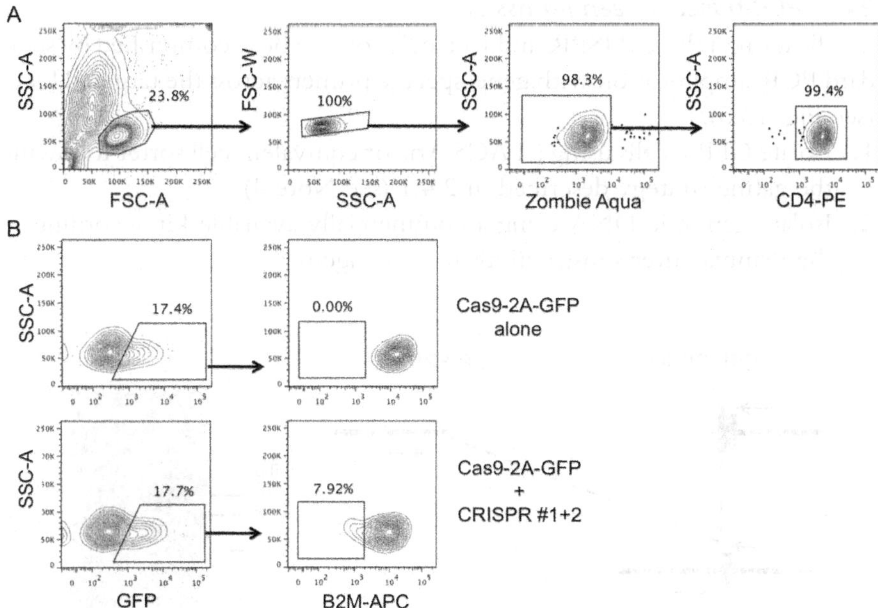


**Figure 13.4** Loss of B2M surface staining as a measure of successful targeting. (A) Gating strategy: dead cells are excluded by their distinctive smaller size (FSC-A) and higher granularity (SSC-A). In addition, dead cells can be excluded by gating on the Zombie Aqua™ negative population that stains positive for CD4-PE. (B) Based on the Cas9-2A-GFP fluorescence, the CD4+ GFP+ population is further analyzed for loss of B2M expression.

6. Resuspend pellet in 100 μl blocking buffer (4% FBS in PBS) and incubate for 30 min on ice.
7. Add 100 μl antibody staining mix and incubate for 30–45 min on ice (see Note 3).
   Antibody staining mix
   1% FBS in PBS
   Anti-CD4-PE (1:100)
   Anti-B2M-APC (1:100) or
   Anti-HLA-A,B,C-Alexa647 (1:100)
8. Wash cells twice with 700 μl staining buffer.
9. Resuspend pellets in 300 μl staining buffer.
10. Keep cells on ice until analysis.
11. FACS data can be recorded using a FACS Calibur (BD Bioscience) or BD LSR II and analyzed with FlowJo software version 7.6 (TreeStar).

### *2.4.2 PCR-based screening assay*

Verification of the CRISPR-induced deletion can be accomplished by standard PCR amplification with gene-specific primers across the targeted locus (see Fig. 13.5).

1. Isolate GFP+ cells using a FACS Aria or equivalent cell sorter following the gating strategy described in 2.4.1 (see Note 4).
2. Isolate genomic DNA using a commercially available kit according to the manufacturer's instructions (e.g., Qiagen).

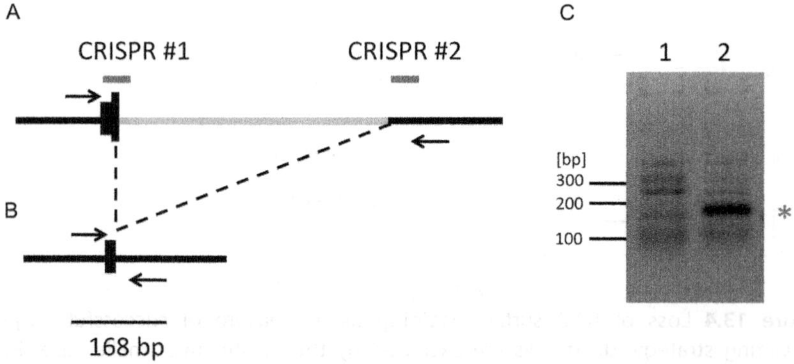


**Figure 13.5** PCR Confirmation of targeted deletion in the *B2M* gene locus. PCR strategy to detect the targeted deletion in the *B2M* locus. (A) WT allele. (B) A product of 168 bp can only be amplified when the 2.2 kb region separating the two primer binding sites (black arrows) is deleted (C, and Lane 2, asterisk). Lane 1: Cas9-2A-GFP only. Lane 2: Cas9-2A-GFP+CRISPR#1+2. The identity of the deletion was verified by Sanger sequencing.

3. PCR amplification is accomplished using standard cycling conditions in a conventional thermocycler (see Note 5).
   1 × PCR reaction mix
   - 50–100 ng genomic DNA
   - 1.0 μl forward primer (10 μ*M*)
   - 1.0 μl reverse primer (10 μ*M*)
   - 4 μl Fusion green GC-rich 5 × buffer
   - 1.0 μl dNTPs
   - 0.2 μl Fusion HS polymerase
   - [fill up to 20 μl with $ddH_20$]

| Cycling conditions | | |
|---|---|---|
| 95 °C | 1 min | |
| 95 °C | 10 s | ×40 |
| 58–62 °C | 10 s | |
| 72 °C | 10 s | |
| 72 °C | 1 min | |
| 16 °C | ∞ | |

4. Following agarose gel electrophoresis, a band of the correct size can be excised, gel purified using a commercially available kit, and sequenced with either the forward or reverse primer to confirm the correct excision on a sequence level.

**Notes**

1. Set some untransfected cells aside for single color and unstained controls, which are required for proper gating and compensation. For the same reason, the sample transfected with Cas9-2A-GFP alone should be split up into two tubes and only one half will be stained. The other half can be used as a single color control in the GFP channel.
2. Keep samples protected from light from here on!
3. Prepare single color controls separately for gating and compensation.
4. A sample transfected with Cas9-2A-GFP alone is required at this step for setting the gates, and also as a negative control for the PCR amplification.
5. Annealing temperature of the primers is best determined using a gradient cycler by varying the annealing temperature.

## 3. TARGETING OF *CCR5* IN HUMAN CD34+ HSPCs USING CRISPR/Cas9

Genetic disruption of chemokine (C-C motif) receptor 5 (CCR5), the major co-receptor of HIV-1 (Trkola et al., 1996), is a well-documented strategy to protect against HIV infection (Catano et al., 2011; Martinson, Chapman, Rees, Liu, & Clegg, 1997; Samson et al., 1996). Encouraged by the naturally occurring delta32 mutation in *CCR5,* which confers resistance to HIV-1 infection, CCR5 has quickly become an intensively studied target of the first wave of genome editing using ZFNs (Holt et al., 2010; Perez et al., 2008). Targeted ablation of *CCR5* in human hematopoietic stem and progenitor cells (HSPCs) gave rise to CD4+ T cells that are resistant to HIV challenge and provided long lasting immunity in mouse models (Holt et al., 2010). CCR5-deficient HSPCs and CD4+ T cells are currently under evaluation in preliminary clinical trials (Tebas et al., 2014).

In this protocol, we explore the applicability of the most recently developed genome-editing tool, the CRISPR/Cas9 system, to target the *CCR5* locus in CD34+ HSPCs. The advantage of CRISPR/Cas9 over ZNFs is their ease of creation and the fact that multiple guides against the same target can be delivered in a single transfection (Hsu et al., 2014). We exploit this multiplexing capacity to introduce specific and predictable deletions, which result in highly efficient homozygous null mutations in *CCR5* (see Fig. 13.6A). For this purpose, CD34+ HSPCs are isolated from either cord

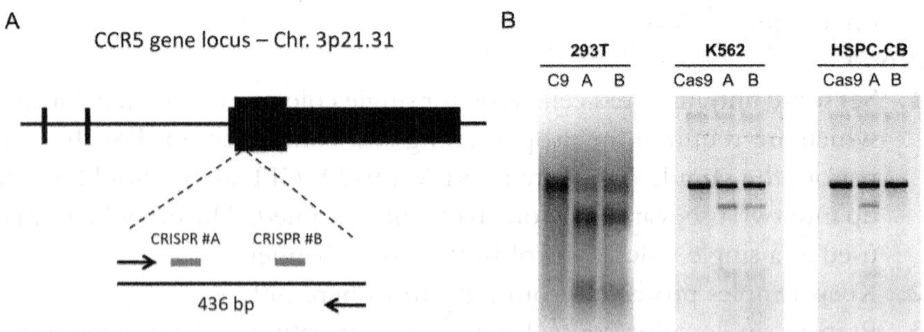


**Figure 13.6** CCR5 ablation in human cells using the CRISPR/Cas9 system. (A) Schematic indicating the locations of guides targeting the human *CCR5* locus. Guides are shown as red (dark gray in the print version) bars. Arrows represent the primers pair used to amplify the targeted region for analysis. (B) CEL assays demonstrating mutational events (indels) introduced at the targeted sites in 293T (left panel), K562 (middle panel), and CD34+ HSPCs isolated from cord blood (HSPC-CB).

blood or mobilized peripheral blood. Cas9-2A-GFP and guide plasmids targeting CCR5, which have been validated by CEL assay in 293T and K562 cells (see Fig. 13.6B), are introduced into CD34+ HSPCs by nucleofection. Twenty-four hours posttransfection, GFP+ cells are isolated via FACS sorting (see Fig. 13.7). FACS-sorted cells are either maintained in liquid culture or plated onto methylcellulose. Two weeks postplating, clonal colonies are counted and can be scored for their contribution to myeloid (granulocyte, macrophage) and erythroid lineages. Subsequently, individual clones are picked and analyzed by a PCR strategy to determine the deletion efficacy at one or both *CCR5* alleles (see Fig. 13.8A). Although we observed significant variation in *CCR5* ablation between individual HSPC donors,

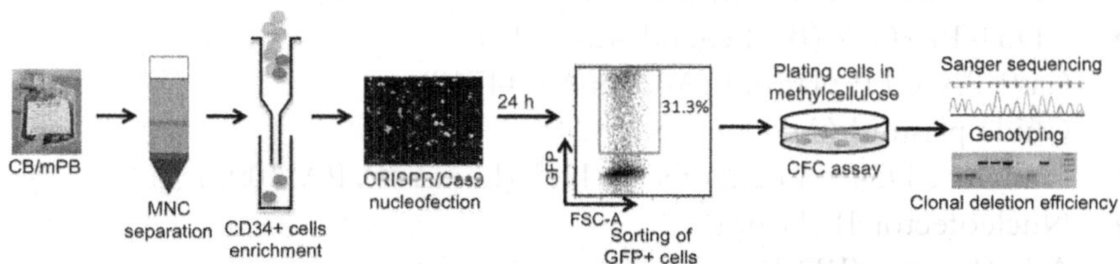


**Figure 13.7** Genome editing in CD34+ HSPCs. Schematic illustrating the workflow of genome editing in CD34+ HSPCs. Mononuclear cells (MNCs) are isolated from cord blood (CB) by Ficoll gradient centrifugation. Subsequently, CD34+ cells are isolated by CD34+ magnetic bead enrichment and are nucleofected with CRISPR/Cas9 plasmids. 24 h posttransfection, GFP-positive cells are sorted by FACS and cultivated in methylcellulose for 2 weeks. Colonies grown in methylcellulose are counted and scored for colony types. DNA isolated from individual colonies can be analyzed by either Sanger sequencing or gel electrophoresis. (See the color plate.)

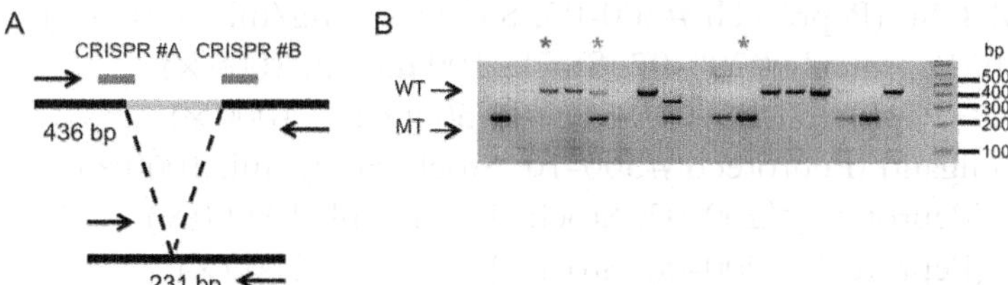


**Figure 13.8** Clonal analysis of *CCR5* deletion in CD34+ HSPCs. (A) Schematic showing the "dual guide" approach for targeting the *CCR5* locus. Arrows indicate the primer pair used to amplify the targeted region. The deleted region (205 bp between the cutting sites) is shown in gray. (B) Gel electrophoresis picture showing the genotype of various clones analyzed by PCR. Wild-type, heterozygous, and null clones are indicated with green (dark gray in the print version), orange (light gray in the print version), and red (dark gray in the print version) asterisks, respectively. DNA isolated from individual clones was analyzed by PCR.

the dual guide approach led to efficacious monoallelic and biallelic deletions in the *CCR5* locus in most cases (see Fig. 13.8B). The exact nature of the deletion can be determined by Sanger sequencing of the PCR product. In conclusion, this protocol provides a straightforward way for efficacious and predictable ablation of clinically relevant genes in primary CD34+ HSPCs using a CRISPR/Cas9-based "dual guide strategy."

## 3.1. Required materials

- Fresh cord blood
- CD34+ cells from mobilized peripheral blood (AllCells #mPB016F)
- Ficoll-Paque™ PLUS (GE Healthcare #17-1440-02)
- CD34+ MicroBead kit (Milteny #130-046-702)
- CD34-PE/Cy7 (BioLegend #343515)
- Cas9-2A-GFP plasmid (Addgene #44719)
- Guide plasmid (Addgene #41824)
- Human CD34+ Nucleofector kit® (Lonza #VPA-1003)
- Nucleofector II (Lonza)
- Aria II sorter (BD Bioscience)
- DMEM/F12 medium (Life technologies #11320-033)
- FBS (Hyclone #SH30070)
- EDTA (Sigma #E7889)
- PBS without $Ca^{2+}$ and $Mg^{2+}$ (Corning Cellgro #21-040-CV)
- β-Mercaptoethanol (Gibco® #21985-023)
- GlutaMax (Gibco® #35050-061)
- Penicillin–Streptomycin (Corning Cellgro #30-002-CI)
- GM-CSF (Peprotech #300-03; Stock: 100 μg/ml, 10,000 ×)
- SCF (Peprotech #300-07; Stock: 100 μg/ml, 1000 ×)
- TPO (Peprotech #300-18; Stock: 50 μg/ml, 1000 ×)
- Flt3 ligand (Peprotech #300-10; Stock: 50 μg/ml, 1000 ×)
- IL3 (Peprotech #200-03; Stock: 100 μg/ml, 10,000 ×)
- IL6 (Peprotech #200-06; Stock: 100 μg/ml, 5000 ×)
- MethoCult™ H4034 Optimum (Stem Cell Technologies #04034)
- Surveyor Mutation detection kit (Transgenomic #706020)
- Direct lysis buffer with detergent and Proteinase K (10 m*M* Tris–HCl, pH 7.6, 50 m*M* NaCl, 6.25 m*M* $MgCl_2$, 0.045% NP40, 0.45% Tween-20. Add Proteinase K 50 μl/ml of lysis buffer)
- Proteinase K (Viagen #501-PK)
- RNase A (Invitrogen #12091-039)

- GoTaq® Green Master Mix (Promega #M7122)
- Sample medium (2 m*M* EDTA, 2% FBS in PBS without $Ca^{2+}$ and $Mg^{2+}$)

## 3.2. Transfection of CD34+ HSPCs

Fresh cord blood samples were obtained from Dana Farber Cancer Institute's cell manipulation lab, and mononuclear cells were purified by the Ficoll-density gradient method. CD34+ HSPCs were isolated by MACS purification using the CD34+ MicroBeads isolation kit (Milteny) as per manufacturer's instructions. CD34+ cells from G-CSF mobilized peripheral blood were purchased from AllCells. CD34+ cells were cotransfected with CRISPR/Cas9 expression plasmids using Amaxa Nucleofection.

### *3.2.1 Isolation of CD34+ HSPCs from cord blood*

1. Dilute cord blood samples with three volumes of PBS without $Ca^{2+}$ and $Mg^{2+}$ (see Note 1).
2. Carefully layer 35 ml of diluted cell suspension over 15 ml of Ficoll-Paque in a 50-ml conical tube.
3. Centrifuge at 400 × *g* for 35 min at 20 °C in a swinging-bucket rotor without brake.
4. Aspirate the upper layer, leaving the buffy coat undisturbed.
5. Carefully transfer the buffy coat to a new 50-ml conical tube, fill the tube with sample medium. Centrifuge at 200 × *g* for 10 min at 20 °C. Low-speed centrifugation removes platelets.
6. Repeat Step 5. Discard the supernatant. Resuspend the cell pellet in 3 ml sample medium and count the cells.
7. Isolate CD34+ cells using CD34+ MicroBead kit (Milteny) as per manufacturer's instructions (see Note 2). Alternatively, thaw frozen vials of CD34+ cells from G-CSF mobilized peripheral blood (AllCells) as per manufacturer's instructions.
8. Cultivate CD34+ cells (either from cord blood or mobilized peripheral blood) in complete HSPC culture medium for 6–8 h at 37 °C at 5% $CO_2$ in incubator prior to transfection.

   HSPC culture medium

   DMEM/F12 medium
   10% FBS
   β-Mercaptoethanol
   Penicillin–Streptomycin
   Minimum nonessential amino acid
   GM-CSF, SCF, TPO, Flt3 ligand, IL3, IL6

### 3.2.2 Nucleofection of CD34+ HSPCs

Here, we describe a protocol for delivering CRISPR and Cas9 expression plasmids to CD34+ cells by nucleofection. Nucleofection is a method of delivering nucleic acid to the cells by creating transient small pores in the cell membrane by applying an electric pulse. Using nucleofection, plasmids can be delivered efficiently to CD34+ HSPCs in a relatively nontoxic manner.

9. Collect the cells in a 15-ml conical tube. Rinse the wells with 2 ml sample medium.
10. Centrifuge cells at 400 × *g* for 5 min at RT. Wash once with sample medium.
11. Resuspend the cell pellet in 1 ml of sample medium. Count the cells.
12. Aliquot $1 \times 10^6$ cells per transfection condition in a 1.7-ml microfuge tube and spin at 400 × *g* for 5 min. Completely remove the sample medium by aspiration (see Note 3).
13. Aliquot 10 μg Cas9-2A-GFP plasmid and 5 μg of each guide RNA in a 1.7-ml microfuge tube.
14. Resuspend the cell pellet in 100 μl of nucleofection buffer. Add plasmid DNA to the cell suspension and mix by pipetting.
15. Transfer the cells with DNA to a cuvette supplied with the nucleofector kit. Proceed immediately to nucleofection (see Note 4).
16. Perform the nucleofection as per manufacturer's instruction using an Amaxa Nucleofector II device.
17. Keep the cells at RT for 10 min.
18. Transfer the cells to 1 ml of HSPC culture medium without Penicillin/Streptomycin in 48-well plate (see Note 5).
19. Transfer the cells to a 37 °C incubator and culture for 24 h.

### 3.2.3 Cell sorting

As the Cas9 protein is expressed from the *CAG* promoter and cotranscribed together with a 2A-peptide coupled eGFP, Cas9 expressing cells can be isolated based on GFP expression by FACS sorting. Co-staining the cells with an anti-CD34 antibody, a very pure population of CRISPR/Cas9 expressing (GFP+) CD34+ cells can be obtained for analysis.

20. 24 h posttransfection, check the cells under a fluorescent microscope for GFP expression.
21. Collect the cells in 1.7-ml microfuge centrifuge. Wash the wells with 500 μl of sample medium.
22. Centrifuge the cells at 400 × *g* for 5 min at 4 °C. Wash once with 1 ml of sample medium.

23. Resuspend cell pellets in 100 μl of sample medium. Add 1 μl of PE-Cy7 conjugated human anti-CD34 (BioLegend) antibody to the cell suspension.
24. Incubate the cells on ice for 20 min in dark.
25. Add 500 μl of sample medium to the cells and centrifuge at 400 × *g* for 5 min at 4 °C. Remove the supernatant.
26. Resuspend the cells in 50 μl of sample medium and keep on ice in the dark.
27. Just before acquisition, add 250 μl of PBS-containing Propidium Iodide (PI). Filter the cell suspension and acquire.
28. Sort live (PI-negative) GFP+ CD34+ cells using a FACS Aria II cell sorter (BD Bioscience) or equivalent instrument.

**Notes**

1. Do not use cord blood older than 6 h for CD34+ cell isolation and transfection.
2. Usually buffy coats are contaminated with erythroid cells but this does not interfere with CD34+ cell enrichment by magnetic beads.
3. Reducing cell number below 500,000 per transfection is not recommended.
4. Do not incubate CD34+ cells in Nucleofector® solution longer than 20 min as this reduces cell viability and transfection efficiency. If handling multiple samples, it is advised to work in a batch of four to five samples at a time. This will reduce the incubation time with nucleofection buffer.
5. Do not cultivate the cells in antibiotics-containing cell culture medium after nucleofection, as this may adversely affect cell viability. After cell sorting, however, cells should be cultured in medium/methylcellulose-containing antibiotics.

## 3.3. Colony-forming cell assay

29. For colony-forming cell (CFC) assay, aliquot 5 ml of methylcellulose (MethoCult™ H4034 Optimum, Stem Cell Technologies) in a 15-ml conical tube (see Notes 1 and 2).
30. Add 5000 GFP+ CD34+ sorted cells.
31. Vortex to mix the cells with methylcellulose. Keep the conical tube in upright position for 10–15 min at RT. This will allow the methylcellulose to settle (see Note 3).
32. With the help of a 3-ml syringe equipped with a 18G needle, plate 1.5 ml of methylcellulose with cells on a 35-mm dish (see Note 4).

33. Tap the dish to disperse methylcellulose uniformly.
34. Incubate the dish in a humidified chamber in a 37 °C incubator. Culture the cells for 2 weeks and count and score the colonies.

**Notes**

1. Methylcellulose should be thawed at 4 °C overnight. After it is thawed completely, smaller aliquots should be made as needed in 15-ml conical tubes and stored at −20 °C.
2. Avoid multiple freeze–thaw cycles. Do not use pipette tips to aliquot methylcellulose.
3. Mix the cells with methylcellulose vigorously to get a uniform distribution of the cells.
4. Do not plate more than 1500 CD34+ cells into a 35-mm culture plate. Higher cell numbers will interfere with picking individual clones.

## 3.4. Clonal analysis

35. Pick colonies grown in MethoCult individually with the help of a p20 pipette.
36. Lyse in 50 μl of direct lysis buffer with detergent and Proteinase K (van der Burg et al., 2011) (see Note 1).
37. Digest samples at 56 °C for 1 h (see Note 2).
38. Inactivate Proteinase K at 95 °C for 15 min.
39. Add 50 μl of water with RNase A to the samples.
40. Use 2 μl of samples for an analytical PCR reaction.
41. Set up PCR reaction (25 μl) using GoTaq® Green Master Mix (Promega) as per manufacturer's instructions.
42. Add 2 μl of DNA samples to PCR mix.
43. Run PCR as per conditions given below.

| Cycling conditions | | |
|---|---|---|
| 95 °C | 1 min | |
| 95 °C | 20 s | ×35 |
| 62 °C | 20 s | ×35 |
| 72 °C | 30 s | ×35 |
| 72 °C | 2 min | |
| 16 °C | ∞ | |

44. For single guide experiments, analyze the PCR products by Sanger sequencing.
45. For dual guide experiments, targeting can be assessed directly by agarose gel electrophoresis.

**Notes**

1. Proteinase K should be added freshly to the lysis buffer before use.
2. Digestion with proteinase K should be carried out in a PCR machine.

## REFERENCES

Banasik, M. B., & McCray, P. B., Jr. (2010). Integrase-defective lentiviral vectors: Progress and applications. *Gene Therapy*, *17*(2), 150–157. http://dx.doi.org/10.1038/gt.2009.135.

Bryder, D., Rossi, D. J., & Weissman, I. L. (2006). Hematopoietic stem cells: The paradigmatic tissue-specific stem cell. *The American Journal of Pathology*, *169*(2), 338–346. http://dx.doi.org/10.2353/ajpath.2006.060312.

Capecchi, M. R. (1989). Altering the genome by homologous recombination. *Science*, *244*(4910), 1288–1292.

Catano, G., Chykarenko, Z. A., Mangano, A., Anaya, J. M., He, W., Smith, A., et al. (2011). Concordance of CCR5 genotypes that influence cell-mediated immunity and HIV-1 disease progression rates. *The Journal of Infectious Diseases*, *203*(2), 263–272. http://dx.doi.org/10.1093/infdis/jiq023.

Challa-Malladi, M., Lieu, Y. K., Califano, O., Holmes, A. B., Bhagat, G., Murty, V. V., et al. (2011). Combined genetic inactivation of beta2-microglobulin and CD58 reveals frequent escape from immune recognition in diffuse large B cell lymphoma. *Cancer Cell*, *20*(6), 728–740. http://dx.doi.org/10.1016/j.ccr.2011.11.006.

Cong, L., Ran, F. A., Cox, D., Lin, S., Barretto, R., Habib, N., et al. (2013). Multiplex genome engineering using CRISPR/Cas systems. *Science*, *339*(6121), 819–823. http://dx.doi.org/10.1126/science.1231143.

Ding, Q., Lee, Y. K., Schaefer, E. A., Peters, D. T., Veres, A., Kim, K., et al. (2013). A TALEN genome-editing system for generating human stem cell-based disease models. *Cell Stem Cell*, *12*(2), 238–251. http://dx.doi.org/10.1016/j.stem.2012.11.011.

Ding, Q., Regan, S. N., Xia, Y., Oostrom, L. A., Cowan, C. A., & Musunuru, K. (2013). Enhanced efficiency of human pluripotent stem cell genome editing through replacing TALENs with CRISPRs. *Cell Stem Cell*, *12*(4), 393–394. http://dx.doi.org/10.1016/j.stem.2013.03.006.

Ding, Q., Strong, A., Patel, K. M., Ng, S. L., Gosis, B. S., Regan, S. N., et al. (2014). Permanent alteration of PCSK9 with in vivo CRISPR-Cas9 genome editing. *Circulation Research*, *115*(5), 488–492. http://dx.doi.org/10.1161/circresaha.115.304351.

D'Urso, C. M., Wang, Z. G., Cao, Y., Tatake, R., Zeff, R. A., & Ferrone, S. (1991). Lack of HLA class I antigen expression by cultured melanoma cells FO-1 due to a defect in B2m gene expression. *The Journal of Clinical Investigation*, *87*(1), 284–292. http://dx.doi.org/10.1172/jci114984.

Genovese, P., Schiroli, G., Escobar, G., Di Tomaso, T., Firrito, C., Calabria, A., et al. (2014). Targeted genome editing in human repopulating haematopoietic stem cells. *Nature*, *510*(7504), 235–240. http://dx.doi.org/10.1038/nature13420.

Gussow, D., Rein, R., Ginjaar, I., Hochstenbach, F., Seemann, G., Kottman, A., et al. (1987). The human beta 2-microglobulin gene. Primary structure and definition of the transcriptional unit. *Journal of Immunology*, *139*(9), 3132–3138.

Holt, N., Wang, J., Kim, K., Friedman, G., Wang, X., Taupin, V., et al. (2010). Human hematopoietic stem/progenitor cells modified by zinc-finger nucleases targeted to CCR5 control HIV-1 in vivo. *Nature Biotechnology*, *28*(8), 839–847. http://dx.doi.org/10.1038/nbt.1663.

Hsu, P. D., Lander, E. S., & Zhang, F. (2014). Development and applications of CRISPR-Cas9 for genome engineering. *Cell*, *157*(6), 1262–1278. http://dx.doi.org/10.1016/j.cell.2014.05.010.

Jinek, M., East, A., Cheng, A., Lin, S., Ma, E., & Doudna, J. (2013). RNA-programmed genome editing in human cells. *eLife*, *2*, e00471. http://dx.doi.org/10.7554/eLife.00471.

June, C., Rosenberg, S. A., Sadelain, M., & Weber, J. S. (2012). T-cell therapy at the threshold. *Nature Biotechnology*, *30*(7), 611–614. http://dx.doi.org/10.1038/nbt.2305.

Kiskinis, E., Sandoe, J., Williams, L. A., Boulting, G. L., Moccia, R., Wainger, B. J., et al. (2014). Pathways disrupted in human ALS motor neurons identified through genetic correction of mutant SOD1. *Cell Stem Cell*, *14*(6), 781–795. http://dx.doi.org/10.1016/j.stem.2014.03.004.

Koller, B. H., Marrack, P., Kappler, J. W., & Smithies, O. (1990). Normal development of mice deficient in beta 2M, MHC class I proteins, and CD8+ T cells. *Science*, *248*(4960), 1227–1230.

Lu, P., Chen, J., He, L., Ren, J., Chen, H., Rao, L., et al. (2013). Generating hypo-immunogenic human embryonic stem cells by the disruption of beta 2-microglobulin. *Stem Cell Reviews*, *9*(6), 806–813. http://dx.doi.org/10.1007/s12015-013-9457-0.

Ma, Y., Shen, B., Zhang, X., Lu, Y., Chen, W., Ma, J., et al. (2014). Heritable multiplex genetic engineering in rats using CRISPR/Cas9. *PLoS One*, *9*(3), e89413. http://dx.doi.org/10.1371/journal.pone.0089413.

Mali, P., Yang, L., Esvelt, K. M., Aach, J., Guell, M., DiCarlo, J. E., et al. (2013). RNA-guided human genome engineering via Cas9. *Science*, *339*(6121), 823–826. http://dx.doi.org/10.1126/science.1232033.

Martinson, J. J., Chapman, N. H., Rees, D. C., Liu, Y. T., & Clegg, J. B. (1997). Global distribution of the CCR5 gene 32-basepair deletion. *Nature Genetics*, *16*(1), 100–103. http://dx.doi.org/10.1038/ng0597-100.

Maus, M. V., Grupp, S. A., Porter, D. L., & June, C. H. (2014). Antibody-modified T cells: CARs take the front seat for hematologic malignancies. *Blood*, *123*(17), 2625–2635. http://dx.doi.org/10.1182/blood-2013-11-492231.

Perez, E. E., Wang, J., Miller, J. C., Jouvenot, Y., Kim, K. A., Liu, O., et al. (2008). Establishment of HIV-1 resistance in CD4+ T cells by genome editing using zinc-finger nucleases. *Nature Biotechnology*, *26*(7), 808–816. http://dx.doi.org/10.1038/nbt1410.

Peters, D. T., Cowan, C. A., & Musunuru, K. (2013). *Genome editing in human pluripotent stem cells. StemBook*. Cambridge (MA): Harvard Stem Cell Institute.

Ran, F. A., Hsu, P. D., Lin, C. Y., Gootenberg, J. S., Konermann, S., Trevino, A. E., et al. (2013). Double nicking by RNA-guided CRISPR Cas9 for enhanced genome editing specificity. *Cell*, *154*(6), 1380–1389. http://dx.doi.org/10.1016/j.cell.2013.08.021.

Ran, F. A., Hsu, P. D., Wright, J., Agarwala, V., Scott, D. A., & Zhang, F. (2013). Genome engineering using the CRISPR-Cas9 system. *Nature Protocols*, *8*(11), 2281–2308. http://dx.doi.org/10.1038/nprot.2013.143.

Riolobos, L., Hirata, R. K., Turtle, C. J., Wang, P. R., Gornalusse, G. G., Zavajlevski, M., et al. (2013). HLA engineering of human pluripotent stem cells. *Molecular Therapy*, *21*(6), 1232–1241. http://dx.doi.org/10.1038/mt.2013.59.

Rosa, F., Berissi, H., Weissenbach, J., Maroteaux, L., Fellous, M., & Revel, M. (1983). The beta2-microglobulin mRNA in human Daudi cells has a mutated initiation codon but is still inducible by interferon. *The EMBO Journal*, *2*(2), 239–243.

Samson, M., Libert, F., Doranz, B. J., Rucker, J., Liesnard, C., Farber, C. M., et al. (1996). Resistance to HIV-1 infection in caucasian individuals bearing mutant alleles of the CCR-5 chemokine receptor gene. *Nature*, *382*(6593), 722–725. http://dx.doi.org/10.1038/382722a0.

Sander, J. D., & Joung, J. K. (2014). CRISPR-Cas systems for editing, regulating and targeting genomes. *Nature Biotechnology*, *32*(4), 347–355. http://dx.doi.org/10.1038/nbt.2842.

Schwank, G., Koo, B. K., Sasselli, V., Dekkers, J. F., Heo, I., Demircan, T., et al. (2013). Functional repair of CFTR by CRISPR/Cas9 in intestinal stem cell organoids of cystic fibrosis patients. *Cell Stem Cell*, *13*(6), 653–658. http://dx.doi.org/10.1016/j.stem.2013.11.002.

Tebas, P., Stein, D., Tang, W. W., Frank, I., Wang, S. Q., Lee, G., et al. (2014). Gene editing of CCR5 in autologous CD4 T cells of persons infected with HIV. *The New England Journal of Medicine*, *370*(10), 901–910. http://dx.doi.org/10.1056/NEJMoa1300662.

Torikai, H., Reik, A., Soldner, F., Warren, E. H., Yuen, C., Zhou, Y., et al. (2013). Toward eliminating HLA class I expression to generate universal cells from allogeneic donors. *Blood*, *122*(8), 1341–1349. http://dx.doi.org/10.1182/blood-2013-03-478255.

Trkola, A., Dragic, T., Arthos, J., Binley, J. M., Olson, W. C., Allaway, G. P., et al. (1996). CD4-dependent, antibody-sensitive interactions between HIV-1 and its co-receptor CCR-5. *Nature*, *384*(6605), 184–187. http://dx.doi.org/10.1038/384184a0.

Tsai, S. Q., Wyvekens, N., Khayter, C., Foden, J. A., Thapar, V., Reyon, D., et al. (2014). Dimeric CRISPR RNA-guided FokI nucleases for highly specific genome editing. *Nature Biotechnology*, *32*(6), 569–576. http://dx.doi.org/10.1038/nbt.2908.

van der Burg, M., Kreyenberg, H., Willasch, A., Barendregt, B. H., Preuner, S., Watzinger, F., et al. (2011). Standardization of DNA isolation from low cell numbers for chimerism analysis by PCR of short tandem repeats. *Leukemia*, *25*(9), 1467–1470. http://dx.doi.org/10.1038/leu.2011.118.

Warren, L., Manos, P. D., Ahfeldt, T., Loh, Y. H., Li, H., Lau, F., et al. (2010). Highly efficient reprogramming to pluripotency and directed differentiation of human cells with synthetic modified mRNA. *Cell Stem Cell*, 7(5), 618–630. http://dx.doi.org/10.1016/j.stem.2010.08.012.

Yin, H., Xue, W., Chen, S., Bogorad, R. L., Benedetti, E., Grompe, M., et al. (2014). Genome editing with Cas9 in adult mice corrects a disease mutation and phenotype. *Nature Biotechnology*, *32*(6), 551–553. http://dx.doi.org/10.1038/nbt.2884.

Zhou, J., Shen, B., Zhang, W., Wang, J., Yang, J., Chen, L., et al. (2014). One-step generation of different immunodeficient mice with multiple gene modifications by CRISPR/Cas9 mediated genome engineering. *The International Journal of Biochemistry & Cell Biology*, *46*, 49–55. http://dx.doi.org/10.1016/j.biocel.2013.10.010.

Zijlstra, M., Bix, M., Simister, N. E., Loring, J. M., Raulet, D. H., & Jaenisch, R. (1990). Beta 2-microglobulin deficient mice lack CD4-8+ cytolytic T cells. *Nature*, *344*(6268), 742–746. http://dx.doi.org/10.1038/344742a0.

CHAPTER FOURTEEN

# Generation of Site-Specific Mutations in the Rat Genome Via CRISPR/Cas9

**Yuting Guan, Yanjiao Shao, Dali Li[1], Mingyao Liu[1]**
Shanghai Key Laboratory of Regulatory Biology, Institute of Biomedical Sciences and School of Life Sciences, East China Normal University, Shanghai, China
[1]Corresponding authors: e-mail address: dlli@bio.ecnu.edu.cn; myliu@bio.ecnu.edu.cn

## Contents



*Methods in Enzymology*, Volume 546
ISSN 0076-6879
http://dx.doi.org/10.1016/B978-0-12-801185-0.00014-3

## Abstract

The laboratory rat is a valuable model organism for basic biological studies and drug development. However, due to the lack of genetic tools for site-specific genetic modification in the rat genome, more and more researchers chose the mouse as their favored mammalian models due to the sophisticated embryonic stem cell-based gene-targeting techniques available. Recently, engineered nucleases, including zinc finger nucleases, transcription activator-like effector nucleases, and CRISPR/Cas9 systems, have been adapted to generate knockout rats efficiently. The purpose of this section is to provide detailed procedures for the generation of site-specific mutations in the rat genome through injection of Cas9/sgRNA into one-cell embryos.

## 1. THEORY

The laboratory rat is an important mammalian model organism widely used by psychologists, pharmacologists, and neurobiologists due to its biological features that in some aspects are more similar to humans than those of mice. Since the establishment of gene targeting by homologous recombination in mouse embryonic stem cells (ESCs), numerous human diseases have been modeled by knockout mice for biomedical studies. The rat ESC culture system was not established until 2008, and the first knockout rat strain was generated through ESC-based gene targeting in 2010 (Tong, Li, Wu, Yan, & Ying, 2010). However, it is an expensive, time-consuming, and laborious technology demanding experienced skills in handling rat ESCs, delaying the adaptation of this technology worldwide. The emergence of engineered DNA nucleases (Gaj, Gersbach, & Barbas, 2013), including zinc finger nucleases, transcription activator-like effector nucleases, engineered meganucleases, and the microbial clustered regularly interspaced short palindromic repeats (CRISPR)/Cas system, is greatly accelerating the development of genetic engineering technologies in the rat and providing benefits to the research community. In this section, we will focus on the CRISPR/Cas system.

The CRISPR/Cas system was originally identified in bacteria and archaea functioning as a RNA-mediated adaptive immune system against the invasion of foreign DNA from viruses and plasmids (Garneau et al., 2010). Typically, there are three major types of the CRISPR system based on the structure of the genomic locus and the signature *cas* gene. Cas9 is the key enzyme in the type II system that has been widely adapted for gene editing (Jinek et al., 2012). In this system, the CRISPR RNA (crRNA) is composed of an array of direct repeats interspaced with variable elements (protospacers) derived from the exogenous target genome. The independently transcribed *trans*-activating crRNA (tracrRNA) forms a duplex with crRNA that guides the Cas9 nuclease to the target DNA through the 5′-most 20 bp of crRNA following the rules of Watson–Crick base pairing. The Cas9 nuclease digests the target DNA to blunt-end double-strand breaks (DSBs) at specific sites upstream of the protospacer adjacent motif (PAM), which may be different depending on the origin of the host species. To simplify Cas9-mediated gene editing, the crRNA/tracrRNA duplex is linked together as a single chimeric RNA, termed a single guide RNA (sgRNA). The most commonly used Cas9 system, derived from *Streptococcus pyogenes* (SpCas9), requires a PAM sequence of NGG (N is any nucleotide). The cleavage activity of Cas9 functions through two nuclease domains and a single mutation in either domain results in a nick rather than a DSB at the target site.

When a DSB is induced by Cas9/sgRNA at specific genomic site, the cell initiates the DNA repair process through homologous recombination (HR) when a donor DNA template is available, or else through error-prone nonhomologous end joining (NHEJ). NHEJ directly ligates the DSB ends and results in random indels, including small deletions, insertions, or substitutions. When the indels are introduced into the coding region of a target gene, frameshift mutations often lead to disruption of gene function. In addition, if two adjacent DSBs are produced, a deletion of the DNA sequence in between will occur resulting in large fragment deletion. DSBs also stimulate the high-fidelity HR pathway when a donor template (either double-strand DNA or single-stranded oligodeoxynucleotides) is present.

We have successfully applied the CRISPR/Cas system for generation of site-specific mutations in mouse and rat (Li et al., 2013; Shao et al., 2014). To generate knockout rats, *in vitro*-transcribed sgRNAs are coinjected with Cas9 mRNA into the cytoplasm of the embryos. For precise genome editing, Cas9/sgRNA and the donor template DNA are injected into the pronucleus of one-cell rat embryos. Injected embryos are transplanted into pseudopregnant females for pregnancy. The genotype of the individual pups is determined for indels by sequencing. Multiple sgRNAs against different genes can be delivered to generate compound genetic mutant rats spontaneously (Shao et al., 2014). To minimize the potential of off-target mutagenesis in the rat genome, the pairing double-nicking strategy is used by injection of Cas9 nickase and two sgRNAs recognizing opposite DNA strands flanking a gap of 4–20 bp (Ran et al., 2013). By following the protocol, the founders of a site-specific modified rat strain can be generated in less than 6 weeks.

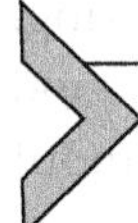

## 2. EQUIPMENT

Water bath or metal bath
1.5-ml microcentrifuge tubes
200-μl microcentrifuge tubes
Agarose gel sets
Power supply
UV imaging system
Centrifuge
Shaker
Spectrophotometer
Thermocycler
Stereomicroscope
Flaming micropipette puller
Microforge
Micromanipulator
Microinjector

Water-jacketed $CO_2$ incubator
Glass needle
Sterilizer
Cauterizer
Bulldog clamp
Light source unit
Forceps
Microscope slides
Petri dishes
Mineral oil
Pipettes and pipette tips

## 3. MATERIALS

Restriction enzyme
Agarose
TAE buffer
Ethidium bromide
*In vitro* transcription SP6 kit (e.g., Invitrogen)
*In vitro* transcriptionT7 kit (e.g., TAKARA)
Nucleotide removal kit (e.g., QIAGEN)
RNase-free water
Ethanol
Phenol/chloroform
Isopropanol
Sodium acetate
DNA polymerase
T7 endonuclease I
Cloning vector
Competent cells
LB medium plate
M2 medium
KSOM medium or mR1ECM
Hyaluronidase
PMSG
hCG
Chloral hydrate

### 3.1. Solutions and buffers

**Step 3** Hyaluronidase solution

---

Dissolve 30 mg hyaluronidase in 10 ml M2 medium. Store at −20°C in 200 μl aliquots.

---

PMSG solution

---

Prepare a stock solution of 500 IU $ml^{-1}$ in DPBS. Store at −20 °C in 100 μl aliquots.

---

hCG solution

---

Prepare a stock solution of 500 IU $ml^{-1}$ in DPBS. Store at −20 °C in 100 μl aliquots.

---

**Step 4** TE microinjection buffer

| Component | Final concentration | Stock | Amount |
|---|---|---|---|
| Tris–HCl, pH 7.4 | 0.01 *M* | 1 *M* | 100 μl |
| EDTA | 0.001 *M* | 0.5 *M* | 20 μl |

Add dd$H_2O$ to a final volume of 10 ml

**Step 5** Lysis buffer

| Component | Final concentration | Stock | Amount |
|---|---|---|---|
| Tris–HCl, pH 8.0 | 0.1 *M* | 1 *M* | 10 ml |
| EDTA | 0.005 *M* | 0.5 *M* | 1 ml |
| NaCl | 0.2 *M* | | 1.17 g |
| SDS | 0.2% (wt/vol) | 10% (wt/vol) | 2 ml |

Add dd$H_2O$ to a final volume of 100 ml

Proteinase K solution

---

Dissolve 20 mg proteinase K in 1 ml dd$H_2O$. Store at −20 °C in 20 μl aliquots.

---

# 4. PROTOCOL

## 4.1. Preparation

Order plasmids and oligonucleotides from the appropriate source. The plasmids used in this protocol can be found in Addgene or can be requested from us.

## 4.2. Duration

| | |
|---|---|
| Preparation | About 5–6 days |
| Protocol | About 4 weeks |

See Fig. 14.1 for the flowchart of the complete protocol.

**Preparation**

**Order plasmids and oligonucleotides from the appropriate source**

**Step 1: *In vitro* transcription of sgRNA target oligonucleotides**

↓

**Step 2: *In vitro* transcription of *Cas9* mRNA**

↓

**Step 3: Preperation of pseudopregnant female rats and one-cell rat embryos**

↓

**Step 4: Microinjection of one-cell embryos and transplanting the embryos into pseudopregnant female rats**

↓

**Step 5: Identification of founder rats**

↓

**Step 6: Production of $F_1$ generation rats**

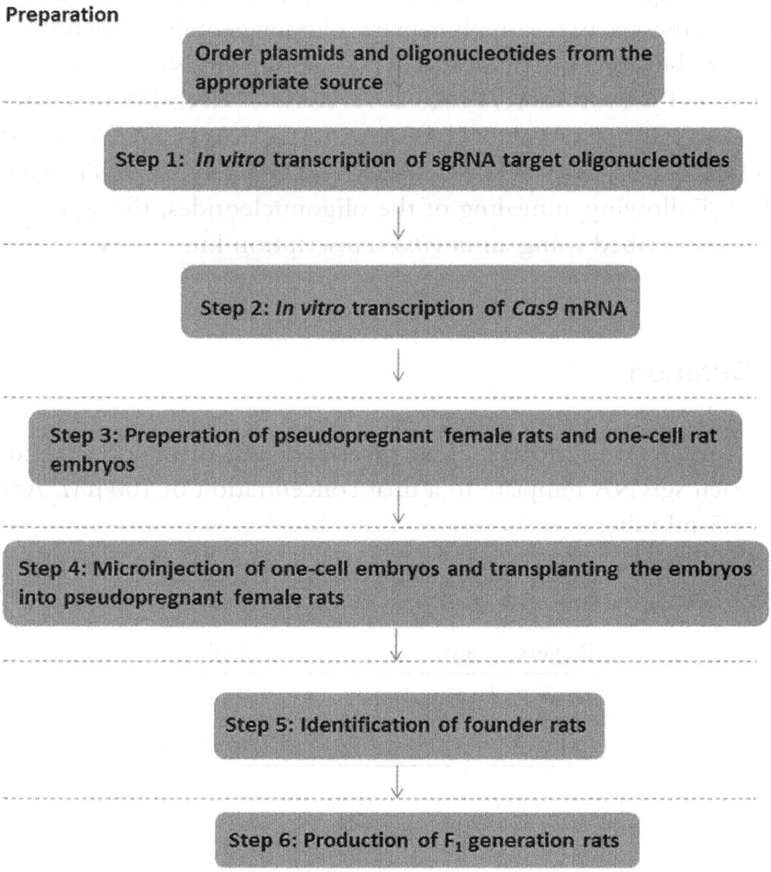

**Figure 14.1** Flowchart of the complete protocol.

### 4.3. Caution

Prior approval for all experimental procedures must be obtained from relevant committees for animal usage.

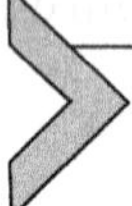

## 5. STEP 1: *IN VITRO* TRANSCRIPTION OF sgRNA TARGET OLIGONUCLEOTIDES

### 5.1. Overview

Select target sites on the genomic sequence of the coding region of the gene. Appropriate target sites can be found through online design tools (e.g., CRISPR Design Tool, http://tools.genome-engineering.org) or manually. The target sequence is the consecutive 20 bp immediately 5′ of the selected PAM. Synthesize two complimentary oligonucleotides containing a T7 promoter, the target site (indicated as N), and the sgRNA scaffold sequence (GATCACTAATACGACTCACTATAGGNNNNNNNNNNNNNNN NNNNNNGTTTTAGAGCTAGAAATAGCAAGTTAAAATAAGGC TAGTCCGTTATCAACTTGAAAAAGTGGCACCGAGTCGGTGC TTTT). Following annealing of the oligonucleotides, the sgRNA template is transcribed using an *in vitro* transcription kit.

### 5.2. Duration

About 5 h

**1.1** Resuspend the forward and reverse strands of oligonucleotides for each sgRNA template to a final concentration of 100 μ*M*. Add to a 1.5-ml tube:

| | |
|---|---|
| Forward oligo | 1 μl |
| Reverse oligo | 1 μl |
| 10 × T4 ligation buffer | 2 μl |
| $ddH_2O$ | to 20 μl |

**1.2** Anneal the oligos in a thermocycler using the following parameters: 95 °C for 5 min; ramp down to 25 °C at 5 °C $min^{-1}$; or in a water bath using the following parameters: 95 °C for 5 min; naturally cool to room temperature.

**1.3** Purify the annealed DNA using a nucleotide removal kit.

**1.4** Transcribe 0.2–1 μg of annealed DNA using an *in vitro* transcription kit (T7). Transfer into a 1.5-ml RNase-free tube:

| | |
|---|---|
| Annealed DNA | 0.2–1 μg |
| 10 × transcription buffer | 2 μl |
| ATP | 2 μl |
| GTP | 2 μl |
| CTP | 2 μl |
| UTP | 2 μl |
| RNase inhibitor | 0.5 μl |
| RNA polymerase | 2 μl |
| RNase-free water | to 20 μl |

**1.5** Incubate reaction at 42 °C for 2 h.

**1.6** Add 2–4 μl of DNase to the reaction to digest the DNA template and incubate the tube at 37 °C for another 30 min.

**1.7** Add 100 μl RNase-free water to stop the reaction.

**1.8** Purify the sgRNA with phenol:chloroform extraction and isopropanol/sodium acetate precipitation.

**1.9** Wash the pellet with 1 ml of 75% RNase-free ethanol.

**1.10** Dissolve the pellet with 10–20 μl RNase-free water, determine RNA concentration, and store it at −80 °C.

## 5.3. Tip

All of the supplies used in the transcription reaction must be RNase free.

See Fig. 14.2 for the flowchart of Step 1.

**Step 1: *In vitro* transcription of sgRNA target oligonucleotides**

1.1 Resuspend the forward and reverse strands of oligonucleotides for each sgRNA to a final concentration of 100 μ*M*:

| | |
|---|---|
| Forward oligo | 1 μl |
| Reverse oligo | 1 μl |
| 10×T4 ligation buffer | 2 μl |
| $ddH_2O$ | to 20 μl |

↓

1.2 Anneal the oligos in a thermocycler by using the following parameters: 95°C for 5 min; ramp down to 25°C at 5°C $min^{-1}$; or in a water bath by using the following parameters: 95°C for 5 min; naturally cool to room temperature

↓

1.3 Purify the annealed reaction using a nucleotide removal kit

↓

1.4 Transcribe 0.2–1 μg of dimerized oligos *in vitro*:

| | |
|---|---|
| 10× transcription buffer | 2 μl |
| ATP | 2 μl |
| GTP | 2 μl |
| CTP | 2 μl |
| UTP | 2 μl |
| Annealed DNA | 0.2–1 μg |
| RNase inhibitor | 0.5 μl |
| RNA polymerase | 2 μl |
| RNase-free water | to 20 μl |

↓

1.5 Incubate the tube at 42°C for 2h

↓

1.6 Add 2–4 μl of DNase to the reaction to digest DNA and incubate the tube at 37°C for another 30 min.

↓

1.7 Add 100 μl RNase-free water to stop the reaction

↓

1.8 Purify the sgRNA with phenol:chloroform extraction and isopropanol/sodium acetate precipitation

↓

1.9 Wash the pellet with 1 ml 75% RNase-free ethanol

↓

1.10 Dissolve the pellet with 10–20 μl RNase-free water and store it at −80°C

**Figure 14.2** Flowchart of Step 1.

## 6. STEP 2: *IN VITRO* TRANSCRIPTION OF *Cas9* mRNA

### 6.1. Overview

The plasmid containing Cas9 cDNA sequence driven by an SP6 (or T7) promoter is linearized with a proper restriction enzyme. The linearized plasmid is transcribed *in vitro* using an *in vitro* mRNA transcription kit (SP6 or T7).

### 6.2. Duration

About 8 h

**2.1** Digest 3 μg of Cas9 expression plasmid containing SP6 (or T7) promoter with the appropriate restriction enzyme. Add to a 1.5-ml tube:

| | |
|---|---|
| Plasmid | *X* μl |
| 10 × enzyme buffer | 5 μl |
| Restriction enzyme | 2 μl |
| $ddH_2O$ | to 50 μl |

**2.2** Incubate the tube at 37 °C or a proper temperature for 3 h.

**2.3** Run 5 μl of the digested reaction on agarose gels with ethidium bromide to confirm the linearization and purify the rest of the linearized plasmid using a nucleotide removal kit (e.g., QIAGEN).

**2.4** Transcribe 0.2–1 μg of digested plasmid using an *in vitro* mRNA transcription SP6 (or T7) kit (e.g., Invitrogen). Add to a 1.5-ml RNase-free tube:

| | |
|---|---|
| 2 × SP6 NTP/CAP | 5 μl |
| 10 × transcription buffer | 2 μl |
| SP6 enzyme | 2 μl |
| Linearized plasmid | *X* μl |
| RNase-free water | to 20 μl |

**2.5** Incubate the tube at 37 °C for 2 h.

**2.6** Add 1 μl of DNase to the reaction to digest DNA and incubate the tube at 37 °C for another 30 min.

**2.7** Add 30 μl RNase-free water and 30 μl 7.5 *M* LiCl to the reaction and incubate the tube at −20 °C for at least 30 min to precipitate RNA.

**2.8** Centrifuge at 12,000 rpm for 10 min at 4 °C and carefully aspirate the supernatant without disturbing the pellet.

**2.9** Wash the pellet with 1 ml of 75% RNase-free ethanol.

**2.10** Dissolve the pellet with 10–20 μl RNase-free water and store it at −80 °C.

## 6.3. Tip

All of the supplies used in the reaction must be RNase free.

See Fig. 14.3 for the flowchart of Step 1.

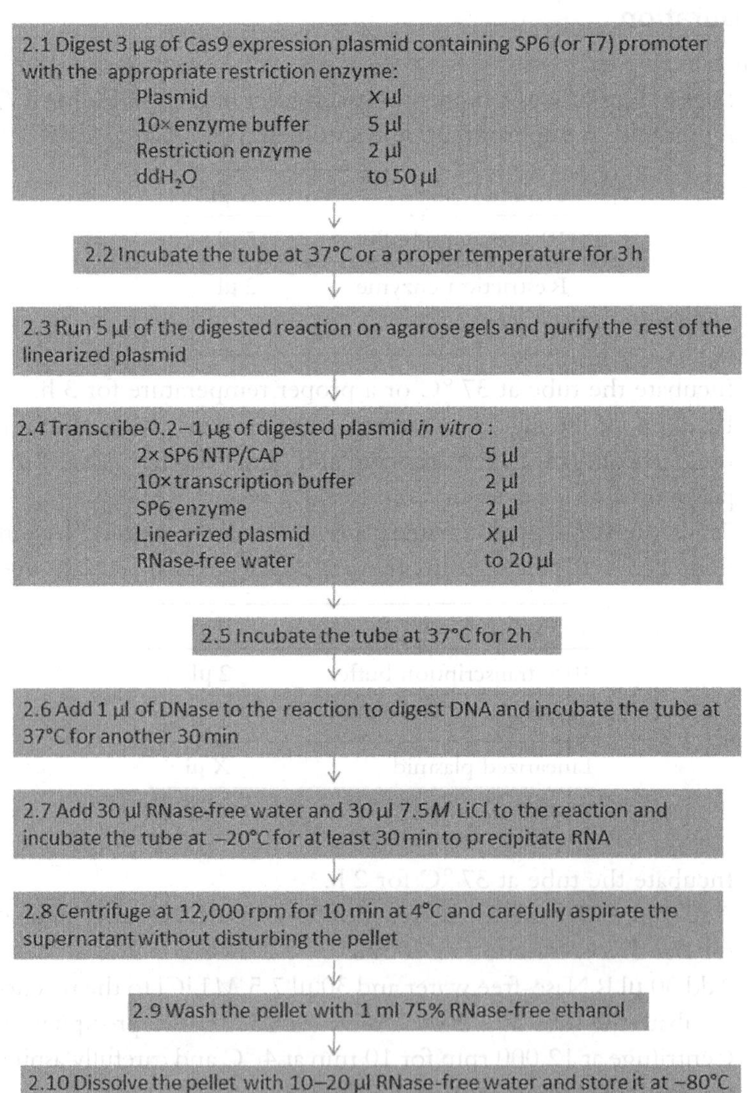


**Figure 14.3** Flowchart of Step 2.

# 7. STEP 3: PREPARATION OF PSEUDOPREGNANT FEMALE RATS AND ONE-CELL RAT EMBRYOS

## 7.1. Overview

The vasectomized male rats are prepared by removal of part of the vas deferens through surgical procedures. Pseudopregnant adult female rats are obtained by crossing with vasectomized males. Adult female rats are treated with PMSG and hCG for superovulation. Hormone-primed female rats are crossed with fertile males and one-cell embryos are collected the next morning.

## 7.2. Duration

About 4 days

- **3.1** Treat adult female rats with 300 μl (30 IU) PMSG at 1–2 p.m. on the first day.
- **3.2** Treat PMSG-treated females with 250 μl (25 IU) hCG at 3–5 p.m. on the third day.
- **3.3** Mate the hormone-treated female rats with adult males overnight following the third day.
- **3.4** Mate adult female rats with vasectomized males overnight at the same day.
- **3.5** Check for the copulatory plug of females mated with vasectomized adult males before 9 a.m. on the fourth day to prepare pseudopregnant female rats. Put the pseudopregnant females in new cages with labeling.
- **3.6** Sacrifice the hormone-treated females with copulatory plugs by $CO_2$ suffocation.
- **3.7** Dissect out and cut off the oviducts from each side of the females and put them in a dish containing prewarm M2 medium. Repeat this step to collect all oviducts.
- **3.8** Tear the ampulla with fine forceps and gently squeeze the embryos out into a dish containing 2 ml prewarmed M2 medium with 40 μl of hyaluronidase to remove the cumulus cells with the help of the mouth pipette under a stereomicroscope.
- **3.9** Wash the embryos with fresh prewarmed M2 medium in a new dish.
- **3.10** Incubate the embryos in drops of KSOM medium covered by mineral oil at 37 °C with 5% $CO_2$ until microinjection.

## 7.3. Tip

Checking the copulatory plug is usually the first job of the day, since it is easy for it to drop out later in the morning.

## 7.4. Tip

M2 medium is used for handling the embryos, and KSOM medium for short-term (less than 24 h) embryo culture. For long-term culture, use mR1ECM medium.

## 7.5. Tip

Do not incubate the embryos in M2 with hyaluronidase more than 5 min, because it will potentially damage the embryos.

See Fig. 14.4 for the flowchart of Step 3.

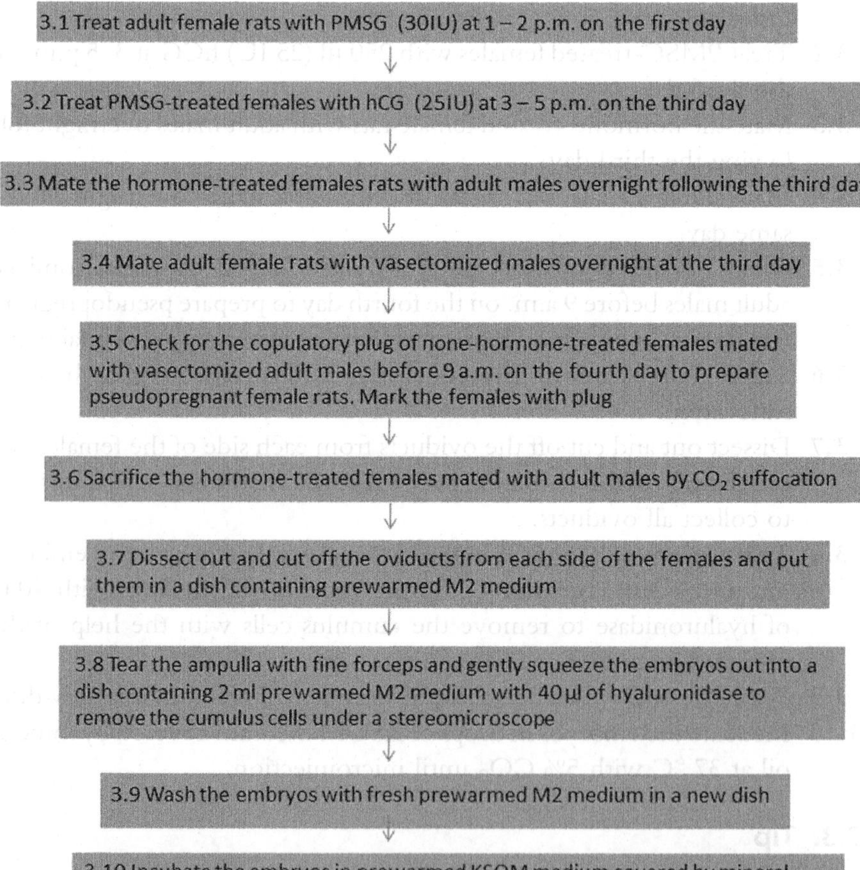


**Figure 14.4** Flowchart of Step 3.

## 8. STEP 4: MICROINJECTION OF ONE-CELL EMBRYOS AND TRANSPLANTING THE EMBRYOS INTO PSEUDOPREGNANT RATS

### 8.1. Overview

Cas9 mRNA and sgRNA are mixed in TE microinjection buffer to prepare the microinjection mix. The mixture is loaded into the injection pipette. The embryos are then injected with the Cas9/sgRNA cytoplasmically through the micromanipulator under the microscope. For precise genome editing, Cas9/sgRNA and donor template DNA are coinjected into the pronucleus of one-cell embryos. The injected embryos are transplanted through the oviduct into the pseudopregnant females.

### 8.2. Duration

About 4–5 h

**4.1** Mix Cas9 mRNA and sgRNA in TE microinjection buffer to a final concentration of 50 and 25 ng/μl, respectively, in 40 μl for each target. For precise mutagenesis, mix donor template DNA with the Cas9/sgRNA to a final concentration of 5 ng/μl in a clean RNase-free 1.5-ml tube.

**4.2** Make an injection needle with the flaming micropipette puller with appropriate parameters.

**4.3** Put the opening of the injection needle into the injection mixture prepared above. The buffer will load to the tip through capillary action.

**4.4** Make a holding pipette through a microforge and insert the pipette into the holder of the microinjector.

**4.5** Insert the injection pipette into the holder of the microinjector.

**4.6** Making an injection slide: Drop 100 μl M2 medium on a glass slide and cover the medium with mineral oil.

**4.7** Transfer 50 embryos onto the injection slide and arrange them in a single vertical line.

**4.8** Break the tip of the injection pipette by scratching it on the holding pipette.

**4.9** Suck an embryo on the holding pipette by applying negative pressure.

**4.10** Penetrate the zona pellucida and the membrane of the embryo and inject the Cas9/sgRNA mixture into the cytoplasm of the embryo, and then withdraw the pipette quickly once a flow of injection solution is observed.

Option: For precise genome editing, penetrate the pronuclear membrane of the embryo with the injection pipette. Quickly and carefully withdraw the pipette when a slight swelling of the pronucleus is observed.

**4.11** Release the injected embryo and repeat the above step.

**4.12** Transfer the injected embryos into prewarmed KSOM medium and put the dish into the incubator.

**4.13** Anesthetize the pseudopregnant female rat with 10% chloral hydrate by IP injection (3 ml/kg).

**4.14** Shave the fur on the back skin (optional); make a skin incision on the back above the ovary. Penetrate the body cavity and pull the fat pad of the ovary out. Fix the ovary and oviduct by clamping the fat pad with a bulldog clamp.

**4.15** Find the oviduct infundibulum under the stereomicroscope with fine forceps.

**4.16** Transfer the injected embryos into prewarmed M2 medium, and place two air bubbles into the transfer pipette and then load 10–15 embryos into the pipette. Seal the pipette with another air bubble.

**4.17** Open the ovarian bursa using fine forceps and expose the oviduct opening. Insert the transfer pipette and transfer the embryos into the oviduct.

**4.18** Repeat Steps 4.13–4.17 to transfer all embryos.

## 8.3. Tip

Mix and place the injection solution on ice all the time to minimize RNA degradation.

## 8.4. Tip

All surgical instruments should be sterilized.

## 8.5. Tip

The injection pipette should be made by fine capillary with filament which will load the solution through capillary action.

## 8.6. Tip

Do no inject too much solution into the embryos or pronucleus.

## 8.7. Tip

If the oviduct opening is hard to find, a small incision on the oviduct ampulla can be made by ophthalmic scissors. The embryos can be transferred in the oviduct through the incision.

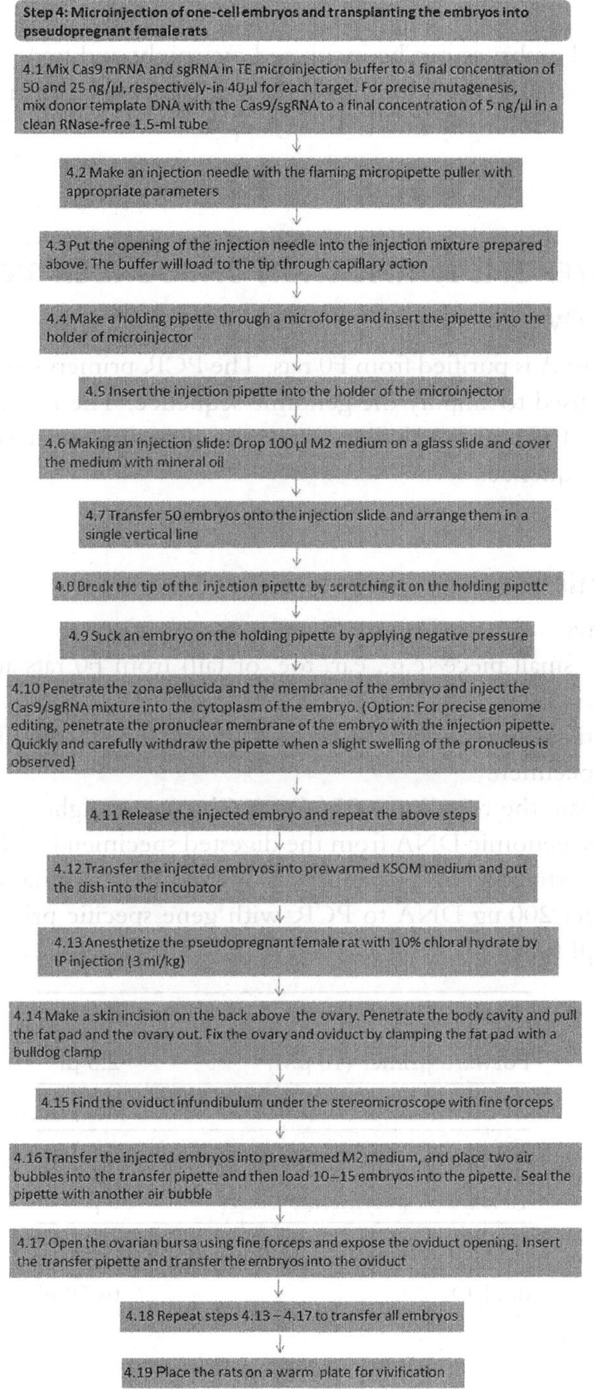


**Figure 14.5** Flowchart of Step 4.

## 8.8. Tip

The injected embryos can be incubated overnight and transplanted on the next day in KSOM medium. For longer culture, mR1ECM is optimal.

See Fig. 14.5 for the flowchart of Step 4.

# 9. STEP 5: IDENTIFICATION OF FOUNDER RATS

## 9.1. Overview

Genomic DNA is purified from F0 rats. The PCR primers spanning the target site are used to amplify the genomic sequence. The PCR products are subjected to T7EI mismatch digestion and sequencing to determine the precise DNA sequence.

## 9.2. Duration

About 4 days

**5.1** Cut a small piece (e.g., ear, toe, or tail) from F0 rats and put it in a 1.5-ml tube.

**5.2** Add tissue lysis buffer with proteinase K to the 1.5-ml tube containing the specimen.

**5.3** Incubate the reaction at 55 °C for 6 h or overnight.

**5.4** Purify genomic DNA from the digested specimen by phenol:chloroform extraction and ethanol/sodium acetate precipitation.

**5.5** Subject 200 ng DNA to PCR with gene specific primers. Add to a 200-μl tube:

| | |
|---|---|
| Genomic DNA (200 ng) | *X* μl |
| Forward primer (10 μ*M*) | 2.5 μl |
| Reverse primer (10 μ*M*) | 2.5 μl |
| dNTP (2.5 m*M* each) | 5 μl |
| 10 × DNA polymerase buffer | 5 μl |
| DNA polymerase | 0.5 μl |
| $ddH_2O$ | to 50 μl |

**5.6** Amplify the desired genomic DNA with a thermocycler using the common parameters according to the instructions of the DNA polymerase used.

**5.7** Anneal the PCR products in a thermocycler using the following parameters: 98 °C for 5 min, ramp down to 35 °C at 1 °C $min^{-1}$.

**5.8** Run 5 μl of each PCR product on a 1.5% (wt/vol) agarose gel and check the PCR product with a UV imaging system.

**5.9** Purify the PCR products using a nucleotide removal kit.

**5.10** Digest 200 ng of purified PCR products with the T7 endonuclease I. Add to a 200-μl tube:

| | |
|---|---|
| Purified PCR product | *X* μl |
| 10× NEBuffer2 | 2 μl |
| T7 endonuclease I | 0.5 μl |
| $ddH_2O$ | to 20 μl |

**5.11** Incubate the tube at 37 °C for 1 h.

**5.12** Run 20 μl of each digested PCR product on a 1.5% (wt/vol) agarose gel and photograph the bands with a UV imaging system. The samples that can be digested by T7EI suggest that mutations occurred in the genome.

**5.13** Ligate the PCR products containing mutations to cloning vectors.

**5.14** Transform the ligated vector to competent cells.

**5.15** Extract the plasmids from 5 to 10 clones. Sequence the plasmids to determine the exact genotype of the founders.

## 9.3. Tip

Use a high-fidelity DNA polymerase to avoid introducing mutations during amplification.

## 9.4. Tip

Sometimes the T7E1 digestion will produce false positive signals, and DNA sequencing must therefore be used for genotyping.

See Fig. 14.6 for the flowchart of Step 5.

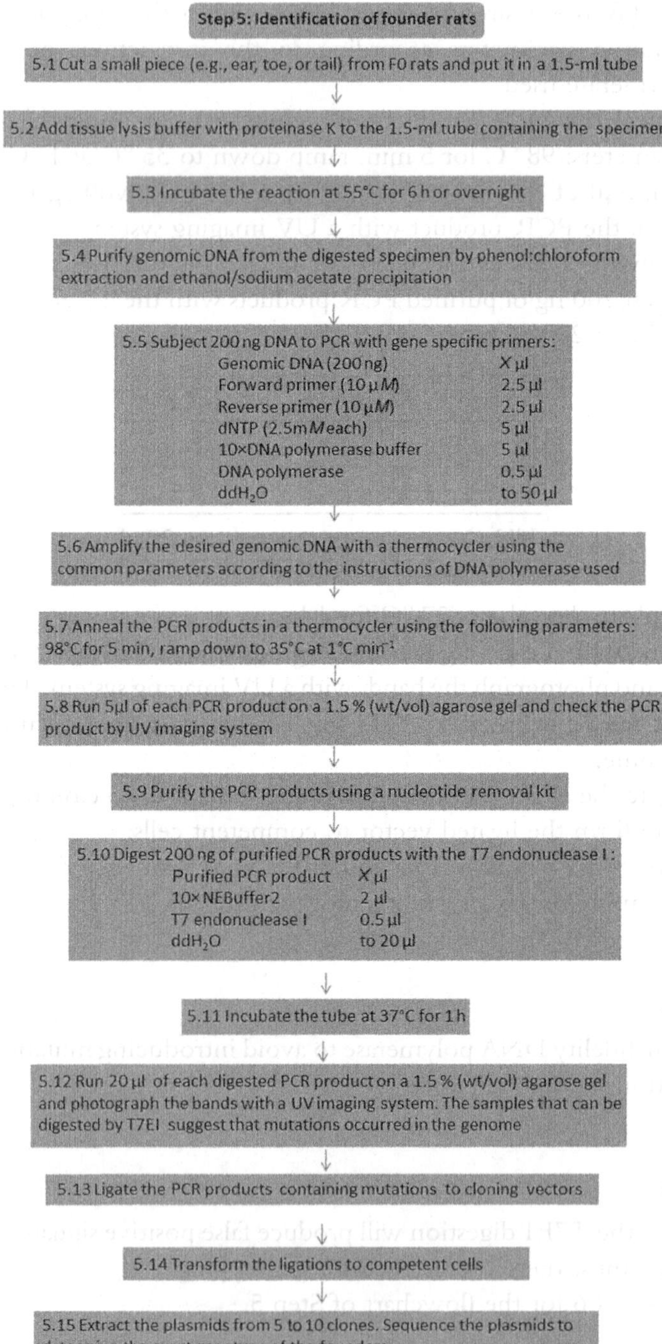


**Figure 14.6** Flowchart of Step 5.

# 10. STEP 6: PRODUCTION OF $F_1$ GENERATION RATS

## 10.1. Overview

Positive founder rats are mated with wild-type rats to produce $F_1$ generation rats. Next, identification of $F_1$ generation rats proceeds as in Step 5.

## 10.2. Duration

About 5 weeks

The procedure is exactly the same as described in Step 5.

## 10.3. Tips

The founders are usually chimeras bearing several kinds of mutations. The exact genomic sequence of the mutations in each $F_1$ rat should be sequenced.

# REFERENCES


Gaj, T., Gersbach, C. A., & Barbas, C. F., 3rd. (2013). ZFN, TALEN, and CRISPR-Cas-based methods for genome engineering. *Trends in Biotechnology*, *31*, 397–405.

Garneau, J. E., Dupuis, M. E., Villion, M., Romero, D. A., Barrangou, R., Boyaval, P., et al. (2010). The CRISPR-Cas bacterial immune system cleaves bacteriophage and plasmid DNA. *Nature*, *468*, 67–71.

Jinek, M., Chylinski, K., Fonfara, I., Hauer, M., Doudna, J. A., & Charpentier, E. (2012). A programmable dual-RNA-guided DNA endonuclease in adaptive bacterial immunity. *Science*, *337*, 816–821.

Li, D., Qiu, Z., Shao, Y., Chen, Y., Guan, Y., Liu, M., et al. (2013). Heritable gene targeting in the mouse and rat using a CRISPR-Cas system. *Nature Biotechnology*, *31*, 681–683.

Ran, F. A., Hsu, P. D., Lin, C. Y., Gootenberg, J. S., Konermann, S., Trevino, A. E., et al. (2013). Double nicking by RNA-guided CRISPR Cas9 for enhanced genome editing specificity. *Cell*, *154*, 1380–1389.

Shao, Y., Guan, Y., Wang, L., Qiu, Z., Liu, M., Li, D., et al. (2014). CRISPR/Cas-mediated genome editing in the rat via direct injection of one-cell embryos. *Nature Protocols*, in press. http://dx.doi.org/10.1038/nprot.2014.171.

Tong, C., Li, P., Wu, N. L., Yan, Y., & Ying, Q. L. (2010). Production of p53 gene knockout rats by homologous recombination in embryonic stem cells. *Nature*, *467*, 211–213.

CHAPTER FIFTEEN

# CRISPR/Cas9-Based Genome Editing in Mice by Single Plasmid Injection

**Yoshitaka Fujihara, Masahito Ikawa[1]**
Research Institute for Microbial Diseases, Osaka University, Suita, Japan
[1]Corresponding author: e-mail address: ikawa@biken.osaka-u.ac.jp

## Contents



## Abstract

CRISPR/Cas-mediated genome modification has opened a new era for elucidating gene function. Gene knockout mice can be generated by injecting humanized Cas9 (hCas9) mRNA and guide RNA (sgRNA) into fertilized eggs. However, delivery of RNA instead of DNA to the fertilized oocyte requires extra preparation and extra care with storage. To simplify the method of delivery, we injected the circular pX330 plasmids expressing both hCas9 and sgRNA and found that mutant mice were generated as efficiently as with RNA injection. Different from the linearized plasmid, the circular plasmid decreased the chance of integration into the host genome. We also developed the pCAG-EGxxFP reporter plasmid for evaluating the sgRNA activity by observing EGFP fluorescence in HEK293T cells. The combination of these techniques allowed us to develop a rapid, easy,

*Methods in Enzymology*, Volume 546
ISSN 0076-6879
http://dx.doi.org/10.1016/B978-0-12-801185-0.00015-5

and reproducible strategy for targeted mutagenesis in living mice. This chapter provides an experimental protocol for the design of sgRNAs, the construction of pX330-sgRNA and pCAG-EGxxFP-target plasmids, the validation of cleavage efficiency *in vitro*, and the generation of targeted gene mutant mice. These mice can be generated within a month.

## 1. INTRODUCTION

Gene knockout (KO) mice are powerful tools for studying biological science and genetic diseases of humans (Skarnes et al., 2011). The most commonly used method consists of three major steps: (i) construction of the gene targeting vector, (ii) homologous recombination (HR) in embryonic stem (ES) cells, and (iii) generation of chimeric mice. The gene targeting vectors contain the drug-resistant gene cassette in the center of an ~10 kbp genomic fragment encompassing the target locus. The negative selection cassette is usually added at the outside of the homology arms in the targeting vector. Then, the ES cells transfected with the targeting vector are selected with drugs and screened to obtain HR clones. After the clonal expansion of the HR clones with normal karyotypes, the ES cells are aggregated within preimplantation embryos to generate chimeric mice. If the mutation is transmitted into the next generation (F1), intercrosses between heterozygous F1 mutant mice generate homozygous gene KO mice. There are significant advances in the techniques performed in each step, but overall this method is still expensive, laborious, and time consuming to perform. Thus, well-trained researchers are required to achieve all the techniques for successful targeted mutation (Fujihara, Kaseda, Inoue, Ikawa, & Okabe, 2013). Moreover, this approach requires germ-line-competent ES cells that have only been established so far in a limited number of organisms (such as mice and rats).

Genome editing using customized nucleases, for instance, zinc-finger nucleases (ZFNs) and transcription activator-like effector nucleases (TALENs), provides a novel approach for targeted mutagenesis in a broad range of cell lines and organisms (Gaj, Gersbach, & Barbas, 2013). ZFNs and TALENs are artificially generated proteins made by fusing a FokI endonuclease with a DNA-recognition domain. These enzymes recognize target DNA sequences by protein–DNA interactions and the FokI endonuclease component induces a DNA double-stranded break (DSB) at the target genomic locus. Thus generated DSBs can be repaired by one of at least two different pathways, nonhomologous end-joining (NHEJ) or homology-directed repair (HDR). NHEJ is error-prone and leads to insertion/deletion mutations (indels) of various sizes. HDR requires a homologous reference sequence to repair the DSB correctly, allowing introduction of designed mutations into the targeted locus via an exogenously supplied reference DNA (single-stranded DNA (ssDNA) or double-stranded DNA (dsDNA)) (Sander & Joung, 2014). Taking advantage of the high efficiency of DSB-mediated gene mutation, gene-targeted mice and rats were generated by injection of the mRNA coding ZFNs/TALENs into the zygote (Carbery et al., 2010; Sung et al., 2013). However, the design and preparation of the DNA-recognition domain of the ZFN/TALEN enzymes has proven highly difficult, ultimately limiting the spread and use of this technique.

Recently, the type II CRISPR (clustered regularly interspaced short palindromic repeat)-Cas (CRISPR-associated) system has been demonstrated to cause DSB-mediated mutation in mammalian cell lines (Cong et al., 2013; Mali et al., 2013). The bacterial Cas protein 9 (Cas9) nuclease from *Streptococcus pyogenes* (SpCas9) is unique and flexible owing to its dependence on RNA as the moiety that targets the nuclease to a desired DNA sequence. In contrast to ZFNs and TALENs, the CRISPR/Cas9 system depends on a simple base-pairing rule between the synthetic guide RNA (sgRNA) and the target DNA sequence, and the Cas9/sgRNA riboprotein complex works as an RNA-guided nuclease (Sander & Joung, 2014). It is highly advantageous that one can target any gene locus by just replacing 20 nucleotides (nts) within the sgRNA sequence.

In this chapter, we outline the use of the plasmid pX330 (Fig. 15.1A) developed by Dr. F. Zhang (Cong et al., 2013) that expresses a humanized Cas9 (hCas9) and sgRNA under the chicken hybrid promoter and human U6 promoter, respectively, to generate gene-modified (GM) mice. We first describe our simple validation system for gene-targeted DSB activity by observing green fluorescence from the reconstituted enhanced green fluorescent protein (EGFP) expression cassette in the reporter plasmid, pCAG-EGxxFP (http://www.addgene.org/50716/) (Fig. 15.1A) (Mashiko et al., 2013). We next describe the one-step generation of mutant mice by injecting circular pX330 plasmids into zygotes (Mashiko et al., 2013, 2014). Whereas mutant mice can be generated by injecting hCas9 mRNA along with sgRNA into zygotes (Fujii, Kawasaki, Sugiura, & Naito, 2013; Wang et al., 2013; Yang et al., 2013), our method allows researchers to skip the technically difficult procedure of RNA synthesis and storage and provides a simple and reproducible method for targeted mutagenesis.

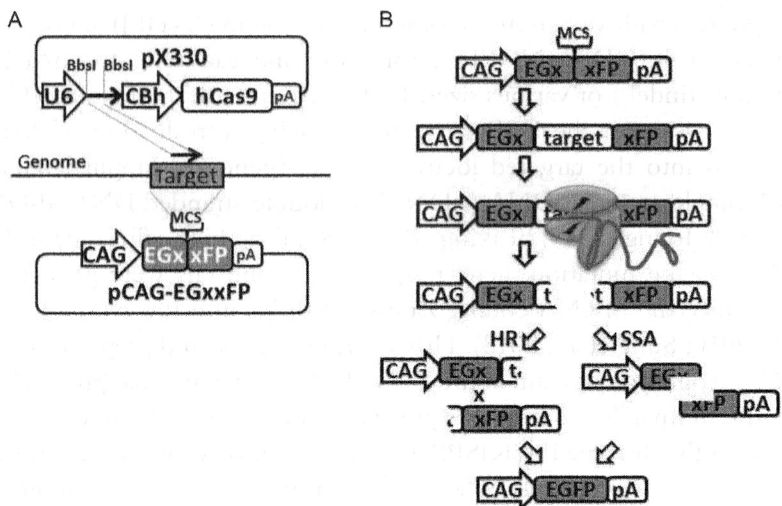


**Figure 15.1** Scheme for CRISPR/Cas9-mediated genome editing. (A) pX330 plasmid and pCAG-EGxxFP plasmid. pX330 plasmid contains two kinds of expression cassettes of both sgRNA and humanized Cas9 (hCas9). The target sgRNA sequence can be cloned directionally into the BbsI site. pCAG-EGxxFP plasmid contains 5′ and 3′ regions overlapping EGFP fragments under the CAG promoter. The 500–1000 bp genomic fragment containing the target sgRNA sequence can be inserted in multicloning site (MCS, *Bam*HI, *Nhe*I, *Pst*I, *Sal*I, *Eco*RI, and *Eco*RV) between EGFP fragments. (B) Scheme of validation for DSB-mediated EGFP expression cassette reconstitution. When the target sequence was digested by sgRNA-guided Cas9 endonuclease, DSB can be repaired and reconstituted the EGFP expression cassette. HR, homologous recombination; SSA, single-strand annealing.

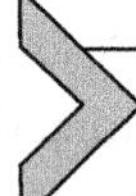

## 2. DESIGN AND CONSTRUCTION OF CRISPR/Cas9 PLASMIDS WITH pX330

The sequence specificity of the SpCas9 nuclease is determined by 20 nts within the sgRNA. The Cas9/sgRNA complex recognizes 20 nts preceding a protospacer-adjacent motif (PAM) sequence at its 3′-end (5′-NNNNNNNNNNNNNNNNNNNN-NGG-3′, N can be A, G, C, or T) and digests between 3 and 4 nts upstream of the PAM sequence of the target DNA. NAG may also function as PAM sequence with one fifth of the targeting efficiency of NGG (Hsu et al., 2013).

### 2.1. Selection and off-target analysis of sgRNA in targeted gene

The human U6 RNA polymerase III promoter prefers a G (guanine) nucleotide at the 5′-end of the sgRNA to initiate translation. However, when we compared sgRNAs inserted into pX330 with or without G at the first position, there was no significant differences in their cleavage activity (Mashiko et al., 2014), and we therefore simply choose 20 nts preceding NGG. However, the addition of an extra G at the 5′-end sometimes increased the activity. It should be noted that more than five continuous T nucleotides (poly-T stretch) may act as a transcriptional termination signal for the U6 promoter.

### *2.1.1 Design of sgRNAs against the target gene: Protocol*

1. Search for the nucleotide sequence NGG, appearing after a translational start site (ATG) in both the sense and antisense strands. If there are any in-frame ATGs within approximately 100 nts, it is preferable to select the NGG downstream of the last ATG.
2. Align 12, 13, and 14 nts at the 3′-end of the target 20 nts plus NGG against the mouse genome (mm9) using the free software Bowtie (http://bowtie-bio.sourceforge.net/index.shtml).

NB: We usually analyze eight candidates with Bowtie and select four candidate sequences with the lowest number of off-target sequences (e.g., less than 5 perfect matches within 13 nts plus NGG). The CRISPR Design Tool (http://crispr.mit.edu/) is also recommended for designing sgRNAs against the target gene (Ran, Hsu, Wright, et al., 2013).

## 2.2. Construction of pX330 with designed sgRNA

For a listing of reagents and equipment required for this protocol refer to Table 15.1. The Taq DNA polymerase for PCR and ligase buffer can be purchased from suppliers other than those mentioned in Table 15.1.

### *2.2.1 Insertion of sgRNA into the pX330 plasmid: Protocol*

1. Once the sgRNAs are designed, prepare a pair of oligos (forward oligo: 5′-cacc + N20-3′; reverse oligo: 5′-aaac + N20-3′) against each sgRNA. Phosphorylation is not required.
2. Resuspend and dilute the oligos to a final concentration of 0.1 μ*M* each in TE buffer. Combine 0.1 μ*M* of forward and reverse oligos in TE buffer. Anneal the forward and reverse oligos in a thermal cycler as follows: one cycle of 95 °C—5 min; 60 °C—5 min; 25 °C—minimum 60 min.
3. Ligate the annealed sgRNA oligos with *Bbs*I-digested pX330 plasmid (alkaline phosphatase treatment is not required) according to the manufacturer's instructions. In brief, set up the ligation reaction for 2 μL of annealed oligos, 1 μL of *Bbs*I-digested pX330 (50–100 ng/μL), and

**Table 15.1** Reagents and equipment for CRISPR/Cas9-mediated mutant mice by plasmid injection

| Reagents | Source and catalog number |
|---|---|
| pX330 | Addgene, Cambridge, MA, #42230 |
| pX330-Cetn1/sgRNA#1 | Addgene, Cambridge, MA, #50718 |
| pCAG-EGxxFP | Addgene, Cambridge, MA, #50716 |
| pCAG-EGxxFP-Cetn1 | Addgene, Cambridge, MA, #50717 |
| *Bbs*I | New England Biolabs, Hitchin, UK, #R0539 |
| *Bam*HI | New England Biolabs, Hitchin, UK, #R0136 |
| *Nhe*I | New England Biolabs, Hitchin, UK, #R0131 |
| *Pst*I | New England Biolabs, Hitchin, UK, #R0140 |
| *Sal*I | New England Biolabs, Hitchin, UK, #R0138 |
| *Eco*RI | New England Biolabs, Hitchin, UK, #R0101 |
| *Eco*RV | New England Biolabs, Hitchin, UK, #R0195 |
| Ligation high Ver. 2 | TOYOBO, Osaka, Japan, #LGK-201 |
| Chemically competent cells (DH5α, Stbl3) | |
| Wizard Plus Minipreps DNA Purification System | Promega, Tokyo, Japan, #A1465 |
| LB plate | MO BIO Laboratories, Carlsbad, CA, #12107 |
| Ampicillin | Sigma, St. Louis, MO, #A0166 |
| 2 × YT medium | Sigma, St. Louis, MO, #Y1003 |
| KOD FX Neo | TOYOBO, Osaka, Japan, #KFX-201 |
| Wizard PCR Preps DNA Purification System | Promega, Tokyo, Japan, #A7170 |
| Cell culture dish | AGC Techno Glass, Tokyo, Japan, #3020-100, #3810-006 |
| DMEM medium | Life Technologies, Carlsbad, CA, #11995 |
| FCS | BioWest, Nuaillé, France |
| 100 × Penicillin–Streptomycin–Glutamine solution | Life Technologies, Carlsbad, CA, #10378016 |
| HEK293T cells | |
| 2.5% Trypsin (10 ×) | Life Technologies, Carlsbad, CA, #15090046 |
| 2 × BBS (pH 6.95) (280 m*M* NaCl, 50 m*M* BES, 1.5 m*M* $Na_2HPO_4$) | |

**Table 15.1** Reagents and equipment for CRISPR/Cas9-mediated mutant mice by plasmid injection—cont'd

| Reagents | Source and catalog number |
|---|---|
| 2.5 *M* $CaCl_2$ | |
| Mouse (sexually mature female and male mice) | |
| Pseudopregnant ICR mouse | |
| Foster mother ICR mouse | |
| PMSG (pregnant mare's serum gonadotropin) | ASKA Pharmaceutical, Tokyo, Japan |
| hCG (human chorionic gonadotropin) | ASKA Pharmaceutical, Tokyo, Japan |
| Mouse embryo culture dish | AGC Techno Glass, Tokyo, Japan, #1010-060 |
| KSOM medium | Merck Millipore, Darmstadt, Germany, #MR-020P-5 F |
| FHM medium | Merck Millipore, Darmstadt, Germany, #MR-024-D |
| Hyaluronidase | Sigma, St. Louis, MO, #H4272 |
| $T_{10}E_{0.1}$ (10 m*M* Tris–HCl (pH 7.4), 0.1 m*M* EDTA in ultrapure distilled water) | |
| Lysis buffer (20 m*M* Tris–HCl (pH 8.0), 5 m*M* EDTA, 400 m*M* NaCl, 0.3% SDS, and 200 μg/mL Proteinase K solution) | |
| Ampdirect Plus | Shimadzu, Kyoto, Japan, #241-08890-92 |
| **Equipment** | |
| Thermal cycler | Life Technologies, Carlsbad, CA, Verti Fast 100 |
| Gel electrophoresis apparatus and power supply | |
| Digital gel imaging system | |
| NanoDrop ND-1000 | Thermo Scientific, Wilmington, DE |
| Fluorescence microscope | |
| Incubator | |
| Microinjection system | |

3 μL of Ligation high Ver. 2. buffer. Incubate the ligation reaction for a total of 1 h at 16 °C.

4. Transform the pX330 with designed sgRNA into a competent *E. coli* strain (e.g., DH5α and Stbl3). Briefly, add 3 μL of the ligation products into 15 μL of ice-cold chemically competent cells, incubate the mixture on ice for 20 min, heat-shock it at 42 °C for 30 s and return it immediately to ice for 2 min. Add 80 μL of SOC medium and plate it onto an LB plate containing 50 μg/mL ampicillin. Incubate plate overnight at 37 °C.
5. Following a minimum of 12 h incubation observe the colonies on each plate. From each plate, pick up several individual colonies to check for the correct integration of the pX330 plasmid containing the designed sgRNA. Inoculate 3 mL of 2 × YT media containing 50 μg/mL ampicillin with a single colony. Incubate the liquid culture and place on a shaker at 37 °C overnight (approximately 14–16 h).
6. Purify the plasmid DNA from liquid cultures by the commercially available "Wizard Plus SV Minipreps DNA Purification System," according to the manufacturer's instructions.
7. Confirm sgRNA sequence of each culture via sanger sequencing using the sequencing primer (5′-TGGACTATCATATGCTTACC-3′).

## 3. VALIDATION OF pX330 *IN VITRO*

Validation of sgRNA cleavage activity was done by cotransfection of HEK293T cells with the pX330-sgRNA and pCAG-EGxxFP-target plasmids as described in the below protocols. The reconstituted EGFP fluorescence was observed under a fluorescence microscope at 48 h after transfection. We used the pCAG-EGxxFP-Cetn1 (http://www.addgene.org/50717/) and pX330-Cetn1/sgRNA#1 (http://www.addgene.org/50718/) as the positive control. These plasmids have been reported to work well both *in vitro* and *in vivo* (Mashiko et al., 2013, 2014).

### 3.1. Construction of pCAG-EGxxFP with the targeted genomic region

In a previous report (Mashiko et al., 2013), we constructed the pCAG-EGxxFP plasmid containing 5′ and 3′ EGFP fragments that share 482 bp under the ubiquitous CAG promoter (Fig. 15.1A). The 500–1000 bp region of the target genome containing the sgRNA target sequence is inserted between the EGFP fragments and this construct is used as a target plasmid.

### 3.1.1 Insertion of the target genomic fragment into pCAG-EGxxFP plasmid: Protocol

1. Design and prepare the primers for amplifying the target genomic region. Amplify the genomic fragment in a thermal cycler by PCR using the KOD FX Neo reagents. The pCAG-EGxxFP plasmid has several multicloning sites (*Bam*HI, *Nhe*I, *Pst*I, *Sal*I, *Eco*RI, and *Eco*RV) between the EGFP fragments. Add the sequence for two of these restriction enzyme sites (plus several nucleotides outside each restriction enzyme site) to the 5′ region of the primers, one for the forward primer and one for the reverse primer (e.g., *Bam*HI sequence is "ggatcc" therefore the forward primer sequence would be 5′-NNggatcc NNNNNNNNNNNNNNNNNN-3′).
2. Check the amplified PCR fragments by gel electrophoresis. When the band can be observed clearly and at the expected size, isolate the PCR products by the commercial kit "Wizard PCR Preps DNA Purification System" according to the manufacturer's instructions.
3. Cut the purified PCR products with the inserted restriction enzymes for 1–2 h at 37 °C. Then purify again by the commercial kit "Wizard PCR Preps DNA Purification System" according to the manufacturer's instructions.
4. Ligate the pCAG-EGxxFP with the PCR fragments. In brief, set up a ligation reaction for 1 μL of digested pCAG-EGxxFP, 2 μL of purified PCR products, and 3 μL of Ligation high Ver. 2. buffer. Incubate the ligation reaction for a total of 1 h at 16 °C.
5. Transform the pCAG-EGxxFP containing the target genomic fragments into a competent *E. coli* strain (e.g., DH5α and Stbl3). Briefly, add 3 μL of the ligation products into 15 μL of ice-cold chemically competent cells, incubate the mixture on ice for 20 min, heat-shock it at 42 °C for 30 s and return it immediately to ice for 2 min. Add 80 μL of SOC medium and plate it onto an LB plate containing 50 μg/mL ampicillin. Incubate overnight at 37 °C.
6. Following a minimum of 12 h incubation at 37 °C, observe the colony growth for each plate. Pick up several colonies to confirm correct integration of the target genomic sequence. Inoculate 3 mL of 2× YT medium supplemented with 50 μg/mL ampicillin with a single colony. Incubate the liquid culture on a shaker at 37 °C overnight.
7. After approximately 14–16 h of incubation, purify the plasmid DNA from liquid cultures using the commercially available "Wizard Plus SV Minipreps DNA Purification System" according to the manufacturer's instructions.

8. Confirm integration of the target genomic sequence via Sanger sequencing using the primers for amplifying the target genomic region or the sequencing primers (5′-GCCTTCTTCTTTTTCCTACAGC-3′ for sequencing from CAG promoter side, 5′-GCCACACCAGCCACCACCTTCTG-3′ for sequencing from polyA side). Do not use the EGFP primers that anneal with both EGFP fragments in pCAG-EGxxFP.

## 3.2. Cotransfection of pX330-sgRNA and pCAG-EGxxFP-target into HEK293T cells

To validate which sgRNA sequence works, we used a simple validation system by the observation of green fluorescence reconstituted by DSB-mediated HDR of cells transfected with an EGFP containing plasmid (Fig. 15.1B) (Mashiko et al., 2013). In this chapter, we will outline the conventional calcium phosphate transfection method using HEK293T cultured cells. We recommend including both positive and negative transfection controls (e.g., positive control: pCAG-EGxxFP-Cetn1 and pX330-Cetn1/sgRNA#1; negative control: pCAG-EGxxFP-target and pX330). It should be noted that pCAG-EGxxFP-Cetn1 and -target plasmids themselves may have some background signals.

### *3.2.1 Cell culture and transfection in HEK293T cells: Protocol*

1. Approximately 6 h before transfection prepare 100-mm plates containing fairly confluent (80–90% or $2 \times 10^7$ cells) cell coverage. Seed the well-dissociated cells onto six-well plates at a density of approximately $1 \times 10^6$ cells per well in a total volume of 2 mL cell culture medium (culture medium; 10% (vol/vol) FCS, 1 × Penicillin–Streptomycin–Glutamine solution in DMEM medium). Cells are maintained and cultured according to the medium manufacturer's protocols.
2. 5–6 h after passaging, transfect cells using the calcium phosphate method according to the manufacturer's instructions. Briefly, prepare transfection reagents and plasmids as follows: 1 μg of pX330-sgRNA, 1 μg of pCAG-EGxxFP-target, and 10 μL of 2.5 *M* $CaCl_2$ in a total volume of 100 μL.
3. Add 100 μL of 2 × BBS (pH 6.95) to each tube, and mix by vortex immediately.
4. Incubate tubes at room temperature for 10 min.
5. Add transfection mix to each well (one pX330 plasmid containing one sgRNA per well). Ensure diffusion of transfection mix by gently shaking and incubate at 37 °C with 5% $CO_2$.

6. Refresh media after 16–24 h transfection.
7. 48 h after transfection, the EGFP fluorescence can be observed under a fluorescence microscope.

## 3.3. Observation of EGFP fluorescence in the transfected cells

pCAG-EGxxFP-Cetn1 and pX330-Cetn1/sgRNA#1 were used as the positive control in this assay. We classified the observed fluorescence intensity into four groups (score-4, brighter than control; score-3, same as control; score-2, darker than control; score-1, very dark; Fig. 15.2A). The validated pX330-sgRNA plasmids with a score of 3 or 4 can be used for the generation of mutant mice by pronuclear injections. If all the validated pX330-sgRNA plasmids have a score of 1 or 2, we recommend redesigning the target sgRNA sequences.

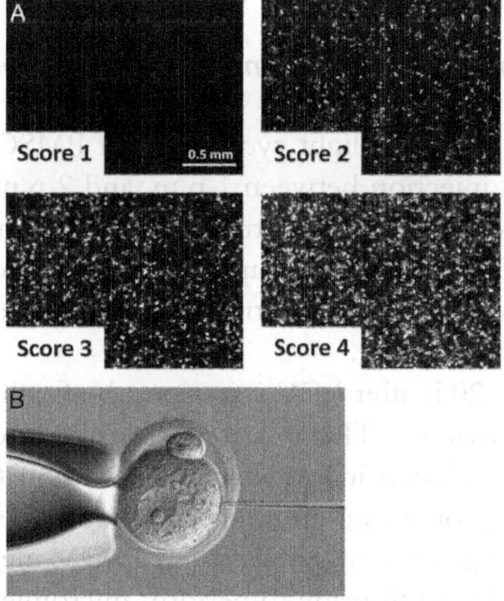


**Figure 15.2** *In vitro* validation system for cleavage activity of pX330-sgRNA plasmid and pronuclear injection of circular pX330-sgRNA plasmid. (A) pCAG-EGxxFP-target plasmid was cotransfected with pX330-sgRNA plasmids into HEK293T cells. The reconstituted EGFP fluorescence was observed under a fluorescence microscope at 48 h after transfection. The fluorescence intensity was classified into four groups (score-4, brighter than control; score-3, same as control (pCAG-EGxxFP-Cetn1/pX330-Cetn1/sgRNA#1); score-2, darker than control; score-1, very dark). (B) The validated pX330-sgRNA plasmids with a score of 3 or 4 were injected into the pronuclei of fertilized eggs. The circular plasmid injection can decrease the risk of integration of the plasmid into the host mouse genome.

# 4. ONE-STEP GENERATION OF MUTANT MICE VIA CIRCULAR PLASMID INJECTION

Following the validation of individual sgRNA cleavage activity in HEK293T cells, the selected pX330-sgRNA plasmid was injected at 5 ng/μL into the pronucleus of the zygote. To reduce the risk of integration of the plasmid into the host mouse genome, we injected pX330-sgRNA plasmids in their circular form.

## 4.1. Collecting the fertilized eggs

Wild-type female B6D2F1 mice are superovulated and mated with wild type B6D2F1 males, and fertilized eggs are collected from the oviduct according to the manual for manipulating the mouse embryo (Nagy, Gertsenstein, Vintersten, & Behringer, 2003). KSOM medium is used for the *in vitro* culture of preimplantation embryos, and FHM medium is used for *in vitro* manipulation outside the incubator.

### *4.1.1 Superovulation treatment and collection of fertilized eggs: Protocol*

1. On an 8 a.m. to 8 p.m. light cycle, 5 IU of PMSG is administered by intraperitoneal injection between 1 p.m. and 2 p.m. followed by 5 IU of the human chorionic gonadotrophin (hCG) administered 48 h later. After the hCG injection, one superovulated female is caged with one adult male. Copulation is confirmed by presence of a vaginal plug in the female next morning.
2. Approximately 20 h after hCG injection, the fertilized eggs are recovered from the oviduct. The collected eggs are surrounded by cumulus cells and are incubated in hyaluronidase solution (final concentration of 300 μg/mL) for 5 min until the cumulus cells can be removed. The fertilized eggs are then transferred to KSOM medium and incubated at 37 °C 5% $CO_2$ until ready to perform microinjection.

NB: There is a 4–6 h window for microinjection to take place; beyond this time the pronuclear stage eggs will develop into zygotes and can no longer be used for this protocol. Injection at later stage may increase the occurrence of mosaicism.

## 4.2. Preparing pX330-sgRNA plasmid for microinjection

To simplify procedures and minimize effort, we directly injected the circular form of pX330-sgRNA plasmids into the pronuclei of fertilized eggs.

The transgenic efficiency with the circular form of the plasmid DNA is approximately 10 times lower than that of the linear form (Mashiko et al., 2014).

#### 4.2.1 Preparation of pX330-sgRNA plasmid for microinjection: Protocol

1. The validated pX330-sgRNA plasmids with a score of 3 or 4 are filtered through spin columns (0.2 μm pore size) via centrifugation at 400 × *g* for 1 min.
2. Dilute the filtered pX330-sgRNA plasmids to a final concentration of 5 ng/μL using $T_{10}E_{0.1}$ buffer. The plasmid DNA can be stored at room temperature until microinjection.

### 4.3. Pronuclear microinjection of circular pX330-sgRNA plasmid

Before experiment, the microinjection systems need to be correctly set up (microscope, micromanipulators, holding, and injecting needles, etc.). These systems are very expensive and must be maintained in optimum conditions. Moreover, it takes practice to handle these systems and be able to generate mutant mice by microinjection (Fig. 15.2B). If you do not have the microinjection systems and/or have had little chance to deal with handling mouse gametes, we recommend ordering this procedure to be done in an animal facility or by companies that routinely generate mutant mice.

#### 4.3.1 Manipulating mouse embryos and microinjection system: Protocol

1. Approximately 24 h after hCG injection the fertilized eggs are ready to be injected with the circular pX330-sgRNA plasmids (5 ng/μL) into the pronucleus. The eggs will remain in the pronuclear stage for 4–6 h.
2. Surviving eggs can be cultured *in vitro* to two-cell stage embryos in KSOM medium at 37 °C 5% $CO_2$ overnight (approximately 20 h).
3. Two-cell stage embryos are then transferred into the oviduct of 0.5 d.p. c. pseudopregnant females.

NB: When mutant mice are generated with precise point mutations or insertions by DSB-mediated HDR, the fertilized eggs are coinjected with the mixture of 5 ng/μL pX330-sgRNA and 100 ng/μL ssODNs (single-stranded oligodeoxynucleotide, HPLC grade) into the pronucleus.

## 5. SCREENING FOR TARGETED MUTATION IN MICE

The pups developed from microinjected eggs are genotyped by PCR and subsequent Sanger sequencing analysis of DNA collected from the tail tips of the potential mutant pups (2–3-week old). The primers used for PCR analysis are the same primers used for the amplification of the target genome used in the construction of the pCAG-EGxxFP plasmid (see Section 3.1). Of the 32 genes tested, approximately half (100/192) of the pups were mutant mice (Mashiko et al., 2014).

### 5.1. Direct sequencing of PCR products: Protocol

1. Place the mouse tail tip into a 1.5-mL tube and add 0.2 mL of lysis buffer. Incubate at 50 °C overnight (minimum 12 h).
2. Following incubation for at least 12 h centrifuge the samples at room temperature for 1 min, 9000 × *g*. Collect 0.5 μL of the supernatant and perform PCR analysis using the Ampdirect Plus reagents according to the manufacturer's instruction.
   NB: This protocol renders the phenol/chloroform purification of the tail DNA unnecessary.
3. Check the size of the amplified PCR fragments by gel electrophoresis. When the band can be observed clearly, purify the PCR products using the commercially available "Wizard PCR Preps DNA Purification System" according to the manufacturer's instructions.
4. Confirm the sequence of purified pup tail DNA via Sanger sequencing using the same primers used for the amplification of the target genome used in the construction of the pCAG-EGxxFP plasmid. The primers for sgRNA insertion into pX330 may also be used. Typically, the majority (~80%) of the indel size in mutant mice were less than 100 bp (Mashiko et al., 2014). If it is difficult to read the direct sequencing of PCR products, it is recommended to subclone the PCR products into a plasmid vector.

## 6. CONCLUDING REMARKS

To examine the efficiency of pX330-mediated genome modification, we developed a validation system in cultured HEK293T cells transfected with pCAG-EGxxFP plasmid containing the target region of the mouse genome (Mashiko et al., 2013). Following validation, the circular form of pX330-sgRNA plasmids were microinjected into the pronuclei of single cell zygotes and mutant mice were obtained. If mice containing pX330 Tg are obtained as well as the desired mutation, the Tg and the mutation are often carried on separate alleles and can be separated by breeding of the F0 generation with wild type mice. This only occurs in very rare instances. Therefore, microinjection of the circular pX330 plasmid into the pronucleus of the single cell zygote is a simple, easy, fast, and viable method to make GM mouse lines within 1 month (Fig. 15.3).

One of the characteristics of the CRISPR/Cas9 system is its ability to induce DSBs at multiple sites in parallel (Wang et al., 2013; Yang et al., 2013). Cointroduction of multiple sgRNAs and the Cas9 nuclease can

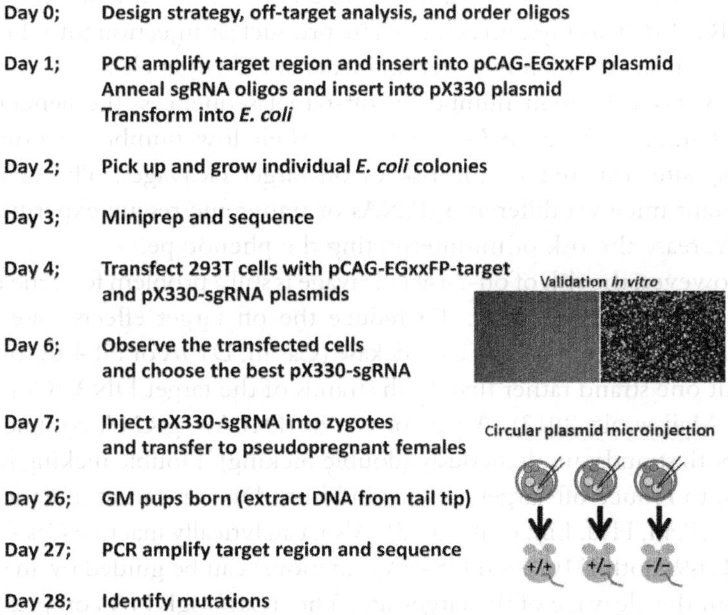

| | |
|---|---|
| **Day 0;** | **Design strategy, off-target analysis, and order oligos** |
| **Day 1;** | **PCR amplify target region and insert into pCAG-EGxxFP plasmid**<br>**Anneal sgRNA oligos and insert into pX330 plasmid**<br>**Transform into *E. coli*** |
| **Day 2;** | **Pick up and grow individual *E. coli* colonies** |
| **Day 3;** | **Miniprep and sequence** |
| **Day 4;** | **Transfect 293T cells with pCAG-EGxxFP-target and pX330-sgRNA plasmids** |
| **Day 6;** | **Observe the transfected cells and choose the best pX330-sgRNA** |
| **Day 7;** | **Inject pX330-sgRNA into zygotes and transfer to pseudopregnant females** |
| **Day 26;** | **GM pups born (extract DNA from tail tip)** |
| **Day 27;** | **PCR amplify target region and sequence** |
| **Day 28;** | **Identify mutations** |



**Figure 15.3** Timeline and generation of mutant mice by single plasmid injection. Conventional embryonic stem (ES) cell-mediated gene targeting requires at least 6 months to obtain homozygous mutant mice. CRISPR/Cas9-mediated gene modification utilizes double-strand breaks (DSBs) and the host cells own DNA repair systems. Cas9 endonuclease-induced DSBs can be repaired by nonhomologous end-joining (NHEJ) or homology-directed repair (HDR) pathway. NHEJ-mediated repair can often lead to the production of small indels (less than 100 bp). HDR-mediated repair can introduce precise point mutations or modification by coinjection of a reference single-stranded oligodeoxynucleotide (ssODN) or double-stranded DNA donor template with pX330-sgRNA plasmid. With this single plasmid injection system, heterozygous/homozygous mutant mice can be generated in 1 month. GM, genetically modified.

induce large deletions or inversions in a specific genomic region (Cong et al., 2013; Fujii et al., 2013). With our protocol, we could also obtain pX330 plasmid injection-mediated mutant mice carrying the complete removal of a region of the genome flanked by sites targeted by two sgRNAs, a deletion of approximately 400 bps (Mashiko et al., 2013). Furthermore, single plasmid injection with ssODN (approximately 120 nts in length) could induce point mutations in the mice (in preparation).

As with the ZFNs and TALENs systems, off-target cleavage events are of concern when using the CRISPR/Cas9 system. To analyze the off-target effects, we performed PCR analysis for each of the off-target sequences that matched the 13 nt seed sequence with an NGG PAM sequence. Only 3 off-target events were found among 382 potential off-target binding sites in 63 mutant mouse lines (Mashiko et al., 2014). The transient expression of the CRISPR/Cas9 system achieved by pronuclear injection into the mouse zygote appears to result in fewer incidents of off-target cleavage. Since each sgRNA has a different number of off-target sequences, the generation of mutant mice with sgRNAs selected for their low number of nondesired binding sites can reduce the risk of off-target cleavages. The generation of mutant mice via different sgRNAs or transgenic rescue experiment can also decrease the risk of misinterpreting the phenotype.

However, the risk of off-target cleavage is still a problem for gene therapy (e.g., human ES/iPS cells). To reduce the off-target effects, one of two nuclease domain-deficient Cas9 nickase (Cas9n; D10A or H840A mutation) can cut one strand rather than both strands of the target DNA (Cong et al., 2013; Mali et al., 2013). A pair of Cas9n (D10A)–sgRNA complexes can nick both strands simultaneously (double nicking). Double nicking has been shown to reduce off-target activity and introduce on-target mutations efficiently (Ran, Hsu, Lin, et al., 2013). Also, catalytically inactive Cas9 (dCas9; dead Cas9, both D10A and H840A mutations) can be guided by an sgRNA without the cleavage of the target site. The dCas9–sgRNA complex specifically binding to the target DNA can efficiently interfere with the expression of the targeted gene (CRISPR interference; CRISPRi) by blocking the binding of other proteins, such as RNA polymerase (Qi et al., 2013). The dCas9 can also be fused to exogenous proteins (e.g., activator, repressor, epigenetic regulator, and fluorescent proteins) to control the genetic and epigenetic modification of target genes (Chen et al., 2013; Gilbert et al., 2013; Maeder et al., 2013; Perez-Pinera et al., 2013). Recently, the human disease HTI (hereditary tyrosinemia type I) was corrected in adult mice following delivery of the CRISPR/Cas9 system via hydrodynamic tail vein injection (Yin et al., 2014). CRISPR/Cas9-based genome editing has emerged as a powerful and efficient method in the biological/biomedical sciences and holds much promise for the future of genetic/mouse research.

## ACKNOWLEDGMENT

We thank Dr. F. Zhang for pX330 plasmid and S.A.M. Young, B.Sc (Hons), for critical reading. Research described in this chapter was supported in part by the MEXT of Japan.

## REFERENCES

Carbery, I. D., Ji, D., Harrington, A., Brown, V., Weinstein, E. J., Liaw, L., et al. (2010). Targeted genome modification in mice using zinc-finger nucleases. *Genetics*, *186*(2), 451–459.

Chen, B., Gilbert, L. A., Cimini, B. A., Schnitzbauer, J., Zhang, W., Li, G. W., et al. (2013). Dynamic imaging of genomic loci in living human cells by an optimized CRISPR-Cas system. *Cell*, *155*(7), 1479–1491.

Cong, L., Ran, F. A., Cox, D., Lin, S., Barretto, R., Habib, N., et al. (2013). Multiplex genome engineering using CRISPR-Cas systems. *Science*, *339*(6121), 819–823.

Fujihara, Y., Kaseda, K., Inoue, N., Ikawa, M., & Okabe, M. (2013). Production of mouse pups from germline transmission-failed knockout chimeras. *Transgenic Research*, *22*(1), 195–200.

Fujii, W., Kawasaki, K., Sugiura, K., & Naito, K. (2013). Efficient generation of large-scale genome-modified mice using sgRNA and CAS9 endonuclease. *Nucleic Acids Research*, *41*(20), e187.

Gaj, T., Gersbach, C. A., & Barbas, C. F., 3rd. (2013). ZFN, TALEN, and CRISPR-Cas-based methods for genome engineering. *Trends in Biotechnology*, *31*(7), 397–405.

Gilbert, L. A., Larson, M. H., Morsut, L., Liu, Z., Brar, G. A., Torres, S. E., et al. (2013). CRISPR-mediated modular RNA-guided regulation of transcription in eukaryotes. *Cell*, *154*(2), 442–451.

Hsu, P. D., Scott, D. A., Weinstein, J. A., Ran, F. A., Konermann, S., Agarwala, V., et al. (2013). DNA targeting specificity of RNA-guided Cas9 nucleases. *Nature Biotechnology*, *31*(9), 827–832.

Maeder, M. L., Linder, S. J., Cascio, V. M., Fu, Y., Ho, Q. H., & Joung, J. K. (2013). CRISPR RNA-guided activation of endogenous human genes. *Nature Methods*, *10*(10), 977–979.

Mali, P., Yang, L., Esvelt, K. M., Aach, J., Guell, M., DiCarlo, J. E., et al. (2013). RNA-guided human genome engineering via Cas9. *Science*, *339*(6121), 823–826.

Mashiko, D., Fujihara, Y., Satouh, Y., Miyata, H., Isotani, A., & Ikawa, M. (2013). Generation of mutant mice by pronuclear injection of circular plasmid expressing Cas9 and single guided RNA. *Scientific Reports*, *3*, 3355.

Mashiko, D., Young, S. A., Muto, M., Kato, H., Nozawa, K., Ogawa, M., et al. (2014). Feasibility for a large scale mouse mutagenesis by injecting CRISPR-Cas plasmid into zygotes. *Development, Growth & Differentiation*, *56*(1), 122–129.

Nagy, A., Gertsenstein, M., Vintersten, K., & Behringer, R. (2003). *Manipulating the mouse embryo: A laboratory manual cold spring harbor*. NY: Cold Spring Harbor Laboratory Press.

Perez-Pinera, P., Kocak, D. D., Vockley, C. M., Adler, A. F., Kabadi, A. M., Polstein, L. R., et al. (2013). RNA-guided gene activation by CRISPR-Cas9-based transcription factors. *Nature Methods*, *10*(10), 973–976.

Qi, L. S., Larson, M. H., Gilbert, L. A., Doudna, J. A., Weissman, J. S., Arkin, A. P., et al. (2013). Repurposing CRISPR as an RNA-guided platform for sequence-specific control of gene expression. *Cell*, *152*(5), 1173–1183.

Ran, F. A., Hsu, P. D., Lin, C. Y., Gootenberg, J. S., Konermann, S., Trevino, A. E., et al. (2013). Double nicking by RNA-guided CRISPR Cas9 for enhanced genome editing specificity. *Cell*, *154*(6), 1380–1389.

Ran, F. A., Hsu, P. D., Wright, J., Agarwala, V., Scott, D. A., & Zhang, F. (2013). Genome engineering using the CRISPR-Cas9 system. *Nature Protocols*, *8*(11), 2281–2308.

Sander, J. D., & Joung, J. K. (2014). CRISPR-Cas systems for editing, regulating and targeting genomes. *Nature Biotechnology*, *32*(4), 347–355.

Skarnes, W. C., Rosen, B., West, A. P., Koutsourakis, M., Bushell, W., Iyer, V., et al. (2011). A conditional knockout resource for the genome-wide study of mouse gene function. *Nature*, *474*(7351), 337–342.

Sung, Y. H., Baek, I. J., Kim, D. H., Jeon, J., Lee, J., Lee, K., et al. (2013). Knockout mice created by TALEN-mediated gene targeting. *Nature Biotechnology*, *31*(1), 23–24.

Wang, H., Yang, H., Shivalila, C. S., Dawlaty, M. M., Cheng, A. W., Zhang, F., et al. (2013). One-step generation of mice carrying mutations in multiple genes by CRISPR-Cas-mediated genome engineering. *Cell*, *153*(4), 910–918.

Yang, H., Wang, H., Shivalila, C. S., Cheng, A. W., Shi, L., & Jaenisch, R. (2013). One-step generation of mice carrying reporter and conditional alleles by CRISPR-Cas-mediated genome engineering. *Cell*, *154*(6), 1370–1379.

Yin, H., Xue, W., Chen, S., Bogorad, R. L., Benedetti, E., Grompe, M., et al. (2014). Genome editing with Cas9 in adult mice corrects a disease mutation and phenotype. *Nature Biotechnology*, *32*(6), 551–553.

CHAPTER SIXTEEN

# Imaging Genomic Elements in Living Cells Using CRISPR/Cas9

**Baohui Chen***, **Bo Huang***,†,1
*Department of Pharmaceutical Chemistry, University of California, San Francisco, California, USA
†Department of Biochemistry and Biophysics, University of California, San Francisco, California, USA
[1]Corresponding author: e-mail address: bo.huang@ucsf.edu

## Contents



## Abstract

In addition to their applications in genome editing and gene expression regulation, programmable DNA recognition systems, including both CRISPR and TALE, have been recently engineered for the visualization of endogenous genomic elements in living cells. This capability greatly helps the study of genome function regulation by its physical organization and interaction with other nuclear structures. This chapter first discusses the general considerations in designing and implementing the imaging system. The subsequent sections provide detailed protocols to use the CRISPR/Cas9 system to label and image specific genomic loci, including the establishment of expression

*Methods in Enzymology*, Volume 546
ISSN 0076-6879
http://dx.doi.org/10.1016/B978-0-12-801185-0.00016-7

systems for dCas9-GFP and sgRNA, the procedure to label repetitive sequences of telomeres and protein-coding genes, the simultaneous expression of many sgRNAs to label a nonrepetitive locus, and the verification of signal specificity by FISH.

# 1. INTRODUCTION

With an enormous amount of information encoded, the total length of genomic DNA in a human cell exceeds 2 m. How these extremely long molecules are packaged into the cell nucleus with a diameter at around 10 μm is an intriguing question. Indeed, it is now widely understood that the spatiotemporal organization of the genome is indispensable for the regulation of its functional output (Misteli, 2007, 2013). To study the physical organization and interactions of genomic elements, a widely used approach is the direct visualization of specific DNA sequences by fluorescent *in situ* hybridization (FISH). However, the requirement of DNA denaturation for probe hybridization makes live-cell FISH impractical in most cases, thus losing the capability to track dynamic processes. On the other hand, live-cell imaging of genomic elements have so far relied on fluorescently tagged DNA-binding proteins. Due to the limited choices of such proteins, their application has been restricted to either special genomic elements such as the telomere (Wang et al., 2008) and centromere (Hellwig et al., 2008), or artificially inserted sequences such as LacO and TetO as tandem arrays (Robinett et al., 1996). It has been a major challenge to visualize arbitrary, endogenous DNA sequences in living cells until the recent demonstration that the genome editing tools of clustered regularly interspaced short palindromic repeats (CRISPR) and transcription activator-like effector (TALE) can be adapted for genome imaging (Anton, Bultmann, Leonhardt, & Markaki, 2014; Chen et al., 2013; Ma, Reyes-Gutierrez, & Pederson, 2013; Miyanari, Ziegler-Birling, & Torres-Padilla, 2013; Thanisch et al., 2014; Yuan, Shermoen, & O'Farrell, 2014).

## 1.1. Choice of target sites and DNA recognition methods

To fluorescently label arbitrary, endogenous genomic elements in a living cell, the basic approach is to introduce a fluorescently tagged, programmable DNA recognition protein. A Cas9 protein in the CRISPR system with its nuclease activity removed (dCas9) (Gilbert et al., 2013; Qi et al.,

2013), as well as the TALE protein without nuclease fusion, serve well for this purpose. By forming a complex with a small guide RNA (sgRNA) complimentary to the target sequence (for dCas9), or by combining the matching base-recognition domains (for TALE), they can stably bind to the target DNA in the nucleus. For live-cell fluorescence imaging, the most straightforward way is to express a fluorescent protein (FP) fusion of either dCas9 or TALE. Alternatively, FP or dye-labeled dCas9 or TALE can be introduced into a living cell through microinjection or electroporation.

With either dCas9 or TALE, the signal level for the target locus is determined by the number of fluorophores at the site. Due to the intrinsic background from the cell autofluorescence and free dCas9-FP or TALE-FP in the nucleoplasm, it has been very challenging to detect a single fluorophore at the target. In order to generate a detectable signal, it is advantageous to target tandem repeats in the genome so that multiple dCas9-FP or TALE-FP can bind the target locus using a single recognition sequence. Such tandem repeats exist throughout the genome of many organisms. For example, the mammalian telomeres consist of large arrays of TTAGGG repeat and have been successfully imaged using either CRISPR (Anton et al., 2014; Chen et al., 2013) or TALE (Ma et al., 2013; Miyanari et al., 2013; Thanisch et al., 2014). Simple tandem repeats are also available for imaging in centromeres (Ma et al., 2013), satellite DNA (Anton et al., 2014; Miyanari et al., 2013; Thanisch et al., 2014; Yuan et al., 2014), and sometimes even in protein-coding regions (Chen et al., 2013).

Although tandem repetitive sequences are convenient to be visualized by either CRISPR or TALE, it is not always possible to find suitable ones near the locus of interest. For these nonrepetitive loci, it is essential to have many targeting sequences simultaneously so that multiple fluorophores can be brought in. In this case, the CRISPR system is much more practical than the TALE system because it only needs to introduce a number of small RNAs, whereas the other must express a different TALE protein for each target sequences. When using the CRISPR/Cas9 system to label nonrepetitive loci, more than 10 FPs should be able to create a reliably detectable fluorescent puncta over the cellular background. In practice, however, because different sgRNAs have highly variable targeting efficiency, it has been reported that a group of at least 30 sgRNAs are necessary to generate good labeling (Chen et al., 2013).

Because of its generality, the rest of this chapter will focus on the use of the CRISPR/Cas9 system for the imaging of genomic elements.

## 1.2. Sensitivity and specificity of genome imaging using CRISPR/Cas9

Because the fluorescence signal level at a genomic locus is limited by the availability of target sites, the fluorescence background from free dCas9-FP in the nucleoplasm is a major practical consideration for imaging experiment. When the binding of dCas9 to the target locus is nearly saturated, increasing their expression level will only result in an increased background but not the signal level, thus decreasing the signal-to-background ratio. Therefore, a weak or inducible expression system is preferred for the expression of dCas9-FP to reduce the background, as long as there is sufficient amount of dCas9-FP to cover the binding sites. Ideally, the nuclear concentration of free dCas9-FP should be comparable to their binding constants, which are yet to be fully characterized and may vary with the target sequence. When a larger number of binding sites are in the nucleus, such as when imaging telomeres, the optimal expression level is higher because it needs more DNA-bound dCas9-FP.

The limited number of recognition sites also means that the fluorescence signal level will not be very high (sometimes just a few copies of FP). Hence, a practical consideration is to use a sensitive detection system. For example, microscopes with high numerical aperture objectives and sensitive cameras (such as the EMCCD or sCMOS cameras) are strongly preferred. In certain cases, widefield microscopes may be more suitable than laser scan confocal microscopes because of the higher detection efficiency. With a more sensitive detection system, not only can photobleaching be minimized with a lower excitation intensity and shorter exposure time, but the phototoxicity associated with long-term imaging can also be reduced. In fact, imaging induced artifacts can easily arise and remain unnoticed. As an example, light-induced DNA damage is often undetected unless DNA damage response reporters are used or cell cycle arrest is observed. On the other hand, the dCas9-FP probes can be made brighter by fusing multiple FPs to dCas9. Although this approach will not improve the signal-to-background ratio because it equally increases the brightness of DNA-bound and free dCas9-FP, it does allow a lower exposure to be used to generate the same quality.

For genome editing, off-target effects, particularly for the CRISPR/Cas9 system (Hsu et al., 2013), are often a major consideration because just a single DNA cleavage event will trigger the DNA damage repair response. In contrast, for genome imaging, multiple copies of fluorophores must be present to constitute a fluorescent signal above the background. Therefore, it is inherently much less prone to off-target effects. Nevertheless, artificial "puncta" which complicate image interpretation can still appear due to nonspecific binding of the DNA targeting protein to nuclear structures. For example, when not forming a complex with sgRNA, dCas9 can be enriched in nucleoli, presumably through nonspecific interactions with the ribosomal RNA. In order to eliminate this nucleoli signal, high expression strength of sgRNA, such as with a lentiviral vector, is required to ensure that all dCas9 are bound to sgRNA. A new version of sgRNA optimized for expression efficiency and dCas9 binding has also been shown to critically improve the efficiency of genome imaging by CRISPR (Chen et al., 2013). Given these potential artifacts, it is always important to perform control experiment using nontargeting sgRNA or two-color imaging with FISH (Section 5.1) to ensure that the observed signal is specific.

In the following chapters, we will use the telomere and mucin gene loci in live human cell lines as an example to illustrate how the CRISPR/Cas9 system can be practically applied to genome imaging. The general workflow is outlined in Fig. 16.1.

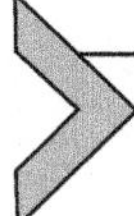

## 2. GENERATION OF CELL LINES STABLY EXPRESSING dCas9-GFP

### 2.1. Generation of dCas9-GFP constructs

To apply the CRISPR/Cas9 system for labeling endogenous genomic sequences, a catalytically inactive form of Cas9 (dCas9 harboring D10A and H840A substitutions), derived from *Streptococcus pyogenes*, is fused with an enhanced green fluorescent protein (GFP). Two copies of a nuclear localization signal (NLS) sequence are used to ensure nuclear localization of dCas9-GFP proteins. To reduce the level of unbound dCas9-GFP which contributes to the background signal, the Tet-On 3G inducible expression system is selected as the expression vector.

1. Design In-Fusion polymerase chain reaction (PCR) primers following the manufacturer's instructions of In-Fusion HD cloning kit (#638909, Clontech). Both dCas9 and GFP fragments will be cloned into the vector in a single reaction. Put NLS sequence in the primers to achieve its different positions within dCas9-GFP insert.
2. Amplify the coding sequences of dCas9 (human codon optimized) and GFP via the PCR separately. Phusion High-Fidelity polymerase DNA kit (#M0530S, NEB) should be used according to manufacturer's instructions.

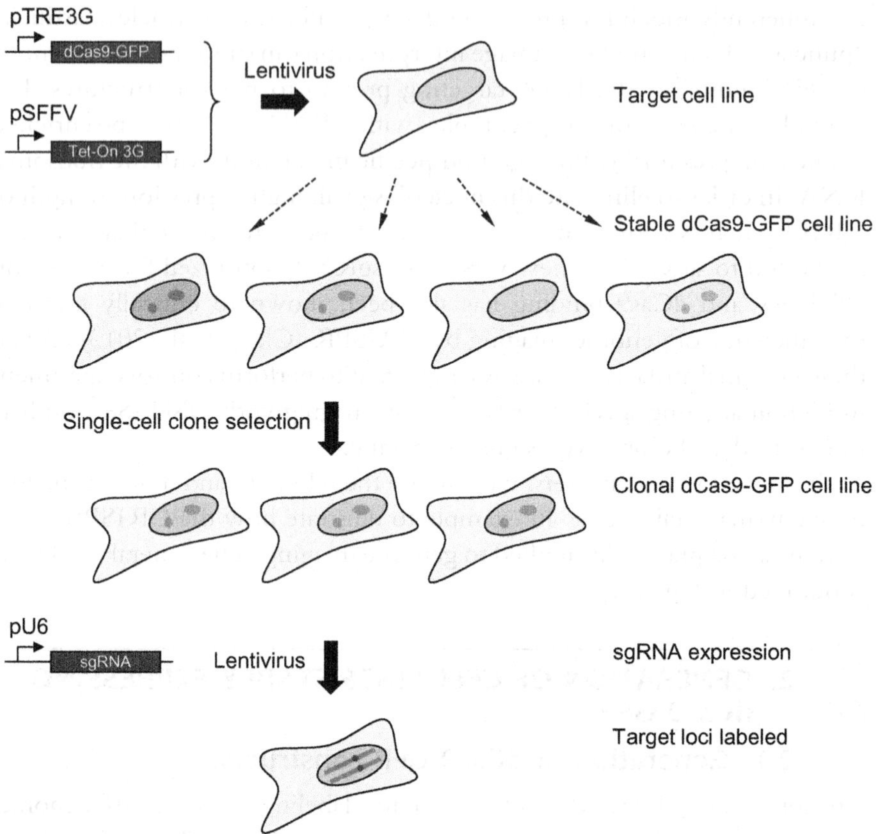


**Figure 16.1** Schematic of the work flow for imaging genomic elements using the CRISPR/Cas9 system. (See the color plate.)

3. Generate a linearized response vector of Lenti-X Tet-On 3G inducible expression system by restriction digestion (*Xma*I/*Not*I).
4. Purify the resulting PCR products of dCas9 and GFP. Ligate them into the vector based on In-Fusion HD cloning protocols.

## 2.2. dCas9-GFP/Tet-On 3G lentiviral production

The CRISPR/Cas9 imaging system can be delivered into mammalian cells via lentivirus which is efficient for infecting a broad variety of mammalian cells. Because lentivirus integrates its genome into the genome of infected cells, the lentiviral vector also facilitates the generation of stable cell lines. It is imperative to fully understand the potential hazards of working with lentivirus. Recombinant lentivirus is listed as a Biosafety Level 2 organism. Any experiments involving this virus have to be performed in a lab with appropriate biosafety level. More biosafety considerations for research with lentiviral vectors can be found at http://www.colorado.edu/ehs/training/recom_dna_lenti.pdf. Here, we briefly describe our procedure to produce lentivirus.

1. One day before transfection, seed 293T cells in the T25 flask containing 6 ml of growth medium, using 90% of Dulbecco's Modified Eagle Medium (DMEM) with high glucose plus 10% of Tet system approved FBS (#631106, Clontech).
2. Maintain cells at 37 °C and 5% $CO_2$ in a humidified incubator.
3. The cells should be around 80% confluent at the time of transfection. Transfect 0.3 μg envelop plasmid pMD2.G, 2.6 μg packaging plasmid pCMV-dR8.91, and 3 μg lentiviral vector (dCas9-GFP or Tet-On 3G) into 293T cells using FuGENE (#E2311, Promega) following the manufacturer's recommended protocol.
4. After 10–12 h, replace the transfection medium with 6 ml fresh growth medium (containing Tet system approved FBS). Incubate the cells at 37 °C for an additional 36–48 h.
5. Harvest the lentiviral supernatants and centrifuge briefly (800 × *g* for 8 min) or filter through a 0.45-μm filter to remove cellular debris.
6. Aliquot the virus (0.5–1 ml) and keep it at −80 °C. Each freeze–thaw cycle may reduce the functional titers by up to two- to four-fold. Freshly harvested virus is recommended for the first try of any new targets.

## 2.3. dCas9-GFP/Tet-On 3G lentiviral infection

This procedure describes the generation of RPE (human retinal pigment epithelium) cells stably coexpressing dCas9-GFP and Tet-On 3G transactivator. To reduce the background fluorescence that comes from unbound dCas9-GFP, we strongly recommend performing imaging experiments with the basal level of dCas9-GFP expression without doxycycline (dox) induction (Fig. 16.2). As the transactivator might also contribute to the basal expression of dCas9-GFP, the cells should still be coinfected with both dCas9-GFP and Tet-On 3G lentiviruses. Any other cell type with flat morphology, such as U2OS (human osteosarcoma) and HeLa cells, are also ideal for CRISPR imaging. For the following protocol, we specify volumes for RPE cells in a 24-well plate.

1. Plate RPE cells into 12 wells of 24-well plate with ~20 to 30% confluence 12–18 h before transduction. Each well contains 0.5 ml complete growth medium (DMEM with GlutaMAX1 plus 10% of Tet system approved FBS).

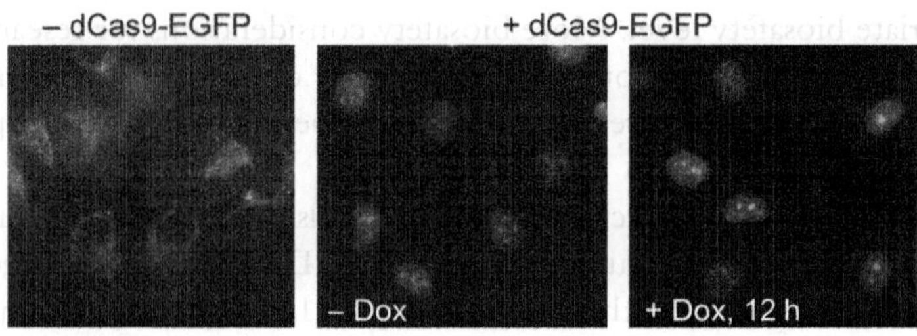


**Figure 16.2** Expression of dCas9-GFP in RPE cells without sgRNA. Only weak cytoplasmic autofluorescence can be detected in cells not expressing dCas9-GFP. For cells carrying a Tet-On 3G inducible expression system for dCas9-GFP, nuclear GFP signal can be observed at the basal expression level without doxycycline induction. At 12 h after doxycycline induction, much stronger (~50-fold) GFP signal can be recorded, with the nucleolar enrichment of dCas9-GFP clearly visible. The three panels are adjusted to different contrasts so that the weak signals are visible. (See the color plate.)

2. Use virus freshly prepared from packaging cells or thaw aliquots of dCas9-GFP and Tet-On 3G lentiviral stocks. Combine the dCas9-GFP and Tet-On 3G lentiviruses in an optimized ratio. If titer values are unknown, use serial dilutions of the viruses mixed at a ratio of 1:1, 1:3, and 3:1. Total volume in each well should be 0.5 ml (e.g., 30 μl dCas9-GFP lentiviruses + 10 μl Tet-On 3G lentiviruses + 460 μl growth medium). Each condition should have two wells. Dox will be added later to one of the well for inducing high expression of dCas9-GFP.
3. Transduce the cells for overnight at 37 °C. Remove and discard the virus-containing medium and replace it with fresh growth medium without Dox.
4. After 36 h, replace the medium again with fresh growth medium with Dox at a concentration of 100 ng/ml. Maintain another well of cells without Dox.
5. Incubate the cells for 12–24 h to allow dCas9-GFP expression to be induced by the presence of Dox.
6. Use fluorescence microscopy to determine if dCas9-GFP is properly localized in the nucleus and the percentage of cells that have dCas9-GFP signal in each condition (different ratio of dCas9-GFP and Tet-On 3G lentiviruses). With the function of NLS, dCas9-GFP is highly enriched in the nucleus, especially in the nucleolus. Because the cells after induction have very strong fluorescence signal, a lower magnification imaging system (10× or 20× objective) can be used so that a large view field can be examined.

7. The condition with 60–70% of cells having dCas9-GFP expression is the best for CRISPR imaging. When the lentiviral dosage is too high, the basal dCas9-GFP expression might exceed the optimal level due to the high multiplicity of infection. Passage the cells with selected condition from the well without adding Dox. Plate the cells in 8-well of chambered coverglass (#155409, Lab-Tek II, Thermo Scientific Nunc) for further imaging and T25 flask for being frozen as stocks.
8. After 12–24 h, image the cells in 8-well chambered coverglass with a high numerical aperture oil immersion objective to examine the basal expression level of dCas9-GFP. Make sure the basal expression level of dCas9-GFP can be visualized.

## 2.4. Selection of clonal cell lines stably expressing dCas9-GFP

Although the stable cell lines created by lentiviral infection can be directly used for CRISPR imaging. The cell-to-cell variation of dCas9-GFP expression level complicates downstream data interpretation. This variation largely results from the different copy number of inserted viral genomes and their site of integration. Therefore, it is desirable to use cells derived from a single cell clone. In addition, in the process of single cell clone selection, clones with different basal expression levels can be identified to match the optimal requirement for different targets. By screening clonal cell lines for a specific repetitive genomic target, we are able to achieve best labeling efficiency and lowest background signal. Here, we briefly describe the procedure to isolate single cell clones of RPE cells stably expressing dCas9-GFP.

1. After allowing dCas9-GFP/Tet-On 3G infected cells to be maintained for 2–3 passages in T25 flask, detach cells by digesting with 1 ml of 0.25% trypsin/EDTA (Invitrogen) for approximately 5 min at 37 °C.
2. Add 6 ml growth medium to suspend the cells and dilute the cells by 1:10, 1:100, 1:1000, and 1:10,000 in the growth medium. Plate 50 μl of each dilution ratio on a glass slide to count the cell number. Select the dilution ratio with 0–1 cell/50 μl for the further experiment.
3. Add 100 μl complete growth medium in each well of 96-well plate. Transfer 50 μl of appropriately diluted RPE cells into each well. Theoretically, there would be 0–1 cell distributed into each well.
4. After 10–14 days, the colonies will be clearly formed. Select the wells with a single colony and detach the colony by digesting with 1 ml of 0.25% trypsin/EDTA (Invitrogen) for 5 min at 37 °C.

5. Transfer the colonies into 24-well plate. Seed each colony in 2-wells. One well of cells are maintained with Dox.
6. After the cells reach >50% confluence, select the colonies with 100% of cells having dCas9-GFP expression.
7. Seed the selected colonies without Dox induction in 8-well chambered coverglass and do sgRNA infection to identify a best clone for labeling a specific target.

## 3. EXPRESSION OF sgRNAs USING LENTIVIRAL VECTOR

Previous work has indicated that sgRNA expression level limits CRISPR/Cas9 function in human cells (Jinek et al., 2013). An optimized sgRNA sequence (Chen et al., 2013) has been created to improve its expression efficiency and assembly with dCas9-GFP. This optimized design enables more efficient labeling of various genomic targets. The delivery of sgRNA expression system using lentiviral vectors also dramatically increased the efficiency. This procedure describes how we clone sgRNAs into lentiviral vector and infect RPE cells with sgRNA lentiviruses.

### 3.1. sgRNA design and cloning

1. To target the template DNA strand, search for 5′-GN$_{(17–24)}$-NGG-3′. The sequence of GN$_{(17–24)}$ will be used directly as the base-paring region (spacer) of the sgRNA. NGG is the PAM sequence recognized by *S. pyogenes* Cas9 protein. To target the nontemplate DNA strand, search for 5′-CCN-N $_{(17–24)}$C-3′. The reverse complementary sequence of N $_{(17–24)}$C will be used as the spacer of the sgRNA. The design principle is the same as that for genome editing (Ran et al., 2013) or gene regulation (Larson et al., 2013).
2. Select 2–4 sequences for a target. The successful rate of sgRNAs design is around 50%.
3. Design forward and reverse primers for a specific target. Forward primer: 5′- ggagaa<u>CCACCTTGTTG</u>**GN$_x$**GTTTAAGAGCTATGCTGG-AAACAGCA-3′. **GN$_x$** (x=17–28) is the base-pairing sequence (spacer) which can be changed for labeling any target. Underlined sequence is BstXI restriction site. Reverse primer: 5′- ctagta <u>CTCGAG</u>AAAAAAAGCACCGACTCGGTGCCAC-3′. Underlined sequence is *Xho*I restriction site.
4. Amplify the optimized sgRNA sequence by using the common reverse primer but unique forward primers containing the specific spacer

sequence. Use any plasmid containing the optimized sgRNA sequence (e.g., Addgene ID: 51024) as the PCR template.

5. Purify the sgRNA PCR products and digest them by *Bst*XI and *Xho*I.
6. Digest the lentiviral U6-based expression vector (Addgene ID: 51024) by *Bst*XI and *Xho*I.
7. Subclone the sgRNA fragment into the U6 vector according to manufacturer's instructions of Quick T4 DNA ligase (#M2200S, NEB).

### 3.2. sgRNA lentiviral infection

The procedure for producing sgRNA lentivirus is the same as dCas9-GFP lentivirus. RPE cells can be efficiently infected with unconcentrated virus stock. Nevertheless, as the labeling efficiency is sgRNA dosage dependent, concentrating viruses for some types of cells with low infection efficiency should be considered.

1. 12 h prior to transduction, seed clonal cells stably expressing dCas9-GFP in 8-well of chambered coverglass.
2. Infect the cells with sgRNA lentivirus (by 1:12, 1:6, or 1:3 dilution) supplemented with 5 μg/ml polybrene (#TR-1003-G, Millipore) in each well. The more target sites, the high sgRNA dosage should be transduced. Incubate the cell with lentivirus for 12 h.
3. Remove the virus-containing medium and replace it with fresh growth medium without phenol red.
4. 48 h posttransduction, the cells are ready for imaging.

## 4. LABELING OF NONREPETITIVE SEQUENCES

To label nonrepetitive sequences, at least 30 sgRNAs should be generated for loci. This is essential for enriching dCas9-GFP signal for detection under conventional fluorescence microscopy. Here, we describe procedures for cloning multiple sgRNAs in a high-throughput way and packaging pooled sgRNA lentiviruses.

### 4.1. Target selection and sgRNA design

When labeling a nonrepetitive sequence, we recommend to design at least 30 sgRNAs. Distance between two adjacent target sequences is critical for dCas9-GFP signal enrichment. Labeling of repetitive regions in mucin genes indicates that 30–60-bp long distance is efficient for visualization. Thus, to label the nonrepetitive region of *MUC4* gene, the distance between two adjacent sgRNAs is kept at 30–50 bp. The sgRNAs are designed by searching for **GN$_x$**NGG (x=17–28) in both template and nontemplate DNA strands of the coding sequence of *MUC4*. NGG is the PAM for the *S. pyogenes* Cas9. The GN$_x$ sequence is the base-pairing region of the sgRNA.

### 4.2. High-throughput sgRNA cloning

This procedure describes cloning multiple sgRNAs into a same U6 vector in a high-throughput way. The protocol is an adaption of published methods to clone multiple sgRNAs for CRISPRi (Larson et al., 2013). Taking cloning 96 sgRNAs simultaneously as an example:

1. Order the 96 forward primers in a 96-well plate for easy handling (e.g., from Integrated DNA Technologies).
2. Pool the forward primers into 12 groups using a multichannel pipet. Thus, each group contains eight primers. Mix the pooled primers with equal amount of the reverse primer to PCR amplify sgRNA fragments using a plasmid (e.g., Addgene ID: 51024) as the template. Perform the PCR reaction using the Phusion High-Fidelity DNA polymerase.
3. Run the PCR products (twelve groups) on a 1.5% agarose gel and purify them by cutting the ~150 bp DNA band.
4. Digest the purified PCR products by *Bst*XI and *Xho*I for 4 h.
5. Purify the digestion reactions and ligate them into digested sgRNA vector (by *Bst*XI/*Xho*I) using Quick T4 DNA ligase for 5 min.
6. Transform the eight groups of ligation reactions into Stellar™ competent cells (#636763, Clontech), respectively. Grow the *Escherichia coli* on agar plates supplemented with 100 μg/ml ampicillin at 37 °C.
7. The second day, randomly pick 24 clones from each plate (a total of 24 × 12 = 288 colonies) and culture them in 1 ml LB + ampicillin in 2 ml deep 96-well plates. Grow the cultures for 6 h. Send 50 μl of each culture for sequencing. Grow the rest of the cultures overnight at 37 °C.
8. Amplify the sequencing-verified cultures (representing correct sgRNAs, we hit 73 correct sgRNAs in our case for labeling *MUC4* nonrepetitive region) in 10 ml LB with ampicillin and grow them for 6 h.
9. Mix cultures of five or six sgRNAs to create a cocktail culture (the total volume will be 50–60 ml). Extract plasmids using the Nucleo Bond Xtra Midi kit (#740420.50, Clontech) for subsequent lentiviral packaging.

### 4.3. Production of pooled sgRNA lentiviruses

1. One day prior to transfection, plate 293T cells in the T25 flask (~40% confluence) containing 6 ml of growth medium.
2. Maintain cells at 37° and 5% $CO_2$ in a humidified incubator.
3. Transfect 0.3 μg envelop plasmid pMD2.G, 2.6 μg packaging plasmid pCMV-dR8.91, and 3 μg pooled lentiviral vectors (including equal amounts of five to six sgRNA plasmids) into 293T cells using FuGENE following the manufacturer's recommended protocol.
4. Replace the transfection medium with 6 ml fresh growth medium after 12 h. Incubate the cells at 37 °C for an additional 36–48 h.
5. Harvest the lentiviral supernatants and centrifuge briefly (800 × *g* for 8 min) or filter through a 0.45-μm filter to remove cellular debris. To label 73 targets in *MUC4* nonrepetitive region, 15 lentiviral cocktails were generated.
6. Concentrate the lentiviruses by using Lenti-X™ Concentrator (#631231, Clontech) according to the manufacturer's direction. For labeling nonrepetitive sequences, this step is strongly recommended. Further optimization is required for delivering more sgRNAs into one cell.

## 5. IMAGING OF GENOMIC LOCI DETECTED BY CRISPR

### 5.1. Verify CRISPR signal by a modified FISH staining protocol

Without sgRNA expression, dCas9-GFP is enriched in nucleolus-like structures. Although such nucleolar signal diminishes and even disappears when dCas9 forms complex with sgRNA, false fluorescent puncta may still arise if sgRNA is not properly expressed or if the original instead of the optimized sgRNA design is used. Therefore, it is important to always verify the specificity of dCas9-GFP signal. A straightforward verification approach is to fix and stain the cells with FISH, and then check the colocalization with the dCas9-GFP signal.

Cas9–RNA complex initiates strand separation to enable base paring between the sgRNA guide sequence and the target DNA strand (Sternberg, Redding, Jinek, Greene, & Doudna, 2014). As the result, oligo-DNA FISH probes are able to bind to the region where the two DNA strands are separated by CRISPR without the DNA denaturation procedure (heating cells at 80 °C) required for regular FISH protocols. Skipping the denaturation step allows the preservation of dCas9-GFP fluorescence for multicolor colocalization imaging. Below is a simplified protocol of FISH to colabel genomic DNA sequence with CRISPR (Fig. 16.3).

1. Select oligo-DNA FISH target sequence. For short repetitive sequence (e.g., TTAGGG of telomere repeat), a similar target sequence with CRISPR targets can be chosen. For long repetitive sequence (more than 40 bp), the sequence close to the CRISPR target can be used for FISH target. Thus, dCas9/sgRNA and FISH probe can bind at regular intervals along the target region (Fig. 16.3).
2. Order oligo-DNA FISH probes with one Cy5 dye attached at the 5′ or 3′ end.
3. Dilute the oligo-DNA FISH probe to 200 ng/μl (stock solution).
4. Fix CRISPR-labeled cells with 4% paraformaldehyde (PFA). Wash the samples three times with PBS.
5. Permeabilize the cells with PBS + 0.5% NP-40 for 10 min.
6. Wash the sample with PBS for 5 min.
7. Incubate the sample with 2 ng/μl oligo-DNA FISH probe in hybridizing solution (10% dextran sulfate, 50% formamide, 500 ng/ml Salmon sperm DNA in 2× SSC buffer) for 12 h in a dark and humidified box (Schmitt et al., 2010).
8. Wash the sample for three times with 2× SSC buffer (#S6639-1 l, Sigma), each for 10 min. Stain nuclei with DAPI if necessary.

## 5.2. Live-cell imaging of genomic loci

We recommend performing the imaging 48 h post-infection of sgRNAs. The low basal level expression of dCas9-GFP is sufficient for the detections of telomeres and mucin genes. To achieve high sensitivity for widefield epifluorescence microscopy, oil immersion objectives with a numerical aperture of at least 1.3 (40× to 100×) should be used, with either an EMCCD camera or a Scientific CMOS (sCMOS) camera. A low excitation intensity (at about 0.1 W/cm$^2$) should be used to avoid photobleaching and phototoxicity, which should generate sufficient signal in an exposure time of as short as 0.2 s. Short-term live imaging within 1 h can be done at room temperature for determining labeling efficiency or nuclear localization; however, to track genomic dynamics, the samples must be maintained at 37 °C using an environmental control chamber.

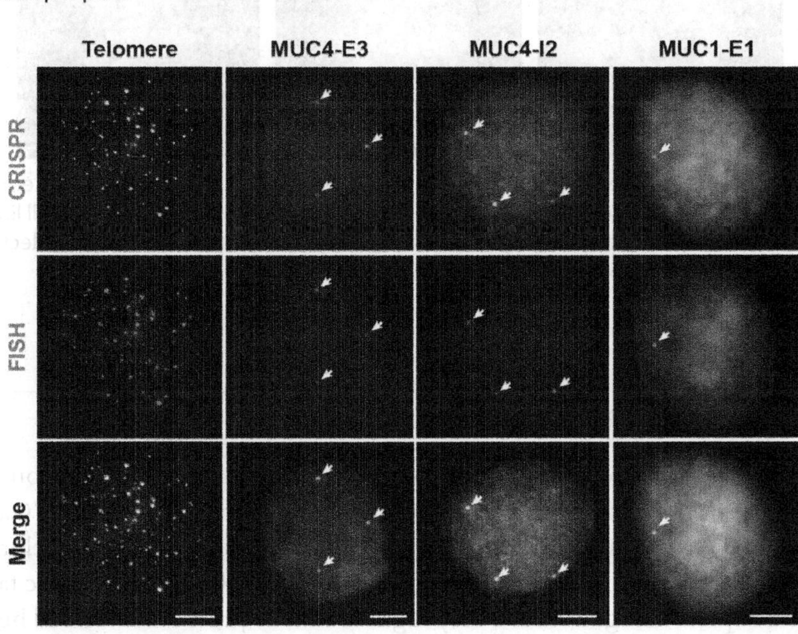


**Figure 16.3** Verification of CRISPR signal specificity by FISH. Four target loci in RPE cells have been tested: telomere, the tandem repeats in exon 2 and intron 2 of the *MUC4* gene, and the tandem repeats in exon 2 of the *MUC1* gene. The recognition sequences for the sgRNAs and oligo-DNA FISH probes are underlined. Complete colocalization of the CRISPR and FISH signal is displayed with the nuclear DNA stained by DAPI. Scale bars: 5 μm. (See the color plate.)

After live-cell imaging, samples can be kept at 4 °C after being fixed with 4% PFA for 15 min if necessary. Figure 16.4 displays fluorescence images of the telomeres and the *MUC4* loci in three different human cell lines labeled with dCas9-GFP.

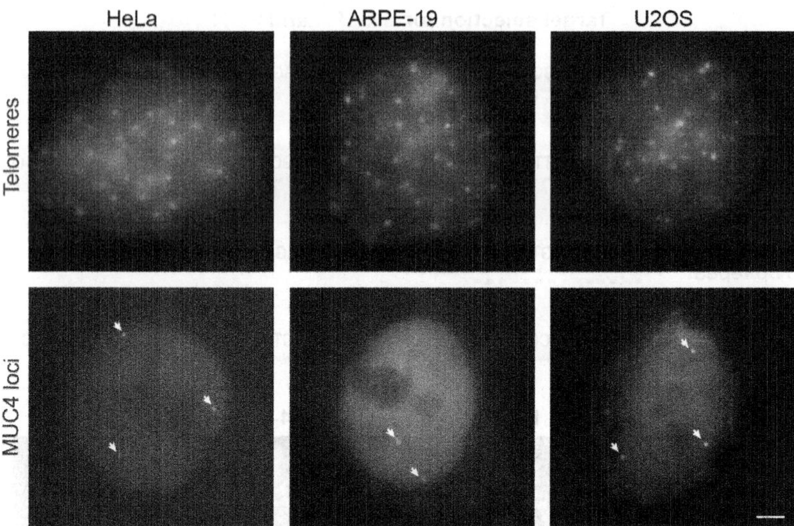


**Figure 16.4** Imaging telomeres and the *MUC4* gene loci using CRISPR/Cas9 system in three different human cell lines: HeLa, ARPE-19, and U2OS. The HeLa and U2OS cell lines contain three copies of the *MUC4* gene due to trisomy chromosome 3, which is reflected in the CRISPR images. Scale bars: 5 μm. (See the color plate.)

## 6. SUMMARY

We have presented procedures to label both repetitive and nonrepetitive genomic sequences in live cells by engineering a CRISPR/Cas9 system. Two key ways to achieve high labeling efficiency are screen clonal cell lines for an appropriate expression level of dCas9-GFP for a specific target and express the sgRNA at a very high level. The protocol described here uses telomeres and mucin genes as examples, and can be applied to many human cell lines including RPE, HeLa, APRE-19, and U2OS cells. Modifications might be necessary for other organisms and cell types. In addition, although CRISPR imaging is relatively robust, the labeling efficiency still varies among different target sequences, making it necessary to screen a number of sgRNAs for a given genomic locus. Further developments need to be conducted to improve the labeling efficiency of nonrepetitive genomic sequences, including strategies to enhance the delivery of multiple sgRNAs into the cells, reduce the free dCas9-FP background, and promote the efficient sgRNA–dCas9 complex formation.

## ACKNOWLEDGMENTS

This work is supported by the W. M. Keck Foundation. The authors thank Christian Covill-Cooke for critical reading of the chapter.

## REFERENCES

Anton, T., Bultmann, S., Leonhardt, H., & Markaki, Y. (2014). Visualization of specific DNA sequences in living mouse embryonic stem cells with a programmable fluorescent CRISPR/Cas system. *Nucleus*, *5*, 163–172.

Chen, B., Gilbert, L. A., Cimini, B. A., Schnitzbauer, J., Zhang, W., Li, G. W., et al. (2013). Dynamic imaging of genomic loci in living human cells by an optimized CRISPR/Cas system. *Cell*, *155*, 1479–1491.

Gilbert, L. A., Larson, M. H., Morsut, L., Liu, Z., Brar, G. A., Torres, S. E., et al. (2013). CRISPR-mediated modular RNA-guided regulation of transcription in eukaryotes. *Cell*, *154*, 442–451.

Hellwig, D., Munch, S., Orthaus, S., Hoischen, C., Hemmerich, P., & Diekmann, S. (2008). Live-cell imaging reveals sustained centromere binding of CENP-T via CENP-A and CENP-B. *Journal of Biophotonics*, *1*, 245–254.

Hsu, P. D., Scott, D. A., Weinstein, J. A., Ran, F. A., Konermann, S., Agarwala, V., et al. (2013). DNA targeting specificity of RNA-guided Cas9 nucleases. *Nature Biotechnology*, *31*, 827–832.

Jinek, M., East, A., Cheng, A., Lin, S., Ma, E., & Doudna, J. (2013). RNA-programmed genome editing in human cells. *eLife*, *2*, e00471.

Larson, M. H., Gilbert, L. A., Wang, X., Lim, W. A., Weissman, J. S., & Qi, L. S. (2013). CRISPR interference (CRISPRi) for sequence-specific control of gene expression. *Nature Protocols*, *8*, 2180–2196.

Ma, H., Reyes-Gutierrez, P., & Pederson, T. (2013). Visualization of repetitive DNA sequences in human chromosomes with transcription activator-like effectors. *Proceedings of the National Academy of Sciences of the United States of America*, *110*, 21048–21053.

Misteli, T. (2007). Beyond the sequence: Cellular organization of genome function. *Cell*, *128*, 787–800.

Misteli, T. (2013). The cell biology of genomes: Bringing the double helix to life. *Cell*, *152*, 1209–1212.

Miyanari, Y., Ziegler-Birling, C., & Torres-Padilla, M. E. (2013). Live visualization of chromatin dynamics with fluorescent TALEs. *Nature Structural and Molecular Biology*, *20*, 1321–1324.

Qi, L. S., Larson, M. H., Gilbert, L. A., Doudna, J. A., Weissman, J. S., Arkin, A. P., et al. (2013). Repurposing CRISPR as an RNA-guided platform for sequence-specific control of gene expression. *Cell*, *152*, 1173–1183.

Ran, F. A., Hsu, P. D., Wright, J., Agarwala, V., Scott, D. A., & Zhang, F. (2013). Genome engineering using the CRISPR-Cas9 system. *Nature Protocols*, *8*, 2281–2308.

Robinett, C. C., Straight, A., Li, G., Willhelm, C., Sudlow, G., Murray, A., et al. (1996). In vivo localization of DNA sequences and visualization of large-scale chromatin organization using lac operator/repressor recognition. *The Journal of Cell Biology*, *135*, 1685–1700.

Schmitt, E., Schwarz-Finsterle, J., Stein, S., Boxler, C., Muller, P., Mokhir, A., et al. (2010). COMBinatorial Oligo FISH: Directed labeling of specific genome domains in differentially fixed cell material and live cells. *Methods in Molecular Biology*, *659*, 185–202.

Sternberg, S. H., Redding, S., Jinek, M., Greene, E. C., & Doudna, J. A. (2014). DNA interrogation by the CRISPR RNA-guided endonuclease Cas9. *Nature*, *507*, 62–67.

Thanisch, K., Schneider, K., Morbitzer, R., Solovei, I., Lahaye, T., Bultmann, S., et al. (2014). Targeting and tracing of specific DNA sequences with dTALEs in living cells. *Nucleic Acids Research*, *42*, e38.

Wang, X., Kam, Z., Carlton, P. M., Xu, L., Sedat, J. W., & Blackburn, E. H. (2008). Rapid telomere motions in live human cells analyzed by highly time-resolved microscopy. *Epigenetics & Chromatin*, *1*, 4.

Yuan, K., Shermoen, A. W., & O'Farrell, P. H. (2014). Illuminating DNA replication during Drosophila development using TALE-lights. *Current Biology*, *24*, R144–R145.

CHAPTER SEVENTEEN

# Cas9-Based Genome Editing in *Xenopus tropicalis*

**Takuya Nakayama*, Ira L. Blitz†, Margaret B. Fish†, Akinleye O. Odeleye*, Sumanth Manohar*, Ken W.Y. Cho†, Robert M. Grainger*,[1]**

*Department of Biology, University of Virginia, Charlottesville, Virginia, USA
†Department of Developmental and Cell Biology, University of California, Irvine, California, USA
[1]Corresponding author: e-mail address: rmg9p@virginia.edu

## Contents



## Abstract

*Xenopus tropicalis* has been developed as a model organism for developmental biology, providing a system offering both modern genetics and classical embryology. Recently, the Clustered Regularly Interspaced Short Palindromic Repeats/CRISPR-associated (CRISPR/Cas) system for genome modification has provided an additional tool for *Xenopus* researchers to achieve simple and efficient targeted mutagenesis. Here, we provide insights into experimental design and procedures permitting successful application of this technique to *Xenopus* researchers, and offer a general strategy for performing loss-of-function assays in F0 and subsequently F1 embryos.

*Methods in Enzymology*, Volume 546
ISSN 0076-6879
http://dx.doi.org/10.1016/B978-0-12-801185-0.00017-9

# 1. INTRODUCTION

*Xenopus* has long been a favored model organism for developmental and cell biology due to its unique combination of advantageous features including: the ability to acquire large numbers of eggs (or oocytes) and embryos; the availability of relatively simple techniques for microinjection of mRNA, DNA, protein, or antisense morpholino oligonucleotides for gain-of-function or loss-of-function (LOF) experiments; ease of transgenesis allowing modern molecular developmental and biochemical studies; and its suitability for classical explant/transplantation embryology.

Although many studies use *Xenopus laevis*, an allotetraploid species with a long generation time, for embryological and cell biological studies, recently *Xenopus tropicalis* has emerged as a new model organism (Harland & Grainger, 2011). Forward and reverse genetic approaches have identified developmental mutants and their causative genes (Abu-Daya, Khokha, & Zimmerman, 2012). However, the number of characterized mutants to date is small. Mutational screens, including directed approaches such as Targeting Induced Local Lesions In Genomes (e.g., Fish et al., 2014) are laborious, and new approaches to efficiently edit the genome are urgently needed.

Recent technological advances have allowed researchers to readily perform targeted gene editing in many organisms. Two major methods that have been used are zinc-finger nucleases (ZFNs) and transcription activator-like effector nucleases (TALENs), both of which have been successfully applied in *Xenopus* (Ishibashi, Cliffe, & Amaya, 2012; Lei et al., 2012; Nakajima, Nakai, Okada, & Yaoita, 2013; Nakajima, Nakajima, Takase, & Yaoita, 2012; Nakajima & Yaoita, 2013; Suzuki et al., 2013; Young et al., 2011). Most recently, Type II CRISPR/Cas (Clustered Regularly Interspaced Short Palindromic Repeats/CRISPR-associated) technology has been developed for genome modification. This system was first identified as part of the naturally occurring bacterial adaptive defense mechanism (Fineran & Dy, 2014; Hsu, Lander, & Zhang, 2014; Terns & Terns, 2014), and now has been successfully applied in numerous organisms (Sander & Joung, 2014) including *X. tropicalis* to effect targeted genome modification (Blitz, Biesinger, Xie, & Cho, 2013; Guo et al., 2014; Nakayama et al., 2013), providing an additional tool for *Xenopus* researchers to achieve simple and efficient targeted mutagenesis. The application of newly available genome engineering tools in the *X. tropicalis* system, with its sequenced diploid genome, high degree of synteny with the human genome, and conservation of key developmental processes, will make this an outstanding model organism for studying human genetic disease and developmental pathologies.

Here, we present a general protocol for CRISPR/Cas9-mediated targeted mutations in *X. tropicalis* including a strategy for performing LOF experiments in F0 mutagenized animals and subsequently F1 animals.

## 2. PRINCIPLE

CRISPR/Cas9 creates genome modifications using a common biological mechanism across taxa, and is described briefly here. The Type II CRISPR/Cas system uses Cas9 (an RNA-guided DNA endonuclease) for genome editing. In bacteria, Cas9 cleaves target DNA by forming a complex with two small RNAs, a CRISPR RNA (crRNA) that has complementary sequence to the target DNA and the *trans*-activating CRISPR RNA (tracrRNA) that base pairs with the crRNA. For efficient cleavage, the target DNA must be followed by a sequence called the protospacer (a complementary sequence targeted by a specific crRNA) adjacent motif (PAM; Fig. 17.1), which varies among bacterial species. *Streptococcus pyogenes* Cas9 is most widely used for genome editing in eukaryotic systems, and its PAM sequence is NGG (where N can be any nucleotide), although NAG can also function at lower efficiency (Anders, Niewoehner, Duerst, & Jinek 2014; Terns &

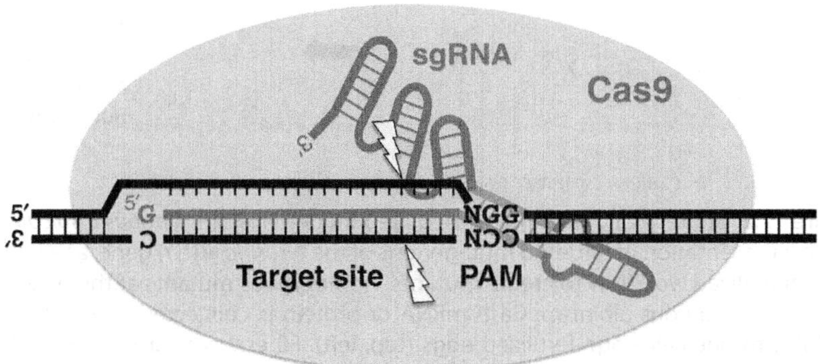


**Figure 17.1** Schematic representation of CRISPR/Cas9-mediated target cleavage. Cas9 protein (oval) and sgRNA together recognize the target sequence. The PAM, NGG, while recognized by Cas9, does not base pair with the sgRNA. Cleavage (indicated by lightning bolts) occurs within the target to create a double-stranded break, which is then repaired by an error-prone NHEJ mechanism.

Terns, 2014). For genome editing applications, a portion of the crRNA has been fused to the tracrRNA to create a cassette for production of synthetic (or single) guide RNAs (sgRNAs) (Hwang et al., 2013; Mali, Yang, et al., 2013). A target sequence (~20 bp) is added to this cassette to create the final sgRNA, which then directs Cas9 to specific sites in the genome for cleavage. In *Xenopus*, sgRNAs are coinjected together with either Cas9 mRNA or protein into fertilized eggs or early embryos (Blitz et al., 2013; Guo et al., 2014; Nakayama et al., 2013) (Fig. 17.2). Following Cas9-mediated scission of the target site, double-strand breaks are often imperfectly repaired by non-homologous end-joining (NHEJ), which frequently leads to insertion and deletion mutations (indels).

The resultant embryos are mosaic; cells bearing combinations of mutant and wild-type (unsuccessfully targeted or repaired perfectly) alleles are

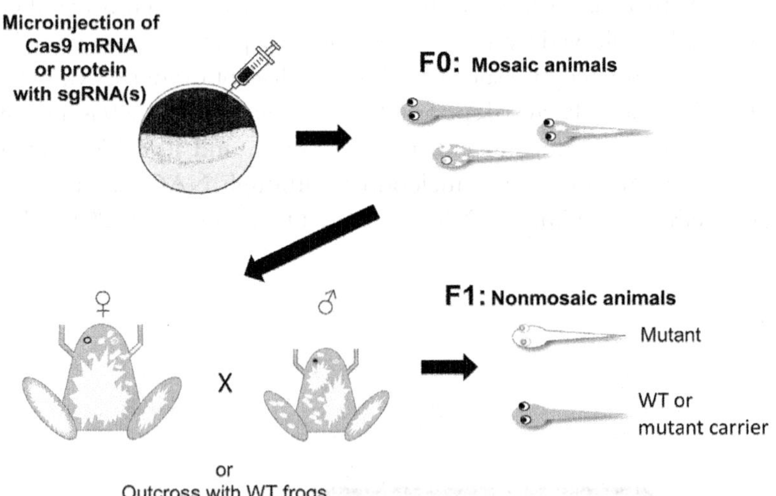


**Figure 17.2** Strategy of CRISPR/Cas9-mediated mutagenesis in *Xenopus tropicalis*. Schematic representation of targeted mutagenesis of the *tyrosinase* (*tyr*) gene, as an example of a generalized workflow to create mutants. Homozygous mutants at the *tyr* locus display oculocutaneous albinism. Cas9 mRNA or protein is coinjected with a *tyr*-specific sgRNA into one-cell stage fertilized eggs (top, left). F0 embryos are mosaics: they are comprised of populations of cells containing different mutant alleles and wild-type tissue. In the case of *tyr* targeting, biallelic loss of gene function results in deficient melanin synthesis, which is observed as a loss of pigmentation in the eyes and skin. Mosaic F0 animals can be intercrossed to produce nonmosaic (compound heterozygous or homozygous) mutants ($tyr^{-/-}$), as well as carriers ($tyr^{+/-}$), and homozygous wild-type F1 progeny.

present in various proportions. The relative amounts of mutant to wild type are likely a function of both the timing of mutagenesis and efficacy of individual sgRNAs to direct Cas9 to target sites. Mutant phenotypes may be scorable in F0 mosaic animals. Intercrosses between F0 adults can be performed to create nonmosaic F1s that are expected to be primarily compound heterozygotes (containing two different mutant allelic variants). Alternatively, in the F2 generation, homozygous mutants can be created that carry a single allelic variant.

# 3. PROTOCOL

## 3.1. Background knowledge and experimental equipment

In this chapter, the authors assume that readers have sufficient experimental knowledge, basic molecular biology skills, and experience handling *Xenopus*, including general embryo manipulation and microinjection technique (Sive, Grainger, & Harland, 2010). All necessary equipment is commonly available in laboratories already equipped for *Xenopus* embryo research and no special equipment is required for the CRISPR-mediated mutagenesis.

## 3.2. sgRNA design

The first step in designing a CRISPR/Cas mutagenesis strategy is to identify possible target sites in the gene of interest. A target-adjacent PAM sequence is an absolute requirement for efficient mutagenesis; it is important to note that the PAM sequence is not included in the sgRNA itself. A further constraint on the target sequence is due to the use of *in vitro* transcription from promoters such as T7, T3, or SP6 to produce sgRNAs, which function optimally with an initiator guanine (G) residue (Fig. 17.3A). If one uses a plasmid template for making sgRNA by *in vitro* transcription as originally described (e.g., Hwang et al., 2013), the target must start with GG due to cohesive end requirements of the cloning strategy. Alternatively, the template can be prepared using a PCR-based method (e.g., Nakayama et al., 2013), which is easier and faster, and allows more flexibility for target sequence selection because it requires only a single 5′ G (+1). The typical length of target sequence included in the design of an sgRNA (i.e., the genomic target sequence before the PAM) is 20 bp, but a recent report (Fu, Sander, Reyon, Cascio, & Joung, 2014) suggests that shorter (i.e., as short as 17 bp) targets function as efficiently as 20 bp and have fewer potential off-target sites. In summary, one needs to find $G(N)_{16-19}$ target sequence

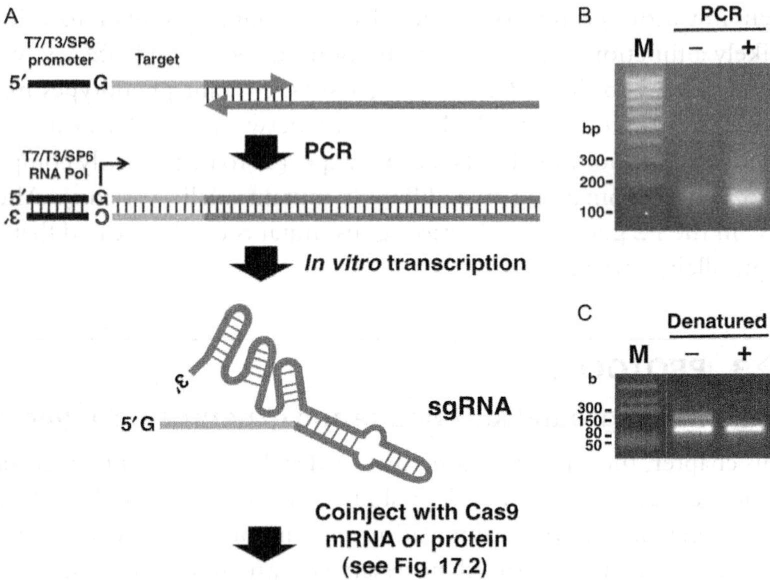


**Figure 17.3** PCR-based template synthesis for sgRNA *in vitro* transcription. (A) Two long, partially overlapping oligonucleotides are annealed and serve as primers for fill-in reactions by a thermostable polymerase. The 5′-oligo (primer) contains the T7 (or T3 or SP6) promoter; the transcription start site (G) corresponds to the first base of the target region (17–20 bp), followed by a portion of the sgRNA backbone. This region is complementary to the 3′-end of the 3′-oligo (primer) that contains the remainder of the sgRNA backbone sequence required for proper RNA folding. (B) Agarose gel (2%) electrophoresis of 2 μl of a PCR reaction (+) compared to no PCR (−) confirms successful sgRNA template synthesis. M, 100-bp DNA ladder. (C) Agarose gel (2%) electrophoresis of ~200 ng sgRNA following *in vitro* transcription, using the DNA template shown in (B). Example of heat-denatured (+) and nondenatured (−) sgRNA is shown. The RNA size marker lane (M) shown here was loaded on the same gel and was digitally spliced to place it adjacent to the sample lanes in this image.

(total length is 17–20 bp), followed by NGG in the genome. Any effects of decreasing the target length have not been rigorously tested in *Xenopus*, and therefore, it remains unknown whether shorter target sequences will work as efficiently in this system.

Potential CRISPR/Cas9 target sequences in the genome can be identified by manually locating PAM sequences within the region of interest. Finding a guanine nucleotide 17–20 bp upstream of a PAM would identify a potential CRISPR target. Recent publications, however, suggest that target sites do not necessarily need to start with G. Any sequence followed by a PAM in the genome may be targeted by simply adding an extra G or GG not encoded in the genome sequence (Ansai & Kinoshita, 2014 and references therein). This approach may be a workable strategy when no other suitable option can be found. Once the target site is chosen, the next step is to determine whether the selected target sequence has similarity to other regions that may lead to off-target mutagenesis. This can be done bioinformatically using the raw genome sequence database (e.g., Blitz et al., 2013), or using a Web-based homology search engine (http://gggenome.dbcls.jp/en/) (Nakayama et al., 2013).

A number of other online tools are now available to search the *X. tropicalis* genome to identify optimal target and putative off-target sites. These include CHOPCHOP (https://chopchop.rc.fas.harvard.edu/index.php), CRISPR/Cas9 target predictor (http://crispr.cos.uni-heidelberg.de), CRISPRdirect (http://crispr.dbcls.jp/), E-CRISP (http://www.e-crisp.org/E-CRISP/designcrispr.html), GT-Scan (http://gt-scan.braembl.org.au/gt-scan/), and Cas-OFFinder (http://www.rgenome.net/cas-offinder/). Each tool offers a variety of input and output options, and depending on the purpose and knowledge of users different tools may be preferable. Notable features offered by many of the tools include: options for PCR primer design to assay CRISPR efficiency in target regions and to check for off-target mutagenesis (along with the inclusion of genomic location information for putative off-target regions); the ability to vary target length and 5′ nucleotide identity; the option to search for alternative PAM sequences; and the ability to target 5′ or 3′ regions of ORFs for the purpose of introducing N- or C-terminal protein modifications by homologous recombination, among others.

### 3.2.1 Considerations in target site choice

The success of a gene targeting strategy will be influenced by location of the mutation within the gene. While repair of double-strand breaks by NHEJ results in indels, most are short deletions. By chance, approximately two-thirds of indels will result in premature termination of protein translation and hence more 5′ target sites might be predicted to yield stronger effects. The remaining indels may or may not result in loss of gene function depending on their effects on protein structure. Therefore, choosing target sites within sequence encoding folded domains may be more advantageous for generating LOF mutations because even in-frame mutations may disrupt proper domain folding, resulting in loss of protein activity or stability. Folded domains are often recognizable as highly conserved sequences across species, or protein families, and protein fold search tools such as Pfam (http://pfam.xfam.org/search) can be used to assist in identification of these regions.

Genes encoding multiple protein isoforms need careful examination to ensure that site choice results in proper targeting of all possible isoforms. However, there may be cases where it is desirable to target specific isoforms while maintaining the function of others. A one-size-fits-all strategy is impossible, and therefore application of these ideas will need to be tailored to each gene's unique characteristics. Since the efficiency of cleavage at different target sites is variable and unpredictable, multiple independent target sites should be explored to identify the best sgRNA (see Section 4.1 for further discussion).

## 3.3. sgRNA template construction

Two sources of template can be used for the *in vitro* synthesis of sgRNAs. One is plasmid-based and requires subcloning (e.g., Guo et al., 2014; Hwang et al., 2013) and the other uses a PCR-based strategy for making linear DNA templates, shown schematically in Fig. 17.3.

### 3.3.1 Template assembly by PCR: Primers

The 5′ primer is unique to each target site and has the form

5′-TAATACGACTCACTATA $G(N)_{16-19}$ GTTTTAGAGCTAGAA ATAGCAAG-3′,

where $G(N)_{16-19}$ is the target sequence of interest, designed as described above, and double underlined is the T7 promoter. T3 or SP6 promoters could be substituted, for example, SP6 would work better for $GA(N)_{15-18}$.

The 3′ primer is common to all sgRNA templates and is as follows:

5′-AAAAGCACCGACTCGGTGCCACTTTTTCAAGTTGATAA CGGACTAGCCTTATTTTAACTTGCTATTTCTAGCTCTAAA AC-3′

Both primers above encode the original backbone published in Hwang et al. (2013). To date, we have tested another backbone, $sgRNA^{(F+E)}$, which has a modified hairpin reported to have stronger activity (Chen et al., 2013). However, we have not observed significant enhancement of mutagenesis activity using the $sgRNA^{(F+E)}$ backbone in *Xenopus* (our unpublished observations).

### 3.3.2 Template assembly by PCR: Assembly conditions

High fidelity polymerase (e.g., Platinum *Pfx* DNA polymerase, Invitrogen) is used to generate PCR-based templates. Assembly reactions are performed in a final volume of 100 μl as follows:

| **PCR assembly reaction** | |
|---|---|
| 10 × Buffer | 10 μl |
| 25 m*M* dNTP mixture | 1.2 μl |
| 50 m*M* $Mg_2SO_4$ | 2 μl |
| 5′ Primer (100 pmol/μl) | 2 μl |
| 3′ Primer (100 pmol/μl) | 2 μl |
| DNA polymerase | 1 μl |
| $H_2O$ (DNAse/RNAse-free) | To 100 μl |

Cycling conditions are:

94 °C for 5 min.

10 cycles (up to 20 cycles) of 94 °C for 20 s, 58 °C for 20 s, 68 °C for 15 s.

68 °C for 5 min.

A typical result is shown in Fig. 17.3B. After confirming that a single product has been synthesized, templates are column-purified using, for example, QIAquick® PCR Purification Kit (Qiagen) or DNA Clean & Concentrator™-5 (Zymo Research) and DNA is eluted with 30–50 μl of RNase-free water. The concentration should be between 30 and 80 ng/μl or more, but, if lower, PCR may need to be repeated to increase yield.

### 3.3.3 In vitro *transcription of sgRNA*

We use the MEGAscript® T7 Transcription Kit (Life Technologies) following the manufacturer's recommended reaction mixture using up to 8 μl of template (0.25–0.6 μg if PCR-generated and 1 μg if using linearized plasmid) in 20 μl final volume. Incubations are 4 h to overnight at 37 °C to maximize yield, followed by DNAse digestion. Subsequent purification of sgRNA can be performed using either the LiCl precipitation (which is usually not recommended for small RNAs, but has been successful for sgRNAs) or by the phenol–chloroform extraction/$NH_4OAc$ precipitation method. sgRNA is dissolved in RNase-free water (5–30 μl depending on pellet size). A minimal RNA concentration of 100 ng/μl is desired to create "cocktails" for microinjection (see Section 3.4.1) and quality is further assessed by gel electrophoresis (see Fig. 17.3C). Secondary structure of sgRNAs can result in multiple bands that are readily resolved by denaturation before electrophoresis. Heat-denaturation can be performed by incubating at 60 °C for 2 min in 60–80% formamide followed by quick cooling on ice for 2 min. The yield of sgRNA is variable and template dependent. Amounts as high as 100 μg of sgRNA have been achieved from a single reaction. However, if yield is unacceptably low, longer incubation times may improve the outcome. If this approach fails, adding an extra 5′ sequence (5′-GCAGC-3′) to the end of the 5′-oligo (Michinori Toriyama, personal communication and our observations, see note added in proof 1) has been shown to enhance T7 polymerase reactions (Baklanov, Golikova, & Malygin, 1996). Other alternatives are to use the plasmid-based strategy for template construction (see Section 3.3), or choose an alternative target site.

Note added in proof 1: We now recommend the use of an improved 5′ primer which ensures higher efficiency of T7 polymerase-mediated transcription under the same experimental conditions as described: 5′-GCAGCTAATACGACTCACTATA*G(N)*$_{16\text{-}19}$GTTTTA-GAGCTAGAAATA-3′.

## 3.4. Procedure for microinjection

Optimal amounts of sgRNA and Cas9 used in embryo microinjections will vary depending both on the target site and experimental goals. For example, F0 analyses will require relatively high doses to observe LOF phenotypes in a high proportion of animals. However, there are scenarios where F0 LOF is either not possible (e.g., F0 knockouts cannot be achieved for maternal RNAs) or insufficient (e.g., phenotypic variability between mosaic LOF animals may not provide consistent results), and creation of lines carrying mutant alleles might be preferable. High doses of sgRNA and Cas9 that cause lethal phenotypes or are otherwise sterile would preclude the possibility of creating lines. Therefore, moderate doses may be needed to create fertile adults for germline transmission. In short, empirical determination of optimal doses will be required.

### *3.4.1 Doses of sgRNA and Cas9*

To achieve mutations at a high efficiency, sgRNA doses ranging from 50 to 500 pg appear to be sufficient in many cases (Blitz et al., 2013; Guo et al., 2014; Nakayama et al., 2013). We recommend testing a range from 50 to 200 pg as increased toxicity can be observed at high doses (Guo et al., 2014).

Cas9 can be microinjected as either mRNA or protein. Three Cas9 plasmid templates have been successfully used in *X. tropicalis* (Blitz et al., 2013; Guo et al., 2014; Nakayama et al., 2013). Since native *S. pyogenes* Cas9 mRNA has weak activity in *Xenopus* (Nakayama et al., 2013), successful Cas9 constructs have incorporated two key modifications: a codon optimization more compatible with vertebrate systems and fusion to nuclear localization signals (NLSs). We recommend using appropriate mMessage mMachine Kits (Life Technologies) and following standard conditions to make capped Cas9 mRNAs. These should be quality-tested by $A_{260}$ quantification and gel electrophoresis.

Table 17.1 summarizes effective mRNA doses that can be used as a starting point for experimental design. Depending on the source, Cas9 mRNAs are generally effective in the dose range from ~300 pg to 3–4 ng levels and researchers will need to empirically determine optimal doses. Although increasing toxicity has been reported with moderate doses of Cas9 prepared from pCS2+ 3xFLAG-NLS-SpCas9-NLS (Guo et al., 2014), this mRNA showed no apparent toxicity at doses up to 2 ng/embryo in our experiments (our unpublished observations).

Recently, a recombinant Cas9 protein (containing an NLS) has become available (PNA Bio, Inc.). We have successfully targeted the *tyrosinase* (*tyr*) gene using this protein with no toxic effects on embryogenesis. Cas9 protein is reconstituted by dissolving 50 μg Cas9 protein in 40 μl of nuclease-free water, creating a stock at 1.25 μg/μl in 20 m*M* HEPES (pH 7.5), 150 m*M* NaCl, and 1% sucrose. This Cas9 solution is stored in small aliquots at −80 °C and diluted in nuclease-free water to create "cocktails" for embryo injections. Cas9 protein is equally efficient in creating mutations

**Table 17.1** Cas9 mRNA and sgRNA dosing for CRISPR/Cas9 mutagenesis in *X. tropicalis*

| Plasmid/protein | Cas9 dose range (per embryo) | sgRNA dose range (per embryo) | References |
|---|---|---|---|
| Cas9 (w/NLS) protein | 900–1200 pg protein | 50–200 pg | This study |
| pXT7-Cas9 | 0.55–3.2 ng mRNA (up to 6 ng[a]) | 25–200 pg (200 pg each[a]) | Nakayama et al. (2013) |
| pCasX | 3–4 ng mRNA | 150 pg | Blitz et al. (2013) |
| pCS2+ 3xFLAG-NLS-SpCas9-NLS | 200–500 pg mRNA | 1–2000 pg | Guo et al. (2014) |
| Our recommendations for RNAs | 300–2000 pg mRNA | 50–200 pg | See text for details |

[a]Cas9 mRNA and sgRNA doses for simultaneous targeting of two different sites.

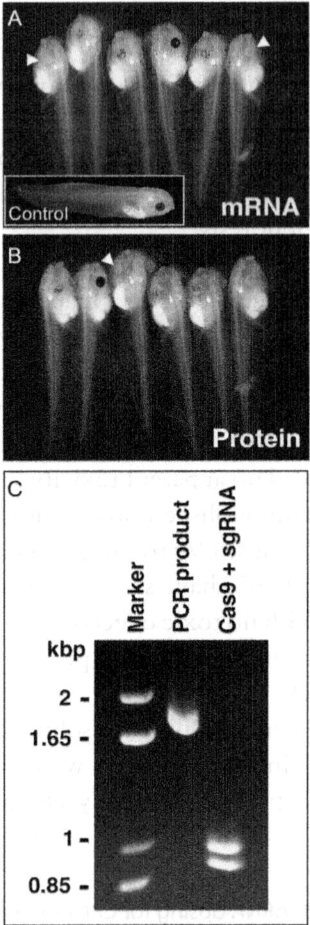


**Figure 17.4** Cas9 protein-mediated mutagenesis and *in vitro* cleavage assays. sgRNA targeting the *tyr* gene was coinjected with either Cas9 mRNA (A) or protein (B) and representative embryos are shown. Inset is an uninjected control embryo. White arrowheads mark embryos displaying complete albinism. (C) The result of an *in vitro* Cas9 cleavage assay performed using a PCR product containing an asymmetrically located *tyr* target site.

at the *tyr* locus as coinjected Cas9 mRNAs transcribed from available plasmids (Fig. 17.4A and B). We find that 900–1200 ng of Cas9 protein coinjected with 200 pg of *tyr* sgRNA efficiently produces albino tadpoles (whereas lower doses of Cas9 protein under this regimen are less effective), at a mutation frequency comparable to 2 ng of Cas9 mRNA.

An additional benefit of the availability of Cas9 protein is the opportunity to test efficiency of newly designed sgRNAs *in vitro* prior to microinjection (see Section 3.4.2).

### 3.4.2 Sidebar: Cas9 protein in vitro cleavage assays

An advantage of using Cas9 protein is that one can assay for sgRNA function using an *in vitro* cleavage assay prior to testing it *in vivo*. It seems reasonable to presume that sgRNAs that fail to efficiently direct Cas9 cleavage *in vitro* would probably also underperform *in vivo*. While success *in vitro* may not forecast success *in vivo*, we anticipate that this assay can serve to help researchers eliminate poor sgRNA candidates.

To assess a sgRNA's function *in vitro*, we have tested the *tyr* sgRNA on a DNA fragment amplified from the *X. tropicalis* genome (Fig. 17.4C). *In vitro* assays are performed in a 10 μl final volume as follows:

**Cas9 *in vitro* cleavage assay reaction**

| | |
|---|---|
| 10 × NEB buffer 3 | 1 μl |
| 10 × NEB BSA (diluted from 100 ×) | 1 μl |
| Target DNA (PCR amplicon) | ~250 ng |
| Cas9 protein | 0.5–1 ng |
| sgRNA | 250 pg |
| $H_2O$ (DNAse/RNAse-free) | To 10 μl |

The reaction is incubated at 37 °C for 1 h. We recommend using the manufacturer's protocol for RNAse digestion and inactivation of the reaction; samples are then analyzed by gel electrophoresis (Fig. 17.4C).

### 3.4.3 Procedure for embryo microinjection

Many factors are critical for successful Cas9-mediated mutagenesis in *X. tropicalis*. *In vitro* fertilizations are preferable to natural matings because they produce synchronous populations of embryos. It is recommended that injections begin shortly after fertilization to maximize the number/extent of animals displaying the strongest phenotypes. We begin de-jellying 10 min postfertilization (mpf; fertilization is defined here as the time of flooding of the eggs with medium after sperm addition; Ogino, McConnell, & Grainger, 2006). We strive to complete de-jellying within 5–10 min and after washing the one-cell embryos, injections can begin approximately

20 mpf. In some batches of embryos, single-site injections produced the strongest phenotypes during the earliest 10 min of injection (30 mpf) and incrementally declined thereafter (our unpublished data). Injections at later time points may have improved outcomes if multiple injection sites are used, but this needs further investigation. Injection regimens can be applied to later stage embryos, which may be useful for achieving a "poor man's" tissue-specific knockout—by taking advantage of the fate map to target specific blastomeres for injection.

Detailed procedures for *in vitro* fertilization, microinjection, and culture of *X. tropicalis* embryos are published in Ogino et al. (2006). We routinely microinject sgRNA/Cas9 cocktails in a volume of 1–4 nl per embryo, with the injection site located in the animal hemisphere. Some researchers prefer to coinject a lineage tracer as a component of their cocktails to permit elimination of embryos that were not properly injected. Inclusion of 5 ng fluorescent dextran/nl (Life Technologies, fluorescein (D-1845) or rhodamine dextrans (D-1818)) into sgRNA/Cas9 cocktails has no deleterious effects on mutagenesis.

## 3.5. Assessment of mutagenesis: Genotyping

A number of methods have been employed for detecting induced indels; examples include the T7 endonuclease I (Kim, Lee, Kim, Cho, & Kim, 2009) and Surveyor assays (Guschin et al., 2010). These detect mismatches occurring between wild-type and mutant DNAs created after denaturation and annealing of mixtures of PCR amplicons from target regions. Another method, the high resolution melting assay, detects differences in melting temperature between wild-type and mutant amplicons (Wittwer, Reed, Gundry, Vandersteen, & Pryor, 2003). Here, we describe the DSP (direct sequencing of PCR amplicons) assay (Nakayama et al., 2013): a rapid, initial screening assay for targeting efficiency in which target regions are PCR amplified from single mosaic F0 embryos and sequenced directly to detect the presence of indels within the population of amplicons.

### *3.5.1 Embryo lysis and PCR*

Individual embryos are transferred to 0.2-ml PCR tubes containing 100 μl of lysis buffer (50 m*M* Tris [pH 8.8], 1 m*M* EDTA, 0.5% Tween 20) containing freshly added proteinase K at a final concentration of 200 μg/ml. Embryos are incubated at 56 °C for 2 h to overnight, followed by incubation at 95 °C for 10 min to inactivate proteinase K. Lysates are centrifuged for 10 min at 4 °C and 1 μl aliquots are used directly in 25 μl

PCR reactions. Individual embryos from blastula stages to approximately stage 40 (Nieuwkoop & Faber, 1967) lysed in this manner generally yield sufficient DNA for many PCR reactions. Similarly prepared samples from later stages typically require 10- to 20-fold dilution prior to use in PCR. The PCR primers ideally should be designed to amplify 300–400 bp genomic regions containing the target sequence. Successful PCR should be confirmed by gel electrophoresis, followed by column-purification of the PCR reaction. Amplicons are then subjected to Sanger sequencing using either primer used for the PCR.

### 3.5.2 Evaluation of sequencing results and subsequent identification of specific indels

If indels have occurred, the PCR product will contain a mixture of heterogeneous fragments with different mutations. The profile will show sequence heterogeneity in peaks in the mutated region (Fig. 17.5A). Subcloning PCR products from the population and sequencing of individual clones confirms the results of the DSP assay (Fig. 17.5B) that successful mutagenesis has

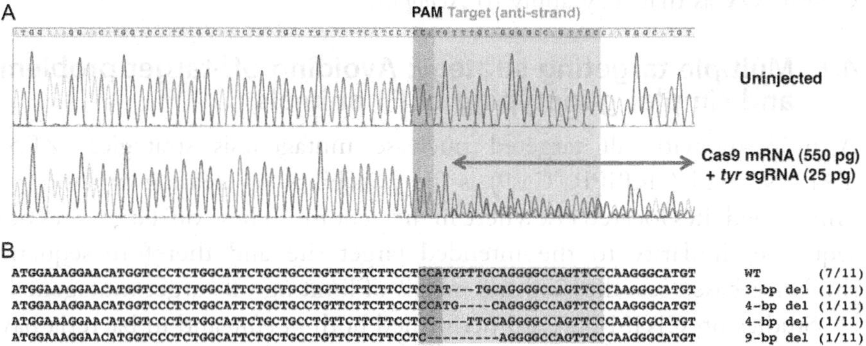


B

```
ATGGAAAGGAACATGGTCCCTCTGGCATTCTGCTGCCTGTTCTTCTTCCTCCATGTTTGCAGGGGCCAGTTCCCAAGGGCATGT    WT         (7/11)
ATGGAAAGGAACATGGTCCCTCTGGCATTCTGCTGCCTGTTCTTCTTCCTCCAT---TGCAGGGGCCAGTTCCCAAGGGCATGT    3-bp del   (1/11)
ATGGAAAGGAACATGGTCCCTCTGGCATTCTGCTGCCTGTTCTTCTTCCTCCATG----CAGGGGCCAGTTCCCAAGGGCATGT    4-bp del   (1/11)
ATGGAAAGGAACATGGTCCCTCTGGCATTCTGCTGCCTGTTCTTCTTCCTCC----TTGCAGGGGCCAGTTCCCAAGGGCATGT    4-bp del   (1/11)
ATGGAAAGGAACATGGTCCCTCTGGCATTCTGCTGCCTGTTCTTCTTCCTC---------AGGGGCCAGTTCCCAAGGGCATGT    9-bp del   (1/11)
```

**Figure 17.5** DSP assay. (A) Representative results of DSP assays. Single embryos (top, uninjected; bottom, injected with indicated RNAs) were lysed and the targeted genomic region was PCR amplified; amplicons were then directly sequenced. Perturbation of peaks (double-headed arrow) on the 3′ side of the PAM region (note that in this specific case, the target is on the antisense strand) seen in the sample of the injected embryo suggests indel events occurred between the target sequence (lightly shaded area where multiple peaks overlap) and the PAM region (darkly shaded). (B) The PCR amplicon from the injected embryo was recloned and sequenced to show the profile of individual mutations found in mosaic individuals (bottom sequence alignments). Dashes (-) indicate gaps. The numbers in parentheses indicate the frequency of each mutation pattern seen in the total number of sequenced clones, suggesting that a mutation frequency of 36% (i.e., 4/11) can be detected as a positive targeting event by the DSP assay. *The figure is adapted and modified from Nakayama et al. (2013).* (See the color plate.)

occurred. The advantages of the DSP assay include its ease of use and the ability to visualize that the genomic region where sequence perturbations are observed corresponds to the intended target site. We have detected mutations by DSP assay in embryos with rates of mutagenesis of ~25% or higher, but generally cannot detect rates at lower levels (our unpublished observations), demonstrating the moderate sensitivity of the assay. Notably, a strength of this assay may be that it is not overly sensitive: since F0 mosaic animals with low rates of germline transmission necessitate laborious screening of F1s, it may be a useful standard to only raise F0 animals for subsequent mating that have mutagenesis levels detectable by DSP assay. This standard would improve one's chances of achieving adult F0 animals with germline transmission rates of at least one in four, which is not too burdensome for genotypic screening of F1s.

## 4. DISCUSSION

We have described basic and essential tips for CRISPR-mediated mutagenesis in *X. tropicalis*. Here, we briefly discuss additional aspects of CRISPR/Cas that may apply to *Xenopus*.

### 4.1. Multiple targeting strategy: Avoiding off-target problems and simpler genotyping of F1 animals

A problem with all targeted nuclease mutagenesis strategies (ZFNs, TALENs, and CRISPR/Cas9) is the possibility of creating mutations at unintended sites located elsewhere in the genome. These off-target sites bear sequence similarity to the intended target site and therefore sequence similarity-based searches are one means of limiting off-target mutagenesis.

Studies on CRISPR/Cas9 performed both *in vitro* and in bacteria have shown that a perfect match between sgRNA and the DNA target in an 8- to 12-base region of the target sequence proximal to the PAM, known as the "seed sequence," is crucial for Cas9-mediated cleavage (see Fineran & Dy, 2014; Hsu et al., 2014; Sander & Joung, 2014 and references therein). Mismatches are tolerated in the target nucleotides that are distal to the seed sequence, which could lead to off-target cleavage. As discussed in Section 3.2, a number of Web-based search tools have been developed that predict both target and off-target sites for sgRNAs in *X. tropicalis*. When choosing between potential sites to target, one should avoid sequences with off-target mismatches that are largely located distal to the seed sequence. Also, one should keep in mind that sequences followed by the NAG

PAM can also be off-target sites, therefore off-target analysis should include both NGG and NAG PAMs.

One possible avenue for reducing off-target mutagenesis is the use of paired nickases (Mali, Aach, et al., 2013; Ran et al., 2013). The Cas9 D10A mutant is incapable of creating double-strand breaks, but instead nicks one strand at the target site. Using a pair of sgRNAs with target sites of appropriate orientation and spacing, in conjunction with Cas9 D10A, permits mutagenesis at the paired site without creating double-strand breaks at sgRNA single (off-target) sites across the genome. Another approach would be to use FokI fusions to dCas9, a doubly mutated Cas9 with no nuclease activity (Guilinger, Thompson, & Liu, 2014; Tsai et al., 2014). This approach similarly involves choosing two nearby target sites to direct Cas9–FokI to the genome, which only functions as a dimer to cleave DNA. Both strategies are limited by the ability to find two nearby target sequences with appropriate directionality and spacing. Neither approach has been reported in *Xenopus* (see note added in proof 2).

Off-target mutagenesis may not be a major concern for LOF assays in *Xenopus*. Studies by Blitz et al. (2013) and Guo et al. (2014) found no evidence for off-target mutagenesis in *X. tropicalis*, similar to studies in mouse embryos (Wang et al., 2013; Yang et al., 2013), but unlike reports using cultured cells (Cradick, Fine, Antico, & Bao, 2013; Fu et al., 2013; Hsu et al., 2013; Pattanayak et al., 2013). This suggests that in whole organisms off-target cleavage may be negligible.

A strategy we recommend when performing F0 analyses to ensure that a phenotype is due to mutagenesis of the targeted gene is to replicate the phenotype using multiple independent sgRNAs targeting the same gene (Nakayama et al., 2013). It is unlikely that multiple sgRNAs would result in mutation of the same off-target gene. An alternative would be to rescue the phenotype by expressing a wild-type mRNA (Guo et al., 2014; Nakayama et al., 2013). These approaches provide strong supporting evidence for an on-target effect and refute the notion that a phenotype is due to off-target mutagenesis.

If the study of F0 mosaic animals is not desirable, then various breeding strategies for creating non-mosaic animals may be employed. Ideally, one would construct a mutant that contains the same lesion at both alleles; however, since multiple generations are needed to achieve this, a time consuming process, analysis of phenotypes may be significantly delayed. Matings between F0 mosaic animals will produce non-mosaic F1s that are (primarily) compound heterozygotes (Fig. 17.2). Because F0 mutagenesis results in a variety of indels in the same target site, F1 offspring will likely be difficult to genotype. These problems can be overcome using two nonoverlapping sgRNA target sites within a gene. A mating between F0 animals mutated at different target sites will produce F1 offspring that can more readily be genotyped. Thus, compound heterozygotes mutated at nonoverlapping target sites permit more rapid phenotypic analyses.

Note added in proof 2: We have confirmed that the nickase version of Cas9 (D10A) can mediate efficient mutagenesis using paired sgRNAs in *Xenopus*.

## 4.2. Further applications of CRISPR-mediated mutagenesis in *Xenopus*

As with morpholino antisense oligo-mediated knockdowns, it may be possible to study the effects of multigene knockouts using CRISPR/Cas. Coinjection of multiple sgRNAs, each targeting a different site or gene, will likely permit such analyses. This is especially useful in situations where redundancy, possibly caused by gene duplications, makes phenotypic assays of gene function problematic.

The use of multiple sgRNAs also permits examining the role of *cis*-regulatory elements. Target sites flanking an element can be used to delete the intervening sequence as has been shown by promoter deletion by Nakayama et al. (2013). This approach in principle can be used to create larger gene deletions though the frequency of successful mutagenesis by this strategy needs to be determined.

Another application is homology-directed recombination (HDR)-dependent gene targeting, i.e., introduction of point mutations, or knocking in larger genetic elements such as sequence tags or transgenes into any site in the genome (see Sander & Joung, 2014 and references therein). This is a future challenge for *Xenopus* researchers.

The CRISPR-mediated mutagenesis strategy outlined herein may provide a means for LOF analysis in the F0 animal. However, the variable mosaicism of F0 embryos may ultimately interfere with some analyses and therefore improvements to the efficiency of CRISPR/Cas in *Xenopus* would be valuable. Major advantages of the *Xenopus* system include the ease of manipulating oocytes and sperm and these techniques may offer new avenues to reduce mosaicism. The host-transfer method (Olson, Hulstrand, & Houston, 2012) allows for *in vitro* culturing and manipulation of oocytes, which are then implanted back into the body cavity of a host female for proper ovulation that is necessary for successful fertilization. Oocytes may be injected with Cas9 protein and sgRNA(s) leading to a mutant gamete with only one gene copy. Fertilization of manipulated eggs with wild-type sperm would result in nonmosaic heterozygotes. Sperm can be similarly manipulated. It may be feasible to incubate decondensed sperm nuclei (Hirsch et al., 2002; Kroll & Amaya, 1996) with Cas9 protein–sgRNA cocktails to create mutations, and to inject these into unfertilized eggs to create nonmosaic heterozygotes. In both cases, further technology development is desirable to achieve HDR. The host-transfer method may be especially useful in this regard because oocytes have naturally high homologous recombination activity (Carroll, Wright, Wolff, Grzesiuk, & Maryon, 1986).

## ACKNOWLEDGMENTS

This work was supported in part by NIH grants EY022954, EY018000, OD010997, and by the Sharon Stewart Aniridia Trust to R. M. G. Funding was also provided by NSF and NIH grants 1147270 and HD073179, respectively, to K. W. Y. C., and NIH grant HD080684 to I. L. B. Fellowship funding that contributed to this work was awarded by the UVA Society of Fellows to M. B. F. Some preliminary experiments were performed during the 2013 and 2014 Cold Spring Harbor Laboratory *Xenopus* Courses, which the authors would like to acknowledge. The authors also thank Amy Sater, Jamina Oomen-Hajagos, Jerry Thomsen, Michinori Toriyama, Doug Houston, and Rob Steele for sharing unpublished information, and past and current members in Grainger Lab, especially Marilyn Fisher and Cristina D'Ancona, for help in experiments and discussion. We also thank Yonglong Chen for pCS2+ 3xFLAG-NLS-SpCas9-NLS and Jianzhong Jeff Xi for pXT7-Cas9. Finally, the authors would also like to gratefully acknowledge the support of Xenbase and the National *Xenopus* Resource and their staffs.

## REFERENCES

Abu-Daya, A., Khokha, M. K., & Zimmerman, L. B. (2012). The Hitchhiker's guide to *Xenopus* genetics. *Genesis, 50*, 164–175.

Anders, C., Niewoehner, O., Duerst, A., & Jinek, M. (2014). Structural basis of PAM-dependent target DNA recognition by the Cas9 endonuclease. *Nature, 513*, 569–573.

Ansai, S., & Kinoshita, M. (2014). Targeted mutagenesis using CRISPR/Cas system in medaka. *Biology Open, 3*, 362–371.

Baklanov, M. M., Golikova, L. N., & Malygin, E. G. (1996). Effect on DNA transcription of nucleotide sequences upstream to T7 promoter. *Nucleic Acids Research, 24*, 3659–3660.

Blitz, I. L., Biesinger, J., Xie, X., & Cho, K. W. Y. (2013). Biallelic genome modification in $F_0$ *Xenopus tropicalis* embryos using the CRISPR/Cas system. *Genesis, 51*, 827–834.

Carroll, D., Wright, S. H., Wolff, R. K., Grzesiuk, E., & Maryon, E. B. (1986). Efficient homologous recombination of linear DNA substrates after injection into *Xenopus laevis* oocytes. *Molecular and Cellular Biology, 6*, 2053–2061.

Chen, B., Gilbert, L. A., Cimini, B. A., Schnitzbauer, J., Zhang, W., Li, G.-W., et al. (2013). Dynamic imaging of genomic loci in living human cells by an optimized CRISPR/Cas system. *Cell, 155*, 1479–1491.

Cradick, T. J., Fine, E. J., Antico, C. J., & Bao, G. (2013). CRISPR/Cas9 systems targeting β-globin and *CCR5* genes have substantial off-target activity. *Nucleic Acids Research, 41*, 9584–9592.

Fineran, P. C., & Dy, R. L. (2014). Gene regulation by engineered CRISPR-Cas systems. *Current Opinion in Microbiology*, *18*, 83–89.

Fish, M. B., Nakayama, T., Fisher, M., Hirsch, N., Cox, A., Reeder, R., et al. (2014). *Xenopus* mutant reveals necessity of *rax* for specifying the eye field which otherwise forms tissue with telencephalic and diencephalic character. *Developmental Biology*, *395*, 317–330.

Fu, Y., Foden, J. A., Khayter, C., Maeder, M. L., Reyon, D., Joung, J. K., et al. (2013). High-frequency off-target mutagenesis induced by CRISPR-Cas nucleases in human cells. *Nature Biotechnology*, *31*, 822–826.

Fu, Y., Sander, J. D., Reyon, D., Cascio, V. M., & Joung, J. K. (2014). Improving CRISPR-Cas nuclease specificity using truncated guide RNAs. *Nature Biotechnology*, *32*, 279–284.

Guilinger, J. P., Thompson, D. B., & Liu, D. R. (2014). Fusion of catalytically inactive Cas9 to Fok1 nuclease improves the specificity of genome modification. *Nature Biotechnology*, *32*, 577–582.

Guo, X., Zhang, T., Hu, Z., Zhang, Y., Shi, Z., Wang, Q., et al. (2014). Efficient RNA/Cas9-mediated genome editing in *Xenopus tropicalis*. *Development*, *141*, 707–714.

Guschin, D. Y., Waite, A. J., Katibah, G. E., Miller, J. C., Holmes, M. C., & Rebar, E. J. (2010). A rapid and general assay for monitoring endogenous gene modification. *Methods in Molecular Biology*, *649*, 247–256.

Harland, R. M., & Grainger, R. M. (2011). *Xenopus* research: Metamorphosed by genetics and genomics. *Trends in Genetics*, *27*, 507–515.

Hirsch, N., Zimmerman, L. B., Gray, J., Chae, J., Curran, K. L., Fisher, M., et al. (2002). *Xenopus tropicalis* transgenic lines and their use in the study of embryonic induction. *Developmental Dynamics*, *225*, 522–535.

Hsu, P. D., Lander, E. S., & Zhang, F. (2014). Development and applications of CRISPR-Cas9 for genome engineering. *Cell*, *157*, 1262–1278.

Hsu, P. D., Scott, D. A., Weinstein, J. A., Ran, F. A., Konermann, S., Agarwala, V., et al. (2013). DNA targeting specificity of RNA-guided Cas9 nucleases. *Nature Biotechnology*, *31*, 827–832.

Hwang, W. Y., Fu, Y., Reyon, D., Maeder, M. L., Tsai, S. Q., Sander, J. D., et al. (2013). Efficient genome editing in zebrafish using a CRISPR-Cas system. *Nature Biotechnology*, *31*, 227–229.

Ishibashi, S., Cliffe, R., & Amaya, E. (2012). Highly efficient bi-allelic mutation rates using TALENs in *Xenopus tropicalis*. *Biology Open*, *1*, 1273–1276.

Kim, H. J., Lee, H. J., Kim, H., Cho, S. W., & Kim, J.-S. (2009). Targeted genome editing in human cells with zinc finger nucleases constructed via modular assembly. *Genome Research*, *19*, 1279–1288.

Kroll, K. L., & Amaya, E. (1996). Transgenic *Xenopus* embryos from sperm nuclear transplantations reveal FGF signaling requirements during gastrulation. *Development*, *122*, 3173–3183.

Lei, Y., Guo, X., Liu, Y., Cao, Y., Deng, Y., Chen, X., et al. (2012). Efficient targeted gene disruption in *Xenopus* embryos using engineered transcription activator-like effector nucleases (TALENs). *Proceedings of the National Academy of Sciences of the United States of America*, *109*, 17484–17489.

Mali, P., Aach, J., Stranges, P. B., Esvelt, K. M., Moosburner, M., Kosuri, S., et al. (2013). CAS9 transcriptional activators for target specificity screening and paired nickases for cooperative genome engineering. *Nature Biotechnology*, *31*, 833–838.

Mali, P., Yang, L., Esvelt, K. M., Aach, J., Guell, M., DiCarlo, J. E., et al. (2013). RNA-guided human genome engineering via Cas9. *Science*, *339*, 823–826.

Nakajima, K., Nakai, Y., Okada, M., & Yaoita, Y. (2013). Targeted gene disruption in the *Xenopus tropicalis* genome using designed TALE nucleases. *Zoological Science*, *30*, 455–460.

Nakajima, K., Nakajima, T., Takase, M., & Yaoita, Y. (2012). Generation of albino *Xenopus tropicalis* using zinc-finger nucleases. *Development, Growth & Differentiation, 54,* 777–784.

Nakajima, K., & Yaoita, Y. (2013). Comparison of TALEN scaffolds in *Xenopus tropicalis. Biology Open, 2,* 1364–1370.

Nakayama, T., Fish, M. B., Fisher, M., Oomen-Hajagos, J., Thomsen, G. H., & Grainger, R. M. (2013). Simple and efficient CRISPR/Cas9-mediated targeted mutagenesis in *Xenopus tropicalis. Genesis, 51,* 835–843.

Nieuwkoop, P. D., & Faber, J. (1967). *Normal table of* Xenopus laevis *(Daudin).* Amsterdam: North-Holland.

Ogino, H., McConnell, W. B., & Grainger, R. M. (2006). High-throughput transgenesis in *Xenopus* using *I-SceI* meganuclease. *Nature Protocols, 1,* 1703–1710.

Olson, D. J., Hulstrand, A. M., & Houston, D. W. (2012). Maternal mRNA knock-down studies: Antisense experiments using the host-transfer technique in *Xenopus laevis* and *Xenopus tropicalis. Methods in Molecular Biology, 917,* 167–182.

Pattanayak, V., Lin, S., Guilinger, J. P., Ma, E., Doudna, J. A., & Liu, D. R. (2013). High-throughput profiling of off-target DNA cleavage reveals RNA-programmed Cas9 nuclease specificity. *Nature Biotechnology, 31,* 839–843.

Ran, F. A., Hsu, P. D., Lin, C.-Y., Gootenberg, J. S., Konermann, S., Trevino, A. E., et al. (2013). Double nicking by RNA-guided CRISPR Cas9 for enhanced genome editing specificity. *Cell, 154,* 1380–1389.

Sander, J. D., & Joung, J. K. (2014). CRISPR-Cas systems for editing, regulating and targeting genomes. *Nature Biotechnology, 32,* 347–355.

Sive, H. L., Grainger, R. M., & Harland, R. M. (2010). *Early development of* Xenopus laevis*: A laboratory manual.* Cold Spring Harbor Press.

Suzuki, K. T., Isoyama, Y., Kashiwagi, K., Sakuma, T., Ochiai, H., Sakamoto, N., et al. (2013). High efficiency TALENs enable F0 functional analysis by targeted gene disruption in *Xenopus laevis* embryos. *Biology Open, 2,* 448–452.

Terns, R. M., & Terns, M. P. (2014). CRISPR-based technologies: Prokaryotic defense weapons repurposed. *Trends in Genetics, 30,* 111–118.

Tsai, S. Q., Wyvekens, N., Khayter, C., Foden, J. A., Thapar, V., Reyon, D., et al. (2014). Dimeric CRISPR RNA-guided FokI nucleases for highly specific genome editing. *Nature Biotechnology, 32,* 569–576.

Wang, H., Yang, H., Shivalila, C. S., Dawlaty, M. M., Cheng, A. W., Zhang, F., et al. (2013). One-step generation of mice carrying mutations in multiple genes by CRISPR/Cas-mediated genome engineering. *Cell, 153,* 910–918.

Wittwer, C. T., Reed, G. H., Gundry, C. N., Vandersteen, J. G., & Pryor, R. J. (2003). High-resolution genotyping by amplicon melting analysis using LCGreen. *Clinical Chemistry, 49,* 853–860.

Yang, H., Wang, H., Shivalila, C. S., Cheng, A. W., Shi, L., & Jaenisch, R. (2013). One-step generation of mice carrying reporter and conditional alleles by CRISPR/Cas-mediated genome engineering. *Cell, 154,* 1370–1379.

Young, J. J., Cherone, J. M., Doyon, Y., Ankoudinova, I., Faraji, F. M., Lee, A. H., et al. (2011). Efficient targeted gene disruption in the soma and germ line of the frog *Xenopus tropicalis* using engineered zinc-finger nucleases. *Proceedings of the National Academy of Sciences of the United States of America, 108,* 7052–7057.

CHAPTER EIGHTEEN

# Cas9-Based Genome Editing in Zebrafish

**Andrew P.W. Gonzales**[*,†], **Jing-Ruey Joanna Yeh**[*,†,1]
[*]Cardiovascular Research Center, Massachusetts General Hospital, Charlestown, Massachusetts, USA
[†]Department of Medicine, Harvard Medical School, Boston, Massachusetts, USA
[1]Corresponding author: e-mail address: jyeh1@mgh.harvard.edu

## Contents



## Abstract

Genome editing using the Cas9 endonuclease of *Streptococcus pyogenes* has demonstrated unprecedented efficacy and facility in a wide variety of biological systems. In zebrafish, specifically, studies have shown that Cas9 can be directed to user-defined genomic target sites via synthetic guide RNAs, enabling random or homology-directed sequence alterations, long-range chromosomal deletions, simultaneous disruption of multiple genes, and targeted integration of several kilobases of DNA. Altogether, these methods are opening new doors for the engineering of knock-outs, conditional alleles, tagged proteins, reporter lines, and disease models. In addition, the ease and high efficiency of generating Cas9-mediated gene knock-outs provides great promise for

*Methods in Enzymology*, Volume 546
ISSN 0076-6879
http://dx.doi.org/10.1016/B978-0-12-801185-0.00018-0

high-throughput functional genomics studies in zebrafish. In this chapter, we briefly review the origin of CRISPR/Cas technology and discuss current Cas9-based genome-editing applications in zebrafish with particular emphasis on their designs and implementations.

# 1. INTRODUCTION

## 1.1. CRISPR/Cas adaptive immunity

In order to persist and thrive within threatening virus-rich environments, prokaryotes over the course of evolutionary history have developed various kinds of defense mechanisms for fending off invading viral genetic elements (Labrie, Samson, & Moineau, 2010), one of which being an immune mechanism mediated by clustered regularly interspaced short palindromic repeat (CRISPR) loci (Barrangou et al., 2007; Ishino, Shinagawa, Makino, Amemura, & Nakata, 1987; Jansen, Embden, Gaastra, & Schouls, 2002). CRISPR loci are common among prokaryotes and have been estimated to be in ~40% and ~90% of all genomically sequenced bacteria and archaea, respectively (Grissa, Vergnaud, & Pourcel, 2007a; Kunin, Sorek, & Hugenholtz, 2007; Sorek, Kunin, & Hugenholtz, 2008). These loci, together with their neighboring CRISPR-associated (Cas) genetic elements, form an unique adaptive immune system called CRISPR/Cas, which utilizes short RNA-guided endonucleases to target, cleave, and degrade specific viral sequences during a recurring infection (Bhaya, Davison, & Barrangou, 2011; Bolotin, Quinquis, Sorokin, & Ehrlich, 2005; Horvath & Barrangou, 2010; Marraffini & Sontheimer, 2010a).

CRISPR loci are characterized by arrays of conserved ~20–50 base-pair (bp) repeats with distinct "spacer" sequences of comparable length interspaced between them (Grissa, Vergnaud, & Pourcel, 2007b; Rousseau, Gonnet, Le Romancer, & Nicolas, 2009). These loci are flanked by a cluster of *cas* genes which encode some of the enzymatic machinery utilized for normal CRISPR/Cas function (Makarova, Grishin, Shabalina, Wolf, & Koonin, 2006). Within a given CRISPR locus, each spacer sequence is unique and derived from fragments of invading viral nucleic acids acquired from a previous pathogenic exposure, thus allowing the prokaryote to generate a genetically stored immunological memory of past infections (Bolotin et al., 2005; Mojica, Diez-Villasenor, Garcia-Martinez, & Soria, 2005). All CRISPR/Cas systems generate this immunological memory by following a common three-step process (Wiedenheft, Sternberg, & Doudna, 2012).

First, during preliminary exposure to a pathogen, invading foreign nucleic acids must be cleaved into fragments called protospacers, which then become integrated as spacers in the CRISPR locus (Barrangou et al., 2007; Garneau et al., 2010). Second, during any subsequent infection, the CRISPR locus is transcribed to produce a single long pre-CRISPR RNA, which is then processed into an active genetic library of many spacer-derived short CRISPR RNAs (crRNAs) (Brouns et al., 2008). Third, crRNAs combined with one or more Cas proteins form RNA-guided endoribonuclease surveillance complexes, which by base-pair interactions between the crRNA spacer region and complementary viral protospacer sequences, allow the complexes to target and cleave invading foreign nucleic acids (Brouns et al., 2008).

## 1.2. The Type II CRISPR/Cas system

Three types of CRISPR/Cas systems are known to exist in nature, each of which uses distinct mechanisms to carry out the aforementioned three-step process to produce CRISPR-mediated adaptive immunity (Makarova, Aravind, Wolf, & Koonin, 2011; Makarova, Haft, et al., 2011; Wiedenheft et al., 2012). Among these three systems, the Type II CRISPR/Cas system is the best characterized and the simplest in certain key aspects. One important difference is that its surveillance complex requires a single Cas9 endonuclease (Chylinski, Makarova, Charpentier, & Koonin, 2014; Sapranauskas et al., 2011), while Type I and Type III systems require several proteins (Makarova, Aravind, Wolf, & Koonin, 2011; Makarova, Haft, et al., 2011).

In addition to the Cas9 protein, the Type II surveillance complex also consists of two RNA components, a crRNA and a transactivating crRNA (tracrRNA) (Deltcheva et al., 2011). The tracrRNA is required for normal crRNA processing and Type II surveillance complex formation, and the crRNAs contain 20-nucleotide (nt) spacer regions derived from the original CRISPR locus (Deltcheva et al., 2011). By complementary base-pair interactions, these crRNAs guide the surveillance complexes to target, bind, and degrade foreign genetic elements that contain protospacer sequences complementary to the spacer, as well as a Cas9-specific protospacer adjacent motif (PAM) directly 3′ to the target protospacer (Gasiunas, Barrangou, Horvath, & Siksnys, 2012).

Having the correct PAM sequence directly adjacent to the protospacer is necessary for DNA interrogation by the Type II surveillance complex and for the triggering of Cas9 cleavage activity. Indeed, mismatches within or nearby the first few nucleotides of the PAM have been shown to inhibit heteroduplex formation and unwinding of the dsDNA target (Sternberg, Redding, Jinek, Greene, & Doudna, 2014). In this manner, the 3′ PAM sequences allow the Type II system to distinguish between sequences belonging to "self" and those that are foreign in order to prevent the destruction of its own CRISPR loci (Horvath et al., 2008; Marraffini & Sontheimer, 2010b). Among Type II Cas9 endonucleases found in various prokaryotic species, PAM sequences vary in complexity, one of the simplest being the 5′-NGG PAM of *Streptococcus pyogenes* Cas9 (Jinek et al., 2012). The natural RNA-guided Cas9 endonuclease from *S. pyogenes* (SpCas9) possesses the ability, in principle, to target any invading protospacer sequence in the form of 5′-$N_{20}$-NGG-3′. Thus, it is both the relatively small number of components required by the Type II system, combined with the flexibility of its required target sequence, which have allowed the recent adaptation of the Type II CRISPR/Cas system as a novel, powerful, and amendable genome-editing platform.

## 1.3. The development of CRISPR/Cas genome-editing technology

Contemporary genome editing relies on the usage of programmable nucleases to artificially produce gene disruptions, DNA insertions, targeted mutations, or chromosomal rearrangements in a predictable and controlled manner (Segal & Meckler, 2013). These engineered nucleases "edit" the genome by introducing targeted double-strand DNA breaks (DSBs), which in turn allow the cell's natural repair mechanisms—e.g., nonhomologous end-joining (NHEJ) mediated repair and homology-directed repair (HDR)—to be co-opted for the purpose of site-specific DNA manipulation (Bibikova, Beumer, Trautman, & Carroll, 2003; Bibikova et al., 2001; Bibikova, Golic, Golic, & Carroll, 2002). The potential applications of such genome-editing technologies are far-reaching, including the bioengineering of disease-resistant, nutrient-rich crops and livestock (Carlson et al., 2012; Li, Liu, Spalding, Weeks, & Yang, 2012), the generation of various animal models and human pluripotent stem cell models that can be used for preclinical drug studies (Brunet et al., 2009; Carbery et al., 2010; Ding et al., 2013; Yang et al., 2013), and even the development of therapies involving the direct delivery of genetically corrected, patient-derived pluripotent stem cells or somatic cells (Schwank et al., 2013; Sebastiano et al., 2011). In light of the various potential benefits programmable nucleases present, the engineering of more facile, precise, and efficient genome-editing platforms is highly sought after and have made the recent development of CRISPR/Cas genome-editing systems all the more valuable.

The first published instance of an engineered CRISPR/Cas system for the purpose of genome editing was in 2012, when researchers adapted the Type II CRISPR/Cas system of *S. pyogenes* and demonstrated that SpCas9 could be guided by a programmable chimeric dual-RNA to target and cleave various DNA sites *in vitro* (Jinek et al., 2012). In this study, the authors simplified the system even further to create a genome-editing platform that required only two components—SpCas9 and a synthetic single guide RNA (sgRNA) consisting of a fusion of the essential features of Type II crRNA and tracrRNA (Fig. 18.1). A few months after this CRISPR/Cas platform's initial debut, its utility quickly expanded to a variety of cellular systems, exhibiting its efficacy in introducing targeted mutations within various species of bacteria (Jiang, Bikard, Cox, Zhang, & Marraffini, 2013), as well as, cultured human cancer cell lines and human pluripotent stem cells (Cho, Kim, Kim, & Kim, 2013; Cong et al., 2013; Jinek et al., 2013; Mali, Yang, et al., 2013). Around the same time, our group reported efficient genome editing in zebrafish using CRISPR/Cas, demonstrating its potential in a whole multicellular organism (Hwang, Fu, Reyon, Maeder, Tsai, et al., 2013). Since then, the platform has proven its effectiveness and versatility in editing the genes of various flora and fauna, only a sample of these being

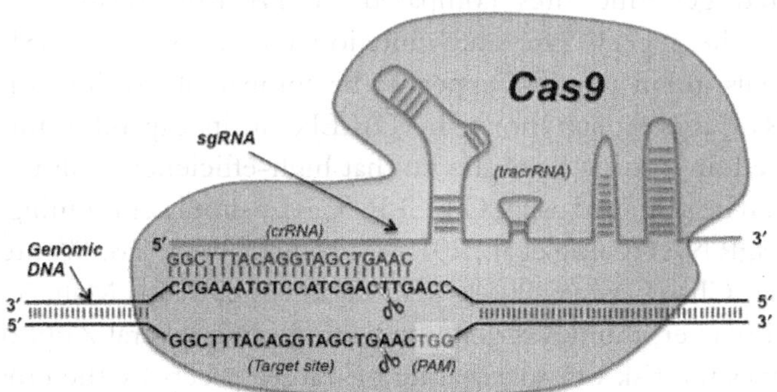


**Figure 18.1** A graphic representation of DNA targeting by sgRNA-guided Cas9. An sgRNA, consisting of a 20-nt crRNA spacer and a tracrRNA tail, guides the *S. pyogenes*-derived Cas9 endonuclease to bind to and unwind a specific 20-nt genomic target site. The target site should be in the form of 5′-N20-NGG, where NGG is the PAM sequence (highlighted in yellow (light gray in the print version)). The top and bottom strands of the genomic DNA are then cleaved by the RuvC-like nuclease domain and the HNH nuclease domain of Cas9 (indicated by the "scissors") to produce a DNA double-strand break (DSB) approximately three base pairs proximal to the PAM.

yeast, rice, wheat, *C. elegans*, silk worms, fruit flies, frogs, mice, and non-human primates (DiCarlo et al., 2013; Friedland et al., 2013; Nakayama et al., 2013; Niu et al., 2014; Shan et al., 2013; Wang, Yang, et al., 2013; Wang, Li, et al., 2013; Yu et al., 2013). It is very rare in biology for a single biotechnology to have the degree of versatility as CRISPR/Cas to work with such effectiveness in the wide scope of organisms that it does, giving this new technology the potential to fulfill many of the research, engineering, and therapeutic goals of the genetic engineering field.

Beyond the extraordinary applicability of CRISPR/Cas, this novel genome-editing platform has also exhibited several key advantages for laboratory use compared to other programmable nuclease systems, such as zinc-finger nucleases (ZFNs) and transcription activator-like effector nucleases (TALENs). The first advantage is the greater ease by which CRISPR/Cas can be designed and implemented. Unlike ZFNs and TALENs which require the complex design of zinc-finger and TALE DNA-binding arrays for every new genomic target site, CRISPR/Cas simply requires changing the 20-nt sgRNA spacer sequence so that it matches the target site (Sander & Joung, 2014). The second advantage of CRISPR/Cas is its comparable or greater genome-editing efficiencies than that of ZFNs or TALENs. In general, CRISPR/Cas functions with greater consistency, efficacy, and less toxicity than lab-produced ZFNs (Cornu et al., 2008; Maeder et al., 2008; Ramirez et al., 2008), and they are likely to be more effective at targeting methylated genomic sites compared to TALENs (Hsu et al., 2013). Although the success rate and mutation efficiency of CRISPR/Cas in human cells and in zebrafish appear to be comparable to those of TALENs, CRISPR/Cas is far superior than TALENs in its capability for multiplex genome editing. It has been shown that high-efficiency multiplex genome editing can be achieved using CRISPR/Cas by simply combining Cas9 with multiple sgRNAs (Cong et al., 2013; Guo et al., 2014; Jao, Wente, & Chen, 2013; Ma, Chang, et al., 2014; Ma, Shen, et al., 2014; Mali, Yang, et al., 2013). However, multiplex genome editing using several ZFN or TALEN pairs carries the risk of exacerbating off-target effects by the cross reaction between nuclease pairs (Sollu et al., 2010). In light of these various advantages of the CRISPR/Cas system over previous programmable nuclease platforms, CRISPR/Cas, also known as RNA-guided nucleases, have rapidly risen to become the flagship of contemporary genome-editing technologies.

## 1.4. The zebrafish animal model and CRISPR/Cas

The zebrafish is a powerful and tractable animal model for functional genomics analysis, the study of human disease pathogenesis, as well as, for the discovery and development of new drugs (Campbell, Hartjes, Nelson, Xu, & Ekker, 2013; Helenius & Yeh, 2012; Lieschke & Currie, 2007). The key strength of the zebrafish model lies in its intermediate evolutionary relationship to humans, between mammalian model systems, such as mice, on one hand, and invertebrate model systems, such as *Drosophila* and *C. elegans*, on the other. The zebrafish has an upper hand over invertebrate models due to its common vertebrate ancestry with humans. This closer ancestry gives the zebrafish greater genetic and anatomical similarity to humans than invertebrates, meaning that orthologous genes carry similar functions as in humans, and most of the organ systems and structures between zebrafish and humans are homologous (Kettleborough et al., 2013; Lieschke & Currie, 2007; Santoriello & Zon, 2012). Due to these genetic and anatomical similarities, various zebrafish models have been developed to study the pathogenesis of human diseases, ranging from genetic disorders such as Duchenne muscular dystrophy and forms of cardiomyopathy (Bassett et al., 2003; Kawahara et al., 2011; Xu et al., 2002), to acquired diseases, such as melanoma and tuberculosis (Cambier et al., 2014; Ceol et al., 2011; Patton et al., 2005; Swaim et al., 2006; White et al., 2011).

Conversely, though the mouse model exhibits greater molecular and anatomical similarity with humans due to their shared mammalian ancestry, the zebrafish carries many key advantages over mouse models due to its nonmammalian features. Because zebrafish reproduce by external fertilization, all stages of zebrafish embryogenesis are accessible to the researcher for study, unlike mammals wherein embryogenesis occurs within the body. This benefit combined with the natural optical transparency of the zebrafish allows for real-time observation of studied processes by fluorescent reporters with great ease (Ignatius & Langenau, 2011; Moro et al., 2013; Pantazis & Supatto, 2014; Weber & Koster, 2013). These observational qualities, in conjunction with the relative size, rapid development, and fecundity of zebrafish compared to mice, enables low animal maintenance and husbandry infrastructure expenses that allow the affordability of high-throughput, whole-animal zebrafish drug screens and reverse-genetic experimentation at a scale simply unfeasible with mouse models (Kari, Rodeck, & Dicker, 2007; Peal, Peterson, & Milan, 2010). Therefore, the zebrafish model serves as an especially prime candidate for the application of state-of-the-art genome-editing technologies such as CRISPR/Cas.

As mentioned above, Hwang et al. was first to demonstrate that the CRISPR/Cas genome-editing platform could be adapted *in vivo* in zebrafish by using it to introduce site-specific insertion/deletion (indel) mutations with mutation frequencies between 24% and 59% at 8 out of 10 tested genes (Hwang, Fu, Reyon, Maeder, Tsai, et al., 2013). Interestingly, some of the successfully mutated target sites were within genomic regions previously untargetable by TALENs. Thus, this pioneer study displayed the robustness and power of the CRISPR/Cas platform in zebrafish. Furthermore, CRISPR/Cas-induced indel mutations were later shown to be heritable with transmission rates up to ~100%, opening the possibility for using CRISPR/Cas to create genetic knock-out lines for specific genes (Hwang, Fu, Reyon, Maeder, Kaini, et al., 2013). The goal of Section 2 of this chapter will be to discuss the methodologies that have been developed since these initial studies for the generation of targeted indel mutations in zebrafish using CRISPR/Cas.

It has recently been shown that the co-injection of several sgRNAs targeting multiple genomic loci can result in the simultaneous generation

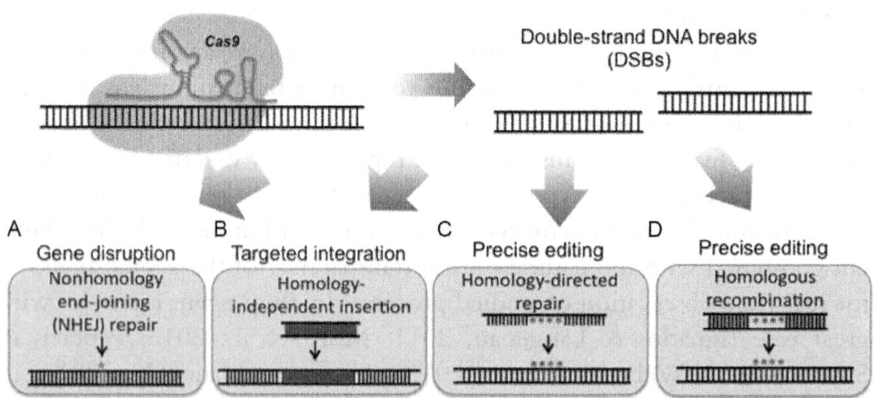


**Figure 18.2** Cas9-mediated genome editing. The RNA-guided Cas9 endonuclease can induce DSBs at its genomic target site. Subsequently, a DSB may be repaired by a non-homologous end-joining (NHEJ) repair mechanism. This mechanism may cause random-length insertion/deletion (indel) mutations (red asterisk) at the target site (A). Additional approaches can be used to create predetermined sequence modifications. Linear donor DNA fragments containing a desired functional cassette without any sequence homology to the target locus may be inserted into the target site during DNA repair (B). Moreover, donor DNA containing homologous sequences of the target locus, in the form of small single-stranded oligonucleotides (C) or plasmid DNA (D), may be recombined with the genomic DNA and replace the target site sequence. (See the color plate.)

of multigene mutations in zebrafish embryos, demonstrating even further the power by which CRISPR/Cas can generate targeted indel mutations (Jao et al., 2013; Ota, Hisano, Ikawa, & Kawahara, 2014). Nevertheless, CRISPR/Cas can also be used in zebrafish for purposes beyond the generation of indels (Fig. 18.2). CRISPR/Cas has been used to create small but precise sequence modifications such as point mutations, to integrate long DNA fragments at target sites, and to facilitate long-range chromosomal deletions and inversions (Fig. 18.2). As the genome-editing repertoire of the CRISPR/Cas system in zebrafish continues to rapidly grow, it will be the goal of Section 3 of this chapter to discuss the other available genome-editing strategies beyond the introduction of indels.

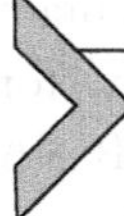

# 2. TARGETED GENERATION OF INDEL MUTATIONS

## 2.1. Cas9 modification and delivery platforms

The most studied and commonly implemented version of Cas9 endonuclease used for CRISPR/Cas genome editing is SpCas9. This is in part due to its short PAM 5′-NGG which is simpler than that of many other Type II Cas9 nucleases (Westra et al., 2012). However, because of the innate differences in cellular contexts between prokaryotic and eukaryotic systems, this bacterial Cas9 must be modified for *in vivo* eukaryotic experimentation.

In the preliminary studies implementing CRISPR/Cas *in vivo* in cultured human cells, SpCas9 was human-codon optimized, allowing the primary structure of SpCas9 to be encoded by codons preferentially used by human cells in order to boost SpCas9 translation efficiency (Cho et al., 2013; Cong et al., 2013; Mali, Yang, et al., 2013). Also in these experiments, one or more commonly used nuclear localization signals (NLS), such as the SV40 NLS, were added to one or both sequence termini of the Cas9 protein to facilitate endonuclease entrance into the eukaryotic nucleus. In our published studies applying CRISPR/Cas to zebrafish, we used an SpCas9 vector (pMLM3613) containing the natural noncodon optimized SpCas9 sequence and an NLS attachment to our construct's C-terminus (Hwang, Fu, Reyon, Maeder, Kaini, et al., 2013; Hwang, Fu, Reyon, Maeder, Tsai, et al., 2013). Though we did not use a codon-optimized version of SpCas9 in our initial studies, our studies demonstrated that a simple NLS attachment to natural SpCas9 suffices for CRISPR/Cas to efficiently generate indel mutation rates up to ~60% in zebrafish (Hwang, Fu, Reyon, Maeder, Tsai, et al., 2013).

Since the time of our initial publications, we have started using a version of SpCas9 that has been optimized for human codon usage. Based upon our experiences, we have consistently seen higher somatic mutation frequencies in zebrafish with this codon-optimized SpCas9 version (pJDS246), and therefore recommend using this version over natural SpCas9. In addition, we have found that the activity of pJDS246 is comparable to a zebrafish codon-optimized SpCas9 (pCS2-nCas9n) (Gonzales, unpublished results), which was reported to produce indel mutagenesis rates up to ~75–99% at various loci (Jao et al., 2013). Last, it is also worth considering the possibility that other modified Type II Cas9 orthologs besides SpCas9 may provide more potent Cas9 options in the future (Esvelt et al., 2013).

All of the Cas9 plasmids mentioned above can be ordered from Addgene (http://www.addgene.org/CRISPR/). After receiving the Cas9-containing plasmid, it should be linearized by the appropriate restriction enzyme and then *in vitro* transcribed to produce Cas9 mRNA. Most Cas9 plasmids contain either a T7 or SP6 promoter upstream of the Cas9 sequence to allow for standard *in vitro* transcription. In order for *in vitro*-transcribed Cas9 mRNA to be translated efficiently in zebrafish embryos, the mRNAs should have a 5′-cap and a 3′-poly(A) tail. Though most published papers to-date implement CRISPR/Cas in zebrafish by co-injecting Cas9 mRNA and sgRNA together into zebrafish embryos, it has recently been argued that the direct injection of preformed Cas9 protein–sgRNA complexes may be more advantageous because it eliminates the need for Cas9 translation before CRISPR/Cas can start functioning. However, results from these experiments are conflicting as to whether this method can more consistently produce efficient site-directed mutagenesis than direct RNA injections (Gagnon et al., 2014; Sung et al., 2014). Nonetheless, the study by Gagnon et al. strongly suggests that the injection of such complexes can raise the indel mutation rates of normally weak sgRNAs up to approximately sixfold.

**Protocol for preparation of SpCas9 mRNA for microinjection**

1. Linearize the human-codon optimized SpCas9 vector, pJDS246 (Addgene, Cambridge, MA), with the *Pme*I restriction enzyme (New England Biolabs, Ipswich, MA) by setting up the following reaction: 5 μg of pJDS246 vector DNA, 10 μL of 10× CutSmart™ Buffer (New England Biolabs), 1 μL of *Pme*I (10 units/μL), and sterile deionized water to a total volume of 100 μL. Add *Pme*I last to the reaction mixture. Incubate the reaction at 37 °C for 3 h to overnight to ensure complete linearization.
2. Purify the *Pme*I-cut vector using Qiagen's QIAquick PCR Purification kit and elute with 25 μL of EB Buffer. Measure the vector DNA

concentration with a spectrometer. Run 100 ng of uncut and cut vector DNA on a 1% wt/vol agarose gel to confirm complete digestion of the vector sample. The purified vector sample can be stored at −20 °C.

**3.** *In vitro* transcribe capped and poly(A)-tailed SpCas9 mRNA using a mMESSAGE mMACHINE® T7 Ultra kit (Life Technologies, Beverly, MA). First, thaw the 2× NTP/ARCA and 10× T7 Reaction Buffer solutions at room temperature while keeping the T7 Enzyme Mix on ice at all times. Put the 2× NTP/ARCA solution on ice as soon as it has thawed. Once the 10× T7 Reaction Buffer has thawed, vortex it to redissolve any precipitate. Next, set up the transcription reaction by mixing the listed reagents in a nuclease-free microfuge tube in the following sequence: 5 μL of 2× NTP/ARCA, 1 μL of 10× T7 Reaction Buffer, 1 μL of T7 Enzyme Mix, and then 3 μL of linearized SpCas9 vector (from Step 2). Gently flick and briefly microfuge the tube to collect the reaction mixture at the bottom of the tube. Incubate the tube at 37 °C for 3 h to overnight for *in vitro* transcription to proceed. After the transcription step, add 1 μL of TURBO DNase to the reaction mixture. Gently flick and briefly microfuge the tube to mix. Incubate the tube at 37 °C for 30 min for DNA removal.

**4.** Prepare the poly(A) tailing reaction master mix by combining the following reagents supplied in the same kit in a nuclease-free microfuge tube: 10 μL of 5× EPAP Buffer, 2.5 μL of 25 m*M* $MnCl_2$, 5 μL of ATP Solution, and 21.5 μL of nuclease-free water. After the TURBO DNase incubation step is complete (end of Step 3), add the poly(A) tailing reaction master mix to the reaction mixture. Gently flick and briefly microfuge the tube to mix. Aliquot 2 μL of this new mixture to a clean nuclease-free tube labeled "−poly(A)" and store this tube at −20 °C for later gel analysis. Next, add 2 μL of E-PAP Enzyme to the reaction mixture. Gently flick and briefly microfuge the tube to mix. Incubate the reaction mixture at 37 °C for 1–2 h for poly(A) tailing reaction to proceed.

**5.** After poly(A) tailing is complete, aliquot 2 μL of the reaction mixture to another clean nuclease-free tube labeled "+poly(A)" and store this tube at −20 °C for later gel analysis. Then, add 25 μL of Lithium Chloride Precipitation Solution to the remaining reaction mixture. The volume of Lithium Chloride Precipitation Solution added should be half the volume of the reaction mixture. Thoroughly mix the solution and incubate it at either 0.5–1 h on dry ice, or preferably overnight at −20 °C for a greater overall mRNA yield.

6. During the mRNA precipitation step, add 5 μL of the Formaldehyde Loading Dye into the 2 μL "−poly(A)" and "+poly(A)" aliquots from before and after the poly(A) tailing reaction (from Steps 4 and 5). Run the samples on a 1% wt/vol agarose gel to check for successful poly(A) tailing by looking for an upshift in the "+poly(A)" sample relative to the "−poly(A)" sample.
7. After the mRNA precipitation step, spin the sample in a microcentrifuge at >10,000 × *g*, 4 °C for 30 min. After the spin, check for an opaque white mRNA pellet at the bottom of the tube. Carefully aspirate the supernatant without dislodging the pellet. Next, add 1 mL of RNase-free 70% ethanol to the tube and wash the pellet by inverting the tube several times. Centrifuge the tube at >10,000 × *g*, 4 °C for 15 min. Again, check for the pellet at the bottom of the tube. Aspirate as much of the supernatant as possible without dislodging the pellet so that the pellet will air-dry quickly. Leave the tube with the lid open in a fume hood until all of the supernatant has evaporated and the pellet becomes dry and translucent.
8. Dissolve the SpCas9 mRNA pellet with 15 μL of non-diethylpyrocarbonate (DEPC)-treated, nuclease-free water. As soon as the mRNA pellet completely dissolves, put the tube on ice. Measure the dissolved SpCas9 mRNA concentration using a spectrometer (the yield is typically 1000–2000 ng/μL). Aliquot the SpCas9 mRNA into multiple nuclease-free microfuge tubes (~1500 ng/tube) to prevent freeze–thaw cycles. Store these tubes at −80 °C.

## 2.2. Single-guide RNA design considerations

The sgRNA design that we use is a ~100-nt sequence in which the first 20 nucleotides interact with the complementary strand of the target site, while the remaining portion interacts with SpCas9 (Hwang, Fu, Reyon, Maeder, Kaini, et al., 2013; Hwang, Fu, Reyon, Maeder, Tsai, et al., 2013). This ~100-nt sgRNA has a longer tracrRNA region compared to the sgRNA first described in the *in vitro* study done by Jinek et al. (2012), and it appears to be more effective *in vivo* compared to sgRNAs that have shorter tracrRNA regions (Jinek et al., 2012, 2013). This sgRNA design is the most common sgRNA design in use (Sander & Joung, 2014), with the same or similar sgRNA design being used in other published zebrafish studies (Auer, Duroure, De Cian, Concordet, & Del Bene, 2014; Chang et al., 2013; Hruscha et al., 2013; Jao et al., 2013).

To express sgRNAs in early stage zebrafish embryos, the sgRNA is usually *in vitro* transcribed and then microinjected. The sgRNA should not have a 5′-cap or a 3′-poly(A) tail, and the sgRNA vectors used for sgRNA production should have a T7 or SP6 promoter. The transcribed sgRNA can recognize any DNA target in a 5′-GG-$N_{18}$-NGG-3′ format, with the 5′-GG required for T7-driven transcription, and the "NGG" being the *S. pyogenes* PAM. Thus, the theoretical targeting range is 1 site for every 128 bp ofDNA (Hwang, Fu, Reyon, Maeder, Tsai, et al., 2013). However, we have shown that sgRNA targeting can often tolerate 2-nt mismatches at its 5′-end, extending the theoretical targeting range to 1 site for every 32 bps (Hwang, Fu, Reyon, Maeder, Kaini, et al., 2013). In addition, we had initially proposed to loosen the constraint of the T7 promoter, which would enable targeting of sequences of the form 5′-(G/A)(G/A)-$N_{18}$-NGG-3′. Nonetheless, a recent report by Gagnon et al. suggests that any change to the 5′-GG can reduce sgRNA efficiency. This result is likely due to T7-driven transcription errors because in all three cases tested, the 5′-GA sgRNAs transcribed by the SP6 promoter showed similar activities to their 5′-GG sgRNA counterparts (Gagnon et al., 2014). Thus, the SP6 promoter may be more flexible than the T7 promoter, tolerating a G→A change at the second position from the transcription start site (Helm, Brule, Giege, & Florentz, 1999; Imburgio, Rong, Ma, & McAllister, 2000; Kuzmine, Gottlieb, & Martin, 2003).

At least two studies have proposed guidelines for choosing sgRNA sequences that are more likely to be effective (Gagnon et al., 2014; Wang, Wei, Sabatini, & Lander, 2014). Together, these studies suggest that the last 1–4 nucleotides of the spacer region should preferably be purines. In addition, the GC content of the spacer region near the PAM should be >50%, but not too high. Although these guidelines may be taken into consideration when there are many sites to choose from, they do not necessarily guarantee success or failure of a sgRNA as shown by the authors and our data (Gagnon et al., 2014; Hwang, Fu, Reyon, Maeder, Tsai, et al., 2013; Wang et al., 2014).

In addition to the above sgRNA design considerations, the location of the sgRNA target site within a gene must also be considered. When designing a sgRNA to produce a gene knock-out, for example, a suitable genome browser (e.g., http://useast.ensembl.org) in combination with free sgRNA design programs (e.g., http://zifit.partners.org/ZiFiT/) should be used to examine the regions of the gene that one wishes to target. Ideally, the sgRNA target site should be as far upstream as possible within the gene's open reading frame to ensure that an introduced indel can disrupt almost the entire protein. However, in addition to this targeting principle, one must also search for alternative splice variants for a given gene, as well as annotated alternative translational start sites. If there are any of these additional confounding factors, the same targeting principle can be applied by designing a sgRNA target site after the most downstream annotated alternative start site within the most upstream exon shared by all alternative transcripts. Upon choosing a sgRNA target site, complementary oligonucleotides containing the designed sgRNA spacer region can be ordered, annealed, and inserted into plasmids for cloning, restriction linearization, and subsequent T7 or SP6 *in vitro* transcription. The sgRNA expression vector used for our previous publications (pDR274) was originally developed by the Joung lab (Hwang, Fu, Reyon, Maeder, Kaini, et al., 2013; Hwang, Fu, Reyon, Maeder, Tsai, et al., 2013). The pDR274 vector and those from other labs are available at the non-profit Addgene (http://www.addgene.org/CRISPR/). In addition, there are now published cloning-free methods for sgRNA synthesis that may make sgRNA production quicker (Cho et al., 2013; Gagnon et al., 2014; Hruscha et al., 2013).

Despite the remarkable success of CRISPR/Cas, a potential drawback of the current platform is the possibility of off-target effects due to sgRNA mistargeting. In published studies within human cells, sgRNA mistargeting has been demonstrated to cause indel frequencies at off-target sites at rates up to ~125% compared to its on-target sites (Fu et al., 2013) and has even been shown to inadvertently induce large chromosomal deletions at several loci (Cradick, Fine, Antico, & Bao, 2013). Although earlier CRISPR/Cas studies had proposed the theory of a seed sequence (Cong et al., 2013; Jiang et al., 2013; Jinek et al., 2012), a ~10–12-bp mismatch intolerant region directly adjacent to the target loci's PAM motif, more recent studies that intentionally investigated sgRNA off-target effects have shown the reality to be much more complicated. Though sequence homology in the first ~10–12 bps directly adjacent to the PAM are relatively more important, all base pairs at a target site, including those in the seed region, confer varying degrees of sgRNA targeting specificity in an sgRNA-specific manner (Fu et al., 2013; Hsu et al., 2013; Pattanayak et al., 2013). Clearly, increasing the number of mismatches reduces the efficiency of target cleavage at a given site, especially as the mismatches become more proximal to the PAM (Fu et al., 2013; Hsu et al., 2013). One study suggests that rC:dC base-pairing causes the greatest destruction to sgRNA–Cas9 targeting activity (Hsu et al., 2013). Another study suggests that shorter sgRNA designs may be less active but more specific than longer sgRNAs (Pattanayak et al., 2013). A couple of studies have also shown that in addition to the 5′-NGG PAM, Cas9 can sometimes target sequences that have a 5′-NAG or 5′-NNGGN PAM (Hsu et al., 2013; Jiang et al., 2013).

Nevertheless, in spite of this progress, exceptions have been demonstrated to these principles. For example, recent zebrafish publications have reported off-target cleavage by sgRNAs at multiple loci, with one study exhibiting detectable cleavage by T7 endonuclease I (T7EI) mismatch assays at an off-target site containing only two mismatches outside the seed region (Jao et al., 2013), while in another study, next-generation sequencing identified indel rates ~1.0–2.5% at off-target sites containing up to six base-pair mismatches at their 5′ most distal ends (Hruscha et al., 2013). On the other hand, not all sgRNAs are promiscuous, and not all potential off-target sites are mistargeted by sgRNAs (Fu et al., 2013; Hruscha et al., 2013; Hsu et al., 2013; Jao et al., 2013; Pattanayak et al., 2013). In light of the complexity regarding sgRNA off-target effects and the lack of *in vivo* genome-wide bioinformatics studies done on this topic, sgRNA targeting zebrafish genomic loci should be designed in conjunction with available software to minimize potential specificity issues (Bae, Park, & Kim, 2014; Hsu et al., 2013; O'Brien & Bailey, 2014; Xiao et al., 2014).

Other recently developed strategies to optimize or go beyond the standard Cas9–sgRNA platform should be considered. One strategy involves simply lowering the concentrations of Cas9 and sgRNA injected into the zebrafish embryo, since it has been shown in human cells that reduced concentrations lower off-target indel rates, although on-target indel rates will also diminish to varying degrees (Fu et al., 2013; Hsu et al., 2013). Another strategy uses truncated sgRNAs (tru-gRNAs) that contain shortened ~17 to 18-nt spacer regions, which have been shown to maintain on-target efficiencies while reducing indel frequencies at off-target sites up to ~5000-fold compared to standard sgRNAs (Fu, Sander, Reyon, Cascio, & Joung, 2014). The implementation of D10A Cas9 nickases guided by paired sgRNAs targeting adjacent genomic sites have likewise been shown to reduce the off-target effects of CRISPR/Cas up to ~1500-fold in human cells without sacrificing on-target activity. These work by doubling the number of base pairs required for double-strand DNA cleavage and by relying on less erroneous DNA repair for any off-target DNA single-strand nicks (Cho et al., 2014; Mali, Aach, et al., 2013; Ran et al., 2013). Recently, two independent groups developed a crRNA-guided FokI nuclease platform, fusing catalytically inactive Cas9 with a FokI nuclease domain. Taking advantage of the RNA-guided capability of Cas9 and the dimerization requirement of the FokI nucleases, this platform recognizes extended target sequences and has been shown to be more specific compared to wild-type Cas9 and paired nickases (Guilinger, Thompson, & Liu, 2014; Tsai et al., 2014). Due to differences in the availability of sgRNA target sites, regardless of which of the options are chosen for generating targeted indels, the varied strategies mentioned should be considered on a case-by-case basis.

**Protocol for preparation of sgRNAs for microinjection:**

1. Linearize the sgRNA vector, pDR274 (Addgene, Cambridge, MA), with the *Bsa*I restriction enzyme (New England Biolabs) by setting up the following reaction: 5 μg of pDR274 vector DNA, 10 μL of 10× CutSmart™ Buffer (New England Biolabs), 1 μL of *Bsa*I (10 units/μL), and sterile deionized water to a total volume of 100 μL. Add *Bsa*I last to the reaction mixture. Incubate the reaction at 37 °C for at least 3 h to overnight to ensure complete digestion.
2. Purify the *Bsa*I-cut vector using Qiagen's QIAquick PCR Purification kit and elute with 25 μL of EB Buffer. Measure the vector DNA concentration with a spectrometer. Run 100 ng of both uncut and cut vector DNA on a 1% wt/vol agarose gel to confirm complete digestion of the vector DNA sample. Dilute the vector sample with EB Buffer to a final concentration of 5–10 ng/μL. Keep a stock of this purified cut vector sample at −20 °C for future use.
3. Design a pair of 22-nt DNA oligos that contain within them complementary sequences that correspond to the 18-bp sequence adjacent to the PAM in a sgRNA target site. These oligos when annealed together by these sequences can be inserted into the pDR274 vector for T7 promoter-driven transcription of a given sgRNA. For insertion into the pDR274 vector, Oligo #1 also contains at its 5′-end a TAGG, which is a part of the T7 promoter sequence, and Oligo #2 contains at its 5′-end an AAAC. These short sequences are necessary to provide the annealed oligo the sticky ends necessary for unidirectional insertion into *Bsa*I-cut pDR274. Investigators can also use free online software ZiFiT Targeter (http://zifit.partners.org/ZiFiT/) to generate the oligo sequences needed for any specified target site.
4. Obtain these DNA oligos from a reliable source. Anneal these DNA oligos together by combining 45 μL of 100 μ*M* Oligo #1, 45 μL of 100 μ*M* Oligo #2, and 10 μL of 10× NEBuffer 2.1 (New England Biolabs) in a microfuge tube. Place the tube in a removable heat block at 95 °C for 5 min. Then, remove the heat block and allow it to

gradually cool until it reaches below 37 °C. The annealed oligos can be stored at −20 °C.

5. Ligate the annealed oligos (from Step 4) into the linearized pDR274 vector (from Step 2) by setting up the following reaction and letting it incubate at room temperature for 1 h or at 4 °C overnight: 1 μL of the annealed oligos, 1 μL of purified *Bsa*I-cut pDR274 vector, 2.5 μL of 2× Rapid Ligation Buffer (Promega, Madison, WI), and 0.5 μL of T4 DNA Ligase (Promega). If you use a different ligase and ligation buffer, follow the ligation condition recommended by the manufacturer. Transform chemically competent bacterial cells with 5 μL of the ligation product. After transformation is complete, spread the cells on a LB/kanamycin plate and incubate it at 37 °C overnight.
6. Pick at least three colonies per transformation and inoculate each colony in a culture tube containing 1.5 mL of LB/kanamycin. Incubate the culture tubes in a shaker at 37 °C overnight. The next day, extract the plasmid DNA using a plasmid DNA miniprep kit. Send the extracted plasmid DNA samples for sequencing using a M13F primer. Confirm whether the sequenced plasmid samples have the correct elements in the appropriate order: (from 5′ to 3′) a T7 promoter, the customized target sequence, and the appropriate sgRNA backbone sequence (Hwang, Fu, Reyon, Maeder, Tsai, et al., 2013).
7. Re-inoculate a bacteria sample (from Step 6) that contains the correct sgRNA vector. Miniprep the sgRNA vector using a gravity-flow column-based QIAGEN Plasmid Mini kit, and measure the vector DNA concentration using a spectrometer. Linearize the sgRNA vector with the *Dra*I restriction enzyme (New England Biolabs) by setting up the following reaction: 5 μg of customized sgRNA vector DNA, 10 μL of 10× CutSmart™ Buffer (New England Biolabs), 1 μL of *Dra*I (10 units/μL), and sterile deionized water to a total volume of 100 μL. Add *Dra*I last to the reaction mixture. Incubate the tube at 37 °C for at least 3 h to overnight to ensure complete digestion.
8. Purify the *Dra*I-cut vector using Qiagen's QIAquick PCR Purification kit and elute with 25 μL of EB Buffer. Measure the vector DNA concentration with a spectrometer. Run 100 ng of both uncut and cut vector DNA on a 3% wt/vol agarose gel to confirm complete digestion of the vector DNA sample. The digested sgRNA vector should exhibit only two fragments at 1.9 kb and 282 bp. The smaller DNA fragment contains the T7 promoter and the customized sgRNA sequence. Gel purification is not necessary. The purified DNA can be stored at −20 °C.

**9.** *In vitro* transcribe the sgRNA using the MAXIscript® T7 kit (Life Technologies) by setting up the following reaction in a nuclease-free microfuge tube: 1 μg of purified *Dra*I-cut sgRNA vector DNA (from Step 8), 2 μL of 10 × Transcription Buffer, 1 μL of each of the ATP, UTP, GTP, CTP 10 mM solutions, 2 μL of T7 Enzyme Mix, and nuclease-free water to a total volume of 20 μL. Incubate the reaction mixture at 37 °C for 2 h to overnight for sgRNA transcription. After sgRNA transcription, add 2 μL of TURBO DNase to the reaction mixture and incubate at 37 °C for 30 min to remove DNA from the sample. Next, add 1 μL of 0.5 *M* EDTA to the reaction mixture. Gently flick and briefly microfuge the tube to stop the reaction. Subsequently, add 30 μL of nuclease-free water and 5 μL of 5 *M* ammonium acetate to the reaction mixture and thoroughly mix. Last, add 150 μL of 100% RNase-free ethanol and thoroughly mix. Incubate the sample on dry ice for 0.5–1 h, or preferably at −20 °C overnight for a greater overall sgRNA yield.

**10.** After the sgRNA precipitation step, spin the sample in a microcentrifuge at >10,000 × *g*, 4 °C for 30 min. After the spin, check for an opaque white sgRNA pellet at the bottom of the tube. Carefully aspirate the supernatant without dislodging the pellet. Next, add 1 mL of RNase-free 70% ethanol to the tube and wash the pellet by inverting the tube several times. Centrifuge the tube at >10,000 × *g*, 4 °C for 15 min. Again, check for the pellet at the bottom of the tube. Aspirate as much of the supernatant as possible without dislodging the pellet so that the pellet will air-dry quickly. Leave the tube with the lid open in a fume hood until all of the supernatant has evaporated and the pellet becomes dry and translucent.

**11.** Dissolve the sgRNA pellet with 11 μL of non-DEPC-treated, nuclease-free water. As soon as the sgRNA pellet completely dissolves, put the tube on ice. Measure the dissolved sgRNA concentration using a spectrometer (the yield is typically 100–200 ng/μL). Aliquot the sgRNA into multiple nuclease-free microfuge tubes (~100 ng/tube) to prevent freeze–thaw cycles. Store the tubes at −80 °C. Check the sgRNA integrity by mixing 1 μL of the sgRNA with 5 μL of Formaldehyde Loading Dye and running it on a 3% wt/vol agarose gel containing 0.2–0.5 μg/mL of ethidium bromide. There should be a distinct sgRNA band without smearing.

## 2.3. Introduction and identification of Cas9–sgRNA-induced indels

After selecting and preparing the appropriate Cas9 and sgRNA(s) for a given experiment, CRISPR/Cas genome editing can be implemented by zebrafish embryo microinjections. Zebrafish embryos must be collected and immediately injected at the one-cell stage before the first round of cellular mitosis to facilitate homogenous distribution of CRISPR/Cas components among all future daughter cells. As previously discussed, the most common practice uses *in vitro*-prepared Cas9 mRNAs for injections, though studies using Cas9 protein–sgRNA complexes have also been reported (Gagnon et al., 2014). Various concentrations of Cas9 mRNA and sgRNAs have been used in different studies. For sgRNAs that show a high on-target mutation rate, reducing the concentration of Cas9 may reduce potential off-target activities. On the other hand, the quality of *in vitro*-transcribed Cas9 mRNA and sgRNA have a direct influence on the observed on-target activity.

Depending on the particular purpose of the study, indel mutation rates may be determined by the T7EI mismatch assay, PCR subcloning and sequencing, high resolution melt analysis (Dahlem et al., 2012), and/or next-generation sequencing. Due to its speed and independence of any sophisticated instrumentation, the T7EI assay is probably the most widely used method. T7EI analysis relies upon the ability of T7EI to recognize and cleave nonperfectly annealed DNA. In this method, PCR amplicons of targeted genomic loci are denatured and gradually cooled to allow for partial hybridization between differentially sized indel-containing single-stranded DNA fragments. These partially hybridized PCR strands will contain mismatched regions at the designed target sites that may be cleaved by T7EI to produce two fragments of predictable lengths. Upon electrophoresis, the percentage of cleaved PCR products can be measured, and the estimated NHEJ percentage rates can be calculated using the calculations given by Guschin et al.: % target site indel rate $= 100 \times (1 - (1 - \text{fraction cleaved})^{1/2})$ (Guschin et al., 2010). Nonetheless, the estimated detection limit of this assay is above ~3% (Hwang, Peterson, & Yeh, 2014). In addition, genetic polymorphisms surrounding the target site may cause false positives and should be carefully controlled. Moreover, if a low amount of or no cleaved product is detected, sequencing of the PCR products should be considered.

# 3. OTHER TARGETED GENOME-EDITING STRATEGIES

## 3.1. Precise sequence modifications mediated by single-stranded oligonucleotides

Previously, it was found that in the presence of DSBs, single-stranded oligonucleotides (ssODNs) flanked by two arms with tens of base pairs of homology to the sequences surrounding the breaks may be co-opted to introduce predesigned sequence modifications in human cells (Chen et al., 2011). Later, a similar strategy was used to introduce small yet precise targeted insertions in conjunction with TALENs (Bedell et al., 2012). We and others have recently shown that small predetermined sequences may be targetedly inserted in zebrafish using CRISPR/Cas and ssODNs containing ~20–50-nt homology arms (Chang et al., 2013; Hruscha et al., 2013; Hwang, Fu, Reyon, Maeder, Kaini, et al., 2013).

Thus far, this ssODN-mediated CRISPR/Cas method has been used to knock-in ~30-nt mloxp sites (Chang et al., 2013), as well as, ~30-nt HA tags at two zebrafish loci (Hruscha et al., 2013). We have used ssODNs to precisely insert up to ~40 nucleotides at different zebrafish loci (Gonzales, unpublished results) and to successfully generate precise point mutations in one zebrafish gene (Hwang, Fu, Reyon, Maeder, Kaini, et al., 2013). It should be noted that even though insertion rates can be as high as ~20%, it has been found in all of these studies, including the aforementioned TALEN study, that only a small portion of the sequence changes are free of additional mutations. These results are very different from the results of studies in human cells and mice, wherein sequence modifications introduced by ssODNs are almost always precise (Chen et al., 2011; Kayali, Bury, Ballard, & Bertoni, 2010; Shen et al., 2013). The exact mechanism of DNA repair in zebrafish that mediates the insertion under these conditions is still unclear.

To implement this ssODN-mediated method, we have noticed that, in some cases, sense or antisense ssODNs containing the sequences either homologous or complementary to the sgRNA target site show varied efficiencies (Hwang, Fu, Reyon, Maeder, Kaini, et al., 2013). Nonetheless, the differences in their efficiencies are locus-dependent, and there is no consensus as to which strand is better to use. Interestingly, a recent study shows that ssODNs containing a stop cassette, which if inserted into a gene provides stop codons at all possible reading frames, has allowed this ssODN-mediated technique to serve as an alternative to producing genetic knock-outs by indel frameshift mutations (Gagnon et al., 2014). Moreover, this study demonstrated the insertion and heritability of these stop cassettes in three zebrafish genes. Therefore, the demonstrated ssODN-mediated insertion strategy provides great versatility to CRISPR/Cas genome editing in zebrafish.

## 3.2. Targeted integration of long DNA fragments

Targeted genome modifications involving insertions of long DNA fragments can be achieved via homologous recombination (HR) or a homology-independent mechanism such as NHEJ (Fig. 18.3). In zebrafish embryos, NHEJ appears to be responsible for most DNA repair activities, whereas HR is estimated to be at least 10-fold less active (Dai, Cui, Zhu, & Hu, 2010; Hagmann et al., 1998; Liu et al., 2012). Zu et al. were the first to successfully demonstrate HR-mediated insertion in zebrafish by

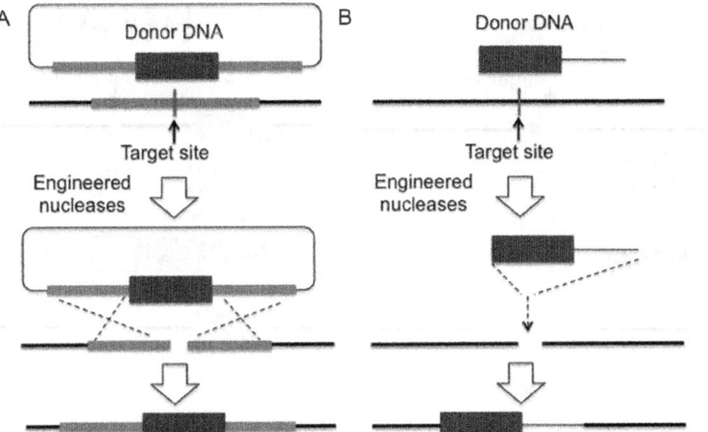


**Figure 18.3** Engineered DNA nucleases can facilitate targeted integration of long DNA fragments via homology-dependent and homology-independent mechanisms. (A) Targeted integration mediated by homologous recombination. In this approach, a plasmid donor DNA containing the DNA cassette (red (dark gray in the print version) box) to be inserted, flanked by several hundred base pairs to a couple kilobase pairs of sequences (gray boxes) upstream and downstream of the genomic target site, is co-injected with the engineered DNA nucleases into zebrafish embryos. The DNA cassette can be inserted into the nuclease target site via homologous recombination. The sequence surrounding the cassette and the joining ends should be precise. (B) Targeted integration mediated by a homology-independent mechanism. In this approach, the donor DNA should be linearized *in vitro* or *in vivo*, and need not have sequence homology to the genomic target locus. The linearized donor DNA can be inserted into the nuclease target site, but the sequences at the joining ends will not be precise.

the use of highly efficient TALEN pairs. In their study, the authors reported targeted insertion of EGFP in three zebrafish genes; however, germline transmission was found for only one gene at a low frequency of 1.5% (Zu et al., 2013). Therefore, although HR-mediated gene targeting affords the ability to generate precise sequence changes, this method may not become a widely used approach unless its efficiency can be improved (Beumer et al., 2008; Liu et al., 2012). To date, an example of Cas9-assisted HR-mediated DNA integration has not been reported in zebrafish.

On the other hand, Auer et al. demonstrated efficient homology-independent targeted integration of >5.7-kilobase (kb) DNA fragments in zebrafish using CRISPR/Cas (Auer et al., 2014). In their study, a donor DNA construct containing a modified *Gal4* gene was co-injected with Cas9

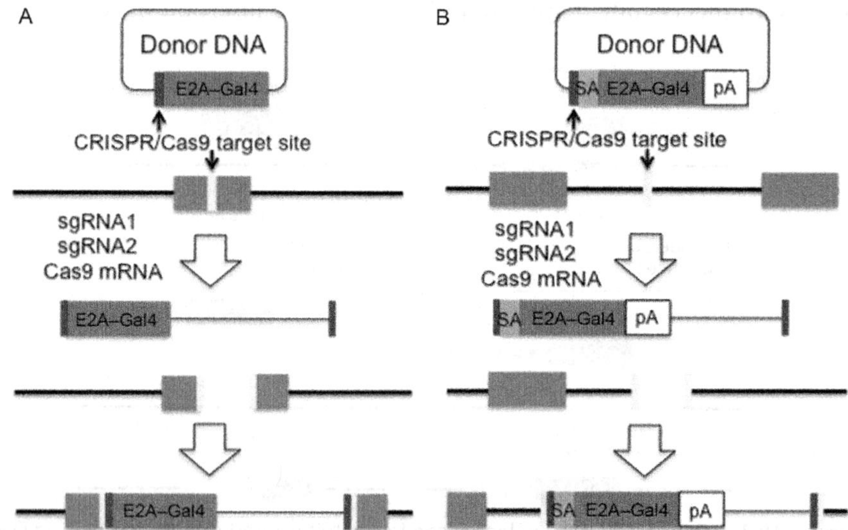


**Figure 18.4** Designs of donor DNAs. A plasmid donor DNA may be linearized *in vivo* by adding a CRISPR/Cas9 target site (blue (dark gray in the print version) vertical line) to it. This target sequence can be the same as or different from the genomic target sequence (yellow (light gray in the print version) vertical line). (A) In this design, Gal4 and a self-processing peptide E2A will be expressed only if the integrated DNA cassette is in the correct orientation and in the right coding frame. Thus, theoretically one-sixth of the integration events will result in the expression of E2A–Gal4. This is the approach utilized by Auer et al. (B) An alternative approach will be to insert a DNA cassette including a splice acceptor site (SA) into an intronic region. In this approach, the reading frame can be predicted because no indel mutation is introduced in the coding region. Thus, approximately one-half of the integration events will result in the expression of E2A–Gal4.

mRNA and a GFP-targeting sgRNA into zebrafish embryos carrying a tissue-specific GFP reporter and the *UAS:RFP* transgene (Fig. 18.4). In theory, one-sixth of the targeted integration events would be in-frame, resulting in Gal4 and subsequent RFP expression. Thus, this experimental setup enabled visual detection of in-frame knock-in events by a green-to-red switch. Detection of RFP+ cells in the expected expression domains, which are the same as the GFP+ domains in the noninjected transgenic embryos, would indicate successful integration at the target site. In contrast, detection of RFP+ cells outside the expected expression domains would suggest off-target or random integration. Moreover, the incorporation of the Gal4–UAS system likely enhanced the sensitivity of this assay.

By using this approach, Auer et al. made a number of important findings (Auer et al., 2014). First, they found that the knock-in efficiency by homology-independent repair can be significantly improved if the donor DNA is linearized in the injected embryos rather than *in vitro* (Fig. 18.4). *In vivo* cleavage of the plasmid donor DNA was accomplished by adding to the donor construct a sgRNA "bait" sequence that could be cleaved by co-injection of a corresponding sgRNA. By co-injecting a sgRNA with a 66% mutation efficiency and a plasmid donor containing the corresponding bait sequence, they observed a near sevenfold increase of knock-in efficiency compared to the use of an *in vitro* linearized donor DNA.

Second, the authors demonstrated that the bait sequence could be made either the same or different from the genomic target sequence. While the former strategy simplifies experimental design due to the need for only one sgRNA, the latter allows the flexibility to use the same donor construct for any given target site (Fig. 18.4). Indeed, this latter strategy may be useful because the Cas9-mediated knock-in efficiency is directly related to the targeting efficiency of a given sgRNA. For example, in the study by Auer et al., there were two cases wherein a single sgRNA was used for both the bait and the target gene, and sgRNAs with 66% and 20% mutation rates were able to induce the green-to-red switch in 75% and 15% of the injected embryos, respectively. Therefore, as shown also by the authors, it may be beneficial when using this knock-in strategy to reuse a sgRNA-bait sequence pair that is known to have a high *in vivo* linearization efficiency in order to increase the knock-in rates at a difficult target site.

Third, Auer et al. found that the germline transmission of the RFP knock-in allele was quite high and that the observed frequencies of identified founder fish could be increased by screening only RFP+ $F_0$ fish. For example, in a case wherein a single sgRNA with a 66% mutation rate was used, Auer et al. observed founder fish frequencies of 10% and 33–40% by screening all of the fish or only RFP+ $F_0$ fish, respectively. Further examination of the $F_1$ fish showed that both single-copy and multiple-copy integrations of the RFP allele could be found. Moreover, despite some low level off-target activities for a single sgRNA they used (probably under 3% at the sites screened), off-target integration was not found in the $F_0$ fish by PCR, Southern blots, or fluorescence.

Overall, this pioneer study significantly extended the power of CRISPR/Cas genome editing and opens the possibility of unprecedented strategies for creating novel functional alleles in zebrafish. Conceivably, one potential means of improving the current system will be to insert DNA cassettes with splice acceptor sites into the intronic regions of target genes (Fig. 18.4). This approach is likely to better guarantee in-frame integrations irrespective of sequence alterations at the joining ends.

## 3.3. Chromosomal deletions and other rearrangements

Functional analysis of noncoding RNAs, transcriptional enhancers, regulatory elements in promoters, gene clusters, and tandem-duplicated genes require methods that, in a targeted fashion, can delete genomic segments ranging from hundreds of base pairs to several hundred kilobase pairs. Consequently, genome-editing tools that can induce large segmental deletions are of particular interest for zebrafish research due to the prevalence of gene duplications in this model organism (Lu, Peatman, Tang, Lewis, & Liu, 2012).

Using two customized pairs of ZFNs or TALENs to target two distant sequences on the same chromosome, deletions of large DNA segments ranging from several kilobase pairs to megabase pairs have been reported in

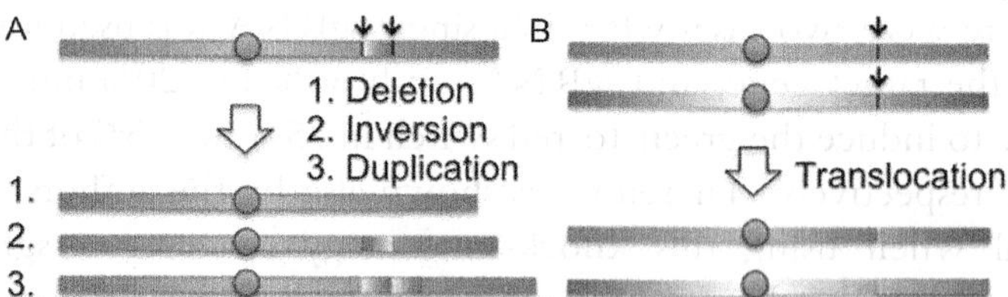


**Figure 18.5** Cas9-mediated chromosomal rearrangements. (A) Chromosomal deletions, inversions, and duplications may be induced by two sgRNA–Cas9 complexes that target two distant sites on the same chromosome. (B) Chromosomal translocations may be induced by two sgRNA–Cas9 complexes that target two different chromosomes. (See the color plate.)

cultured mammalian cells (Carlson et al., 2012; Lee, Kim, & Kim, 2010), silkworms (Ma et al., 2012), plants (Qi et al., 2013), and zebrafish (Gupta et al., 2013; Xiao et al., 2013) (Fig. 18.5). Interestingly, re-insertions of the DNA segments cleaved off by these programmable nucleases have also been observed as inversions and duplications (Gupta et al., 2013; Lee, Kweon, Kim, Kim, & Kim, 2012; Qi et al., 2013; Xiao et al., 2013) (Fig. 18.5). Furthermore, targeted chromosomal translocations have also been successfully created between two ZFN or TALEN sites located on two different chromosomes (Brunet et al., 2009; Piganeau et al., 2013; Simsek et al., 2011) (Fig. 18.5). In light of these studies, ZFN and TALEN-mediated chromosomal rearrangements have been applied to cultured cells for the development of human disease models involving genome rearrangements (Piganeau et al., 2013). Consequently, it has become a very attractive prospect to apply the more facile CRISPR/Cas platform for the generation of such disease models (Choi & Meyerson, 2014).

To date, zebrafish chromosomal deletions and inversions mediated by CRISPR/Cas have been shown by only two groups (Ota et al., 2014; Xiao et al., 2013). In one study, Xiao et al. was able to perform chromosomal deletions of two loci with CRISPR/Cas; however, lower deletion rates were reported by CRISPR/Cas compared to those by TALENs for the tested loci, though the reasons for this discrepancy are unclear (Xiao et al., 2013). More recently, Ota et al. demonstrated germline transmission of a 7.1-kb CRISPR/Cas deletion that had been successfully identified in one of eleven screened potential founder fish (Ota et al., 2014). Because of the ease of use and general robustness of the CRISPR/Cas system, it is likely that Cas9-based approaches will play a more prominent role in future studies in zebrafish involving chromosomal deletions and other rearrangements.

## 4. FUTURE DIRECTIONS

Over the past year-and-a-half, the CRISPR/Cas system has proven itself to be a powerful yet facile and efficient genome-editing platform in zebrafish, demonstrating its broad adaptability by its ability to generate targeted indels, exact point mutations, site-specific insertions of varying lengths, and chromosomal rearrangements in a rapid and low-cost manner. In spite of its quick rise to fame, the CRISPR/Cas platform still has several hurdles to overcome, if it is to advance the field of genetic engineering even further. The two most crucial obstacles that the CRISPR/Cas system faces relate to its targeting range and its specificity. Though the current CRISPR/Cas platform has a relatively widespread targeting range of 1 site in every 8 bps, this is still lower than that of previous programmable nucleases such as TALENs, which has been engineered to theoretically target any sequence in the genome (Joung & Sander, 2013; Lamb, Mercer, & Barbas, 2013). The investigation of other Type II Cas9 orthologs besides that of *S. pyogenes*, coupled with the application of continuous directed evolutionary efforts (Esvelt, Carlson, & Liu, 2011), will likely provide an avenue to expand the CRISPR/Cas targeting range. With regard to the variable targeting specificity of CRISPR/Cas, global genome-wide off-target studies will be needed to accurately test the effectiveness of current tru-sgRNA, paired Cas9 nickase, and other strategies. Off-target studies will also be needed for the potential development of more reliable methods to increase CRISPR/Cas specificity.

Finally, CRISPR/Cas possesses untapped potential for future zebrafish research. From a genetic engineering standpoint, CRISPR/Cas shows great promise to be used for conditional, tissue-specific genome editing. The development of such a spatially and temporally controlled genome-editing application has yet to be demonstrated by CRISPR/Cas, but its development is certainly underway and will be of tremendous benefit to future disease and functional genomics research when applied to zebrafish. Beyond genome editing, CRISPR/Cas has already been adapted in human cells as a system for transcriptional regulation (Gilbert et al., 2013) and as a means for dynamic imaging of genetic structures (Chen et al., 2013). These adaptations are made possible by deactivating the nuclease activity of Cas9 without removing its sgRNA-mediated targeting function, and in principle, these methods can be extended to zebrafish. It is logical that such engineering could allow CRISPR/Cas to be adapted for purposes of epigenetic editing (Maeder et al., 2013) and modulation of genomic architectures (Deng et al., 2012). Also, a recent publication has shown that orthogonal Cas9 enzymes could be used in bacteria to allow for simultaneous indel targeting and transcriptional repression at two different targets, thus opening up the possibility of combining several CRISPR/Cas capabilities, in parallel, within a single organism (Esvelt et al., 2013). Such a report points to the enormous potential CRISPR/Cas has to one day become a universal, all-purpose biomolecular engineering platform that can revolutionize not only zebrafish research, but could influence all future biomedical pursuits and medical treatments.

## ACKNOWLEDGMENTS

This work is supported by the National Institutes of Health (R01 GM088040 and R01 CA140188) and by the Hassenfeld Scholar's Award.

## REFERENCES

Auer, T. O., Duroure, K., De Cian, A., Concordet, J. P., & Del Bene, F. (2014). Highly efficient CRISPR/Cas9-mediated knock-in in zebrafish by homology-independent DNA repair. *Genome Research*, *24*(1), 142–153. http://dx.doi.org/10.1101/gr.161638.113 (Research Support, Non-U.S. Gov't).

Bae, S., Park, J., & Kim, J. S. (2014). Cas-OFFinder: A fast and versatile algorithm that searches for potential off-target sites of Cas9 RNA-guided endonucleases. *Bioinformatics*, *30*(10), 1473–1475. http://dx.doi.org/10.1093/bioinformatics/btu048.

Barrangou, R., Fremaux, C., Deveau, H., Richards, M., Boyaval, P., Moineau, S., et al. (2007). CRISPR provides acquired resistance against viruses in prokaryotes. *Science*, *315*(5819), 1709–1712. http://dx.doi.org/10.1126/science.1138140 (Research Support, Non-U.S. Gov't).

Bassett, D. I., Bryson-Richardson, R. J., Daggett, D. F., Gautier, P., Keenan, D. G., & Currie, P. D. (2003). Dystrophin is required for the formation of stable muscle attachments in the zebrafish embryo. *Development*, *130*(23), 5851–5860. http://dx.doi.org/10.1242/dev.00799 (Research Support, Non-U.S. Gov't).

Bedell, V. M., Wang, Y., Campbell, J. M., Poshusta, T. L., Starker, C. G., Krug, R. G., 2nd, et al. (2012). In vivo genome editing using a high-efficiency TALEN system. *Nature*, *491*(7422), 114–118. http://dx.doi.org/10.1038/nature11537 (Research Support, N.I.H., Extramural Research Support, Non-U.S. Gov't Research Support, U.S. Gov't, Non-P.H.S.).

Beumer, K. J., Trautman, J. K., Bozas, A., Liu, J. L., Rutter, J., Gall, J. G., et al. (2008). Efficient gene targeting in Drosophila by direct embryo injection with zinc-finger nucleases. *Proceedings of the National Academy of Sciences of the United States of America*, *105*(50), 19821–19826. http://dx.doi.org/10.1073/pnas.0810475105 (Research Support, N.I.H., Extramural Research Support, Non-U.S. Gov't).

Bhaya, D., Davison, M., & Barrangou, R. (2011). CRISPR-Cas systems in bacteria and archaea: Versatile small RNAs for adaptive defense and regulation. *Annual Review of Genetics*, *45*, 273–297. http://dx.doi.org/10.1146/annurev-genet-110410-132430 (Research Support, Non-U.S. Gov't Research Support, U.S. Gov't, Non-P.H.S. Review).

Bibikova, M., Beumer, K., Trautman, J. K., & Carroll, D. (2003). Enhancing gene targeting with designed zinc finger nucleases. *Science*, *300*(5620), 764. http://dx.doi.org/10.1126/science.1079512 (Research Support, U.S. Gov't, P.H.S.).

Bibikova, M., Carroll, D., Segal, D. J., Trautman, J. K., Smith, J., Kim, Y. G., et al. (2001). Stimulation of homologous recombination through targeted cleavage by chimeric nucleases. *Molecular and Cellular Biology*, *21*(1), 289–297. http://dx.doi.org/10.1128/MCB.21.1.289-297.2001 (Research Support, Non-U.S. Gov't Research Support, U.S. Gov't, Non-P.H.S. Research Support, U.S. Gov't, P.H.S.).

Bibikova, M., Golic, M., Golic, K. G., & Carroll, D. (2002). Targeted chromosomal cleavage and mutagenesis in drosophila using zinc-finger nucleases. *Genetics*, *161*(3), 1169–1175 (Research Support, Non-U.S. Gov't Research Support, U.S. Gov't, P.H.S.).

Bolotin, A., Quinquis, B., Sorokin, A., & Ehrlich, S. D. (2005). Clustered regularly interspaced short palindrome repeats (CRISPRs) have spacers of extrachromosomal origin. *Microbiology*, *151*(Pt. 8), 2551–2561. http://dx.doi.org/10.1099/mic.0.28048-0.

Brouns, S. J., Jore, M. M., Lundgren, M., Westra, E. R., Slijkhuis, R. J., Snijders, A. P., et al. (2008). Small CRISPR RNAs guide antiviral defense in prokaryotes. *Science*, *321*(5891), 960–964. http://dx.doi.org/10.1126/science.1159689 (Research Support, Non-U.S. Gov't).

Brunet, E., Simsek, D., Tomishima, M., DeKelver, R., Choi, V. M., Gregory, P., et al. (2009). Chromosomal translocations induced at specified loci in human stem cells. *Proceedings of the National Academy of Sciences of the United States of America*, *106*(26), 10620–10625. http://dx.doi.org/10.1073/pnas.0902076106 (Research Support, N.I.H., Extramural Research Support, Non-U.S. Gov't).

Cambier, C. J., Takaki, K. K., Larson, R. P., Hernandez, R. E., Tobin, D. M., Urdahl, K. B., et al. (2014). Mycobacteria manipulate macrophage recruitment through coordinated use of membrane lipids. *Nature*, *505*(7482), 218–222. http://dx.doi.org/10.1038/nature12799 (Research Support, N.I.H., Extramural Research Support, Non-U.S. Gov't Research Support, U.S. Gov't, Non-P.H.S.).

Campbell, J. M., Hartjes, K. A., Nelson, T. J., Xu, X., & Ekker, S. C. (2013). New and TALENted genome engineering toolbox. *Circulation Research*, *113*(5), 571–587. http://dx.doi.org/10.1161/CIRCRESAHA.113.301765 (Research Support, N.I.H., Extramural Research Support, Non-U.S. Gov't Review).

Carbery, I. D., Ji, D., Harrington, A., Brown, V., Weinstein, E. J., Liaw, L., et al. (2010). Targeted genome modification in mice using zinc-finger nucleases. *Genetics*, *186*(2), 451–459. http://dx.doi.org/10.1534/genetics.110.117002.

Carlson, D. F., Tan, W., Lillico, S. G., Stverakova, D., Proudfoot, C., Christian, M., et al. (2012). Efficient TALEN-mediated gene knockout in livestock. *Proceedings of the National Academy of Sciences of the United States of America*, *109*(43), 17382–17387. http://dx.doi.org/10.1073/pnas.1211446109 (Research Support, N.I.H., Extramural Research Support, Non-U.S. Gov't).

Ceol, C. J., Houvras, Y., Jane-Valbuena, J., Bilodeau, S., Orlando, D. A., Battisti, V., et al. (2011). The histone methyltransferase SETDB1 is recurrently amplified in melanoma and accelerates its onset. *Nature*, *471*(7339), 513–517. http://dx.doi.org/10.1038/nature09806 (Research Support, N.I.H., Extramural Research Support, Non-U.S. Gov't).

Chang, N., Sun, C., Gao, L., Zhu, D., Xu, X., Zhu, X., et al. (2013). Genome editing with RNA-guided Cas9 nuclease in zebrafish embryos. *Cell Research*, *23*(4), 465–472. http://dx.doi.org/10.1038/cr.2013.45 (Research Support, Non-U.S. Gov't).

Chen, B., Gilbert, L. A., Cimini, B. A., Schnitzbauer, J., Zhang, W., Li, G. W., et al. (2013). Dynamic imaging of genomic loci in living human cells by an optimized CRISPR/Cas system. *Cell*, *155*(7), 1479–1491. http://dx.doi.org/10.1016/j.cell.2013.12.001 (Research Support, N.I.H., Extramural Research Support, Non-U.S. Gov't).

Chen, F., Pruett-Miller, S. M., Huang, Y., Gjoka, M., Duda, K., Taunton, J., et al. (2011). High-frequency genome editing using ssDNA oligonucleotides with zinc-finger nucleases. *Nature Methods*, *8*(9), 753–755. http://dx.doi.org/10.1038/nmeth.1653 (Research Support, N.I.H., Extramural Research Support, Non-U.S. Gov't).

Cho, S. W., Kim, S., Kim, J. M., & Kim, J. S. (2013). Targeted genome engineering in human cells with the Cas9 RNA-guided endonuclease. *Nature Biotechnology*, *31*(3), 230–232. http://dx.doi.org/10.1038/nbt.2507 (Research Support, Non-U.S. Gov't).

Cho, S. W., Kim, S., Kim, Y., Kweon, J., Kim, H. S., Bae, S., et al. (2014). Analysis of off-target effects of CRISPR/Cas-derived RNA-guided endonucleases and nickases. *Genome Research*, *24*(1), 132–141. http://dx.doi.org/10.1101/gr.162339.113 (Research Support, Non-U.S. Gov't).

Choi, P. S., & Meyerson, M. (2014). Targeted genomic rearrangements using CRISPR/Cas technology. *Nature Communications*, *5*, 3728. http://dx.doi.org/10.1038/ncomms4728 (Research Support, N.I.H., Extramural Research Support, U.S. Gov't, Non-P.H.S.).

Chylinski, K., Makarova, K. S., Charpentier, E., & Koonin, E. V. (2014). Classification and evolution of type II CRISPR-Cas systems. *Nucleic Acids Research*, *42*(10), 6091–6105. http://dx.doi.org/10.1093/nar/gku241.

Cong, L., Ran, F. A., Cox, D., Lin, S., Barretto, R., Habib, N., et al. (2013). Multiplex genome engineering using CRISPR/Cas systems. *Science*, *339*(6121), 819–823. http://dx.doi.org/10.1126/science.1231143 (Research Support, N.I.H., Extramural Research Support, Non-U.S. Gov't).

Cornu, T. I., Thibodeau-Beganny, S., Guhl, E., Alwin, S., Eichtinger, M., Joung, J. K., et al. (2008). DNA-binding specificity is a major determinant of the activity and toxicity of zinc-finger nucleases. *Molecular Therapy*, *16*(2), 352–358. http://dx.doi.org/10.1038/sj.mt.6300357 (Research Support, N.I.H., Extramural Research Support, Non-U.S. Gov't).

Cradick, T. J., Fine, E. J., Antico, C. J., & Bao, G. (2013). CRISPR/Cas9 systems targeting beta-globin and CCR5 genes have substantial off-target activity. *Nucleic Acids Research*, *41*(20), 9584–9592. http://dx.doi.org/10.1093/nar/gkt714 (Research Support, N.I.H., Extramural Research Support, Non-U.S. Gov't).

Dahlem, T. J., Hoshijima, K., Jurynec, M. J., Gunther, D., Starker, C. G., Locke, A. S., et al. (2012). Simple methods for generating and detecting locus-specific mutations induced with TALENs in the zebrafish genome. *PLoS Genetics*, *8*(8), e1002861. http://dx.doi.org/10.1371/journal.pgen.1002861 (Research Support, N.I.H., Extramural Research Support, Non-U.S. Gov't).

Dai, J., Cui, X., Zhu, Z., & Hu, W. (2010). Non-homologous end joining plays a key role in transgene concatemer formation in transgenic zebrafish embryos. *International Journal of Biological Sciences*, *6*(7), 756–768 (Research Support, Non-U.S. Gov't).

Deltcheva, E., Chylinski, K., Sharma, C. M., Gonzales, K., Chao, Y., Pirzada, Z. A., et al. (2011). CRISPR RNA maturation by trans-encoded small RNA and host factor RNase III. *Nature*, *471*(7340), 602–607. http://dx.doi.org/10.1038/nature09886 (Research Support, Non-U.S. Gov't).

Deng, W., Lee, J., Wang, H., Miller, J., Reik, A., Gregory, P. D., et al. (2012). Controlling long-range genomic interactions at a native locus by targeted tethering of a looping factor. *Cell*, *149*(6), 1233–1244. http://dx.doi.org/10.1016/j.cell.2012.03.051 (Research Support, N.I.H., Extramural Research Support, N.I.H., Intramural).

DiCarlo, J. E., Norville, J. E., Mali, P., Rios, X., Aach, J., & Church, G. M. (2013). Genome engineering in saccharomyces cerevisiae using CRISPR-Cas systems. *Nucleic Acids Research*, *41*(7), 4336–4343. http://dx.doi.org/10.1093/nar/gkt135 (Research Support, N.I.H., Extramural Research Support, U.S. Gov't, Non-P.H.S.).

Ding, Q., Regan, S. N., Xia, Y., Oostrom, L. A., Cowan, C. A., & Musunuru, K. (2013). Enhanced efficiency of human pluripotent stem cell genome editing through replacing TALENs with CRISPRs. *Cell Stem Cell*, *12*(4), 393–394. http://dx.doi.org/10.1016/j.stem.2013.03.006 (Letter Research Support, N.I.H., Extramural Research Support, Non-U.S. Gov't).

Esvelt, K. M., Carlson, J. C., & Liu, D. R. (2011). A system for the continuous directed evolution of biomolecules. *Nature*, *472*(7344), 499–503. http://dx.doi.org/10.1038/nature09929 (Research Support, N.I.H., Extramural Research Support, Non-U.S. Gov't).

Esvelt, K. M., Mali, P., Braff, J. L., Moosburner, M., Yaung, S. J., & Church, G. M. (2013). Orthogonal Cas9 proteins for RNA-guided gene regulation and editing. *Nature Methods*, *10*(11), 1116–1121. http://dx.doi.org/10.1038/nmeth.2681 (Research Support, N.I.H., Extramural Research Support, U.S. Gov't, Non-P.H.S.).

Friedland, A. E., Tzur, Y. B., Esvelt, K. M., Colaiacovo, M. P., Church, G. M., & Calarco, J. A. (2013). Heritable genome editing in C. elegans via a CRISPR-Cas9 system. *Nature Methods*, *10*(8), 741–743. http://dx.doi.org/10.1038/nmeth.2532 (Research Support, N.I.H., Extramural Research Support, Non-U.S. Gov't).

Fu, Y., Foden, J. A., Khayter, C., Maeder, M. L., Reyon, D., Joung, J. K., et al. (2013). High-frequency off-target mutagenesis induced by CRISPR-Cas nucleases in human cells. *Nature Biotechnology*, *31*(9), 822–826. http://dx.doi.org/10.1038/nbt.2623 (Research Support, N.I.H., Extramural Research Support, Non-U.S. Gov't Research Support, U.S. Gov't, Non-P.H.S.).

Fu, Y., Sander, J. D., Reyon, D., Cascio, V. M., & Joung, J. K. (2014). Improving CRISPR-Cas nuclease specificity using truncated guide RNAs. *Nature Biotechnology*, *32*(3), 279–284. http://dx.doi.org/10.1038/nbt.2808 (Research Support, N.I.H., Extramural Research Support, Non-U.S. Gov't Research Support, U.S. Gov't, Non-P.H.S.).

Gagnon, J. A., Valen, E., Thyme, S. B., Huang, P., Ahkmetova, L., Pauli, A., et al. (2014). Efficient mutagenesis by Cas9 protein-mediated oligonucleotide insertion and large-scale assessment of single-guide RNAs. *PLoS One*, *9*(5), e98186. http://dx.doi.org/10.1371/journal.pone.0098186.

Garneau, J. E., Dupuis, M. E., Villion, M., Romero, D. A., Barrangou, R., Boyaval, P., et al. (2010). The CRISPR/Cas bacterial immune system cleaves bacteriophage and plasmid DNA. *Nature*, *468*(7320), 67–71. http://dx.doi.org/10.1038/nature09523 (Research Support, Non-U.S. Gov't).

Gasiunas, G., Barrangou, R., Horvath, P., & Siksnys, V. (2012). Cas9-crRNA ribonucleoprotein complex mediates specific DNA cleavage for adaptive immunity in bacteria. *Proceedings of the National Academy of Sciences of the United States of America*, *109*(39), E2579–E2586. http://dx.doi.org/10.1073/pnas.1208507109 (Research Support, Non-U.S. Gov't).

Gilbert, L. A., Larson, M. H., Morsut, L., Liu, Z., Brar, G. A., Torres, S. E., et al. (2013). CRISPR-mediated modular RNA-guided regulation of transcription in eukaryotes. *Cell*, *154*(2), 442–451. http://dx.doi.org/10.1016/j.cell.2013.06.044 (Research Support, N.I.H., Extramural Research Support, Non-U.S. Gov't Research Support, U.S. Gov't, Non-P.H.S.).

Grissa, I., Vergnaud, G., & Pourcel, C. (2007a). The CRISPRdb database and tools to display CRISPRs and to generate dictionaries of spacers and repeats. *BMC Bioinformatics*, *8*, 172. http://dx.doi.org/10.1186/1471-2105-8-172.

Grissa, I., Vergnaud, G., & Pourcel, C. (2007b). CRISPRFinder: A web tool to identify clustered regularly interspaced short palindromic repeats. *Nucleic Acids Research*, *35*(Web Server issue), W52–W57. http://dx.doi.org/10.1093/nar/gkm360 (Research Support, Non-U.S. Gov't).

Guilinger, J. P., Thompson, D. B., & Liu, D. R. (2014). Fusion of catalytically inactive Cas9 to FokI nuclease improves the specificity of genome modification. *Nature Biotechnology*, *32*(6), 577–582. http://dx.doi.org/10.1038/nbt.2909.

Guo, X., Zhang, T., Hu, Z., Zhang, Y., Shi, Z., Wang, Q., et al. (2014). Efficient RNA/Cas9-mediated genome editing in xenopus tropicalis. *Development*, *141*(3), 707–714. http://dx.doi.org/10.1242/dev.099853 (Research Support, Non-U.S. Gov't).

Gupta, A., Hall, V. L., Kok, F. O., Shin, M., McNulty, J. C., Lawson, N. D., et al. (2013). Targeted chromosomal deletions and inversions in zebrafish. *Genome Research*, *23*(6), 1008–1017. http://dx.doi.org/10.1101/gr.154070.112 (Research Support, N.I.H., Extramural).

Guschin, D. Y., Waite, A. J., Katibah, G. E., Miller, J. C., Holmes, M. C., & Rebar, E. J. (2010). A rapid and general assay for monitoring endogenous gene modification. *Methods in Molecular Biology*, *649*, 247–256. http://dx.doi.org/10.1007/978-1-60761-753-2_15.

Hagmann, M., Bruggmann, R., Xue, L., Georgiev, O., Schaffner, W., Rungger, D., et al. (1998). Homologous recombination and DNA-end joining reactions in zygotes and early embryos of zebrafish (danio rerio) and drosophila melanogaster. *Biological Chemistry*, *379*(6), 673–681 (Research Support, Non-U.S. Gov't).

Helenius, I. T., & Yeh, J. R. (2012). Small zebrafish in a big chemical pond. *Journal of Cellular Biochemistry*, *113*(7), 2208–2216. http://dx.doi.org/10.1002/jcb.24120 (Research Support, N.I.H., Extramural Research Support, Non-U.S. Gov't Review).

Helm, M., Brule, H., Giege, R., & Florentz, C. (1999). More mistakes by T7 RNA polymerase at the 5' ends of in vitro-transcribed RNAs. *RNA*, *5*(5), 618–621 (Letter Research Support, Non-U.S. Gov't).

Horvath, P., & Barrangou, R. (2010). CRISPR/Cas, the immune system of bacteria and archaea. *Science*, *327*(5962), 167–170. http://dx.doi.org/10.1126/science.1179555 (Research Support, Non-U.S. Gov't Review).

Horvath, P., Romero, D. A., Coute-Monvoisin, A. C., Richards, M., Deveau, H., Moineau, S., et al. (2008). Diversity, activity, and evolution of CRISPR loci in streptococcus thermophilus. *Journal of Bacteriology*, *190*(4), 1401–1412. http://dx.doi.org/10.1128/JB.01415-07 (Research Support, Non-U.S. Gov't).

Hruscha, A., Krawitz, P., Rechenberg, A., Heinrich, V., Hecht, J., Haass, C., et al. (2013). Efficient CRISPR/Cas9 genome editing with low off-target effects in zebrafish. *Development*, *140*(24), 4982–4987. http://dx.doi.org/10.1242/dev.099085 (Research Support, Non-U.S. Gov't).

Hsu, P. D., Scott, D. A., Weinstein, J. A., Ran, F. A., Konermann, S., Agarwala, V., et al. (2013). DNA targeting specificity of RNA-guided Cas9 nucleases. *Nature Biotechnology*, *31*(9), 827–832. http://dx.doi.org/10.1038/nbt.2647 (Research Support, N.I.H., Extramural Research Support, Non-U.S. Gov't).

Hwang, W. Y., Fu, Y., Reyon, D., Maeder, M. L., Kaini, P., Sander, J. D., et al. (2013). Heritable and precise zebrafish genome editing using a CRISPR-Cas system. *PLoS One*, *8*(7), e68708. http://dx.doi.org/10.1371/journal.pone.0068708 (Research Support, N.I.H., Extramural Research Support, Non-U.S. Gov't Research Support, U.S. Gov't, Non-P.H.S.).

Hwang, W. Y., Fu, Y., Reyon, D., Maeder, M. L., Tsai, S. Q., Sander, J. D., et al. (2013). Efficient genome editing in zebrafish using a CRISPR-Cas system. *Nature Biotechnology*, *31*(3), 227–229. http://dx.doi.org/10.1038/nbt.2501 (Research Support, N.I.H., Extramural Research Support, Non-U.S. Gov't).

Hwang, W. Y., Peterson, R. T., & Yeh, J. R. (2014). Methods for targeted mutagenesis in zebrafish using TALENs. *Methods*. http://dx.doi.org/10.1016/j.ymeth.2014.04.009.

Ignatius, M. S., & Langenau, D. M. (2011). Fluorescent imaging of cancer in zebrafish. *Methods in Cell Biology*, *105*, 437–459. http://dx.doi.org/10.1016/B978-0-12-381320-6.00019-9 (Research Support, N.I.H., Extramural).

Imburgio, D., Rong, M., Ma, K., & McAllister, W. T. (2000). Studies of promoter recognition and start site selection by T7 RNA polymerase using a comprehensive collection of promoter variants. *Biochemistry*, *39*(34), 10419–10430 (Research Support, U.S. Gov't, P.H.S.).

Ishino, Y., Shinagawa, H., Makino, K., Amemura, M., & Nakata, A. (1987). Nucleotide sequence of the iap gene, responsible for alkaline phosphatase isozyme conversion in Escherichia coli, and identification of the gene product. *Journal of Bacteriology*, *169*(12), 5429–5433 (Research Support, Non-U.S. Gov't).

Jansen, R., Embden, J. D., Gaastra, W., & Schouls, L. M. (2002). Identification of genes that are associated with DNA repeats in prokaryotes. *Molecular Microbiology*, *43*(6), 1565–1575 (Research Support, Non-U.S. Gov't).

Jao, L. E., Wente, S. R., & Chen, W. (2013). Efficient multiplex biallelic zebrafish genome editing using a CRISPR nuclease system. *Proceedings of the National Academy of Sciences of the United States of America*, *110*(34), 13904–13909. http://dx.doi.org/10.1073/pnas.1308335110 (Research Support, N.I.H., Extramural Research Support, Non-U.S. Gov't).

Jiang, W., Bikard, D., Cox, D., Zhang, F., & Marraffini, L. A. (2013). RNA-guided editing of bacterial genomes using CRISPR-Cas systems. *Nature Biotechnology*, *31*(3), 233–239. http://dx.doi.org/10.1038/nbt.2508 (Research Support, N.I.H., Extramural Research Support, Non-U.S. Gov't).

Jinek, M., Chylinski, K., Fonfara, I., Hauer, M., Doudna, J. A., & Charpentier, E. (2012). A programmable dual-RNA-guided DNA endonuclease in adaptive bacterial immunity. *Science*, *337*(6096), 816–821. http://dx.doi.org/10.1126/science.1225829 (Research Support, Non-U.S. Gov't).

Jinek, M., East, A., Cheng, A., Lin, S., Ma, E., & Doudna, J. (2013). RNA-programmed genome editing in human cells. *elife*, *2*, e00471. http://dx.doi.org/10.7554/eLife.00471.

Joung, J. K., & Sander, J. D. (2013). TALENs: A widely applicable technology for targeted genome editing. *Nature Reviews Molecular Cell Biology*, *14*(1), 49–55. http://dx.doi.org/10.1038/nrm3486 (Research Support, N.I.H., Extramural Review).

Kari, G., Rodeck, U., & Dicker, A. P. (2007). Zebrafish: An emerging model system for human disease and drug discovery. *Clinical Pharmacology and Therapeutics*, *82*(1), 70–80. http://dx.doi.org/10.1038/sj.clpt.6100223 (Research Support, N.I.H., Extramural Research Support, Non-U.S. Gov't Review).

Kawahara, G., Karpf, J. A., Myers, J. A., Alexander, M. S., Guyon, J. R., & Kunkel, L. M. (2011). Drug screening in a zebrafish model of duchenne muscular dystrophy. *Proceedings of the National Academy of Sciences of the United States of America*, *108*(13), 5331–5336. http://dx.doi.org/10.1073/pnas.1102116108 (Research Support, N.I.H., Extramural Research Support, Non-U.S. Gov't).

Kayali, R., Bury, F., Ballard, M., & Bertoni, C. (2010). Site-directed gene repair of the dystrophin gene mediated by PNA-ssODNs. *Human Molecular Genetics*, *19*(16), 3266–3281. http://dx.doi.org/10.1093/hmg/ddq235 (Research Support, Non-U.S. Gov't).

Kettleborough, R. N., Busch-Nentwich, E. M., Harvey, S. A., Dooley, C. M., de Bruijn, E., van Eeden, F., et al. (2013). A systematic genome-wide analysis of zebrafish protein-coding gene function. *Nature*, *496*(7446), 494–497. http://dx.doi.org/10.1038/nature11992 (Research Support, N.I.H., Extramural Research Support, Non-U.S. Gov't).

Kunin, V., Sorek, R., & Hugenholtz, P. (2007). Evolutionary conservation of sequence and secondary structures in CRISPR repeats. *Genome Biology*, *8*(4), R61. http://dx.doi.org/10.1186/gb-2007-8-4-r61 (Research Support, U.S. Gov't, Non-P.H.S.).

Kuzmine, I., Gottlieb, P. A., & Martin, C. T. (2003). Binding of the priming nucleotide in the initiation of transcription by T7 RNA polymerase. *The Journal of Biological Chemistry*, *278*(5), 2819–2823. http://dx.doi.org/10.1074/jbc.M208405200 (Research Support, U.S. Gov't, Non-P.H.S. Research Support, U.S. Gov't, P.H.S.).

Labrie, S. J., Samson, J. E., & Moineau, S. (2010). Bacteriophage resistance mechanisms. *Nature Reviews Microbiology*, *8*(5), 317–327. http://dx.doi.org/10.1038/nrmicro2315 (Research Support, Non-U.S. Gov't Review).

Lamb, B. M., Mercer, A. C., & Barbas, C. F., 3rd. (2013). Directed evolution of the TALE N-terminal domain for recognition of all 5' bases. *Nucleic Acids Research*, *41*(21), 9779–9785. http://dx.doi.org/10.1093/nar/gkt754 (Research Support, N.I.H., Extramural).

Lee, H. J., Kim, E., & Kim, J. S. (2010). Targeted chromosomal deletions in human cells using zinc finger nucleases. *Genome Research*, *20*(1), 81–89. http://dx.doi.org/10.1101/gr.099747.109 (Evaluation Studies Research Support, Non-U.S. Gov't).

Lee, H. J., Kweon, J., Kim, E., Kim, S., & Kim, J. S. (2012). Targeted chromosomal duplications and inversions in the human genome using zinc finger nucleases. *Genome Research*, *22*(3), 539–548. http://dx.doi.org/10.1101/gr.129635.111 (Research Support, Non-U.S. Gov't).

Li, T., Liu, B., Spalding, M. H., Weeks, D. P., & Yang, B. (2012). High-efficiency TALEN-based gene editing produces disease-resistant rice. *Nature Biotechnology*, *30*(5), 390–392. http://dx.doi.org/10.1038/nbt.2199 (Letter Research Support, U.S. Gov't, Non-P.H.S.).

Lieschke, G. J., & Currie, P. D. (2007). Animal models of human disease: Zebrafish swim into view. *Nature Reviews Genetics*, *8*(5), 353–367. http://dx.doi.org/10.1038/nrg2091 (Research Support, N.I.H., Extramural Research Support, Non-U.S. Gov't Review).

Liu, J., Gong, L., Chang, C., Liu, C., Peng, J., & Chen, J. (2012). Development of novel visual-plus quantitative analysis systems for studying DNA double-strand break repairs in zebrafish. *Journal of Genetics and Genomics*, *39*(9), 489–502. http://dx.doi.org/10.1016/j.jgg.2012.07.009 (Research Support, Non-U.S. Gov't).

Lu, J., Peatman, E., Tang, H., Lewis, J., & Liu, Z. (2012). Profiling of gene duplication patterns of sequenced teleost genomes: Evidence for rapid lineage-specific genome expansion mediated by recent tandem duplications. *BMC Genomics*, *13*, 246. http://dx.doi.org/10.1186/1471-2164-13-246 (Research Support, Non-U.S. Gov't Research Support, U.S. Gov't, Non-P.H.S.).

Ma, S., Chang, J., Wang, X., Liu, Y., Zhang, J., Lu, W., et al. (2014). CRISPR/Cas9 mediated multiplex genome editing and heritable mutagenesis of BmKu70 in bombyx Mori. *Scientific Reports*, *4*, 4489. http://dx.doi.org/10.1038/srep04489 (Research Support, Non-U.S. Gov't).

Ma, Y., Shen, B., Zhang, X., Lu, Y., Chen, W., Ma, J., et al. (2014). Heritable multiplex genetic engineering in rats using CRISPR/Cas9. *PLoS One*, *9*(3), e89413. http://dx.doi.org/10.1371/journal.pone.0089413 (Research Support, Non-U.S. Gov't).

Ma, S., Zhang, S., Wang, F., Liu, Y., Xu, H., Liu, C., et al. (2012). Highly efficient and specific genome editing in silkworm using custom TALENs. *PLoS One*, 7(9), e45035. http://dx.doi.org/10.1371/journal.pone.0045035 (Research Support, Non-U.S. Gov't).

Maeder, M. L., Angstman, J. F., Richardson, M. E., Linder, S. J., Cascio, V. M., Tsai, S. Q., et al. (2013). Targeted DNA demethylation and activation of endogenous genes using programmable TALE-TET1 fusion proteins. *Nature Biotechnology*, *31*(12), 1137–1142. http://dx.doi.org/10.1038/nbt.2726 (Research Support, N.I.H., Extramural Research Support, Non-U.S. Gov't Research Support, U.S. Gov't, Non-P.H.S.).

Maeder, M. L., Thibodeau-Beganny, S., Osiak, A., Wright, D. A., Anthony, R. M., Eichtinger, M., et al. (2008). Rapid "open-source" engineering of customized zinc-finger nucleases for highly efficient gene modification. *Molecular Cell*, *31*(2), 294–301. http://dx.doi.org/10.1016/j.molcel.2008.06.016 (Research Support, N.I.H., Extramural Research Support, Non-U.S. Gov't Research Support, U.S. Gov't, Non-P.H.S.).

Makarova, K. S., Aravind, L., Wolf, Y. I., & Koonin, E. V. (2011). Unification of Cas protein families and a simple scenario for the origin and evolution of CRISPR-Cas systems. *Biology Direct*, *6*, 38. http://dx.doi.org/10.1186/1745-6150-6-38 (Comparative Study Research Support, N.I.H., Intramural).

Makarova, K. S., Grishin, N. V., Shabalina, S. A., Wolf, Y. I., & Koonin, E. V. (2006). A putative RNA-interference-based immune system in prokaryotes: Computational analysis of the predicted enzymatic machinery, functional analogies with eukaryotic RNAi, and hypothetical mechanisms of action. *Biology Direct*, *1*, 7. http://dx.doi.org/10.1186/1745-6150-1-7.

Makarova, K. S., Haft, D. H., Barrangou, R., Brouns, S. J., Charpentier, E., Horvath, P., et al. (2011). Evolution and classification of the CRISPR-Cas systems. *Nature Reviews Microbiology*, *9*(6), 467–477. http://dx.doi.org/10.1038/nrmicro2577 (Research Support, N.I.H., Extramural Research Support, Non-U.S. Gov't).

Mali, P., Aach, J., Stranges, P. B., Esvelt, K. M., Moosburner, M., Kosuri, S., et al. (2013). CAS9 transcriptional activators for target specificity screening and paired nickases for cooperative genome engineering. *Nature Biotechnology*, *31*(9), 833–838. http://dx.doi.

org/10.1038/nbt.2675 (Research Support, N.I.H., Extramural Research Support, U.S. Gov't, Non-P.H.S.).

Mali, P., Yang, L., Esvelt, K. M., Aach, J., Guell, M., DiCarlo, J. E., et al. (2013). RNA-guided human genome engineering via Cas9. *Science*, *339*(6121), 823–826. http://dx.doi.org/10.1126/science.1232033 (Research Support, N.I.H., Extramural).

Marraffini, L. A., & Sontheimer, E. J. (2010a). CRISPR interference: RNA-directed adaptive immunity in bacteria and archaea. *Nature Reviews Genetics*, *11*(3), 181–190. http://dx.doi.org/10.1038/nrg2749 (Research Support, N.I.H., Extramural Research Support, Non-U.S. Gov't Review).

Marraffini, L. A., & Sontheimer, E. J. (2010b). Self versus non-self discrimination during CRISPR RNA-directed immunity. *Nature*, *463*(7280), 568–571. http://dx.doi.org/10.1038/nature08703 (Research Support, N.I.H., Extramural).

Mojica, F. J., Diez-Villasenor, C., Garcia-Martinez, J., & Soria, E. (2005). Intervening sequences of regularly spaced prokaryotic repeats derive from foreign genetic elements. *Journal of Molecular Evolution*, *60*(2), 174–182. http://dx.doi.org/10.1007/s00239-004-0046-3 (Research Support, Non-U.S. Gov't).

Moro, E., Vettori, A., Porazzi, P., Schiavone, M., Rampazzo, E., Casari, A., et al. (2013). Generation and application of signaling pathway reporter lines in zebrafish. *Molecular Genetics and Genomics*, *288*(5–6), 231–242. http://dx.doi.org/10.1007/s00438-013-0750-z (Research Support, Non-U.S. Gov't Review).

Nakayama, T., Fish, M. B., Fisher, M., Oomen-Hajagos, J., Thomsen, G. H., & Grainger, R. M. (2013). Simple and efficient CRISPR/Cas9-mediated targeted mutagenesis in xenopus tropicalis. *Genesis*, *51*(12), 835–843. http://dx.doi.org/10.1002/dvg.22720 (Research Support, N.I.H., Extramural Research Support, Non-U.S. Gov't).

Niu, Y., Shen, B., Cui, Y., Chen, Y., Wang, J., Wang, L., et al. (2014). Generation of gene-modified cynomolgus monkey via Cas9/RNA-mediated gene targeting in one-cell embryos. *Cell*, *156*(4), 836–843. http://dx.doi.org/10.1016/j.cell.2014.01.027 (Research Support, Non-U.S. Gov't).

O'Brien, A., & Bailey, T. L. (2014). GT-scan: Identifying unique genomic targets. *Bioinformatics*. http://dx.doi.org/10.1093/bioinformatics/btu354.

Ota, S., Hisano, Y., Ikawa, Y., & Kawahara, A. (2014). Multiple genome modifications by the CRISPR/Cas9 system in zebrafish. *Genes to Cells*. http://dx.doi.org/10.1111/gtc.12154.

Pantazis, P., & Supatto, W. (2014). Advances in whole-embryo imaging: A quantitative transition is underway. *Nature Reviews Molecular Cell Biology*, *15*(5), 327–339. http://dx.doi.org/10.1038/nrm3786 (Review).

Pattanayak, V., Lin, S., Guilinger, J. P., Ma, E., Doudna, J. A., & Liu, D. R. (2013). High-throughput profiling of off-target DNA cleavage reveals RNA-programmed Cas9 nuclease specificity. *Nature Biotechnology*, *31*(9), 839–843. http://dx.doi.org/10.1038/nbt.2673 (Research Support, N.I.H., Extramural Research Support, Non-U.S. Gov't Research Support, U.S. Gov't, Non-P.H.S.).

Patton, E. E., Widlund, H. R., Kutok, J. L., Kopani, K. R., Amatruda, J. F., Murphey, R. D., et al. (2005). BRAF mutations are sufficient to promote nevi formation and cooperate with p53 in the genesis of melanoma. *Current Biology*, *15*(3), 249–254. http://dx.doi.org/10.1016/j.cub.2005.01.031 (Comparative Study Research Support, Non-U.S. Gov't).

Peal, D. S., Peterson, R. T., & Milan, D. (2010). Small molecule screening in zebrafish. *Journal of Cardiovascular Translational Research*, *3*(5), 454–460. http://dx.doi.org/10.1007/s12265-010-9212-8 (Review).

Piganeau, M., Ghezraoui, H., De Cian, A., Guittat, L., Tomishima, M., Perrouault, L., et al. (2013). Cancer translocations in human cells induced by zinc finger and TALE nucleases. *Genome Research*, *23*(7), 1182–1193. http://dx.doi.org/10.1101/gr.147314.112 (Research Support, N.I.H., Extramural Research Support, Non-U.S. Gov't).

Qi, Y., Li, X., Zhang, Y., Starker, C. G., Baltes, N. J., Zhang, F., et al. (2013). Targeted deletion and inversion of tandemly arrayed genes in Arabidopsis thaliana using zinc finger nucleases. *G3 (Bethesda)*, *3*(10), 1707–1715. http://dx.doi.org/10.1534/g3.113.006270 (Research Support, N.I.H., Extramural Research Support, U.S. Gov't, Non-P.H.S.).

Ramirez, C. L., Foley, J. E., Wright, D. A., Muller-Lerch, F., Rahman, S. H., Cornu, T. I., et al. (2008). Unexpected failure rates for modular assembly of engineered zinc fingers. *Nature Methods*, *5*(5), 374–375. http://dx.doi.org/10.1038/nmeth0508-374 (Letter Research Support, N.I.H., Extramural Research Support, Non-U.S. Gov't Research Support, U.S. Gov't, Non-P.H.S.).

Ran, F. A., Hsu, P. D., Lin, C. Y., Gootenberg, J. S., Konermann, S., Trevino, A. E., et al. (2013). Double nicking by RNA-guided CRISPR Cas9 for enhanced genome editing specificity. *Cell*, *154*(6), 1380–1389. http://dx.doi.org/10.1016/j.cell.2013.08.021 (Research Support, N.I.H., Extramural Research Support, Non-U.S. Gov't Research Support, U.S. Gov't, Non-P.H.S.).

Rousseau, C., Gonnet, M., Le Romancer, M., & Nicolas, J. (2009). CRISPI: A CRISPR interactive database. *Bioinformatics*, *25*(24), 3317–3318. http://dx.doi.org/10.1093/bioinformatics/btp586 (Research Support, Non-U.S. Gov't).

Sander, J. D., & Joung, J. K. (2014). CRISPR-Cas systems for editing, regulating and targeting genomes. *Nature Biotechnology*, *32*(4), 347–355. http://dx.doi.org/10.1038/nbt.2842 (Research Support, N.I.H., Extramural Research Support, Non-U.S. Gov't Research Support, U.S. Gov't, Non-P.H.S.).

Santoriello, C., & Zon, L. I. (2012). Hooked! modeling human disease in zebrafish. *The Journal of Clinical Investigation*, *122*(7), 2337–2343. http://dx.doi.org/10.1172/JCI60434.

Sapranauskas, R., Gasiunas, G., Fremaux, C., Barrangou, R., Horvath, P., & Siksnys, V. (2011). The Streptococcus thermophilus CRISPR/Cas system provides immunity in Escherichia coli. *Nucleic Acids Research*, *39*(21), 9275–9282. http://dx.doi.org/10.1093/nar/gkr606 (Research Support, Non-U.S. Gov't).

Schwank, G., Koo, B. K., Sasselli, V., Dekkers, J. F., Heo, I., Demircan, T., et al. (2013). Functional repair of CFTR by CRISPR/Cas9 in intestinal stem cell organoids of cystic fibrosis patients. *Cell Stem Cell*, *13*(6), 653–658. http://dx.doi.org/10.1016/j.stem.2013.11.002 (Research Support, Non-U.S. Gov't).

Sebastiano, V., Maeder, M. L., Angstman, J. F., Haddad, B., Khayter, C., Yeo, D. T., et al. (2011). In situ genetic correction of the sickle cell anemia mutation in human induced pluripotent stem cells using engineered zinc finger nucleases. *Stem Cells*, *29*(11), 1717–1726. http://dx.doi.org/10.1002/stem.718 (Research Support, N.I.H., Extramural Research Support, Non-U.S. Gov't Research Support, U.S. Gov't, Non-P.H.S.).

Segal, D. J., & Meckler, J. F. (2013). Genome engineering at the dawn of the golden age. *Annual Review of Genomics and Human Genetics*, *14*, 135–158. http://dx.doi.org/10.1146/annurev-genom-091212-153435 (Research Support, N.I.H., Extramural Research Support, Non-U.S. Gov't Review).

Shan, Q., Wang, Y., Li, J., Zhang, Y., Chen, K., Liang, Z., et al. (2013). Targeted genome modification of crop plants using a CRISPR-Cas system. *Nature Biotechnology*, *31*(8), 686–688. http://dx.doi.org/10.1038/nbt.2650 (Letter Research Support, Non-U.S. Gov't).

Shen, B., Zhang, X., Du, Y., Wang, J., Gong, J., Tate, P. H., et al. (2013). Efficient knockin mouse generation by ssDNA oligonucleotides and zinc-finger nuclease assisted homologous recombination in zygotes. *PLoS One*, *8*(10), e77696. http://dx.doi.org/10.1371/journal.pone.0077696 (Research Support, Non-U.S. Gov't).

Simsek, D., Brunet, E., Wong, S. Y., Katyal, S., Gao, Y., McKinnon, P. J., et al. (2011). DNA ligase III promotes alternative nonhomologous end-joining during chromosomal translocation formation. *PLoS Genetics*, 7(6), e1002080. http://dx.doi.org/10.1371/journal.pgen.1002080 (Research Support, N.I.H., Extramural).

Sollu, C., Pars, K., Cornu, T. I., Thibodeau-Beganny, S., Maeder, M. L., Joung, J. K., et al. (2010). Autonomous zinc-finger nuclease pairs for targeted chromosomal deletion. *Nucleic Acids Research*, *38*(22), 8269–8276. http://dx.doi.org/10.1093/nar/gkq720 (Research Support, N.I.H., Extramural Research Support, Non-U.S. Gov't Research Support, U.S. Gov't, Non-P.H.S.).

Sorek, R., Kunin, V., & Hugenholtz, P. (2008). CRISPR–a widespread system that provides acquired resistance against phages in bacteria and archaea. *Nature Reviews Microbiology*, *6*(3), 181–186. http://dx.doi.org/10.1038/nrmicro1793 (Research Support, U.S. Gov't, Non-P.H.S. Review).

Sternberg, S. H., Redding, S., Jinek, M., Greene, E. C., & Doudna, J. A. (2014). DNA interrogation by the CRISPR RNA-guided endonuclease Cas9. *Nature*, *507*(7490), 62–67. http://dx.doi.org/10.1038/nature13011 (Research Support, N.I.H., Extramural Research Support, Non-U.S. Gov't Research Support, U.S. Gov't, Non-P.H.S.).

Sung, Y. H., Kim, J. M., Kim, H. T., Lee, J., Jeon, J., Jin, Y., et al. (2014). Highly efficient gene knockout in mice and zebrafish with RNA-guided endonucleases. *Genome Research*, *24*(1), 125–131. http://dx.doi.org/10.1101/gr.163394.113 (Research Support, Non-U.S. Gov't).

Swaim, L. E., Connolly, L. E., Volkman, H. E., Humbert, O., Born, D. E., & Ramakrishnan, L. (2006). Mycobacterium marinum infection of adult zebrafish causes caseating granulomatous tuberculosis and is moderated by adaptive immunity. *Infection and Immunity*, *74*(11), 6108–6117. http://dx.doi.org/10.1128/IAI.00887-06 (Research Support, N.I.H., Extramural Research Support, Non-U.S. Gov't).

Tsai, S. Q., Wyvekens, N., Khayter, C., Foden, J. A., Thapar, V., Reyon, D., et al. (2014). Dimeric CRISPR RNA-guided FokI nucleases for highly specific genome editing. *Nature Biotechnology*, *32*(6), 569–576. http://dx.doi.org/10.1038/nbt.2908.

Wang, Y., Li, Z., Xu, J., Zeng, B., Ling, L., You, L., et al. (2013). The CRISPR/Cas system mediates efficient genome engineering in Bombyx Mori. *Cell Research*, *23*(12), 1414–1416. http://dx.doi.org/10.1038/cr.2013.146 (Letter Research Support, Non-U.S. Gov't).

Wang, T., Wei, J. J., Sabatini, D. M., & Lander, E. S. (2014). Genetic screens in human cells using the CRISPR-Cas9 system. *Science*, *343*(6166), 80–84. http://dx.doi.org/10.1126/science.1246981 (Evaluation Studies Research Support, N.I.H., Extramural Research Support, Non-U.S. Gov't Research Support, U.S. Gov't, Non-P.H.S.).

Wang, H., Yang, H., Shivalila, C. S., Dawlaty, M. M., Cheng, A. W., Zhang, F., et al. (2013). One-step generation of mice carrying mutations in multiple genes by CRISPR/Cas-mediated genome engineering. *Cell*, *153*(4), 910–918. http://dx.doi.org/10.1016/j.cell.2013.04.025 (Research Support, N.I.H., Extramural Research Support, Non-U.S. Gov't).

Weber, T., & Koster, R. (2013). Genetic tools for multicolor imaging in zebrafish larvae. *Methods*, *62*(3), 279–291. http://dx.doi.org/10.1016/j.ymeth.2013.07.028 (Review).

Westra, E. R., Swarts, D. C., Staals, R. H., Jore, M. M., Brouns, S. J., & van der Oost, J. (2012). The CRISPRs, they are a-changin': How prokaryotes generate adaptive immunity. *Annual Review of Genetics*, *46*, 311–339. http://dx.doi.org/10.1146/annurev-genet-110711-155447 (Research Support, Non-U.S. Gov't Review).

White, R. M., Cech, J., Ratanasirintrawoot, S., Lin, C. Y., Rahl, P. B., Burke, C. J., et al. (2011). DHODH modulates transcriptional elongation in the neural crest and melanoma. *Nature*, *471*(7339), 518–522. http://dx.doi.org/10.1038/nature09882 (Research Support, N.I.H., Extramural Research Support, Non-U.S. Gov't).

Wiedenheft, B., Sternberg, S. H., & Doudna, J. A. (2012). RNA-guided genetic silencing systems in bacteria and archaea. *Nature*, *482*(7385), 331–338. http://dx.doi.org/

10.1038/nature10886 (Research support, Non-U.S. Gov't research support, U.S. Gov't, Non-P.H.S. Review).

Xiao, A., Cheng, Z., Kong, L., Zhu, Z., Lin, S., & Gao, G. (2014). CasOT: A genome-wide Cas9/gRNA off-target searching tool. *Bioinformatics*. http://dx.doi.org/10.1093/bioinformatics/btt764.

Xiao, A., Wang, Z., Hu, Y., Wu, Y., Luo, Z., Yang, Z., et al. (2013). Chromosomal deletions and inversions mediated by TALENs and CRISPR/Cas in zebrafish. *Nucleic Acids Research*, *41*(14), e141. http://dx.doi.org/10.1093/nar/gkt464 (Research Support, N.I.H., Extramural).

Xu, X., Meiler, S. E., Zhong, T. P., Mohideen, M., Crossley, D. A., Burggren, W. W., et al. (2002). Cardiomyopathy in zebrafish due to mutation in an alternatively spliced exon of titin. *Nature Genetics*, *30*(2), 205–209. http://dx.doi.org/10.1038/ng816 (Research Support, Non-U.S. Gov't Research Support, U.S. Gov't, P.H.S.).

Yang, H., Wang, H., Shivalila, C. S., Cheng, A. W., Shi, L., & Jaenisch, R. (2013). One-step generation of mice carrying reporter and conditional alleles by CRISPR/Cas-mediated genome engineering. *Cell*, *154*(6), 1370–1379. http://dx.doi.org/10.1016/j.cell.2013.08.022.

Yu, Z., Ren, M., Wang, Z., Zhang, B., Rong, Y. S., Jiao, R., et al. (2013). Highly efficient genome modifications mediated by CRISPR/Cas9 in drosophila. *Genetics*, *195*(1), 289–291. http://dx.doi.org/10.1534/genetics.113.153825 (Research Support, N.I.H., Intramural Research Support, Non-U.S. Gov't).

Zu, Y., Tong, X., Wang, Z., Liu, D., Pan, R., Li, Z., et al. (2013). TALEN-mediated precise genome modification by homologous recombination in zebrafish. *Nature Methods*, *10*(4), 329–331. http://dx.doi.org/10.1038/nmeth.2374 (Research Support, N.I.H., Extramural Research Support, Non-U.S. Gov't).

CHAPTER NINETEEN

# Cas9-Based Genome Editing in *Drosophila*

**Benjamin E. Housden*[,1], Shuailiang Lin*, Norbert Perrimon*[,†,1]**
*Department of Genetics, Harvard Medical School, Boston, Massachusetts, USA
†Howard Hughes Medical Institute, Harvard Medical School, Boston, Massachusetts, USA
[1]Corresponding authors: e-mail address: bhousden@genetics.med.harvard.edu; perrimon@receptor.med.harvard.edu

## Contents



## Abstract

Our ability to modify the *Drosophila* genome has recently been revolutionized by the development of the CRISPR system. The simplicity and high efficiency of this system allows its widespread use for many different applications, greatly increasing the range of genome modification experiments that can be performed. Here, we first discuss some general design principles for genome engineering experiments in *Drosophila* and then present detailed protocols for the production of CRISPR reagents and screening strategies to detect successful genome modification events in both tissue culture cells and animals.

## 1. INTRODUCTION

The development of genome engineering technologies such as zinc finger nucleases (ZFNs), transcription activator-like effector nucleases

*Methods in Enzymology*, Volume 546
ISSN 0076-6879
http://dx.doi.org/10.1016/B978-0-12-801185-0.00019-2

(TALENs), and clustered regularly interspaced short palindromic repeats (CRISPR) has revolutionized our ability to modify endogenous genomic sequences in *Drosophila* both in cultured cells and *in vivo* (Bassett & Liu, 2014; Bassett, Tibbit, Ponting, & Liu, 2013, 2014; Beumer, Bhattacharyya, Bibikova, Trautman, & Carroll, 2006; Beumer & Carroll, 2014; Bottcher et al., 2014; Gratz et al., 2013; Gratz, Wildonger, Harrison, & O'Connor-Giles, 2013; Gratz et al., 2014; Kondo, 2014; Kondo & Ueda, 2013; Ren et al., 2013; Sebo, Lee, Peng, & Guo, 2014; Yu et al., 2013, 2014). ZFNs, TALENs, and CRISPR all function with similar mechanisms whereby a nonspecific nuclease is combined with a sequence-specific DNA binding element to generate a targeted double-strand break (DSB). The DSB is then repaired using either the non-homologous end joining (NHEJ) pathway or the homologous recombination (HR) pathway (Bibikova, Golic, Golic, & Carroll, 2002; Chapman, Taylor, & Boulton, 2012). The generation of a DSB in a coding region and repair by NHEJ can lead to small insertions or deletions (indel mutations) and therefore generate a knockout of a specific gene. In contrast to NHEJ, HR generally uses the homologous chromosome as a template and repairs the DSB with no sequence alterations. However, this mechanism can be exploited by including a donor construct with "arms" homologous to the target region. At some frequency, the donor will be used by the HR machinery as a template instead of the homologous chromosome, leading to a precise modification of the target site (Bottcher et al., 2014). Depending on the nature of the donor construct, this could be an insertion of exogenous sequence (e.g., GFP), introduction of a mutant allele, etc. Such insertions are generally referred to as knock-ins.

Although all three genome engineering technologies have been used successfully to produce genomic changes, CRISPR appears to function with considerably higher efficiency than ZFNs or TALENs (Beumer, Trautman, Christian, et al., 2013; Bibikova et al., 2002; Yu et al., 2013). Furthermore, generation of the required reagents is considerably simpler, making CRISPR the method of choice in most situations. The CRISPR system requires two components. The first is Cas9, a nonspecific nuclease protein, and the second is a single-guide RNA (sgRNA) molecule, which provides sequence specificity by base pairing with the target genomic sequence (Cong et al., 2013; Mali, Yang, et al., 2013). By altering the sequence of the sgRNA, highly specific DSBs can be generated at defined loci.

In order to take advantage of the CRISPR system in *Drosophila*, several factors must be considered and the approach taken must match the experimental goals. For example, depending on the desired genomic modification (gene knockout, precise sequence modification, gene tagging, etc.), the sgRNA target site must be positioned differently. Furthermore, off-target effects, mutation efficiency, use of a donor construct, and method of reagent delivery must all be considered to achieve the intended result with high specificity and efficiency.

## 2. APPLICATIONS AND DESIGN CONSIDERATIONS FOR CRISPR-BASED GENOME EDITING

CRISPR can be used to generate a diverse range of genomic modifications including small random changes, insertions, deletions, and substitutions. In order to achieve the desired outcome, different approaches must be taken and several aspects of sgRNA design should be considered. Here we describe the most common applications and some general approaches to achieve them.

1. *Random mutations at a given target site*: In the absence of a donor construct, DSBs generated with CRISPR will be repaired primarily by NHEJ, leading to small indel mutations at the target site (Chapman et al., 2012). This approach is somewhat limited due to the lack of control over the mutations produced and the small region of sequence affected. Therefore, NHEJ is not the best approach for deletion of large regions of sequence or disruption of poorly characterized elements such as regulatory sequences. NHEJ is however very effective at generating frameshift mutations in coding sequences (Bibikova et al., 2002) and so is the approach of choice for gene disruption.

   By targeting a sgRNA to the coding sequence of the gene of interest, frameshifts can be produced with high efficiency (Bassett et al., 2013; Cong et al., 2013; Gratz, Cummings, et al., 2013; Kondo & Ueda, 2013; Mali, Yang, et al., 2013; Ren et al., 2013; Sebo et al., 2014), leading to truncation of the encoded protein. An optimal sgRNA design for this application would target a genomic site close to the 5′ end of the coding sequence and in an exon common to all transcripts in order to maximize the chance of ablating protein function.
2. *Insertion of exogenous sequences*: In contrast to knockout of a gene via frameshift-inducing indels, insertion of exogenous sequences, such as for generation of GFP-tagged proteins, requires precise sequence alteration. To achieve this, CRISPR must be used in combination with a donor construct. Donor constructs consist of the sequence to be inserted,

flanked on either side by "arms" with sequences homologous to the target site. Once a DSB has been generated, a subset will be repaired by HR using the donor construct as a template and therefore insert the exogenous sequence into the target site (Auer, Duroure, De Cian, Concordet, & Del Bene, 2014; Bassett et al., 2014; Dickinson, Ward, Reiner, & Goldstein, 2013; Gratz et al., 2014; Xue et al., 2014; Yang et al., 2013). For this application, it is unlikely that an sgRNA target site will be available exactly at the point of insertion so sequences should be selected as close to this as possible to maximize efficiency.

Longer homology arms have been associated with higher efficiency of HR but only to a certain extent and no further improvement is seen past ~1 kb (Beumer, Trautman, Mukherjee, & Carroll, 2013; Bottcher et al., 2014; Urnov et al., 2005). We therefore design all homology arms to be roughly 1 kb in length. In addition, knocking down the *ligase4* gene, a component of the NHEJ repair pathway, biases repair toward HR and can improve efficiency of insertion. This method has been shown to be effective at increasing the rate of HR both *in vivo* and in cultured cells (Beumer et al., 2008; Bottcher et al., 2014; Bozas, Beumer, Trautman, & Carroll, 2009; Gratz et al., 2014).

Note that for some applications, such as generation of a point mutant allele, it may be possible to use a single-stranded DNA oligo as the donor, avoiding the need to generate a longer donor construct. However, this approach is limited to very small insertions (Gratz, Cummings, et al., 2013).

3. *Specific deletions and substitutions*: Similar to insertions, the generation of a deletion or substitution requires precise sequence alteration and so a donor construct should be used in combination with CRISPR. To generate a deletion, homology arms should be designed flanking the sequence to be deleted with no additional sequences cloned in between. In this case, the sgRNA target site can be anywhere within the sequence to be deleted.

   Substitutions are produced in a similar manner to insertions with a donor containing the sequence to be inserted flanked by homology arms. The difference in this case is that the homology arms induce recombination at sites that are not directly adjacent but separated by the sequence to be deleted. Again, the sgRNA target should be within the sequence to be replaced.

## 2.1. Selection of sgRNA target sites

One of the advantages of the CRISPR system over other existing genome editing technologies is the relative lack of sequence limitations in the targeted sites. The only requirement is the presence of a PAM sequence at the 3′ end of the target site. For Cas9 derived from *Streptococcus pyogenes* (SpCas9), the optimal PAM sequence is NGG (or NAG although this leads to lower efficiency) (Jiang, Bikard, Cox, Zhang, & Marraffini, 2013), which occurs often throughout the *Drosophila* genome (every 10.4 bp on average). However, in some cases, such as modification of very precise regions, it may be difficult to find an appropriate PAM sequence. In this case, a more distant sgRNA target site can be used with an HR-based approach.

As described above, the target site of sgRNAs should be positioned based on the type of modification desired. Common to all of these approaches, a second consideration in the selection of a suitable sgRNA target site is the possibility of DSB generation at off-target sites. In mammalian systems, several reports suggest that off-target mutations may be a significant issue associated with the use of the CRISPR system (Fu et al., 2013; Hsu et al., 2013; Mali, Aach, et al., 2013; Pattanayak et al., 2013). Likely due in part to the lower complexity of the *Drosophila* genome, off-target events appear to be much less of a concern in this system (Ren et al., 2013). Indeed, no publications have yet reported detection of off-target effects. In addition, the presence of off-target sites may not be a problem for some applications. For example, when performing genome alterations *in vivo*, off-target mutations on nontarget chromosomes can be tolerated because they can be crossed out of the initial stock. In contrast, off-target events anywhere in the genome may be of concern in cultured cells.

Although it is clear from several studies that sgRNAs have widely varying efficiencies, little is currently known about the factors affecting efficiency. It is therefore difficult to predict how well a specific sgRNA will function prior to testing. For this reason, it is often sensible to test the efficiency of several sgRNAs targeting the region of interest in cell culture to determine which are the most likely to generate DSBs at high efficiency before making the investment of time and effort involved with *in vivo* genome engineering. Moreover, in some situations, it may be advantageous to use an sgRNA with lower efficiency, such as when a homozygous mutation is cell lethal. In such cases, sgRNAs with lower efficiency may result in higher recovery of mutant lines due to an increase in heterozygous compared to homozygous mutants.

## 2.2. Tools facilitating sgRNA design

To aid the design of sgRNAs, we recently developed an online tool allowing the user to browse all possible SpCas9 sgRNA targets in the *Drosophila* genome (Ren et al., 2013) (http://www.flyrnai.org/crispr2). sgRNAs can be filtered based on their genomic location (intron, CDS, UTR, intergenic, etc.), predicted off-target annotation with customizable stringency, and PAM sequence type (NGG or NAG). Individual target sites can then be selected from the genome browser interface to access more detailed annotation. This includes details of any potential off-target sites (genomic location, number and position of mismatches, etc.), whether the sgRNA sequence contains features that may prevent activity (such as a U6 terminator sequence), restriction enzyme sites that could be used to screen for mutations, and a score predicting the likely mutation efficiency at the intended target site. Using these predictions, we estimate that 97% of *Drosophila* protein-coding genes can be mutated with no predicted off-target mutations.

Several other tools are also available to facilitate sgRNA design (Mohr, Hu, Kim, Housden, & Perrimon, 2014) (Table 19.1). For example, targetFinder (Gratz et al., 2014) (http://tools.flycrispr.molbio.wisc.edu/targetFinder/) can be used to design sgRNAs for many different *Drosophila* species. With e-CRISP (Heigwer et al., 2014) (http://www.e-crisp.org/E-CRISP), a specific purpose such as gene knockout or protein tagging can be indicated to aid selection of the most appropriate sgRNA target sites.

**Table 19.1** Tools for sgRNA design

| Lab | Web site | Reference |
|---|---|---|
| DRSC | http://www.flyrnai.org/crispr/ | Ren et al. (2013) |
| O'Connor-Giles | http://tools.flycrispr.molbio.wisc.edu/targetFinder/ | Gratz et al. (2014) |
| DKFZ/Boutros | http://www.e-crisp.org/E-CRISP/designcrispr.html | Heigwer, Kerr, and Boutros (2014) |
| NIG-FLY/Ueda | www.shigen.nig.ac.jp/fly/nigfly/cas9/index.jsp | Kondo and Ueda (2013) |
| Center for Bioinformatics, PKU | http://cas9.cbi.pku.edu.cn/ | Ma, Ye, Zheng, and Kong (2013) |
| Zhang | http://crispr.mit.edu/ | Hsu et al. (2013) |
| Joung | http://zifit.partners.org/ZiFiT/ | Hwang et al. (2013) |

## 3. DELIVERY OF CRISPR COMPONENTS

To generate a genomic modification using the CRISPR system, it is vital that both Cas9 and one or more sgRNAs are delivered efficiently to the cells of interest (usually the germ line). Various methods have been developed to deliver these components and each is associated with advantages and disadvantages. One option is to generate RNA for both Cas9 and the sgRNA and directly inject these into embryos (Bassett et al., 2013; Yu et al., 2013). This approach is attractive because it does not require any cloning steps. It is also possible to inject purified Cas9 protein (Lee et al., 2014). However, compared to other delivery methods, injection of RNA or protein appears to lead to relatively low mutation rates (Bassett & Liu, 2014; Beumer & Carroll, 2014; Gratz, Wildonger, Harrison, & O'Connor-Giles, 2013; Lee et al., 2014).

An alternative approach is to use *Drosophila* stocks that express Cas9 in the germ line (Table 19.2). Several such stocks have been generated using either *vasa* or *nanos* regulatory sequences to drive SpCas9 expression specifically in the germ cells (Kondo & Ueda, 2013; Ren et al., 2013; Sebo et al., 2014; Xue et al., 2014). This offers the advantages of increased efficiency and that viability effects due to somatic mutations in the injected flies are unlikely, aiding the recovery of deleterious mutations through the germ line.

Using these lines means that Cas9 delivery is no longer a consideration, significantly reducing the effort required to prepare CRISPR components. Delivery of sgRNA into Cas9-expressing flies can be achieved using several different methods. As discussed above, the sgRNA can be generated *in vitro* and injected into Cas9-expressing embryos. Alternatively, the sgRNA can be encoded into an expression vector, usually containing a constitutive promoter such as U6, to drive expression of the RNA. While this requires the greater effort of cloning the sequence into a vector, it generally produces higher mutation efficiency than direct delivery of RNA (Kondo & Ueda, 2013; Ren et al., 2013). A final option is to generate fly lines expressing sgRNA, which can then be crossed to Cas9-expressing flies to generate mutant offspring (Kondo, 2014; Kondo & Ueda, 2013). This approach produces the highest efficiency but is a lengthy process due to the need to establish a new fly stock for every sgRNA.

We have found that the best compromise between mutation efficiency, effort, and time required is achieved by injecting an sgRNA expression plasmid into embryos expressing Cas9 in the germ line.

**Table 19.2** CRISPR-related fly lines

| Source | Genotype | Description | Reference |
|---|---|---|---|
| BDSC 51323 | y1 M{vas-Cas9}ZH-2A w1118/ FM7c | Expresses Cas9 from vasa promoter | Gratz et al. (2014) |
| BDSC 51324 | w1118; PBac{vas-Cas9} VK00027 | Expresses Cas9 from vasa promoter | Gratz et al. (2014) |
| BDSC 55821 | y1 M{vas-Cas9.RFP-}ZH-2A w1118/FM7a, P{Tb1}FM7-A | Expresses Cas9 from vasa promoter, marked with RFP | Gratz et al. (2014) |
| BDSC 52669 | y1 M{vas-Cas9.S}ZH-2A w1118 | Expresses Cas9 from vasa promoter | Sebo et al. (2014) |
| BDSC 54590 | y1 M{Act5C-Cas9.P}ZH-2A w* | Expresses Cas9 from Act5c promoter | CRISPR fly design project |
| BDSC 54591 | y1 M{nos-Cas9.P}ZH-2A w* | Expresses Cas9 from nanos promoter | CRISPR fly design project |
| BDSC 54592 | P{hsFLP}1, y1 w1118; P{UAS-Cas9.P}attP2 | Expresses Cas9 from UAS promoter | CRISPR fly design project |
| BDSC 54593 | P{hsFLP}1, y1 w1118; P{UAS-Cas9.P}attP2 P{GAL4::VP16-nos.UTR}CG6325MVD1 | Expresses Cas9 from UAS promoter and Gal4 from nanos promoter | CRISPR fly design project |
| BDSC 54594 | P{hsFLP}1, y1 w1118; P{UAS-Cas9.P}attP40 | Expresses Cas9 from UAS promoter | CRISPR fly design project |
| BDSC 54595 | w1118; P{UAS-Cas9.C}attP2 | Expresses Cas9 from UAS promoter | CRISPR fly design project |
| BDSC 54596 | w1118; P{UAS-Cas9.D10A} attP2 | Expresses Cas9 (nickase) from UAS promoter | CRISPR fly design project |
| NIG-Fly CAS-0001 | *y2 cho2 v1; attP40{nos-Cas9}/ CyO* | Expresses Cas9 from nanos promoter | Kondo and Ueda (2013) |

**Table 19.2** CRISPR-related fly lines—cont'd

| Source | Genotype | Description | Reference |
|---|---|---|---|
| NIG-Fly CAS-0002 | *y2 cho2 v1 P{nos-Cas9, y+, v+} 1A/FM7c, KrGAL4 UAS-GFP* | Expresses Cas9 from nanos promoter | Kondo and Ueda (2013) |
| NIG-Fly CAS-0003 | *y2 cho2 v1; P{nos-Cas9, y+, v+} 3A/TM6C, Sb Tb* | Expresses Cas9 from nanos promoter | Kondo and Ueda (2013) |
| NIG-Fly CAS-0004 | *y2 cho2 v1; Sp/CyO, P{nos-Cas9, y+, v+}2A* | Expresses Cas9 from nanos promoter | Kondo and Ueda (2013) |

For delivery of CRISPR components in cell culture, there are fewer options. As for *in vivo*, the components can be delivered either as *in vitro*-generated RNA or in expression plasmids. For example, an expression vector encoding SpCas9 and sgRNA that can be transfected into cultured cells was recently reported (Bassett et al., 2014) and we have developed a similar plasmid (pL018) (Housden et al., unpublished) (Table 19.3).

The major issue associated with genome engineering in cell lines is currently the inability to generate alterations with 100% efficiency. The rate of alterations is limited by both sgRNA efficiency and transfection efficiency, which is generally low in *Drosophila* cell lines (Bassett et al., 2014). A recent report introduced a selection cassette with the CRISPR components, which increased the mutation rate considerably (Bottcher et al., 2014). The persistence of wild-type sequences, however, results in selection against any unfavorable mutations (e.g., that slow growth) and, over time, reversion of the population to wild type. Until methods are developed to overcome these problems, the use of CRISPR in cell culture is limited to the generation of unstable, mixed populations.

## 4. GENERATION OF CRISPR REAGENTS

As discussed above, there are several methods available to deliver CRISPR components either *in vivo* or in cultured cells. Therefore, the procedures involved in generation of the relevant reagents will depend on the approach taken. Here we will focus on the generation of expression plasmids for delivery of sgRNA into Cas9-expressing flies or for transfection into cultured cells.

**Table 19.3** Cas9 and sgRNA expression plasmids and donor vectors

| Source | Plasmid name | Plasmid purpose | Reference |
|---|---|---|---|
| Perrimon lab | pL018 | Expression of Cas9 (codon optimized) under Act5c promoter and sgRNA under fly U6 promoter | Unpublished |
| Addgene #49330 | pAc-sgRNA-Cas9 | Expression of sgRNA and Cas9-Puro in cell culture | Bassett et al. (2014) |
| Addgene #49408 | pCFD1-dU6:1gRNA | Expression of sgRNA under control of the *Drosophila* U6:1 promoter | CRISPR fly design project |
| Addgene #49409 | pCFD2-dU6:2gRNA | Expression of sgRNA under control of the *Drosophila* U6:2 promoter | CRISPR fly design project |
| Addgene #49410 | pCFD3-U6:3gRNA | Expression of sgRNA under control of the *Drosophila* U6:3 promoter | CRISPR fly design project |
| Addgene #49411 | pCFD4-U6:1_U6:3-tandemgRNAs | Expression of two sgRNAs from *Drosophila* U6:1 and U6:3 promoters | CRISPR fly design project |
| Addgene #45946 | pU6-*Bbs*I-chiRNA | Plasmid for expression of chiRNA under the control of the *Drosophila* snRNA:U6:96Ab promoter | Gratz, Cummings, et al. (2013) |
| Addgene #45945 | pHsp70-Cas9 | Expression of Cas9 (codon optimized) under control of Hsp70 promoter | Gratz, Cummings, et al. (2013) |
| Addgene #46294 | pBS-Hsp70-Cas9 | Expression of Cas9 (codon optimized) under control of Hsp70 promoter | Gratz, Cummings, et al. (2013) |
| NIG-Fly | pBFv-nosP-Cas9 | Expression of Cas9 from the nanos promoter | Kondo and Ueda (2013) |
| NIG-Fly | pBFv-U6.2 | sgRNA expression vector with attB | Kondo and Ueda (2013) |
| Perrimon lab | pBH-donor | Vector for generation of donor constructs | Unpublished |

## 4.1. Cloning of sgRNAs into expression vectors

Few vectors are currently available for expression of sgRNAs. However, those that are available are generally compatible with similar cloning protocols using type IIs restriction enzymes (Table 19.3). For example, the procedure described below is based on one previously developed for mammalian CRISPR vectors (Ran et al., 2013) but can be used with pL018 (Housden et al., unpublished), pU6-*Bbs*I-chiRNA (Gratz, Cummings, et al., 2013), U6b-sgRNA-short (Ren et al., 2013), pAc-sgRNA-Cas9 (note that this plasmid is compatible with *Bsp*QI instead of *Bbs*I) (Bassett et al., 2014), and pBFv-U6.2 (Kondo & Ueda, 2013) *Drosophila* plasmids. Here we will focus on pL018 but the procedure can be easily modified for other plasmids. When designing sgRNA oligos, be sure to include the relevant 4-bp overhangs to allow ligation into the digested vector and when using a U6 expression plasmid, also include an additional G at the start of the sgRNA sequence to initiate transcription.

Materials

- Complementary oligos carrying sgRNA target sequence (not including PAM)
- Suitable plasmid for the desired application
- *Bbs*I restriction enzyme (Thermo Scientific)
- T4 ligase buffer (NEB)
- T4 PNK enzyme (NEB)
- T7 ligase and buffer (Enzymatics)
- FastDigest Buffer (Thermo Scientific)
- FastAP enzyme (Thermo Scientific)

Protocol

1. Set up a restriction digest reaction as shown below and incubate at 37 °C for 30 min:
   1 μg pL018 (or other suitable plasmid)
   2 μl 10 × FastDigest Buffer
   1 μl FastAP
   1 μl *Bbs*I or other suitable enzyme
   Water to 20 μl total volume
2. Purify reaction products using a PCR purification kit and normalize concentration to 10 ng/ μl.
3. Resuspend sense and antisense sgRNA oligos to 100 μ*M* and anneal using the following reaction mixture:
   1 μl 100 μ*M* sense sgRNA oligo
   1 μl 100 μ*M* antisense sgRNA oligo
   1 μl 10 × T4 ligation buffer
   0.5 μl T4 PNK (NEB)
   6.5 μl water
   *Note*: In this step, use T4 ligation buffer with the PNK enzyme as this contains ATP required for phosphorylation of the annealed oligos.

Use a thermocycler with the following program to phosphorylate and anneal oligos:

- 37 °C—30 min
- 95 °C—5 min
- Ramp from 95 °C to 25 °C at 5 °C/min

4. Dilute the annealed oligos from step 3 by 200-fold using water (to 50 n*M*). If the concentration is too high, multiple copies may be inserted into the vector.
5. Ligate annealed oligos into digested vector as follows:

   1 μl digested plasmid (from step 2)
   1 μl diluted annealed oligos (from step 4)
   5 μl 2 × Quick Ligase Buffer
   0.5 μl T7 ligase
   2.5 μl water
   Incubate at room temperature for 5 min.

   Note: While 5 min of ligation is generally sufficient, longer incubation periods may be used to increase colony numbers if necessary.

   When using new vectors, we recommend performing negative controls in parallel using the same conditions but omitting the annealed oligos from the ligation reaction.
6. Transform 2 μl of ligation product into chemically competent *E. coli* using standard procedures and spread on LB plates containing ampicillin. Incubate the plates at 37 °C overnight.
7. Culture and miniprep single colonies and sequence to confirm successful cloning. In general, we have very high success rates for screening single colonies although more can be tested if necessary.

## 4.2. Cloning of donor constructs

As described above, single-stranded DNA oligos can be used as donors, making the insertion of small sequences very simple. However, due to the limit on the length of oligos that can be reliably generated, this approach can only be used to make small changes to the genome.

Production of a double-stranded donor construct requires the ligation of three or four components; two homology arms, an insert (for most but not all applications), and a backbone vector. These constructs can be produced using standard restriction digests followed by four-way ligation, although this approach is generally in efficient and time consuming. Instead, more advanced cloning procedures can be used to generate the construct in a single step. Here, as an example, we describe a detailed protocol to produce a donor construct for insertion of an exogenous sequence using golden gate cloning (Engler, Kandzia, & Marillonnet, 2008; Engler & Marillonnet, 2014). Note that other approaches are also possible to generate these constructs, including Gibson assembly (Gibson et al., 2009).

When designing homology arms for golden gate cloning, it is important to ensure that the restriction enzymes used for construction of the donor plasmid do not cut within these sequences. To facilitate this, the donor vector we use (pBH-donor) is compatible with three type IIs restriction enzymes: *Bsa*I, *Bbs*I, and *Bsm*BI (Housden et al., unpublished).

Type IIs restriction enzymes cut outside their recognition sequences. A single enzyme can therefore be used to generate multiple different sticky ends in a single digest reaction. Furthermore, the enzyme recognition sequence is cleaved from the DNA molecule to be cloned during the digest reaction. When the fragments are subsequently ligated together, the restriction enzyme recognition sites will not be present and so the molecule cannot be recut. If the small fragment cleaved from the end of the molecule religates, however, the restriction site will be restored and the molecule can therefore be redigested. Golden gate cloning works on the principle of cycling between digest and ligation conditions in the presence of both the restriction and ligation enzymes. Iterative rounds of digest and ligation therefore drive the accumulation of correctly ligated products even when multiple fragments are present. By including a backbone vector in the reaction, it is possible to transform the products directly without the need for additional cloning steps. Using this easily scalable approach, we generally obtain greater than 80% of constructs from a single reaction by screening only one resulting colony. Screening additional colonies increases the success rate to above 95%.

Materials

- High-fidelity polymerase (e.g., Phusion high-fidelity polymerase from NEB)
- Gel extraction kit (e.g., QIAGEN gel purification kit)
- 10 × BSA
- 10 m*M* ATP
- NEB buffer 4
- T7 ligase (Enzymatics)
- Type IIs restriction enzyme (e.g., *Bsa*I, *Bsm*BI, or *Bbs*I)
- Thermal cycler

- Chemically competent bacteria
- Miniprep kit
- Restriction enzymes for test digest
- Oligos for sequencing
- pBH-donor or other suitable vector

Protocol

1. Design oligos for PCR amplification of two homology arms and insert fragment. These oligos should add type IIs restriction enzyme cut sites to the PCR products required for later cloning steps.
2. PCR amplify each of the homology arms using a high-fidelity polymerase.
3. Run PCR products on a gel to check the sizes of the bands. We recommend using 1 kb for all homology arms.
4. Gel purify homology arms from the gel using standard kits according to manufacturer's instructions.
5. PCR amplify insert sequence and gel purify if necessary (if using a short sequence, this can also be produced as complementary oligos annealed together).
6. Set up a golden gate reaction using the gel-purified homology arms, insert fragment and backbone vector:
   10 ng each homology arm
   10 ng donor vector
   10 ng insert fragment
   1 μl 10 × BSA
   1 μl 10 m*M* ATP
   1 μl NEB buffer 4
   0.5 μl T7 ligase
   0.5 μl type IIs restriction enzyme
   Water to 10 μl
7. Place samples in a thermal cycler and run the following program:
   1. 37 °C—2 min
   2. 20 °C—3 min
   3. Repeat steps 1 and 2 a further nine times
   4. 37 °C—2 min
   5. 95 °C—5 min
8. Transform 5 μl of reaction product into chemically competent *E. coli* using standard procedures and spread onto a kanamycin plate. Incubate overnight at 37 °C.
9. Culture two of the resulting colonies overnight in selective media.

10. Miniprep samples using a standard kit according to manufacturer's instructions.
11. Send samples for sequencing using suitable primers for the plasmid used. For pBH-donor, sequence with the primer 5′-GAATCGCAGACCGATACCAG-3′.

    Using this cloning approach, we generally find that a high proportion of clones carry the desired components, correctly assembled, and so screening one or two clones is sufficient.

As an alternative approach, it is not necessary to include the backbone vector in the golden gate reaction. The homology arms and insert can be assembled by golden gate cloning and then reamplified by PCR using a high-fidelity polymerase before cloning into a vector of choice using standard procedures. This can be a useful alternative approach if none of the restriction enzymes compatible with the donor vector are appropriate for the homology arms being generated.

### 4.3. Isolation of *In vivo* genome modifications

Once the relevant reagents have been generated and injected into fly embryos, the next stage is the identification and recovery of the desired genome modification events. As described below, there are various methods that can be used to detect modifications but the injected G0 flies must first be crossed to obtain nonmosaic animals before screening can be performed (Fig. 19.1). In order to do this, we generally cross G0 flies to a line with balancers on the chromosome of interest. The resulting F1 flies can then be collected and screened with one of the methods described below before recrossing to the same balancer line to isolate the stock.

## 5. DETECTION OF MUTATIONS

Several methods are available to detect genome alterations induced using CRISPR. For most cases involving insertions, deletions or substitutions, customized methods must be used to detect the change. One option for insertions or substitutions is to include a visible marker such as miniwhite or 3 × P3-dsRed in the inserted sequence. This then allows simple selection of flies carrying the desired modification (Gratz et al., 2014). However, in some cases it may be undesirable to insert the additional sequences associated with these markers, in which case PCR-based screening approaches may be more appropriate.

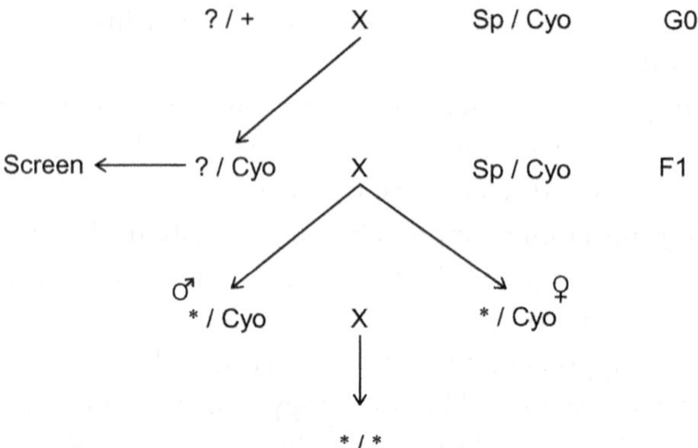


**Figure 19.1** Crossing scheme to generate modified *Drosophila* lines. Following injection of CRISPR components into embryos, the modified chromosome must then be detected and isolated. This can be done by first crossing the injected G0 flies to a suitable balancer stock (second chromosome balancers are used as an example: e.g., *Sp/Cyo*) and isolating individual F1 flies for screening by restriction profiling, surveyor, HRMA, or other suitable approach. Screening can be performed using nonlethal methods and so flies carrying the desired modification (*) can then be recrossed to balancers to isolate the chromosome in a stable stock.

Detection of indels caused by NHEJ is more difficult due to the general lack of visible phenotypes and unreliable effects on PCR-based assays. Many mutations caused by NHEJ are very small (1-bp insertions or deletions are relatively common) (Cong et al., 2013; Mali, Yang, et al., 2013; Ren et al., 2013) and so will not necessarily affect amplification with an overlapping PCR primer. Therefore, alternative screening methods must be used.

Several methods are available, including restriction profiling, endonuclease assays and high-resolution melt assays (HRMAs) (Bassett et al., 2013; Cong et al., 2013; Wang et al., 2013). Each of these methods has advantages and disadvantages and should be chosen based on the number of samples to screen and the availability of suitable reagents. Note that following all of these screening methods, it is recommended to sequence the target site to confirm sequence alteration and determine the nature of the mutation.

## 5.1. Preparation of genomic DNA from fly wings

In order to screen flies prior to establishing stocks, it is possible to extract genomic DNA from a single wing for screening without killing the flies

(Carvalho, Ja, & Benzer, 2009). This allows selection of the correct F1 adults prior to crossing and therefore significantly reduces the workload compared to whole fly screening, which requires all crosses to be established first. All three of the screening methods detailed below require the preparation of genomic DNA. Note that for preparation of genomic DNA from cells, a similar protocol can be used in which the cells are resuspended in squishing buffer and lysed in a thermocycler as described.

Materials

- Squishing buffer (10 m*M* Tris–HCl pH 8.2, 1 m*M* EDTA, 25 m*M* NaCl, 400 μg/ml Proteinase K (add fresh from 50 × stock stored at −20 °C))
- Blender or homogenizer pestles
- Thermocycler

Protocol

1. Remove one wing from fly (tear the wing close to the hinge but avoiding damage to the thorax) using forceps and place in a 1.5-ml microcentrifuge tube. The exact position of the tear can vary as long as the thorax is not damaged.
2. Add 20 μl of squishing buffer and homogenize well in the tube using a pestle or blender.
3. Transfer to a PCR tube and run the following program in a thermocycler:
   - 50 °C—1 h
   - 98 °C—10 min
   - 10 °C—hold

### *5.1.1 Restriction profiling*

This screening approach relies on the disruption of a genomic restriction enzyme recognition sequence by an NHEJ-induced mutation. Particularly when generating gene knockouts, there are often several possible sgRNA targets that can be used, allowing selection of one that overlaps with such a restriction site. Note that these restriction sites are annotated in the DRSC sgRNA design tool described above. In order to detect mutations, a fragment surrounding the target site must first be amplified by PCR from the genomic DNA prepared as described above from F1 generation flies. Next, the PCR product is digested with the relevant restriction enzyme and visualized on a gel. Any alteration of wild-type sequence that disrupts the restriction site will change the band pattern produced, therefore indicating the presence of a mutation.

### *5.1.2 Surveyor assay to detect indels*

It will often be the case that no suitable restriction sites are present at the sgRNA target locus for screening. One alternative possibility is to use endonuclease assays. These work by first amplifying a fragment from genomic DNA containing the sgRNA target site and then melting and reannealing to form homoduplexes and heteroduplexes between wild-type and mutant sequences. An endonuclease enzyme is then used that specifically cuts mismatches in the heteroduplexed molecules, resulting in a change in the band pattern when the samples are visualized on a gel (Fig. 19.2). We generally use SURVEYOR Mutation Detection Kit (Transgenomics) to detect indels generated by NHEJ, although other enzymes can also be used (e.g., T7 nuclease).

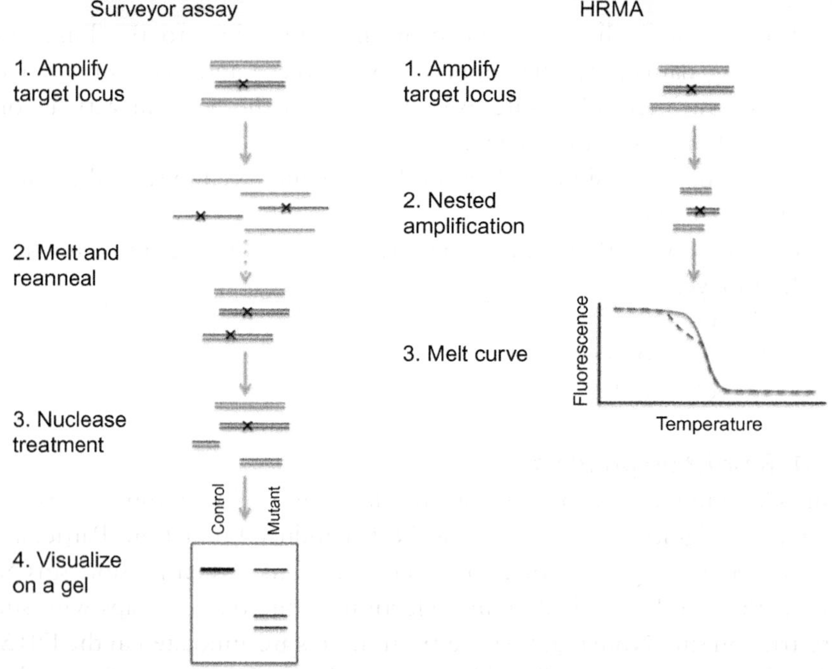


**Figure 19.2** Surveyor and HRMA screening protocols. Surveyor assays rely on the ability of the surveyor endonuclease to cleave mismatched DNA strands. Melting and reannealing fragments amplified from a mixture of wild-type and mutant alleles lead to the formation of heteroduplexes, which are then cleaved by the surveyor enzyme to alter the band pattern when the products are visualized on a gel. HRMA measures differences in the melt curves between amplified fragments. These differences may be small depending on the sequence change and so specialized software may be required to detect mutated samples. (See the color plate.)

Materials

- Suitable primers for PCR amplification
- Standard PCR reagents including a high-fidelity polymerase enzyme
- PCR purification kit (e.g., QIAquick PCR purification kit from QIAGEN)
- Thermocycler
- Taq PCR buffer
- Surveyor mutation detection kit (Transgenomic)

Protocol

1. Extract genomic DNA as described above, either from whole flies or individual wings.
2. Design and optimize surveyor primers such that a single, strong band is produced by PCR and amplify fragments from genomic DNA using optimized PCR conditions and a high-fidelity polymerase. The optimal fragment length is around 500 bp as this allows reliable amplification and easy visualization of changes in band sizes following nuclease treatment.

   It is important to use a high-fidelity polymerase for this step to prevent the introduction of sequence differences due to PCR errors. These would be detected by the surveyor nuclease and generate false-positive results.
3. Purify PCR products using a PCR purification kit and normalize to 20 ng/μl with water.

   Note that primer optimization is a key factor and generating a specific PCR product is very important for surveyor success. Including a negative control consisting of unmutated genomic DNA is also recommended.
4. Melt and reanneal PCR products using the following conditions:

   Reaction mixture:

   2.5 μl 10 × Taq PCR buffer
   22.5 μl 20 ng/μl purified PCR product

   Place samples in a thermocycler and run the following program:

   95 °C—10 min
   Ramp from 95 °C to 85 °C (−2.0 C/s)
   85 °C—1 min
   Ramp from 85 °C to 75 °C (−0.3 °C/s)
   75 °C—1 min
   Ramp from 75 °C to 65 °C (−0.3 °C/s)
   65 °C—1 min
   Ramp from 65 °C to 55 °C (−0.3 °C/s)

55 °C—1 min
Ramp from 55 °C to 45 °C (−0.3 °C/s)
45 °C—1 min
Ramp from 45 °C to 35 °C (−0.3 °C/s)
35 °C—1 min
Ramp from 35 °C to 25 °C (−0.3 °C/s)
25 °C—1 min
4 °C—hold

5. Set up surveyor digest reactions as shown below and incubate for 30 min at 42 °C:
   25 μl annealed product from step 4
   3 μl 0.15 *M* $MgCl_2$
   1 μl SURVEYOR nuclease S
   1 μl SURVEYOR enhancer S
6. Add 2 μl Stop Solution from the kit and visualize on an agarose gel to detect mutations.

When screening cell-based samples, it can be useful to estimate the proportion of mutated alleles in the population. This can be done as follows:

1. Measure the integrated intensity of the uncleaved band A and cleaved bands B and C using ImageJ or other gel quantification software.
2. Calculate $f_{cut}$ as: $f_{cut} = (B + C)/(A + B + C)$.
3. Estimate the indel occurrence by: $\text{indel}(\%) = 100 \times \left(1 - \sqrt{(1 - f_{cut})}\right)$

### 5.1.3 Detection of mutations using HRMA

A cost-effective and scalable approach to mutation detection is to use HRMAs (Bassett et al., 2013). Many companies sell RT-PCR machines with built-in modules for HRMA analysis but the reactions can be performed on almost any RT-PCR machine as long as a melt curve can be performed with fluorescence reads at intervals of 0.1 °C.

The principle of HRMA is that a small fragment is first amplified from the genomic locus potentially containing a mutation. This is then slowly melted and the amount of double-stranded DNA measured throughout the melting process. This produces a melt curve similar to those produced as quality controls in standard RT-PCR assays but with higher resolution. Changes in the sequence of the DNA fragments will alter the shape of this curve thereby allowing detection of samples containing mutations by comparison with curves generated from wild-type samples (Fig. 19.2). Such changes are often very small, especially when using samples containing many different mutations, as is often the case in cell culture, so specialized software is required to detect them.

This assay is also more sensitive than the other methods, meaning that flies can be screened at the G0 generation and therefore reducing the amount of work required to isolate mutant lines. To generate genomic DNA from the G0 generation, it is recommended that crosses are set up to produce the F1 generation for all G0s and whole flies are used for DNA preps once it is clear that the crosses will produce progeny. Note that wing DNA preps cannot be used in this case because mutations are likely to be in the germ line.

Materials

- Standard PCR reagents including a high-fidelity polymerase enzyme
- Suitable primers for nested amplification
- Precision melt supermix (Bio-Rad) or other similar reaction mix
- RT-PCR machine with high-resolution melt ability

Protocol

1. Prepare genomic DNA as described above for wings or using 50–100 μl squishing buffer for whole flies.
2. PCR amplify a fragment (300–600 bp) around the sgRNA target site using the following reaction mixture and PCR program:

   Reaction mixture:

   2 μl DNA
   10 μl buffer
   1 μl dNTPs (25 m*M* each)
   1 μl primers (10 μ*M* each)
   0.5 μl Phusion polymerase
   1.25 μl $MgCl_2$ (100 m*M*)
   34.25 μl water

   PCR program:

   98 °C—3 min
   98 °C—30 s
   50 °C—30 s
   72 °C—30 s
   Goto step 2—34 times
   10° C hold
3. Run 5 μl of reaction products on a gel to determine whether the correct size fragment has been produced. It is not necessary to obtain a specific product because nonspecific bands will not be reamplified in the following steps.

4. Dilute PCR product 1:10,000 using water.
5. Set up a nested PCR and melt assay in an RT-PCR machine as follows:
   Reaction mixture:
      1 μl DNA template
      5 μl precision melt supermix
      0.3 μl left primer (10 μ*M*)
      0.3 μl right primer (10 μ*M*)
      3.4 μl water
   HRMA program:
      95 °C—3 min
      95 °C—18 s
      50 °C—30 s
      Fluorescence read
      Repeat 50 times
      95 °C—2 min
      25 °C—2 min
      4 °C—2 min
      Melt curve from 55 °C to 95 °C with fluorescence reads every 0.1 °C

Even small amounts of nonspecific product can affect the results of the HRMA. The method described above uses nested PCR in order to avoid the need to optimize the original PCR. However, it is possible to skip the first PCR amplification step when highly specific primers can be designed.

## 5.2. Analysis of HRMA data

Many RT-PCR machines come with commercial HRMA analysis software, which can be used to identify samples carrying mutations following the manufacturer's instructions. However, if such software is not available, an online tool can be used. For example, we recently developed HRMAnalyzer (http://www.flyrnai.org/HRMA) (Housden, Flockhart, & Perrimon, unpublished), which can be used to identify samples carrying mutations either using a clustering-based approach or via statistical comparison with control samples.

## ACKNOWLEDGMENTS

We would like to thank David Doupé and Stephanie Mohr for helpful discussion in the preparation of this chapter. Work in the Perrimon lab is supported by the NIH and HHMI.

## REFERENCES

Auer, T. O., Duroure, K., De Cian, A., Concordet, J. P., & Del Bene, F. (2014). Highly efficient CRISPR/Cas9-mediated knock-in in zebrafish by homology-independent DNA repair. *Genome Research*, *24*(1), 142–153. http://dx.doi.org/10.1101/gr.161638.113.

Bassett, A. R., & Liu, J. L. (2014). CRISPR/Cas9 and genome editing in Drosophila. *Journal of Genetics and Genomics*, *41*(1), 7–19. http://dx.doi.org/10.1016/j.jgg.2013.12.004.

Bassett, A. R., Tibbit, C., Ponting, C. P., & Liu, J. L. (2013). Highly efficient targeted mutagenesis of Drosophila with the CRISPR/Cas9 system. *Cell Reports*, *4*(1), 220–228. http://dx.doi.org/10.1016/j.celrep.2013.06.020.

Bassett, A. R., Tibbit, C., Ponting, C. P., & Liu, J. L. (2014). Mutagenesis and homologous recombination in Drosophila cell lines using CRISPR/Cas9. *Biology Open*, *3*(1), 42–49. http://dx.doi.org/10.1242/bio.20137120.

Beumer, K., Bhattacharyya, G., Bibikova, M., Trautman, J. K., & Carroll, D. (2006). Efficient gene targeting in Drosophila with zinc-finger nucleases. *Genetics*, *172*(4), 2391–2403. http://dx.doi.org/10.1534/genetics.105.052829.

Beumer, K. J., & Carroll, D. (2014). Targeted genome engineering techniques in Drosophila. *Methods*, *68*(1), 29–37. http://dx.doi.org/10.1016/j.ymeth.2013.12.002.

Beumer, K. J., Trautman, J. K., Bozas, A., Liu, J. L., Rutter, J., Gall, J. G., et al. (2008). Efficient gene targeting in Drosophila by direct embryo injection with zinc-finger nucleases. *Proceedings of the National Academy of Sciences of the United States of America*, *105*(50), 19821–19826. http://dx.doi.org/10.1073/pnas.0810475105.

Beumer, K. J., Trautman, J. K., Christian, M., Dahlem, T. J., Lake, C. M., Hawley, R. S., et al. (2013). Comparing zinc finger nucleases and transcription activator-like effector nucleases for gene targeting in Drosophila. *G3*, *3*(10), 1717–1725. http://dx.doi.org/10.1534/g3.113.007260.

Beumer, K. J., Trautman, J. K., Mukherjee, K., & Carroll, D. (2013). Donor DNA utilization during gene targeting with zinc-finger nucleases. *G3*, *3*(4), 657–664. http://dx.doi.org/10.1534/g3.112.005439.

Bibikova, M., Golic, M., Golic, K. G., & Carroll, D. (2002). Targeted chromosomal cleavage and mutagenesis in Drosophila using zinc-finger nucleases. *Genetics*, *161*(3), 1169–1175.

Bottcher, R., Hollmann, M., Merk, K., Nitschko, V., Obermaier, C., Philippou-Massier, J., et al. (2014). Efficient chromosomal gene modification with CRISPR/cas9 and PCR-based homologous recombination donors in cultured Drosophila cells. *Nucleic Acids Research*, *42*(11), e89. http://dx.doi.org/10.1093/nar/gku289.

Bozas, A., Beumer, K. J., Trautman, J. K., & Carroll, D. (2009). Genetic analysis of zinc-finger nuclease-induced gene targeting in Drosophila. *Genetics*, *182*(3), 641–651. http://dx.doi.org/10.1534/genetics.109.101329.

Carvalho, G. B., Ja, W. W., & Benzer, S. (2009). Non-lethal PCR genotyping of single Drosophila. *BioTechniques*, *46*(4), 312–314. http://dx.doi.org/10.2144/000113088.

Chapman, J. R., Taylor, M. R., & Boulton, S. J. (2012). Playing the end game: DNA double-strand break repair pathway choice. *Molecular Cell*, *47*(4), 497–510. http://dx.doi.org/10.1016/j.molcel.2012.07.029.

Cong, L., Ran, F. A., Cox, D., Lin, S., Barretto, R., Habib, N., et al. (2013). Multiplex genome engineering using CRISPR/Cas systems. *Science*, *339*(6121), 819–823. http://dx.doi.org/10.1126/science.1231143.

Dickinson, D. J., Ward, J. D., Reiner, D. J., & Goldstein, B. (2013). Engineering the Caenorhabditis elegans genome using Cas9-triggered homologous recombination. *Nature Methods*, *10*(10), 1028–1034. http://dx.doi.org/10.1038/nmeth.2641.

Engler, C., Kandzia, R., & Marillonnet, S. (2008). A one pot, one step, precision cloning method with high throughput capability. *PLoS One*, *3*(11), e3647. http://dx.doi.org/10.1371/journal.pone.0003647.

Engler, C., & Marillonnet, S. (2014). Golden Gate cloning. *Methods in Molecular Biology*, *1116*, 119–131. http://dx.doi.org/10.1007/978-1-62703-764-8_9.

Fu, Y., Foden, J. A., Khayter, C., Maeder, M. L., Reyon, D., Joung, J. K., et al. (2013). High-frequency off-target mutagenesis induced by CRISPR-Cas nucleases in human cells. *Nature Biotechnology*, *31*(9), 822–826. http://dx.doi.org/10.1038/nbt.2623.

Gibson, D. G., Young, L., Chuang, R. Y., Venter, J. C., Hutchison, C. A., 3rd., & Smith, H. O. (2009). Enzymatic assembly of DNA molecules up to several hundred kilobases. *Nature Methods*, *6*(5), 343–345. http://dx.doi.org/10.1038/nmeth.1318.

Gratz, S. J., Cummings, A. M., Nguyen, J. N., Hamm, D. C., Donohue, L. K., Harrison, M. M., et al. (2013). Genome engineering of Drosophila with the CRISPR RNA-guided Cas9 nuclease. *Genetics*, *194*(4), 1029–1035. http://dx.doi.org/10.1534/genetics.113.152710.

Gratz, S. J., Ukken, F. P., Rubinstein, C. D., Thiede, G., Donohue, L. K., Cummings, A. M., et al. (2014). Highly specific and efficient CRISPR/Cas9-catalyzed homology-directed repair in Drosophila. *Genetics*, *196*(4), 961–971. http://dx.doi.org/10.1534/genetics.113.160713.

Gratz, S. J., Wildonger, J., Harrison, M. M., & O'Connor-Giles, K. M. (2013). CRISPR/Cas9-mediated genome engineering and the promise of designer flies on demand. *Fly*, 7(4), 249–255. http://dx.doi.org/10.4161/fly.26566.

Heigwer, F., Kerr, G., & Boutros, M. (2014). E-CRISP: Fast CRISPR target site identification. *Nature Methods*, *11*(2), 122–123. http://dx.doi.org/10.1038/nmeth.2812.

Hsu, P. D., Scott, D. A., Weinstein, J. A., Ran, F. A., Konermann, S., Agarwala, V., et al. (2013). DNA targeting specificity of RNA-guided Cas9 nucleases. *Nature Biotechnology*, *31*(9), 827–832. http://dx.doi.org/10.1038/nbt.2647.

Hwang, W. Y., Fu, Y., Reyon, D., Maeder, M. L., Tsai, S. Q., Sander, J. D., et al. (2013). Efficient genome editing in zebrafish using a CRISPR-Cas system. Research Support, N.I.H., Extramural.

Jiang, W., Bikard, D., Cox, D., Zhang, F., & Marraffini, L. A. (2013). RNA-guided editing of bacterial genomes using CRISPR-Cas systems. *Nature Biotechnology*, *31*(3), 233–239. http://dx.doi.org/10.1038/nbt.2508.

Kondo, S. (2014). New horizons in genome engineering of Drosophila melanogaster. *Genes & Genetic Systems*, *89*(1), 3–8.

Kondo, S., & Ueda, R. (2013). Highly improved gene targeting by germline-specific Cas9 expression in Drosophila. *Genetics*, *195*(3), 715–721. http://dx.doi.org/10.1534/genetics.113.156737.

Lee, J. S., Kwak, S. J., Kim, J., Noh, H. M., Kim, J. S., & Yu, K. (2014). RNA-guided genome editing in drosophila with the purified cas9 protein. *G3*, *4*(7), 1291–1295. http://dx.doi.org/10.1534/g3.114.012179.

Ma, M., Ye, A. Y., Zheng, W., & Kong, L. (2013). A guide RNA sequence design platform for the CRISPR/Cas9 system for model organism genomes. *BioMed Research International*, *2013*, 270805. http://dx.doi.org/10.1155/2013/270805.

Mali, P., Aach, J., Stranges, P. B., Esvelt, K. M., Moosburner, M., Kosuri, S., et al. (2013). CAS9 transcriptional activators for target specificity screening and paired nickases for cooperative genome engineering. *Nature Biotechnology*, *31*(9), 833–838. http://dx.doi.org/10.1038/nbt.2675.

Mali, P., Yang, L., Esvelt, K. M., Aach, J., Guell, M., DiCarlo, J. E., et al. (2013). RNA-guided human genome engineering via Cas9. *Science*, *339*(6121), 823–826. http://dx.doi.org/10.1126/science.1232033.

Mohr, S. E., Hu, Y., Kim, K., Housden, B. E., & Perrimon, N. (2014). Resources for functional genomics studies in Drosophila melanogaster. *Genetics*, *197*(1), 1–18. http://dx.doi.org/10.1534/genetics.113.154344.

Pattanayak, V., Lin, S., Guilinger, J. P., Ma, E., Doudna, J. A., & Liu, D. R. (2013). High-throughput profiling of off-target DNA cleavage reveals RNA-programmed Cas9 nuclease specificity. *Nature Biotechnology*, *31*(9), 839–843. http://dx.doi.org/10.1038/nbt.2673.

Ran, F. A., Hsu, P. D., Wright, J., Agarwala, V., Scott, D. A., & Zhang, F. (2013). Genome engineering using the CRISPR-Cas9 system. *Nature Protocols*, *8*(11), 2281–2308. http://dx.doi.org/10.1038/nprot.2013.143.

Ren, X., Sun, J., Housden, B. E., Hu, Y., Roesel, C., Lin, S., et al. (2013). Optimized gene editing technology for Drosophila melanogaster using germ line-specific Cas9. *Proceedings of the National Academy of Sciences of the United States of America*, *110*(47), 19012–19017. http://dx.doi.org/10.1073/pnas.1318481110.

Sebo, Z. L., Lee, H. B., Peng, Y., & Guo, Y. (2014). A simplified and efficient germline-specific CRISPR/Cas9 system for Drosophila genomic engineering. *Fly*, *8*(1), 52–57. http://dx.doi.org/10.4161/fly.26828.

Urnov, F. D., Miller, J. C., Lee, Y. L., Beausejour, C. M., Rock, J. M., Augustus, S., et al. (2005). Highly efficient endogenous human gene correction using designed zinc-finger nucleases. *Nature*, *435*(7042), 646–651. http://dx.doi.org/10.1038/nature03556.

Wang, H., Yang, H., Shivalila, C. S., Dawlaty, M. M., Cheng, A. W., Zhang, F., et al. (2013). One-step generation of mice carrying mutations in multiple genes by CRISPR/Cas-mediated genome engineering. *Cell*, *153*(4), 910–918. http://dx.doi.org/10.1016/j.cell.2013.04.025.

Xue, Z., Ren, M., Wu, M., Dai, J., Rong, Y. S., & Gao, G. (2014). Efficient gene knock-out and knock-in with transgenic Cas9 in Drosophila. *G3*, *4*(5), 925–929. http://dx.doi.org/10.1534/g3.114.010496.

Yang, L., Guell, M., Byrne, S., Yang, J. L., De Los Angeles, A., Mali, P., et al. (2013). Optimization of scarless human stem cell genome editing. *Nucleic Acids Research*, *41*(19), 9049–9061. http://dx.doi.org/10.1093/nar/gkt555.

Yu, Z., Chen, H., Liu, J., Zhang, H., Yan, Y., Zhu, N., et al. (2014). Various applications of TALEN- and CRISPR/Cas9-mediated homologous recombination to modify the Drosophila genome. *Biology Open*, *3*(4), 271–280. http://dx.doi.org/10.1242/bio.20147682.

Yu, Z., Ren, M., Wang, Z., Zhang, B., Rong, Y. S., Jiao, R., et al. (2013). Highly efficient genome modifications mediated by CRISPR/Cas9 in Drosophila. *Genetics*, *195*(1), 289–291. http://dx.doi.org/10.1534/genetics.113.153825.

CHAPTER TWENTY

# Transgene-Free Genome Editing by Germline Injection of CRISPR/Cas RNA

**Hillel T Schwartz**[*,†], **Paul W Sternberg**[*,†,1]
[*]Division of Biology and Biological Engineering, California Institute of Technology, Pasadena, California, USA
[†]Howard Hughes Medical Institute, Pasadena, California, USA
[1]Corresponding author: e-mail address: pws@caltech.edu

## Contents



*Methods in Enzymology*, Volume 546
ISSN 0076-6879
http://dx.doi.org/10.1016/B978-0-12-801185-0.00021-0

## Abstract

Genome modification by CRISPR/Cas offers its users the ability to target endogenous sites in the genome for cleavage and for engineering precise genomic changes using template-directed repair, all with unprecedented ease and flexibility of targeting. As such, CRISPR/Cas is just part of a set of recently developed and rapidly improving tools that offer great potential for researchers to functionally access the genomes of organisms that have not previously been extensively used in a laboratory setting. We describe in detail protocols for using CRISPR/Cas to target genes of experimental organisms, in a manner that does not require transformation to obtain transgenic lines and that should be readily applicable to a wide range of previously little-studied species.

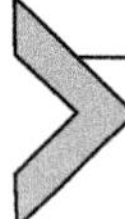

## 1. THEORY, PHILOSOPHY, AND PRACTICAL CONSIDERATIONS

### 1.1. Overview

CRISPR/Cas provides a method for the generation of double-strand DNA breaks at sites that the user can select with great efficiency and a high degree of freedom (Gaj, Gersbach, & Barbas, 2013; Kim & Kim, 2014; Mali, Esvelt, & Church, 2013). It should not be necessary in this context to discuss in detail the origins and nature of nucleases with engineered specificity, or of CRISPR/Cas, in particular; many others have done this in great detail, and anyone reading this chapter is likely familiar with what CRISPR/Cas can do for them. Instead, we wish to explain the particular issues and potential advantages associated with transgene-free delivery of CRISPR/Cas, and to explain in detail how this can be performed, both in general and with specific application to the nematode worm *Caenorhabditis elegans*, or indeed to other animals possessing a similarly accessible germline.

The first work toward the creation of modern standard laboratory model organisms began a little more than a century ago in T. H. Morgan's laboratory at Columbia University (Sturtevant, 2001). Since those beginnings, worldwide communities of researchers have combined their efforts to build the resources that have made it highly rewarding to perform genetic and later molecular studies in standard laboratory model organisms, including the fruit fly *Drosophila melanogaster*, the roundworm *C. elegans*, the zebrafish *Danio rerio*, the mouse *Mus musculus*, and the thale cress *Arabidopsis thaliana*. The global communities focusing on each of these organisms shared their mutant collections, developed and optimized specialized protocols for mutagenesis and for transgenesis in each species, and eventually had access to completely sequenced genomes. Emerging technologies promise to make it feasible for researchers interested in organisms not previously well studied and lacking such a worldwide network of collaborators to establish these animals as powerful molecular genetic experimental systems: an individual researcher with limited resources can generate useable draft genome assemblies by means of high-throughput sequencing (Schatz, Delcher, & Salzberg, 2010); the use of molecular markers has made access to rich collections of visibly phenotypic mutations unnecessary for mapping and for strain construction (Rounsley & Last, 2010; Wicks, Yeh, Gish, Waterston, & Plasterk, 2001), and RNAi and now engineered nucleases make it possible to study gene function (Boutros & Ahringer, 2008; Frokjaer-Jensen, 2013; Gaj et al., 2013; Selkirk, Huang, Knox, & Britton, 2012). Application of these technologies has made it possible to engineer gene knock-outs in organisms that have only been studied sufficiently to generate on the order of 100 published papers listed in PubMed (Lo et al., 2013; Zantke, Bannister, Rajan, Raible, & Tessmar-Raible, 2014). In particular, the transgene-free delivery of CRISPR/Cas activity by direct injection of *in vitro*-synthesized RNAs (Chiu, Schwartz, Antoshechkin, & Sternberg, 2013; Katic & Grosshans, 2013; Lo et al., 2013) makes it possible to engineer the genomes of new species even without access to reliable protocols for DNA transformation in the species of interest.

## 1.2. When to use or not to use transgenes for delivery of CRISPR/Cas

When available protocols make the generation of transgenes straightforward and efficient, as is the case for *C. elegans*, CRISPR/Cas using transformation with DNA should be considered the strongly favored approach: it avoids the need to generate or to store reagents for injection as relatively unstable RNA, and the highest efficiencies reported using DNA transgenes to deliver CRISPR/Cas activity are better than the highest efficiencies reported using injection of *Cas9* mRNA or protein (Frokjaer-Jensen, 2013). This situation, however, results from the advantages of *C. elegans* as an established research organism; in other species there may be no reported efficient method of generating DNA transgenes, and experience suggests that some considerable effort may be required to develop transgenesis for new species, even given a similar anatomy and reproductive mechanism to that of *C. elegans* (Schlager, Wang, Braach, & Sommer, 2009). The transgene-free nature of the direct injection of CRISPR/Cas reagents is thus likely to offer a powerful tool in organisms in which transgenesis has not been attempted, or is known to be challenging.

There have been recent developments in the application of CRISPR technology that make use of the easily programmable DNA binding of CRISPR but that use this binding to target the recruitment of transcriptional modification machinery rather than to induce double-strand DNA breaks (Gilbert et al., 2013). These methods are very exciting in their potential, but require for their use the persistent expression of the modified CRISPR reagents involved. The transient delivery of CRISPR reagents by direct injection of Cas9 mRNA is therefore unlikely to be well suited to these new technologies, transformation with DNA transgenes will be required.

### 1.3. Altered mutation profile from transgene-free treatment with CRISPR/Cas

Many groups have now reported on their experience using various methodologies to achieve genome modification by CRISPR/Cas in *C. elegans*; these approaches have included expressing the CRISPR/Cas reagents from a transgene, typically for a full generation; delivering the reagents as mRNA and guide RNA; or even direct injection of the guide RNA with Cas9 protein (Chiu et al., 2013; Cho, Lee, Carroll, & Kim, 2013; Dickinson, Ward, Reiner, & Goldstein, 2013; Friedland et al., 2013; Katic & Grosshans, 2013; Liu et al., 2014; Lo et al., 2013; Waaijers et al., 2013; Zhao, Zhang, Ke, Yue, & Xue, 2014). From these reports, it has become apparent that there are important differences between the use of transformation and transgene-free approaches for performing CRISPR/Cas regarding the types of mutations most frequently recovered.

Considering only those mutations recovered on the basis of a visible phenotype, rather than identified molecularly on the basis of altered sequence at the targeted site, most of the mutations induced in *C. elegans* by CRISPR/Cas cleavage using DNA transgenes were very small insertions or deletions (Friedland et al., 2013; Liu et al., 2014; Waaijers et al., 2013). By contrast, several groups have reported that the mutations induced by transient application of CRISPR/Cas reagents, whether by direct injection of guide RNAs with *Cas9* mRNA or by direct injection of Cas9 protein, showed a strong preference for very large deletions or even chromosomal rearrangements (Chiu et al., 2013; Cho et al., 2013; Lo et al., 2013); others reported mixed results or reported finding only small deletions and insertions (Katic & Grosshans, 2013; Liu et al., 2014). These large

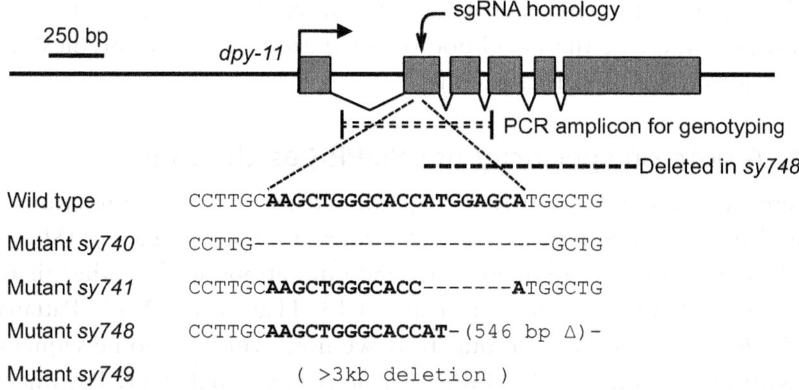


**Figure 20.1** A schematic representation of the *dpy-11* locus of *C. elegans*. The position of the site selected for cleavage by targeted CRISPR/Cas activity is indicated. Note that the targeted site was within an early exon of the gene, to make it more likely that any resulting mutations would strongly disrupt gene function. Partial sequences of the wild type and of selected mutants isolated after CRISPR/Cas treatment targeting this site are shown (Chiu et al., 2013). In these sequences, the 20 nucleotides immediately prior to the **NGG** motif that were incorporated into the sgRNA used to target this locus are bolded. Although in this instance mutants were isolated on the basis of their mutant phenotype, it would have been possible to screen for molecular changes at this site, for example, by using PCR to amplify a small (600 nucleotide) region centered on the targeted site and testing for cleavage using *Nco*I (recognition sequence CCATGG). Note that this approach would only have recovered one of the four mutants shown, *sy740*; *sy748* and *sy749* mutant chromosomes would lack at least one of the primer binding sites used in the PCR amplification and would not be represented in the PCR product (compare the hypothetical PCR amplicon and the extent of the deletion *sy748* in the figure), and the seven nucleotide deletion in *sy741* destroys the endogenous *Nco*I site but the resulting sequence change generates a new *Nco*I site.

alterations have the advantage of generating mutations that are extremely likely to be molecular nulls, but these mutations are much more difficult to detect molecularly, if they cannot be recovered on the basis of a predictable viable visible phenotype. Molecular approaches to detecting changes induced by CRISPR/Cas rely on PCR amplification of the locus so that alterations can be detected using mismatch detection, by altered restriction digestion, or by a difference in amplicon size; large deletions of the sort preferentially seen with transgene-free CRISPR/Cas in *C. elegans* usually lack the primer-binding sites required for this amplification, and so such mutations are likely to be missed (Fig. 20.1). When genes are being targeted mutants of which cannot be screened for phenotypically, it may be helpful to use available template-directed repair knock-in strategies (Chen, Fenk, &

de Bono, 2013; Dickinson et al., 2013; Tzur et al., 2013; Zhao et al., 2014) in order to increase the likelihood of generating a lesion of predicable and easily detected molecular structure.

## 1.4. A note on specificity of CRISPR/Cas cleavage

There is considerable controversy regarding the specificity of mutations generated using CRISPR/Cas. Several studies have suggested CRISPR/Cas nuclease activity may frequently generate mutations at sites other than the one targeted for cleavage (Fu et al., 2013; Hsu et al., 2013; Pattanayak et al., 2013). Conversely, in our study we used whole-genome sequencing to assess the specificity of CRISPR/Cas activity when delivered by injection of *in vitro* synthesized RNAs in *C. elegans*, and did not observe frequent off-site changes (Chiu et al., 2013). It is conceivable that *C. elegans* for unknown reasons displays high specificity for CRISPR/Cas, or that the induction of mutations using transient delivery of CRISPR/Cas reagents inclines toward high specificity; even if so, off-target cleavage events might easily be more common in other species. If extremely high specificity is required, the researcher may wish to look into cleaving with two CRISPR/Cas complexes, each targeted very close to the other and targeting single-strand nickase activity or dimerizing Fok1 activity, such that binding at both sites is required to effect a double-strand break (Cho et al., 2014; Mali, Aach, et al., 2013; Ran et al., 2013; Tsai et al., 2014).

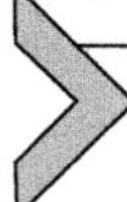

# 2. EQUIPMENT

Oligonucleotide synthesis service
DNA sequencing service
Microcentrifuge
UV spectrophotometer (e.g., NanoDrop)
Water bath
37 °C Incubator
−20 °C Freezer
−80 °C Freezer
PCR machine
Dissecting microscope
Bunsen burner
Needle puller
Inverted microscope with microinjection equipment

Agarose slab gel apparatus
Power supply
UV transilluminator

## 3. MATERIALS

Addgene plasmid 47911 SP6-hCas9-Ce-mRNA
Addgene plasmid 47912 SP6-sgRNA-scaffold
Restriction enzymes *Afl*II and *Kpn*I-HF, and accompanying incubation buffers
Phusion DNA polymerase (NEB)
QIAquick PCR Purification Kit (Qiagen)
QIAquick Gel Extraction Kit (Qiagen)
Gibson cloning kit (NEB)
MAXIscript SP6 Transcription Kit (Life Technologies)
mMESSAGE mMACHINE SP6 Transcription Kit (Life Technologies)
PolyA Tailing Kit (Life Technologies)
TURBO DNase (Life Technologies)
Ampicillin-sensitive transformation-competent *E. coli* (e.g., $CaCl_2$-competent DH5α)
Deoxynucleotide triphosphates (dNTPs)
RNase-free water
NGM agar
LB broth, LB agar
Agarose
Ethidium bromide (or alternative DNA visualization reagent)
Micropipettors
Micropipette tips
0.5 and 1.5-ml microcentrifuge tubes
250-μl PCR tubes
6 and 10 cm diameter Petri plates
Microcapillaries for the preparation of microinjection needles
Cover glasses for microinjection (24 × 50 mm)
20-μl micropipettes with mouth pipettor

## 4. IDENTIFYING A TARGET SEQUENCE

Look in your gene for suitable candidate target sites; when using the *Streptococcus pyogenes* Cas9 enzyme, these should be a 23 nucleotide sequence ending with **NGG**. The site can be on either strand, so also seek out 23mer sequences starting with **CCN**. Cleavage will happen close to the end of the 20mer that precedes the **NGG** (after position 17). To ensure that mutations are likely to be strong loss-of-function alleles you should look within the coding sequence, ideally near to the 5′-end. The eventual sequence will be transcribed *in vitro* using SP6 RNA polymerase, which requires that the sequence start with a **G** (preferably with a **GA** or **GG**); however, this is not a consideration in identifying a target side, as if the endogenous 5′-end of the 23 nucleotide sequence is not a **G** then it can be replaced with a **G** or a **G** can be appended to the 5′-end.

Having compiled a list of candidates, use BLAST homology searches to identify target sites that do not have excessive homology to other, off-target sites in the genome. Note that it may be necessary to adjust the settings of your BLAST query for low stringency so that it will return results with appropriately weak homology; if running BLAST from the command line, include the argument "-word_size 7." Even the targeted site will only have an E-value on the order of e-05; any candidate off-target sites with an E-value less than 1 should be examined according to the criteria below. Target recognition and cleavage by CRISPR/Cas is most strongly influenced by sequences close to the 3′-end of the 20 nucleotides preceding the **NGG**, and the terminal GG sequence is required. Similar sequences elsewhere in the genome are common, with perfect matches of 15 nucleotides in a row being a common occurrence, but it should be readily feasible to identify a site for which such extended stretches of homology do not include nucleotides at or close to the 3′-end of the 20 nucleotides preceding the **NGG**, or that have high similarity but are not followed by **NGG**, indicating that these are not strong candidates for off-target cleavage by CRISPR/Cas. Note that there is another method to maximize specificity, instead of seeking sites with minimal sequence identity to other sites in the genome: as mentioned above, it is possible to modify the CRISPR/Cas cleavage protocol to replace the double-strand DNA cleavage activity with single-strand nickases or dimerizing FokI nuclease, such that two sites in close proximity must be recognized for a double-strand break to occur.

Another consideration is your ability to screen molecularly for any mutations you generate. If you can identify a cleavage site that is within or extremely close to the recognition sequence for a restriction enzyme (and one that does not also cleave again very close to the target site), PCR amplification of the target region followed by digestion with that restriction enzyme should readily detect lesions that destroy the recognition site.

# 5. GENERATING YOUR sgRNA CONSTRUCT

## 5.1. Oligonucleotide design

You will be using Gibson cloning to insert the first 20 nucleotides of your 23 nucleotide target ending in **NGG** into the vector SP6-sgRNA-scaffold. To do this, you will first order the synthesis of two oligos:

*Forward oligo*: Examine the first 20 nucleotides of your 23 nucleotide target site: this, or a modified version of this, will be incorporated into an oligonucleotide and cloned into an sgRNA expression vector—the final three nucleotides present in the genomic target site, the **NGG**, will not be included in this oligonucleotide or in this construct. If the first two of these 20 nucleotides are not either **GA** or **GG**, either replace these two nucleotides with **GA** or add a **G** or **GA** to the 5′-end of this 20 nucleotide sequence. Now, order the synthesis of an oligonucleotide that has this 20, 21, or 22 nucleotide sequence appended to the 3′-end of the sequence provided immediately below:

GATCCCCCGGGCTGCAGGAATTCATTTAGGTGACACTATA

*Reverse oligo*: Determine the reverse-complement of the sequence you appended to the 3′-end of the sequence provided above. Order the synthesis of an oligonucleotide that has this reverse-complement sequence appended to the 3′-end of the sequence immediately below:

GACTAGCCTTATTTTAACTTGCTATTTCTAGCTCTAAAAC

## 5.2. Insert generation

Mix the following in a 250-μl PCR tube:

10.0 μl of 5 × Phusion buffer
15.0 μl of 1 m*M* dNTPs
2.5 μl of 10 μ*M* forward primer
2.5 μl of 10 μ*M* reverse primer
19.5 μl $H_2O$
0.5 μl Phusion polymerase

Perform PCR according to the following program:

**(1)** 94.5 °C—3′
**(2)** 94.5 °C—20″
**(3)** 47 °C—25″
**(4)** 72 °C—25″
**(5)** Goto step 2, four times

**(6)** 94.5 °C—20″
**(7)** 50 °C—25″
**(8)** 72 °C—25″
**(9)** Goto step 6, four times
**(10)** 94.5 °C—20″
**(11)** 53 °C—25″
**(12)** 72 °C—25″
**(13)** Goto step 10, 19 times
**(14)** 72 °C—10′
**(15)** 4 °C—4″
**(16)** 15 °C—until stopped

Use QIAquick PCR Purification Kit to clean up the product according to manufacturer's protocol. At the end of the process, apply 10 μl dd$H_2O$ heated to 70 °C to the column, centrifuge to collect eluate, and repeat once, collecting the eluate in the same tube. Store at −20 °C until linearized vector is available.

## 5.3. Preparation of linearized vector for the sgRNA construct

Mix the following in a 1.5-ml microcentrifuge tube:

4 μl of 10 × NEB CutSmart buffer
5 μl of 500 ng/μl SP6-sgRNA-scaffold
29 μl $H_2O$
2 μl *Afl*II restriction enzyme (20 units/μl)

Incubate in 37 °C water bath for 4 h. Run product on a 0.8% agarose gel and purify using QIAquick Gel Extraction Kit according to manufacturer's protocol. At the end of the process, apply 16 μl dd$H_2O$ heated to 70 °C to the column, centrifuge to collect eluate, and repeat once, collecting the eluate in the same tube. Store at −20 °C and use aliquots as needed.

## 5.4. Construction and identification of sgRNA synthesis plasmid

Mix the following in a 250 μl PCR tube:

3 μl of 2 × Gibson reaction mix
1 μl Purified *Afl*II-digested SP6-sgRNA-scaffold (see Section 5.3)
2 μl Purified Phusion PCR product (see Section 5.2)

Incubate 1 h at 50 °C using a PCR machine with heated lid. It may be desirable to do a control Gibson reaction with no insert. Use at least half of the Gibson product to transform competent *E. coli* (for example, $CaCl_2$-

competent DH5α) according to standard methods (Seidman, Struhl, Sheen, & Jessen, 2001). Plate transformants on LB plates containing ampicillin or carbenicillin and grow overnight at 37 °C. Pick several (at least six) individual colonies and grow in 2–5 ml LB broth containing ampicillin or carbenicillin. Prep plasmid DNA minipreps by alkaline lysis and ethanol precipitation (Engebrecht, Brent, & Kaderbhai, 2001). Perform test digests using *Afl*II by mixing the following in a 1.5-ml microcentrifuge tube or a 250-μl PCR tube:

2.0 μl of 10 × NEB CutSmart buffer
1.0 μl (~500 ng) Miniprep DNA
16.8 μl $H_2O$
0.2 μl *Afl*II restriction enzyme (20 units/μl)

Incubate in 37 °C water bath, heat block, or PCR machine for 2 h. Run the product on 0.8% agarose slab gel to identify colonies whose minipreps were not linearized by *Afl*II digest; these will be the colonies that have incorporated your insert and can be used to synthesize sgRNA. Errors are infrequently introduced in the amplification or cloning process; it may be desirable to confirm the DNA sequence of the clone you use prior to performing CRISPR/Cas treatment. Sequencing can be done using a T3 sequencing primer, keeping in mind that the minipreps produced by alkaline lysis were pure enough for restriction digestion but that DNA should be further purified for sequencing. Sequencing can wait until the end of Section 6.1.

# 6. *IN VITRO* SYNTHESIS OF sgRNA

## 6.1. Linearization of sgRNA template plasmid

By the end of step 5, you will have generated and miniprepped plasmids for *in vitro* transcription of your sgRNA. Before transcribing from these plasmids you should linearize them. If you have not yet confirmed the DNA sequence of your finished sgRNA template plasmid, you may wish to linearize more than one clone. You will later require linearized SP6-hCas9-Ce-mRNA plasmid, which can be prepared identically and simultaneously (see Section 7.1).

Mix the following in a 1.5-ml microcentrifuge tube:

5 μl of 10 × NEB CutSmart buffer
12 μl (~6 μg) Miniprep DNA (from clone that was not linearized by *Afl*II)
31 μl $H_2O$
2 μl *Kpn*I-HF restriction enzyme (20 units/μl)

Incubate in 37 °C water bath for 4 h. Purify with QIAquick PCR Purification Kit or run on a 0.8% agarose slab gel and purify with QIAquick Gel Extraction Kit. At the end of the process, apply 9 μl RNase-free water heated to 70 °C to the column, centrifuge to collect eluate, and repeat once, collecting the eluate in the same tube. Product can be stored at −20 °C. Determine DNA concentration using an UV spectrophotometer. If the DNA sequence of the template plasmid has not been determined, submit part of the purified sample(s) for sequence determination prior to the next step.

### 6.2. *In vitro* transcription to generate sgRNA

Use MAXIscript SP6 Transcription Kit according to manufacturer's protocol; briefly, combine the following in a 1.5-ml microcentrifuge tube:

250 ng linearized sgRNA template DNA, plus RNase-free water, totaling 8 μl
2 μl of 10 × Reaction buffer
2 μl each ATP, CTP, GTP, and UTP solutions
2 μl Enzyme mix

Incubate in 37 °C water bath for 3 h. Optionally, treat reaction mix with DNase according to manufacturer's protocol. Proceed immediately to next step.

### 6.3. Purification of *in vitro*-transcribed sgRNA

Use MEGAclear Transcription Clean-Up Kit according to manufacturer's protocol. Elute with 30 μl RNase-free water and repeat once, collecting eluate in same tube. Determine RNA concentration using UV spectrophotometer. Store in aliquots in a −80 °C freezer.

## 7. *IN VITRO* SYNTHESIS OF *hCas9* mRNA

### 7.1. Linearization of SP6-hCas9-Ce-mRNA plasmid

If you have not already linearized the SP6-hCas9-Ce-mRNA plasmid, do so now, similarly to the procedure in Section 6.1. Mix the following in a 1.5-ml microcentrifuge tube:

5 μl of 10 × NEB CutSmart buffer
8 μl (500 ng/μl) SP6-hCas9-Ce-mRNA plasmid
35 μl $H_2O$
2 μl *Kpn*I-HF restriction enzyme (20 units/μl)

Incubate in a 37 °C water bath for 4 h. Purify with QIAquick PCR Purification Kit or run on an 0.8% agarose slab gel and purify with QIAquick Gel Extraction Kit. At the end of the process, apply 9 μl RNase-free water heated to 70 °C to the column, centrifuge to collect the eluate, and repeat once, collecting the eluate in the same tube. Product can be stored at -20 °C.

### 7.2. *In vitro* transcription of *hCas9* mRNA

Use mMESSAGE mMACHINE SP6 Transcription Kit according to the manufacturer's protocol. Briefly, mix in a 1.5-ml microcentrifuge tube:

6 μl Purified linearized SP6-hCas9-Ce-mRNA plasmid
10 μl NTP/CAP solution
2 μl of 10 × Buffer
2 μl Enzyme mix

Incubate in 37 °C water bath for 4 h. Add 1 μl TURBO DNase and incubate in a 37 °C water bath for 15 min. Immediately proceed to the next step.

### 7.3. Polyadenylation of *in vitro*-transcribed *hCas9* mRNA

Use the PolyA Tailing Kit according to the manufacturer's protocol. Briefly, mix in a 1.5-ml microcentrifuge tube:

20 μl DNase-treated reaction mix
36 μl RNase-free water
20 μl of 5 × *E*-PAP buffer
10 μl of 25 m*M* $MnCl_2$
10 μl of 10 m*M* ATP
4 μl *E*-PAP enzyme

Incubate in 37 °C water bath for 1 h. Proceed immediately to next step.

### 7.4. Purification of *in vitro*-transcribed, polyadenylated *hCas9* mRNA

Use MEGAclear Transcription Clean-Up Kit according to manufacturer's protocol. Elute with 30 μl RNase-free water and repeat once, collecting eluate in the same tube. Determine the RNA concentration using a UV spectrophotometer. Store in aliquots in a −80 °C freezer.

## 8. INJECTION OF sgRNA AND mRNA

You will prepare a mixture of *hCas9* mRNA and sgRNA. Note that in order to enhance your ability to molecularly identify any resulting mutation events, it may be desirable to include a template for homology-directed repair of the double-strand breaks induced by CRISPR/Cas activity; several groups have published relevant methods (Chen et al., 2013; Dickinson et al., 2013; Tzur et al., 2013; Zhao et al., 2014).

In preparing your mixture, we recommend a ratio of sgRNA concentration to mRNA concentration of approximately 1:4 to 1:8, and have thus far found the best results when injecting the highest concentration we could readily achieve—concentrations on the order of 100 ng/μl sgRNA and 800 ng/μl mRNA. Some DNA microinjection protocols call for the use of a microinjection buffer (Mello & Fire, 1995); we did not find this to be necessary. Having prepared the mixture, spin at maximum speed in a microcentrifuge (~13,000 rcf) for at least five minutes to clear the supernatant of any particulate matter, then heat the mixture briefly to 95 °C and place it on ice. The mixture should remain on ice until it is loaded into injection needles.

For *C. elegans* and similar nematodes, follow standard microinjection procedures (Mello & Fire, 1995), injecting to flood the germline syncytia of young adults as is done for DNA transformation.

## 9. RECOVERY OF MUTANTS GENERATED USING CRISPR/Cas

### 9.1. Recovery and plating of injected animals

After being injected, animals can be floated in M9 buffer and transferred using a 20 μl micropipette and a mouth pipettor to Petri plates containing NGM that have been seeded with a bacterial food source. After the animals have visibly recovered, individual injected animals or a small number per Petri plate should be transferred to new seeded NGM Petri plates in preparation for screening their progeny. Mutations induced by injection of CRISPR/Cas mRNA and sgRNA in *C. elegans* will be found at highest frequency among the progeny produced between 8 and 16 h postinjection, and essentially none will be recovered among the progeny produced more than 24 h postinjection (Katic & Grosshans, 2013; Liu et al., 2014). It is therefore recommended that injected animals should be transferred from their recovery plate up to 8 h after being injected, and they should be discarded between 16 and 24 h postinjection, leaving behind the progeny they produced during their time on the plate.

## 9.2. Identification of animals carrying mutations induced by CRISPR/Cas

Induced visible mutants of *C. elegans* can be recovered by examining the $F_2$ progeny of injected animals for expression of the expected phenotype. If no visible phenotype can be predicted, or if homozygosity and expression of the visible phenotype is expected to be associated with an inviable phenotype such as sterility, it will be necessary to clone out individual $F_1$ progeny of injected animals, permit them to produce progeny, and to screen molecularly for mutations at the targeted site. This protocol will not describe in detail the methods involved in molecular detection of mutation events; briefly, the options include targeting for mutation an endogenous restriction site, such that any alteration will result in a PCR product that cannot be cleaved by the corresponding restriction enzyme (Friedland et al., 2013); mismatch detection methods such as the CEL-I Surveyor method (Colbert et al., 2001); recombinant CRISPR/Cas targeting the wild-type sequence (Kim, Kim, Kim, & Kim, 2014); or the inclusion of a template for homology-directed repair of the induced double strand DNA break, such that repair events will produce a predictable and readily detectable sequence change. The addition of a second sgRNA previously demonstrated to reliably induce mutations in another gene that cause a visible phenotype, and screening for mutations in your targeted site only among animals displaying that visible phenotype, can enrich for the presence of mutations in the targeted site (Kim, Ishidate, et al., 2014).

Regardless of how mutations are detected, care should be taken to track all candidates back to the injected animal or small pool of injected animals that gave rise to them. If two mutants are recovered derived from the same injected animal or animals, these may not represent independent mutation events.

## REFERENCES


Boutros, M., & Ahringer, J. (2008). The art and design of genetic screens: RNA interference. *Nature Reviews Genetics*, *9*, 554–566.

Chen, C., Fenk, L. A., & de Bono, M. (2013). Efficient genome editing in *Caenorhabditis elegans* by CRISPR-targeted homologous recombination. *Nucleic Acids Research*, *41*, e193.

Chiu, H., Schwartz, H. T., Antoshechkin, I., & Sternberg, P. W. (2013). Transgene-free genome editing in *Caenorhabditis elegans* using CRISPR-Cas. *Genetics*, *195*, 1167–1171.

Cho, S. W., Kim, S., Kim, Y., Kweon, J., Kim, H. S., Bae, S., et al. (2014). Analysis of off-target effects of CRISPR/Cas-derived RNA-guided endonucleases and nickases. *Genome Research*, *24*, 132–141.

Cho, S. W., Lee, J., Carroll, D., & Kim, J. S. (2013). Heritable gene knockout in *Caenorhabditis elegans* by direct injection of Cas9-sgRNA ribonucleoproteins. *Genetics*, *195*, 1177–1180.

Colbert, T., Till, B. J., Tompa, R., Reynolds, S., Steine, M. N., Yeung, A. T., et al. (2001). High-throughput screening for induced point mutations. *Plant Physiology*, *126*, 480–484.

Dickinson, D. J., Ward, J. D., Reiner, D. J., & Goldstein, B. (2013). Engineering the *Caenorhabditis elegans* genome using Cas9-triggered homologous recombination. *Nature Methods*, *10*, 1028–1034.

Engebrecht, J., Brent, R., & Kaderbhai, M. A. (2001). Minipreps of plasmid DNA. *Current Protocols in Molecular Biology*, *15*, 1.6.1–1.6.10, edited by Frederick M Ausubel et al..

Friedland, A. E., Tzur, Y. B., Esvelt, K. M., Colaiacovo, M. P., Church, G. M., & Calarco, J. A. (2013). Heritable genome editing in *C. elegans* via a CRISPR-Cas9 system. *Nature Methods*, *10*, 741–743.

Frokjaer-Jensen, C. (2013). Exciting prospects for precise engineering of *Caenorhabditis elegans* genomes with CRISPR/Cas9. *Genetics*, *195*, 635–642.

Fu, Y., Foden, J. A., Khayter, C., Maeder, M. L., Reyon, D., Joung, J. K., et al. (2013). High-frequency off-target mutagenesis induced by CRISPR-Cas nucleases in human cells. *Nature Biotechnology*, *31*, 822–826.

Gaj, T., Gersbach, C. A., & Barbas, C. F., 3rd. (2013). ZFN, TALEN, and CRISPR/Cas-based methods for genome engineering. *Trends in Biotechnology*, *31*, 397–405.

Gilbert, L. A., Larson, M. H., Morsut, L., Liu, Z., Brar, G. A., Torres, S. E., et al. (2013). CRISPR-mediated modular RNA-guided regulation of transcription in eukaryotes. *Cell*, *154*, 442–451.

Hsu, P. D., Scott, D. A., Weinstein, J. A., Ran, F. A., Konermann, S., Agarwala, V., et al. (2013). DNA targeting specificity of RNA-guided Cas9 nucleases. *Nature Biotechnology*, *31*, 827–832.

Katic, I., & Grosshans, H. (2013). Targeted heritable mutation and gene conversion by Cas9-CRISPR in *Caenorhabditis elegans*. *Genetics*, *195*, 1173–1176.

Kim, H., Ishidate, T., Ghanta, K. S., Seth, M., Conte, D., Jr., Shirayama, M., et al. (2014). A Co-CRISPR strategy for efficient genome editing in *Caenorhabditis elegans*. *Genetics*. http://dx.doi.org/10.1534/genetics.1114.166389.

Kim, H., & Kim, J. S. (2014). A guide to genome engineering with programmable nucleases. *Nature Reviews Genetics*, *15*, 321–334.

Kim, J. M., Kim, D., Kim, S., & Kim, J. S. (2014). Genotyping with CRISPR-Cas-derived RNA-guided endonucleases. *Nature Communications*, *5*, 3157.

Liu, P., Long, L., Xiong, K., Yu, B., Chang, N., Xiong, J.-W., et al. (2014). Heritable/conditional genome editing in *C. elegans* using a CRISPR-Cas9 feeding system. *Cell Research*, *24*(7), 886–889.

Lo, T. W., Pickle, C. S., Lin, S., Ralston, E. J., Gurling, M., Schartner, C. M., et al. (2013). Precise and heritable genome editing in evolutionarily diverse nematodes using TALENs and CRISPR/Cas9 to engineer insertions and deletions. *Genetics*, *195*, 331–348.

Mali, P., Aach, J., Stranges, P. B., Esvelt, K. M., Moosburner, M., Kosuri, S., et al. (2013). CAS9 transcriptional activators for target specificity screening and paired nickases for cooperative genome engineering. *Nature Biotechnology*, *31*, 833–838.

Mali, P., Esvelt, K. M., & Church, G. M. (2013). Cas9 as a versatile tool for engineering biology. *Nature Methods*, *10*, 957–963.

Mello, C., & Fire, A. (1995). DNA transformation. *Methods in Cell Biology*, *48*, 451–482.

Pattanayak, V., Lin, S., Guilinger, J. P., Ma, E., Doudna, J. A., & Liu, D. R. (2013). High-throughput profiling of off-target DNA cleavage reveals RNA-programmed Cas9 nuclease specificity. *Nature Biotechnology*, *31*, 839–843.

Ran, F. A., Hsu, P. D., Lin, C. Y., Gootenberg, J. S., Konermann, S., Trevino, A. E., et al. (2013). Double nicking by RNA-guided CRISPR Cas9 for enhanced genome editing specificity. *Cell*, *154*, 1380–1389.

Rounsley, S. D., & Last, R. L. (2010). Shotguns and SNPs: How fast and cheap sequencing is revolutionizing plant biology. *The Plant Journal*, *61*, 922–927.

Schatz, M. C., Delcher, A. L., & Salzberg, S. L. (2010). Assembly of large genomes using second-generation sequencing. *Genome Research*, *20*, 1165–1173.

Schlager, B., Wang, X., Braach, G., & Sommer, R. J. (2009). Molecular cloning of a dominant roller mutant and establishment of DNA-mediated transformation in the nematode *Pristionchus pacificus*. *Genesis*, *47*, 300–304.

Seidman, C. E., Struhl, K., Sheen, J., & Jessen, T. (2001). Introduction of plasmid DNA into cells. *Current Protocols in Molecular Biology*, *37*, 1.8.1–1.8.10, edited by Frederick M Ausubel et al.

Selkirk, M. E., Huang, S. C., Knox, D. P., & Britton, C. (2012). The development of RNA interference (RNAi) in gastrointestinal nematodes. *Parasitology*, *139*, 605–612.

Sturtevant, A. H. (2001). *A history of genetics*. Cold Spring Harbor, N.Y.: Cold Spring Harbor Laboratory Press.

Tsai, S. Q., Wyvekens, N., Khayter, C., Foden, J. A., Thapar, V., Reyon, D., et al. (2014). Dimeric CRISPR RNA-guided FokI nucleases for highly specific genome editing. *Nature Biotechnology*, *32*(6), 569–577.

Tzur, Y. B., Friedland, A. E., Nadarajan, S., Church, G. M., Calarco, J. A., & Colaiacovo, M. P. (2013). Heritable custom genomic modifications in *Caenorhabditis elegans* via a CRISPR-Cas9 system. *Genetics*, *195*, 1181–1185.

Waaijers, S., Portegijs, V., Kerver, J., Lemmens, B. B., Tijsterman, M., van den Heuvel, S., et al. (2013). CRISPR/Cas9-targeted mutagenesis in *Caenorhabditis elegans*. *Genetics*, *195*, 1187–1191.

Wicks, S. R., Yeh, R. T., Gish, W. R., Waterston, R. H., & Plasterk, R. H. (2001). Rapid gene mapping in *Caenorhabditis elegans* using a high density polymorphism map. *Nature Genetics*, *28*, 160–164.

Zantke, J., Bannister, S., Rajan, V. B., Raible, F., & Tessmar-Raible, K. (2014). Genetic and genomic tools for the marine annelid *Platynereis dumerilii*. *Genetics*, *197*, 19–31.

Zhao, P., Zhang, Z., Ke, H., Yue, Y., & Xue, D. (2014). Oligonucleotide-based targeted gene editing in *C. elegans* via the CRISPR/Cas9 system. *Cell Research*, *24*, 247–250.

CHAPTER TWENTY-ONE

# Cas9-Based Genome Editing in *Arabidopsis* and Tobacco

**Jian-Feng Li*,†,1, Dandan Zhang*,†, Jen Sheen*,†**

*Department of Molecular Biology, Centre for Computational and Integrative Biology, Massachusetts General Hospital, Boston, Massachusetts, USA

†Department of Genetics, Harvard Medical School, Boston, Massachusetts, USA

[1]Corresponding author: e-mail address: lijfeng3@mail.sysu.edu.cn

## Contents



## Abstract

Targeted modification of plant genome is key to elucidating and manipulating gene functions in plant research and biotechnology. The clustered regularly interspaced short palindromic repeats (CRISPR)/CRISPR-associated protein (Cas) technology is emerging as a powerful genome-editing method in diverse plants that traditionally lacked facile and versatile tools for targeted genetic engineering. This technology utilizes easily reprogrammable guide RNAs (sgRNAs) to direct *Streptococcus pyogenes* Cas9 endonuclease to generate DNA double-stranded breaks in targeted genome sequences, which facilitates efficient mutagenesis by error-prone nonhomologous end-joining (NHEJ) or sequence replacement by homology-directed repair (HDR). In this chapter, we describe the procedure to design and evaluate dual sgRNAs for plant codon-optimized Cas9-mediated genome editing using mesophyll protoplasts as model cell systems in *Arabidopsis thaliana* and *Nicotiana benthamiana*. We also discuss future directions in sgRNA/Cas9 applications for generating targeted genome modifications and gene regulations in plants.

*Methods in Enzymology*, Volume 546
ISSN 0076-6879
http://dx.doi.org/10.1016/B978-0-12-801185-0.00022-2

# 1. INTRODUCTION

The CRISPR/Cas9 technology is derived from the bacterial type-II CRISPR/Cas adaptive immune system (Jinek et al., 2012). The technology uses a single chimeric guide RNA (sgRNA) containing a 20-nt guide sequence to direct coexpressed *Streptococcus pyogenes* Cas9 endonuclease to an intended genomic $N_{20}$NGG sequence through base pairing. Two separate nuclease domains of Cas9 each cleave one DNA strand to generate a DSB in the targeted sequence. During the DSB repair, site-specific gene mutagenesis or replacement can be obtained via the NHEJ pathway or homologous recombination pathway, the later depending on the availability of a DNA repair template (Cong et al., 2013; Li et al., 2013; Mali et al., 2013). Among the designer nucleases for genome editing, the CRISPR/Cas9 system exhibits unparalleled simplicity and multiplexibility in genome editing because sgRNAs can be easily modified to achieve new DNA binding specificities and multiple sgRNAs can work simultaneously with the same Cas9 nuclease on many different target sites (Gaj, Gersbach, & Barbas, 2013; Li et al., 2013; Sander & Joung, 2014).

Effective delivery of genome-editing reagents, including Cas9 nucleases, sgRNAs, and homologous recombination DNA donors, is key to the high efficiency of targeted genome modification, which remains challenging for most plant cells that are enclosed in cell walls. In this chapter, we describe the detailed procedure for designing and evaluating constructs using the CRISPR/Cas9 system for genome editing in *Arabidopsis thaliana* and tobacco (*Nicotiana benthamiana*) mesophyll protoplasts (Fig. 21.1), which support highly efficient DNA transfection and RNA and protein expression (Li, Zhang, & Sheen, 2014; Yoo, Cho, & Sheen, 2007). The procedure is potentially adaptable to diverse plant species that are amenable to protoplast isolation and transfection (Li et al., 2014). Plant protoplasts offer a valuable system for rapidly evaluating the performance of a given combination of sgRNA and Cas9 at the genomic target site. To enhance the rate of generating null mutations, dual sgRNAs are designed and evaluated. We discuss promising strategies to apply the CRISPR/Cas system for generating targeted and inheritable genome modifications in plants. The CRISPR/Cas system has the potential to generate loss-of-function mutations or desirable modifications and regulations in virtually any plant genes and sequences to elucidate their functions and regulatory mechanisms. The new technologies also offer powerful genetic engineering tools to inactivate or modify desired plant genes and traits for agricultural improvement.

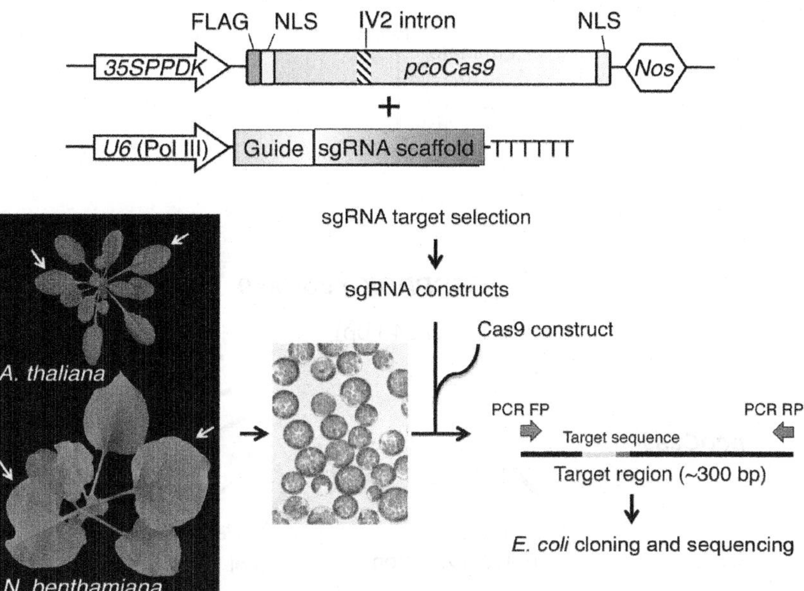


**Figure 21.1** Unbiased sgRNA/Cas9-mediated genome editing in plant protoplasts. The expression cassettes of Cas9 and sgRNA are shown. Plant codon-optimized Cas9 (pcoCas9) is fused to dual nuclear localization sequences (NLSs) and FLAG tags. The constitutive *35SPPDK* promoter and the *Arabidopsis U6-1* promoter were used to express pcoCas9 and sgRNA, respectively, in protoplasts. NGG, the protospacer adjacent motif (PAM), in the target sequence is highlighted in red. The diagram illustrates the key procedure to generate and evaluate Cas9/sgRNA-mediated genome editing in *Arabidopsis* and tobacco protoplasts. Yellow arrows indicate the leaves at optimal developmental stage for protoplast isolation from 4-week-old plants. Scale bar = 2 cm. In the target region, the target sequence of $N_{20}$ and NGG (the PAM) are represented in cyan and red, respectively. Genomic DNA from protoplasts was PCR amplified and cloned into a sequencing vector. *E. coli* colonies were picked randomly for PCR amplification and sequencing. (See the color plate.)

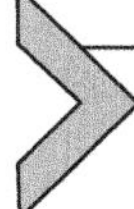

## 2. Cas9 AND sgRNA EXPRESSION

1. *p35SPPDK-pcoCas9*: a plant transient expression plasmid for expressing the plant codon-optimized *Streptococcus pyogenes Cas9* (*pcoCas9*) gene (Li et al., 2013) under the constitutive and strong hybrid *35SPPDK* promoter (Fig. 21.2A). This hybrid plant promoter (Sheen, 1993) and potentially the potato IV2 intron alleviated problems associated with cloning of the *pcoCas9* coding sequence in *Escherichia coli*. This plasmid is available at Addgene (www.addgene.org; Plasmid #52254).

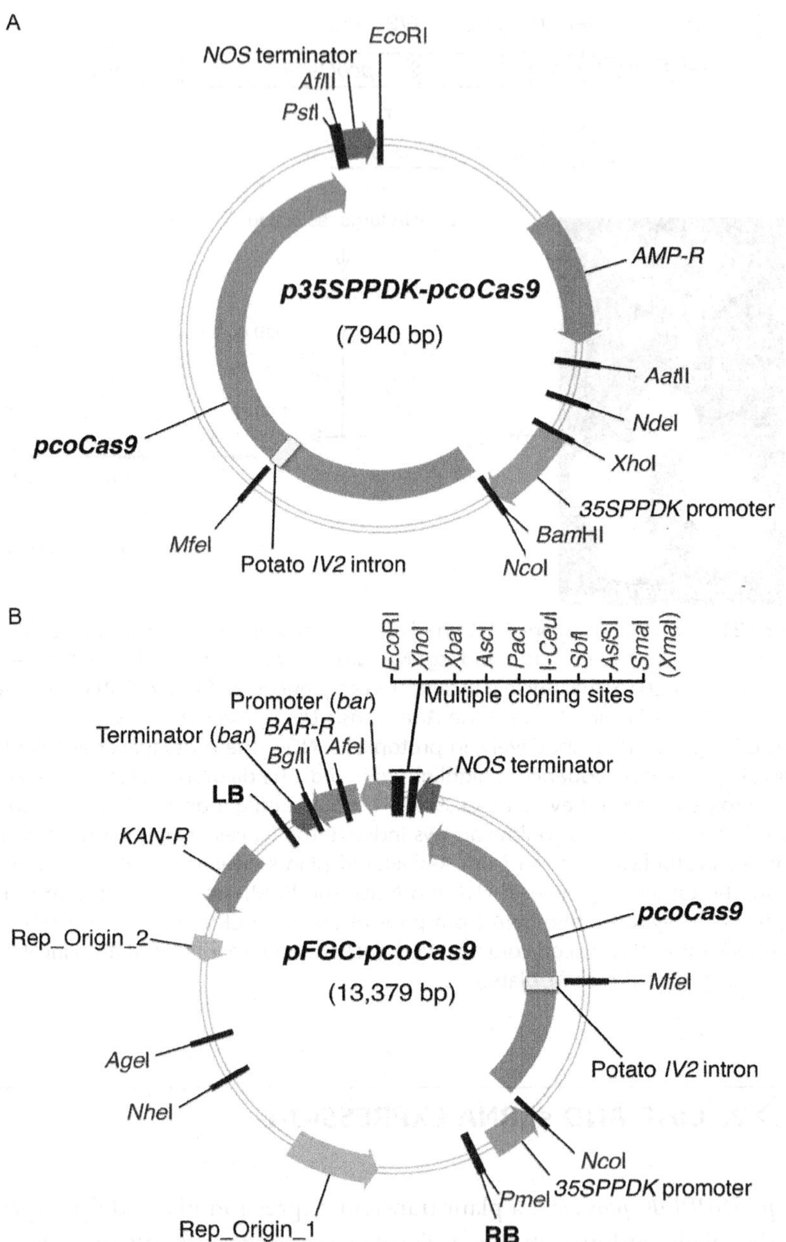


**Figure 21.2** Expression plasmid maps. (A) *p35SPPDK-pcoCas9* plasmid for protoplast transient expression. (B) Binary plasmid *pFGC-pcoCas9* for *Agrobacterium*-mediated stable or transient expression analyses.

2. *pUC119-sgRNA*: the plasmid serves as the PCR template to assemble the expression cassette of a new sgRNA with desired DNA targeting specificity. It harbors the *Arabidopsis U6-1* promoter (Li et al., 2007; Waibel & Filipowicz, 1990), an RNA polymerase III promoter required for sgRNA expression, a sgRNA targeting to the *Arabidopsis PDS3* gene (target site: 5′ GGACTTTTGCCAGCCATGGT**CGG** 3′), and a "TTTTTT" transcription terminator (Li et al., 2013). This plasmid is available at Addgene (Plasmid #52255).
3. *pFGC-pcoCas9*: a binary plasmid expressing *pcoCas9* under the *35SPPDK* promoter and containing multiple cloning sites (MCSs) for inserting single or multiple sgRNA expression cassettes (Fig. 21.2B). This plasmid is designed for *Agrobacterium*-mediated DNA delivery to the plant nuclei and available at Addgene (Plasmid #52256). Sequencing primer (sequencing from *Eco*RI toward *Sma*I): 5′ AATAAAAACTG ACTCGGA 3′.

# 3. DUAL sgRNA-GUIDED GENOME EDITING

## 3.1. Designing and constructing dual sgRNAs

1. Select a pair of closely located sgRNA targets in an *Arabidopsis* gene of interest (see *Note 1*) by referring to a preexisting database of *Arabidopsis* gene-specific sgRNA targets (Li et al., 2013) or a sgRNA target list generated upon request via the CRISPR-Plant web server (Xie, Zhang, & Yang, 2014, www.genome.arizona.edu/crispr/CRISPRsearch.htmL; see *Note 2*).
2. Design PCR primers for PCR-based seamLess assembly of new sgRNA expression cassettes (Li et al., 2013; see *Notes 3 and 4*).
3. Generate expression cassettes of sgRNAs, including the *U6-1* promoter, sgRNA, and the terminator, by an overlapping PCR strategy (see *Note 5*) using Phusion high-fidelity DNA polymerase (Li et al., 2013).
4. Insert one sgRNA expression cassette into any MCSs of a vector (e.g., *pUC119-MCS*, Addgene Plasmid #58807) to obtain the *pUC119-one-sgRNA* plasmid by restriction digestion of both the vector and the final PCR products with the same restriction enzyme(s) and subsequent ligation. The MCS are EcoRI, XhoI, BamHI, XbaI, AscI, EcoRV, SacI, PacI, I-CeuI, PstI, KpnI, SmaI, SalI, StuI, HindIII, and AscI.
5. Transform *E. coli* and inoculate a few single colonies from ampicillin-containing LB solid medium for plasmid miniprep.

6. Verify sequence accuracy of the cloned sgRNA expression cassette by Sanger sequencing.
7. Insert a second sgRNA expression cassette into the MCSs of the *pUC119-one-sgRNA* plasmid to obtain the *pUC119-dual-sgRNA* plasmid by restriction digestion and subsequent ligation (see *Note 6*).
8. Transform *E. coli* and inoculate a few single colonies on ampicillin-containing LB solid medium for plasmid miniprep.
9. Verify sequence accuracy of the second sgRNA expression cassettes in the *pUC119-dual-sgRNA* plasmid by Sanger sequencing.
10. To obtain high plasmid DNA yield, retransform *E. coli* with the *pUC119-dual-sgRNA* plasmid and the *p35SPPDK-pcoCas9* plasmid (Addgene plasmid #52254), respectively.
11. Scrape off overnight grown bacteria from the ampicillin-containing LB plate into 200 mL of Terrific broth with ampicillin using a sterile disposable inoculating loop and shake the culture vigorously at 37 °C for 8 h.
12. Maxiprep the plasmid DNA of both constructs (see *Note 7*).

## 3.2. Transfecting and expressing Cas9/sgRNAs in protoplasts

1. Mix 10 μL of the *p35PPDK-pcoCas9* plasmid (2 μg/μL) and 10 μL of the *pUC119-dual-sgRNA* plasmid (2 μg/μL) in a 2-mL round-bottom microcentrifuge tube (see *Note 8*).
2. Add 200 μL of protoplasts (40,000 cells) to the microcentrifuge tube containing the DNA cocktail. *Arabidopsis* and tobacco mesophyll protoplasts are isolated by the established protocol (Yoo et al., 2007).
3. Add 220 μL of PEG4000 solution (40% PEG4000, v/v, 0.2 *M* mannitol, 100 m*M* $CaCl_2$; Yoo et al., 2007) and gently tap the bottom of the tube a few times to completely mix DNA, protoplasts and PEG solution.
4. Incubate the transfection mixture at room temperature for 5 min.
5. Stop transfection by gently adding 800 μL of W5 solution (154 m*M* NaCl, 125 m*M* $CaCl_2$, 5 m*M* KCl, 2 m*M* MES, pH 5.7; Yoo et al., 2007) to the tube and inverting the tube twice.
6. Centrifuge the tubes at 100 × *g* for 2 min using a CL2 clinical centrifuge and remove the supernatant without disturbing the protoplast pellet (see *Note 9*).
7. Add 100 μL of W5 solution to resuspend the protoplasts.
8. Coat a 6-well culture plate with 5% bovine calf serum, remove the serum and add 1 mL of W5 or WI solution (0.5 *M* mannitol, 4 m*M* MES, pH 5.7, 20 m*M* KCl; Yoo et al., 2007) to each well.

9. Transfer transfected protoplasts to one well of the 6-well plate and mix well with the W5 or WI solution.
10. Incubate transfected protoplasts in the dark at 23–25 °C up to 36 h by covering the plate with aluminum foil.

## 3.3. Evaluating the frequency of targeted genome modifications

1. Design and synthesize a pair of genomic PCR primers (PCR FP and PCR RP, Fig. 21.1) for amplifying a ~300-bp genomic region covering the two sgRNA target sites in the target gene and introduce restriction sites into the forward primer and the reverse primer, respectively (see *Note 10*).
2. Transfer protoplasts from the 6-well plate to a 1.5 mL microcentrifuge tube and harvest protoplasts by centrifugation at 100 × *g* for 2 min using a CL2 clinical centrifuge and subsequent removal of the supernatant.
3. Freeze protoplasts immediately in liquid nitrogen.
4. Add 50 μL of sterile water to resuspend protoplasts by vortexing.
5. Heat resuspended protoplasts at 95 °C for 10 min.
6. Take 2 μL of heated protoplast suspension as the PCR template to amplify the genomic target region in a 50 μL volume using Phusion high-fidelity DNA polymerase.
7. Purify PCR products corresponding to the expected genomic amplicons and digest the PCR products with restriction enzymes at 37 °C for 1–3 h before cloning into any sequencing vector.
8. Transform *E. coli* and the next day randomly select 20–30 single colonies from ampicillin-containing LB solid medium for plasmid miniprep.
9. Conduct Sanger sequencing for plasmids extracted from individual colonies.
10. Visualize genome modifications in the target sequence by aligning DNA sequencing results to the native genomic target sequence (Fig. 21.3).
11. Calculate genome modification frequency using the following formula: genome modification frequency = (number of mutant colonies/number of total sequenced colonies) × 100%.
12. After evaluation of the editing efficacy mediated by several different pairs of sgRNAs for the target gene of interest in *Arabidopsis* and tobacco protoplasts, the most efficient sgRNA pair can be further used for generating targeted modifications in the desired genes in *Arabidopsis* and tobacco plants to obtain inheritable mutations (Fauser, Schiml, &

**Figure 21.3** Representative results of dual sgRNA/Cas9-mediated genome editing in protoplasts. Dual sgRNA-induced mutagenesis in the *AtBON1* and *NbPDS* genes in *Arabidopsis* and tobacco protoplasts, respectively. A black line marks each target sequence in the *AtBON1* and *NbPDS* genes. The protospacer adjacent motif "NGG" is in red (gray in the print version). Nucleotide deletions and substitution are shown in red (gray in the print version) as dashes and lower case letter, respectively.

Puchta, 2014; Feng et al., 2014; Nekrasov, Staskawwicz, Weigel, Jones, & Kamoun, 2013). A commonly used strategy is to clone the Cas9 and sgRNA expression cassettes into a single binary vector and then generate transgenic *Arabidopsis* plants stably expressing Cas9 and two sgRNAs using the *Agrobacterium*-mediated floral-dip transformation method (Fauser et al., 2014; Feng et al., 2014). The T1 transgenic *Arabidopsis* will express Cas9 and two sgRNAs to facilitate mutagenesis in the target gene predominantly in somatic cells and occasionally in shoot apical meristem cells and germ line cells, and the latter can eventually lead to heritable homozygous mutations in the target gene in some of the T2 transgenic *Arabidopsis* (Fauser et al., 2014; Feng et al., 2014). A DNA repair donor with homology to the target region can also be codelivered into transgenic *Arabidopsis* via the same binary plasmid (De Pater, Pinas, Hooykaas, & van der Zaal, 2013) to facilitate homologous recombination-mediated genome modifications in transgenic *Arabidopsis*. Currently, the entire procedure to generate and screen targeted homozygous mutatants is time and labor consuming. Integration of Cas9 and sgRNA expression cassettes into the *Arabidopsis* genome and constant production of these genome-editing reagents, even after the generation of intended site-specific mutagenesis, may increase risk of off targets but could be genetically segregated.

## 4. PERSPECTIVES

Rapid advances in less than a year have demonstrated that the CRISPR/Cas9 technology is applicable in protoplasts, callus tissues and intact plants in diverse plant species (Baltes, Gil-Humanes, Cermak, Atkins, & Voytas, 2014; Fauser et al., 2014; Feng et al., 2014; Li et al., 2013; Miao et al., 2013; Nekrasov et al., 2013; Shan et al., 2013; Sugano et al., 2014; Xie et al., 2014). It is conceivable that the new genetic engineering tools could be established in all plant species amenable to transient or stable gene expression manipulations. The available data suggest that mutagenesis rates appear to be much higher in tobacco and rice protoplasts with higher deletion events than in *Arabidopsis* protoplasts using similar pcoCas9 and sgRNA designs (Li et al., 2013; Shan et al., 2013). Although homozygous mutants have been obtained in transgenic *Arabidopsis* plants (Fauser et al., 2014; Feng et al., 2014), it is possible to further enhance the mutagenesis rates using dual sgRNAs demonstrated here (20% in *Arabidopsis* and 63% in tobacco; Fig. 21.3; Li et al., 2013). Manipulation of DNA repair pathways (Qi et al., 2013) and the introduction of geminivirus-based DNA replicons expressing Cas9, sgRNAs, and donor DNA templates offer promising strategies to further enhance mutagenesis rates and HDR-based gene replacement (Baltes et al., 2014). Recent improvement in tissue culture methods is promising in converting *Arabidopsis* protoplasts harboring targeted genome modifications into plants through regeneration (Chupeau et al., 2013). Coexpression of sgRNA and Cas9 by DNA or RNA bombardment and agroinfiltration in regenerating tissues, meristems, embryos or germ cells may potentially broaden the plant range accessible to genome editing.

Several issues remain to be addressed to achieve robustness, versatility and specificity in targeted genome editing and gene expression manipulation using the sgRNA/Cas9 system and its derivatives as transcription activators and repressors, chromosomal locators, and epigenome regulators. Although off-target mutations do not appear to be prevailing based on the limited cases examined in plant cells (Feng et al., 2014; Nekrasov et al., 2013; Shan et al., 2013) and can potentially be outcrossed, genome-wide sequencing in targeted mutants remains the most thorough and comprehensive option to precisely detect and critically evaluate off-target sites in each plant species. To improve specificity, it is necessary to systematically evaluate the "seed" sequences of sgRNAs and test truncated sgRNA designs and paired nickases (Sander & Joung, 2014). The effects of sgRNA sequences and target sites, paired sgRNA configurations (Fig. 21.3), protospacer adjacent motif (PAM) numbers, distance and locations in the genome, alternative PAM sequences, as well as chromatin structures and modifications may all contribute to the efficiency and specificity. It is unexplored regarding the nuclear retention, stability and sgRNA/Cas9 efficacy in different cell-types, organs, developmental stages, and plant species.

One of the most exciting applications of the sgRNA/Cas9-based genome-editing tools is the realization of simple and efficient homologous recombination-based gene or sequence replacement, or creation of novel plant genome designs that was out of reach in most plant species in the past. As shown in tobacco protoplasts, short homologous sequences flanking the sgRNA target site enabled a relatively high rate of gene replacement specifically in the presence of a DNA donor template (Li et al., 2013). Further improvement and refinement of the sgRNA/Cas9 technology will promise unprecedented opportunities and innovations in plant research, breeding and agriculture.

## 5. NOTES

1. Although targeting an *Arabidopsis* gene with a single sgRNA may be sufficient in triggering loss-of-function mutagenesis in some cases, we generally recommend using two closely targeting sgRNAs for a single gene to trigger genomic deletion to ensure the disruption of target gene function. However, single sgRNA may generate different missense or dominant gain-of-function mutations. As different sgRNAs targeting to the same gene may work with variable efficiency due to unknown factors, it is most desirable to evaluate three to four pairs of sgRNAs for targeting the same gene using the simple and rapid protoplast transient expression system (Li et al., 2013; Yoo et al., 2007). An optimal pair of sgRNAs can be rapidly identified within a week for the target gene prior to the time- and labor-consuming endeavor of generating CRISPR/Cas-mediated mutagenesis in plants with inheritable and homozygous mutations. For targeted homologous recombination,

we recommend the use of a single sgRNA whose target sequence is overlapping with or closest to the intended genomic modification site to reduce mutagenesis via NHEJ DNA repair.

2. The priority in sgRNA target selection should be given to the 5′ exons of target gene because mutagenesis in 3′ exons or all the introns may not lead to null mutations. There is currently no database or web server to aid the prediction for gene-specific sgRNA target sites in *N. benthamiana*. Genomic $N_{20}$NGG sequences can be manually identified from a tobacco gene of interest as the sgRNA target sites based on the draft genome sequence for *N. benthamiana* (http://solgenomics.net/organism/Nicotiana_benthamiana/genome).
3. The RNA polymerase III promoter (e.g., *Arabidopsis U6-1* promoter; Waibel & Filipowicz, 1990) is required to drive sgRNA transcription. Optimal transcription by the *Arabidopsis U6-1* promoter is initiated with "G". Therefore, if the selected sgRNA target sequence ($N_{20}$NGG) is not initiated with "G" ($N_1$ as "C", "A" or "T"), an additional "G" should be introduced behind the *Arabidopsis U6-1* promoter through the primer R1 using a sequence of 5′ the reverse complement of $N_{20}$ + CAATCACTACTTCGTCTCT 3′ (Fig. 21.3B). The *Arabidopsis U6-26* promoter has been used successfully in transgenic plants to obtain inheritable homozygous mutations in T2 generation (Fauser et al., 2014; Feng et al., 2014).
4. Restriction sites of *Sac*I, *Pac*I, *Pst*I, *Kpn*I, *Sma*I, or *Hin*dIII in the *pUC119-MCS* vector as cloning sites for multiple sgRNAs flanked by two *Asc*I sites are highly recommended, as sgRNAs can be easily subcloned into the binary plasmid *pFGC-pcoCas9* through *Asc*I digestion and insertion (Fig. 21.2B). Avoid using *Stu*I in sgRNA cloning because the *Arabidopsis U6-1* promoter contains an internal *Stu*I site.
5. A sgRNA expression cassette from the *Arabidopsis U6-1* promoter to the TTTTTT terminator flanked by desired restriction sites can also be synthesized as a gBlocks Gene Fragment at Integrated DNA Technologies (www.idtdna.com), despite with much increased time and cost. A more convenient *U6-26* promoter plasmid (*pChimera*) based on type II restriction enzyme BbsI cloning is recently published (Fauser et al., 2014).
6. One can also clone individual sgRNA expression cassettes into the *pUC119-MCS* vector to obtain separate sgRNA expression plasmids and then achieve sgRNA coexpression by protoplast cotransfection

with two different sgRNA expression plasmids. However, cloning a pair of sgRNA expression cassettes into the same *pUC119-MCS* vector better ensures coexpression of two sgRNAs in transfected protoplasts.

7. High quality and concentrated (2 μg/μL) plasmid DNA is key for high protoplast transfection efficiency. It is highly recommended to use CsCl gradient ultracentrifugation method to purify plasmid DNA by following the protocol on the Sheen laboratory website (http://molbio.mgh.harvard.edu/sheenweb/protocols_reg.htmL). Plasmid DNA purified by commercial DNA maxiprep kits is acceptable but may lead to lower protoplast transfection efficiency.
8. In the case of obtaining targeted homologous recombination in protoplasts, 20 μL of DNA transfection cocktail is composed of 8 μL of the *p35SPPDK-pcoCas9* plasmid (2 μg/μL), 8 μL of the *pU6-sgRNA* plasmid (2 μg/μL) and 4 μL of DNA repair template (~2 μg/μL), which can be double-stranded DNA (e.g., PCR products) containing a desired mutation flanked by two homology arms, each with at least 100 bp identical to the genomic target region (Li et al., 2013). Longer homology arms are likely to promote the efficiency of homologous recombination.
9. After centrifugation, transfected tobacco protoplasts are not pelleted as tightly as the *Arabidopsis* protoplasts, so removal of the supernatant should be conducted with caution and ~30 μL supernatant can be kept in the tube so that the pellet will not be disturbed. Tobacco protoplasts tend to aggregate during incubation.
10. Design of genomic PCR amplicons with sizes around 300 bp (Fig. 21.1) allows efficient PCR amplification using crudely prepared genomic DNA as template and makes PCR products clearly distinguishable from possible primer dimers. In addition, keeping the PCR amplicons short minimizes the possibility of PCR-introduced DNA mutagenesis.

## ACKNOWLEDGMENTS

The authors thank the Church lab at Harvard Medical School for generating the *Arabidopsis* sgRNA target database. This research was supported by the National Science Foundation grant ISO-0843244 and the National Institutes of Health grants R01 GM60493 and R01 GM70567 to J. S.

## REFERENCES

Baltes, N. J., Gil-Humanes, J., Cermak, T., Atkins, P. A., & Voytas, D. F. (2014). DNA replicons for plant genome engineering. *Plant Cell*, *26*, 151–163.

Chupeau, M. C., Granier, F., Pichon, O., Renou, J. P., Gaudin, V., & Chupeau, Y. (2013). Characterization of the early events leading to totipotency in an Arabidopsis protoplast liquid culture by temporal transcript profiling. *Plant Cell*, *25*, 2444–2463.

Cong, L., Ran, F. A., Cox, D., Lin, S., Barretto, R., Habib, N., et al. (2013). Multiplex genome engineering using CRISPR/Cas systems. *Science*, *339*, 819–823.

De Pater, S., Pinas, J. E., Hooykaas, P. J., & van der Zaal, B. J. (2013). ZFN-mediated gene targeting of the Arabidopsis *protoporphyrinogen oxidase* gene through *Agrobacterium*-mediated floral dip transformation. *Plant Biotechnology Journal*, *11*, 510–515.

Fauser, F., Schiml, S., & Puchta, H. (2014). Both CRISPR/Cas-based nucleases and nickases can be used efficiently for genome engineering in *Arabidopsis thaliana*. *Plant Journal*, *79*(2), 348–359. http://dx.doi.org/10.1111/tpj.12554.

Feng, Z., Mao, Y., Xu, N., Zhang, B., Wei, P., Yang, D. L., et al. (2014). Multigeneration analysis reveals the inheritance, specificity, and patterns of CRISPR/Cas-induced gene modifications in *Arabidopsis*. *Proceedings of the National Academy of Sciences of the United States of America*, *111*, 4632–4637.

Gaj, T., Gersbach, C. A., & Barbas, C. F. (2013). ZFN, TALEN, and CRISPR/Cas-based methods for genome engineering. *Trends in Biotechnology*, *31*, 397–405.

Jinek, M., Chylinski, K., Fonfara, I., Hauer, M., Doudna, J. A., & Charpentier, E. (2012). A programmable dual-RNA-guided DNA endonuclease in adaptive bacterial immunity. *Science*, *337*, 816–821.

Li, X., Jiang, D. H., Yong, K., & Zhang, D. B. (2007). Varied transcriptional efficiencies of multiple *Arabidopsis U6* small nuclear RNA genes. *Journal of Integrative Plant Biology*, *49*, 222–229.

Li, J. F., Norville, J. E., Aach, J., McCormack, M., Zhang, D., Bush, J., et al. (2013). Multiplex and homologous recombination-mediated genome editing in *Arabidopsis* and *Nicotiana benthamiana* using guide RNA and Cas9. *Nature Biotechnology*, *31*, 688–691.

Li, J. F., Zhang, D., & Sheen, J. (2014). Epitope-tagged protein-based artificial microRNA screens for optimized gene silencing in plants. *Nature Protocols*, *9*, 939–949.

Mali, P., Yang, L., Esvelt, K. M., Aach, J., Guell, M., DiCarlo, J. E., et al. (2013). RNA-guided human genome engineering via Cas9. *Science*, *339*, 823–826.

Miao, J., Guo, D., Zhang, J., Huang, Q., Qi, G., Zhang, X., et al. (2013). Targeted mutagenesis in rice using CRISPR-Cas system. *Cell Research*, *12*, 1233–1236.

Nekrasov, V., Staskawwicz, B., Weigel, D., Jones, J. D., & Kamoun, S. (2013). Targeted mutagenesis in the model plant Nicotiana benthamiana using Cas9 RNA-guided endonuclease. *Nature Biotechnology*, *31*, 691–693.

Qi, Y., Zhang, Y., Zhang, F., Baller, J. A., Cleland, S. C., Ryu, Y., et al. (2013). Increasing frequencies of site-specific mutagenesis and gene targeting in Arabidopsis by manipulating DNA repair pathways. *Genome Research*, *23*, 547–554.

Sander, J. D., & Joung, J. K. (2014). CRISPR-Cas systems for editing, regulating and targeting genomes. *Nature Biotechnology*, *32*, 347–355.

Shan, Q., Wang, Y., Li, J., Zhang, Y., Chen, K., Liang, Z., et al. (2013). Targeted genome modification of crop plants using a CRISPR-Cas system. *Nature Biotechnology*, *31*, 686–688.

Sheen, J. (1993). Protein phosphatase activity is required for light-inducible gene expression in maize. *EMBO Journal*, *12*, 3497–3505.

Sugano, S., Shirakawa, M., Takagi, J., Matsuda, Y., Shimada, T., Hara-Nishimura, I., et al. (2014). CRISPR/Cas9-mediated targeted mutagenesis in the liverwort *Marchantia polymorpha* L. *Plant & Cell Physiology*, *55*, 475–481.

Waibel, F., & Filipowicz, W. (1990). U6 snRNA genes of *Arabidopsis* are transcribed by RNA polymerase III but contain the same two upstream promoter elements as RNA polymerase II-transcribed U-snRNA genes. *Nucleic Acids Research*, *18*, 3451–3458.

Xie, K., Zhang, J., & Yang, Y. (2014). Genome-wide prediction of highly specific guide RNA spacers for CRISPR-Cas9-mediated genome editing in model plants and major crops. *Molecular Plant*, 7, 923–926.
Yoo, S. D., Cho, Y. H., & Sheen, J. (2007). Arabidopsis mesophyll protoplasts: A versatile cell system for transient gene expression analysis. *Nature Protocols*, *2*, 1565–1572.

CHAPTER TWENTY-TWO

# Multiplex Engineering of Industrial Yeast Genomes Using CRISPRm

**Owen W. Ryan*, Jamie H.D. Cate*,†,‡,§,1**

*Energy Biosciences Institute, University of California, Berkeley, California, USA
†Department of Molecular and Cell Biology, University of California, Berkeley, California, USA
‡Department of Chemistry, University of California, Berkeley, California, USA
§Physical Biosciences Division, Lawrence Berkeley National Laboratory, Berkeley, California, USA
[1]Corresponding author: e-mail address: jcate@lbl.gov

## Contents



## Abstract

Global demand has driven the use of industrial strains of the yeast *Saccharomyces cerevisiae* for large-scale production of biofuels and renewable chemicals. However, the genetic basis of desired domestication traits is poorly understood because robust genetic tools do not exist for industrial hosts. We present an efficient, marker-free, high-throughput, and multiplexed genome editing platform for industrial strains of *S. cerevisiae* that uses plasmid-based expression of the CRISPR/Cas9 endonuclease and multiple ribozyme-protected single guide RNAs. With this multiplex CRISPR (CRISPRm) system, it is possible to integrate DNA libraries into the chromosome for evolution experiments, and to engineer multiple loci simultaneously. The CRISPRm tools should therefore find use in many higher-order synthetic biology applications to accelerate improvements in industrial microorganisms.

*Methods in Enzymology*, Volume 546
ISSN 0076-6879
http://dx.doi.org/10.1016/B978-0-12-801185-0.00023-4

# 1. INTRODUCTION

For thousands of years, humans have domesticated Baker's yeast *Saccharomyces cerevisiae* for the production of alcohol and bread. More recently, global demand has driven the use of industrial strains of *S. cerevisiae* for large-scale production of biofuels and renewable chemicals (Farrell et al., 2006; Rubin, 2008). However, the genetic basis of desired domestication traits is poorly understood because robust genetic tools do not exist for industrial production hosts. Industrial *S. cerevisiae* strains are more stress tolerant and produce much higher yields of desired biofuel or renewable chemical end products than laboratory strains. However, linking genotypes of industrial yeasts to their phenotypes remains difficult because these strains are often polyploid with low-efficiency mating and sporulation. The standard genetic tool of integrating linear DNA into the genome by homologous recombination (HR) is too inefficient for the creation of loss-of-function phenotypes in these strains, and current technologies rely on dominant selectable markers for chromosomal integrations or plasmid maintenance. Since only a small number of markers exist, deciphering and improving important complex multigenic phenotypes in industrial strains remains a challenge.

Due to the lack of genetic tools, most industrially relevant phenotypes must be tested in haploid derivatives of the industrial strains or in lab strains. These haploid derivatives may not cosegregate the alleles required for the relevant phenotype, particularly if the phenotype is complex, so in many cases they are not ideal surrogates for industrial isolates. Phenotypes observed in one segregant may not be similar to the others. Therefore, phenotypes are best tested within the relevant industrial strain in the exact state to which it is found in the industrial process. Further, lab strains do not act as good proxies for strain-specific phenotypes, as even the most straightforward phenotypes such as essentiality in rich medium can differ substantially between two laboratory strains due to complex genetics with unpredictable allelic combinations (Dowell et al., 2010).

Present technologies for heterologous gene expression by integrating genes into yeast chromosomes require the recombination and cointegration of the gene to be expressed and a dominant selectable marker to identify cells with the integrated DNA. The efficiency of chromosomal integration is low and homozygous integrations need to be made by iteratively incorporating the gene with a different selectable marker. Further complicating matters, the gene integrated second may replace the first integration by recombination. Finally, at the end of the process, the selectable markers need to be removed before the engineered yeast can be used in an industrial setting (Solis-Escalante, Kuijpers, van der Linden, Pronk, & Daran-Lapujade, 2014). Therefore, it is difficult, time-consuming and labor-intensive to generate homozygous integrations in diploid (or higher ploidy) yeast strains. Ideally, an experimenter needs a targeting method that does not require an integrated marker and precisely cuts all chromosomes without the requirement of any premade genetic modifications to the cell, such as auxotrophic markers. A system such as this would be ready for use in any industrial, wild or unmodified isolate, including those with higher chromosome copy number.

Bacterial type II CRISPR/Cas9 genome editing has been used successfully in several eukaryotic organisms but has not been adapted for genome-wide studies or for heterologous protein engineering in industrially important eukaryotic microbes. CRISPR/Cas systems require a Cas9 endonuclease that is targeted to specific DNA sequences by a noncoding single guide RNA (sgRNA) (Jinek et al., 2012). The Cas9–sgRNA ribonucleoprotein complex precisely generates double-strand breaks (DSBs) in eukaryotic genomes at sites specified by a twenty-nucleotide guide sequence at the 5′ end of the sgRNA that base pairs with the protospacer DNA sequence preceding a genomic Protospacer adjacent motif (PAM) (Sternberg, Redding, Jinek, Greene, & Doudna, 2014). Repair by nonhomologous end joining results in small deletions or insertions in the genome 5′ of the PAM motif (Cong et al., 2013; Mali, Esvelt, & Church, 2013; Mali, Yang, et al., 2013). Alternatively, the presence of the Cas9-produced DSB in genomic DNA can increase the rate of HR with linear DNA at the DSB locus by several thousand-fold (DiCarlo et al., 2013), potentially enabling high-throughput genetic studies.

CRISPR gene targeting lends itself to the efficient targeting of yeast genomes, for both loss-of-function analysis and heterologous gene expression. In the yeast system developed in our lab, the CRISPR/Cas9 endonuclease is coexpressed with multiple ribozyme-protected sgRNAs (Fig. 22.1). This system enables efficient, marker-free, single step, and multiplexed genome editing in industrial strains of *S. cerevisiae.* The power of multiplex CRISPR (CRISPRm) can be used to accelerate discoveries of the genetic and molecular determinants of improved industrial microorganisms. Further, CRISPRm can be used in any prototrophic yeast isolate, making the system plug-and-play ready. Unlike other systems, CRISPRm requires

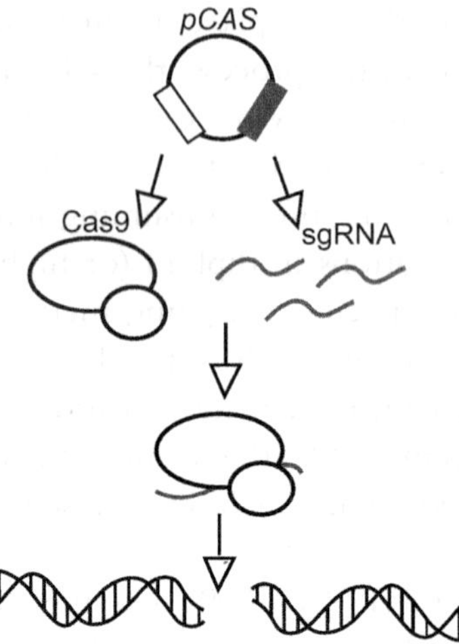


**Figure 22.1** Cas9–sgRNA coexpression. Cas9 protein and the tRNA–HDV–sgRNA are coexpressed from a single plasmid in yeast cells. Cas9 and the sgRNA form a ribonucleoprotein complex within the nucleus. Cas9 is directed to cut genomic DNA by the 20 bp target sequence encoded within the sgRNA.

no previous genetic modifications (Wingler & Cornish, 2011) and the drug-resistant plasmid used to coexpress the Cas9 protein and sgRNA confers dominant drug resistance. Because there is no fitness advantage in non-selective conditions the Cas9 plasmid (pCAS) is readily lost in rich medium shortly after the drug has been removed from the medium. CRISPRm results in homozygous mutants of yeast cells with diploid (or higher) copy number; to date we have not recovered heterozygous mutants. This is likely the case because Cas9 protein will cut all of the targeted chromosomes. We suggest that the linear DNA is used as a homology directed repair template to correct one chromosome and then that repaired chromosome is then used as DNA template to repair the other chromosomes by HR. With respect to the genome, CRISPRm is marker-free and iterative, enabling much more complex genome editing practices in lab and industrial yeast.

## 2. PLASMID DESIGN

The first step in creating a coexpression system is to build a plasmid that can be: (A) stably maintained in the expression host and (B) propagated in bacteria. A plasmid that can be maintained in both bacteria and yeast requires a species-specific origin of replication and a dual function selection marker. Coexpressing the two components of the CRISPRm system (Cas9, sgRNAs) from a single plasmid has the advantage of not needing to be cotransformed, coinherited, or coexpressed and requires only one marker selection in the medium. For stable propagation and inheritance in bacteria and yeast, the plasmid contains a bacterial pUC origin of replication, a yeast 2 μ origin of replication and a dominant selectable marker that works in both yeast and bacteria, such as G418 resistance in eukaryotes versus kanamycin in prokaryotes. The drugs nourseothricin and hygromycin also work in both yeast and bacteria so they can be used as well as selection markers. Cas9 is a large protein and there is a limit to the size of extrachomosomal DNA that can be correctly replicated and inherited by yeast cells. Therefore, we made the smallest possible plasmid with minimal extraneous DNA sequences (Fig. 22.2).

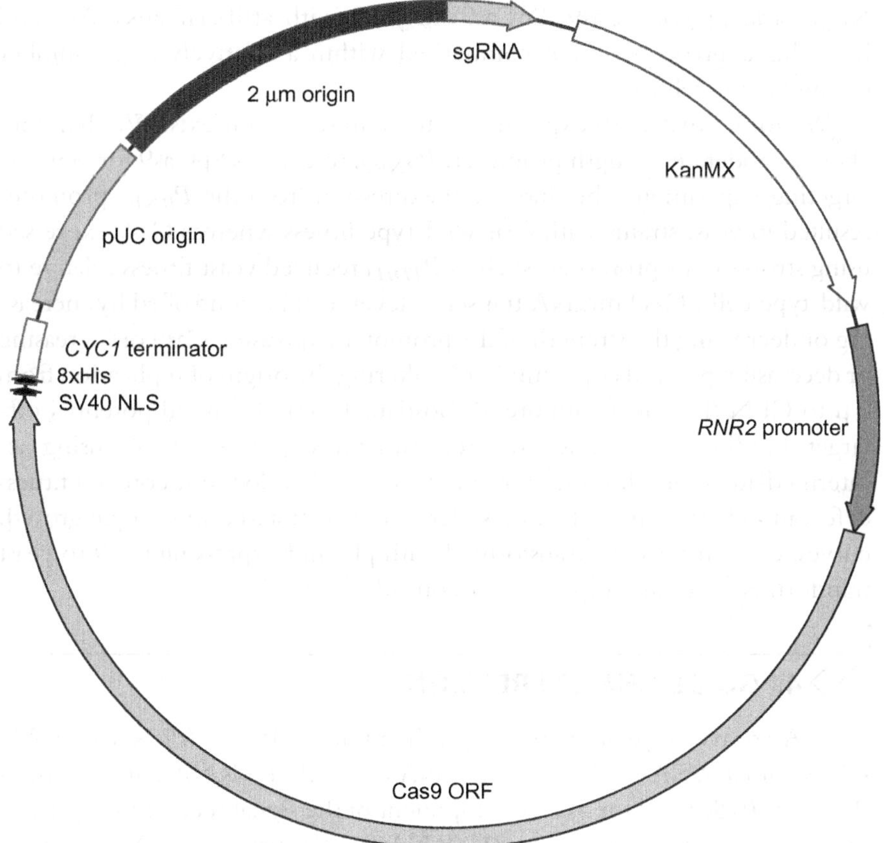


**Figure 22.2** The structure of pCAS. The pCAS plasmid contains a bacterial origin of replication (pUC), a yeast high copy origin of replication (2 μ), the tRNA–HDV–sgRNA expression module, a dual-purpose dominant marker (KANMX cassette), and the Cas9 expression module, which uses the *RNR2* promoter used to drive the expression of Cas9 and a *CYC1* terminator.

## 3. Cas9 EXPRESSION

Cas9 genome editing requires the coexpression of the Cas9 endonuclease and the guide RNA (Jinek et al., 2012). Correctly expressed, the guide RNA binds to Cas9 and forms a functional ribonucleoprotein, equipped with a precise targeting sequence within the guide RNA (Jinek et al., 2014). *In vivo*, this means that both the protein and RNA components need to be expressed at physiologically tolerable (nontoxic) levels by the cell, colocalized and correctly folded. Overexpression of proteins can be toxic to yeast (Sopko et al., 2006) and the heterologous expression of physiologically active RNAs in yeast is relatively unexplored. The goal is to express *Streptococcus pyogenes* Cas9 (SpCas9) in yeast with artificial sgRNA(s) and have this expression system coinherited within a relatively large number of competent cells.

We found that Cas9 expression is toxic in some contexts. We therefore chose a moderate strength promoter, $P_{RNR2}$, to express SpCas9 for genome targeting experiments, because Cas9 expression from the $P_{RNR2}$ promoter resulted in yeast strains with near wild-type fitness whereas Cas9 expressed using strong yeast promoters such as $P_{TDH3}$ reduced yeast fitness relative to wild-type cells. Cas9 mRNA transcript levels can be controlled by increasing or decreasing the strength of the promoter expressing Cas9 or increasing or decreasing plasmid copy number by altering the origin of replication from 2 μ to CEN (Parent, Fenimore, & Bostian, 1985). To avoid potential off-target binding or cutting, we recommend expressing Cas9 using an intermediate-strength promoter or at least one that does not confer a fitness defect in rich medium. The fitness defect can be tested using simple growth curves, comparing yeast transformed with plasmid expressing Cas9 to yeast transformed with an empty vector control.

## 4. GUIDE RNA EXPRESSION

Ascomycete yeasts express all of their transfer RNAs (tRNAs), the U6 spliceosomal RNA *SNR6*, the snoRNA *SNR52*, the RNA component of RNase P *RPR1* and the RNA component of the signal recognition particle *SCR1* using RNA Polymerase III (RNA Pol III) promoters (Marck et al., 2006). These RNA Pol III transcripts have varying architectures but all contain the essential components for RNA Pol III transcription initiation. They contain A Box and B Box binding domains and a TATA Box binding domain. Only one transcript, *SNR52*, resembles a canonical RNA Polymerase II promoter with the polymerase binding motifs (A and B Boxes) 5′ of the TATA box and *SNR52* RNA coding sequence. Transfer RNAs, by contrast, have the A and B Box motifs within the mature tRNA sequence. All RNA Polymerase III transcription termination occurs by the same mechanism, with RNA Pol III transcription terminated by a string of poly-U nucleotides, six in yeast and five in higher eukaryotes (Marck et al., 2006). This means that screening promoters for the expression of the sgRNA is directly comparable and not confounded by promoter–terminator effects found with RNA Pol II promoters (Curran, Karim, Gupta, & Alper, 2013). The sgRNA expression systems developed in higher eukaryotes use the promoter from the U6 snRNA gene as the RNA Pol III promoter to express the sgRNA, an approach that has been used for RNA interference experiments for several years (Mali, Esvelt, et al., 2013; Mali, Yang, et al., 2013).

In mammalian cells, cellular levels of sgRNAs correlate with the efficiency of Cas9-mediated genome targeting (Hsu et al., 2013), raising the possibility that sgRNA abundance is rate limiting for *in vivo* CRISPR mediated genome targeting. It is therefore important to consider ways to increase the transcript levels of the sgRNAs. One way to increase the cellular abundance is to protect the sgRNA from the intracellular RNA degradation machinery. This can be done by adding self-cleaving ribozymes (or any RNA) that could be physically connected to the sgRNA to increase *in vivo* stability and regulation of expression or processing. However, the added RNA must not interfere with the structure or expression of the sgRNA so that the sgRNA is in a conformational state that permits binding to Cas9 within the cell. To increase sgRNA levels in yeast, we engineered a new sgRNA architecture by fusing the sgRNA(+85) (Mali, Esvelt, et al., 2013; Mali, Yang, et al., 2013) to the 3′ end of the self-cleaving hepatitis delta virus (HDV) ribozyme, to protect the sgRNA from 5′-exonucleolytic activities in the cell. Naturally occurring self-cleaving ribozymes are noncoding RNAs widespread in nature (Webb, Riccitelli, Ruminski, & Luptak, 2009), and HDV-like ribozymes have been identified in a multitude of eukaryotes. We chose to use the HDV ribozyme, because its cleavage leaves a clean 5′-end on the RNA with no extraneous nucleotides 5′ of the HDV ribozyme (Ke, Ding, Batchelor, & Doudna, 2007). Furthermore, the clean 5′ end may also aid in nuclear retention of the RNA (Kohler & Hurt, 2007). HDV-like ribozymes are highly conserved, forming a double-pseudoknot secondary structure, and the nucleotides that are essential for its enzymatic activity have been mapped (Ke et al., 2007). Although it should be possible to use an inactive HDV ribozyme or other structured but noncatalytic "protection" RNAs, we selected the active form of the HDV ribozyme because its cleavage would remove any structured or unstructured RNA used as a promoter, leaving the HDV ribozyme covalently fused 5′ to the sgRNA (Fig. 22.3A).

We measured the relative cellular abundance of sgRNAs expressed by a yeast RNA Pol III promoter with and without the 5′ ribozyme, by reverse transcription quantitative polymerase chain reaction (PCR) and found that the presence of the ribozyme increases the intracellular abundance of the sgRNAs by sixfold. This is consistent with our hypothesis that the structure of the HDV ribozyme serves as protection at the 5′ end of the sgRNA from 5′ exonucleases (Houseley & Tollervey, 2009). Notably, in the absence of the HDV ribozyme, we found that RNA Polymerase II promoters resulted in highly expressed sgRNA molecules in yeast but they were physiologically

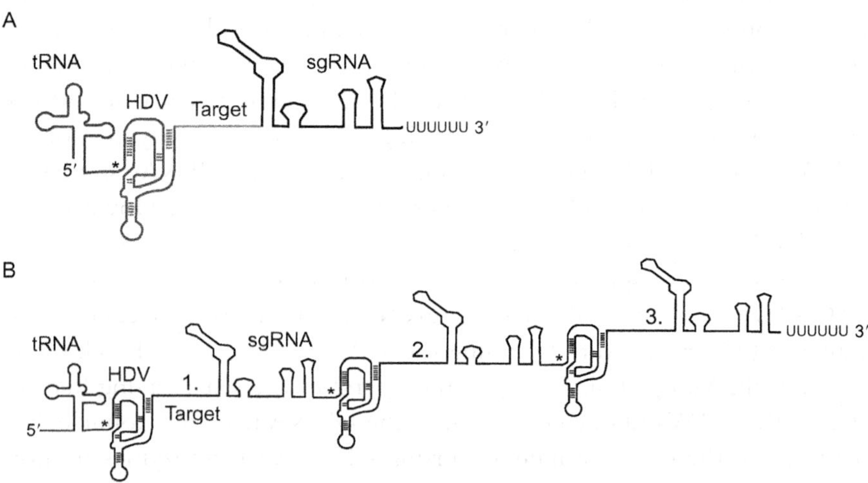


**Figure 22.3** The RNA structure of the sgRNA expression module. (A) The sgRNA is expressed using a tRNA as an RNA Pol III promoter. The HDV folds into its catalytically active form and removes the 5′ tRNA sequence from the mature sgRNA (cleavage site marked by an asterisk). The target sequence is protected between the HDV and sgRNA. RNA Pol III expression is terminated by a series of six or more uridine nucleotides. (B) Engineering sgRNAs for improved coexpression. Future polycistronic sgRNAs could be expressed using a single RNA PolIII promoter (tRNA) and processed internally by their catalytically active HDV (cleavage sites are marked by *). This panel contains three sgRNAs in tandem arrays. (See the color plate.)

inactive, resulting in very low efficiency ($\ll$1%) Cas9-targeting. Although we did not characterize these sgRNAs further, we propose that the lack of sgRNA activity could be due to the failure of the RNA Pol II terminators to cleanly terminate transcription, resulting in 3′ RNA sequences that interfere with the sgRNA folding, or sgRNA nuclear localization. Therefore, we propose that the total abundance of sgRNA is not limiting but rather a total abundance of correctly folded and localized sgRNAs. In summary, we have shown that the sgRNA is readily engineered in a modular fashion that incorporates structured RNAs 5′ to the sgRNA guide sequence. In the future, it could be possible to incorporate several HDV-like ribozyme-sgRNA chimeras in series to produce multiple sgRNAs with increased genome editing functionality. Expressing the sgRNA chimeras using a single promoter and allowing them to be posttranscriptionally modified could control the coexpression of multiple guides used in higher-order multiplex editing (Fig. 22.3B). In addition, the ribozyme enables the modification of the 5′ leader sequence (promoter) so future promoters may be engineered to adjust the expression of the sgRNA. Further, it is possible to add 3′ sequences to the sgRNA to protect them from 3′ exonucleases (Hsu et al., 2013; Jinek et al., 2013). There remain many options for the engineering of the sgRNA so that the expression level divide between Cas9 and the rate limiting sgRNA may be bridged. We anticipate that by improving the coexpression stoichiometry of the protein and RNA components of CRISPR/Cas9 systems will be of great value in more complex editing experiments such as higher-order multiplexing.

## 5. SCREENING METHOD

For CRISPRm to become a common practice in nonspecialized laboratories it is important that the protocol and reagent sets are simple, cost efficient and rely on well-established protocols. The pCAS plasmid developed in our lab can be used in any *S. cerevisiae* strain, including prototrophic isolates. Only one modification to pCAS is required for genome targeting. The researcher only needs to clone the target (protospacer) 20-mer sequence into the sgRNA encoded in the pCAS plasmid.

Screening consists of two steps: (1) cloning the desired guide into an sgRNA expression system within a universal plasmid and (2) genome targeting by cotransforming the plasmid with linear repair DNA in yeast by chemical transformation.

### 5.1. Cloning the target sequence into pCAS

Restriction-free (RF) cloning (van den Ent & Lowe, 2006) is efficient for cloning the target 20-mer sequences into pCAS. Restriction free cloning (ligation-independent cloning) uses PCR followed by treatment with the enzyme DpnI to incorporate scarless clones into any DNA fragment. In the case of cloning target sequences into pCAS, the primers are 60-mer oligonucleotides. These (complementary) 60-mer sequences code for the target RNA sequence (20 bp) flanked by 20 bp of homology to the ribozyme (5′ of the target) and the sgRNA (3′ of the target). The two homology sequences used for cloning are never altered because in pCAS, the target sequence is always preceded by the ribozyme and followed by the sgRNA, which are termed Left and Right homology sequences, respectively. The RF cloning reaction inserts the target sequence between the homology regions, resulting in a functional sgRNA construct within pCAS.

Insert the target 20-mer in between the L and R homology sequences. Order two oligonucleotides: (1) the 60-mer of [L + target + R] and (2) the reverse complement of [L + target + R] to be used as the primers for the RF reaction (Table 22.1).

Homology L (HDV) = CGGGTGGCGAATGGGACTTT

Homology R (sgRNA) = GTTTTAGAGCTAGAAATAGC

The cloned 60-mer sequence is:

5′-CGGGTGGCGAATGGGACTTTXXXXXXXXXXXXXXXX
XXXXGTTTTAGAGCTAGAAATAGC-3′

Perform a DpnI reaction to ensure that all methylated (plasmid, non-PCR generated) DNA is degraded. It is most efficient to run the DpnI reaction overnight at 37 °C and heat-inactivate DpnI at 65 °C for 15 min.

Transform the clone into *E. coli* using 5–10 μL of the DpnI-treated reaction in 50 μL of competent bacterial cells by standard bacterial transformation. Pick 3–5 bacterial transformants to verify the sRNA sequence by using the sgRNA sequencing primer.

The sgRNA sequencing primer = 5′-CGGAATAGGAACTTCAAA GCG-3′

### 5.2. Double-stranded linear DNA repair oligos

To create a markerless yet traceable integration site in the yeast genome, we developed a modified barcode system (Giaever, Chu, et al., 2002). This barcoded DNA used for genome integration is assembled in a modular fashion by PCR. Three 60-mer oligonucleotides are purchased and assembled

**Table 22.1** RF cloning reaction

| |
|---|
| $H_2O$ = to 50 μL |
| 5 × buffer = 10 μL |
| dNTP = 1 μL |
| Insert (target DNA) = 0.1 μL + 0.1 μL (rev. complement) |
| pCAS plasmid = 40 ng |
| Pfusion polymerase = 1 μL |
| **In a thermal cycler, run the following reaction for 30 cycles:** |
| 98 °C—1:00 |
| 98 °C—0:30 |
| 58 °C—1:00 |
| 72 °C—10:00 |
| 72 °C—10:00 |
| 4 °C (end) |

into a single 140-mer double-stranded DNA molecule (Fig. 22.4). There are two reasons for assembling the repair DNA within the laboratory: (A) the repair molecule is modular and can be assembled to fit the experiment and (B) cost of the respective oligonucleotides.

## 5.3. CRISPRm screening consists of the cotransformation of pCAS and the double-stranded linear DNA homologous repair template

To create a stock of yeast competent cells, grow a preculture of yeast cells to saturation. Subculture the saturated yeast culture in rich medium and grow to mid-logarithmic phase with an optical density of 1.0 ($OD_{600}$ = 1.0). Pellet and resuspend in 600 μL of equal parts 40% glycerol and PLATE (Polyethyleneglycol 2000 (PEG 2000), 0.1 *M* Lithium acetate—0.05 *M* Tris–HCl EDTA) in centrifuge tubes. Transfer cells to a −80 °C freezer. Cells can be stored indefinitely.

The Cas9 transformation mix consisted of 90 μL yeast competent cell mix ($OD_{600}$ = 1.0), 10.0 μL ssDNA (10 m*M*) 1.0 μg pCAS plasmid, 5.0 μg of linear repair DNA and 900 μL Polyethyleneglycol 2000 (PEG 2000), 0.1 *M* Lithium acetate—0.05 *M* Tris–HCl EDTA (Table 22.2). To measure

**Table 22.2** CRISPR/Cas9 screening protocol

| Screening reagents: |
|---|
| 1. 90 μL competent cells |
| 2. 10.0 μL ssDNA (boil 5 min, ice 5 min) |
| 3. 1.0 μg pCAS plasmid DNA + 5.0 μg of linear barcode (BC) DNA |
| 4. 0.25 μg of pOR1.1 + 5.0 μg of linear barcode (BC) DNA |
| 5. 900 μL PLATE |
| **Protocol:** |
| 1. Pipette mix for each sample:<br>a. pCAS + Barcode (any linear DNA with homology)<br>b. pOR1.1 + Barcode (linear DNA)<br>c. Negative control (no DNA) |
| 2. Incubate 30 min at 30 °C |
| 3. Shake tubes using the tube holding block |
| 4. Heat shock 42 °C for 17 min |
| 5. Centrifuge 5 K for 2 min |
| 6. Remove PLATE (dump, then pipette aspirate) |
| 7. Resuspend cells in 250 μL YPD |
| 8. Recover for 2.0 h at 30 °C |
| 9. Plate entire contents onto YPD + G418 |
| 10. Grow 48 h at 37 °C |
| 11. Replica plate to confirm mutant phenotype |
| 12. Perform gDNA isolation |
| 13. PCR to confirm barcode integration |
| 14. Sequence PCR amplicon to confirm barcode sequence |

**Figure 22.4** The integrated barcode repair DNA. Each barcode contains a unique 20-mer sequence flanked by common primer sites and a 5′ STOP codon (60-mer). The barcode is amplified using two 60-mer primers with 10 bp of homology to the barcode oligonucleotide and 50 bp homology to the genome. The resulting homology directed repair dsDNA is 140 bp in length.

Cas9 independent integration, the linear DNA can be cotransformed with a plasmid lacking the Cas9 protein and sgRNA (i.e., plasmid pOR1.1). Cells are incubated 30 min at 30 °C, and then subjected to heat shock at 42 °C for 17 min. Following heat shock, cells are resuspended in 250 μL YPD at 30 °C for 2 h and then the entire contents were plated onto YPD + G418 plates (20 g/L Peptone, 10 g/L Yeast Extract, 20 g/L Agar, 0.15 g/L Adenine hemisulfate, 20 g/L Glucose, and G418 at 200 mg/L). Cells are grown for 48 h at 37 °C and replica plated onto phenotype-selective media. Genomic DNA is isolated, PCR amplified and sequenced to confirm barcode sequence in the amplicon. The experimenter can expect >95% efficiency of correctly targeted repair DNA integration in colonies transformed with pCAS, i.e., that are G418 resistant. All screens are performed using a linear DNA background control with an empty vector, which we named pOR1.1. pOR1.1 lacks the Cas9 and sgRNA of pCAS. pOR1.1 is used to measure the CRISPRm-independent integration of linear DNA, which in our experience has always equaled zero.

## 5.4. Industrial yeast

We have tested the barcode targeting efficiency in a polyploid industrial strain, ATCC4124, which was isolated from a molasses distillery. We found that the efficiency of homozygosis in these strains was nearly 100% using a tRNA as a promoter to drive sgRNA expression, but was weak (~5%) when using the non-tRNA promoter $P_{SNR52}$ (Fig. 22.5A). Thus, there is a major advantage to using tRNAs as sgRNA promoters for the efficient creation of homozygous null mutants in industrial yeast isolates, and it may be necessary to test different tRNA promoters to find the optimal ones for sgRNA expression.

## 5.5. Markerless gene assembly in the yeast chromosome

To determine whether CRISPRm could be used for *in vivo* yeast gene assembly for chromosomal integrations, we tested the efficiency of assembly of a functional nourseothricin-resistance (Nat$^R$) gene from multiple PCR products. The correct assembly and insertion of three overlapping PCR products that encode a transcription promoter, protein open reading frame (ORF) and transcription terminator, result in expression of the Nat$^R$ gene and confer nourseothricin resistance (Boeke, LaCroute, & Fink, 1984; Krugel, Fiedler, Haut, Sarfert, & Simon, 1988) (Fig. 22.5B).

PCR amplification of DNA is used for *in vivo* assembly. We used PCR to amplify genomic DNA containing 50 bp of overlapping DNA on the 5′ end of the primers. We cotransformed three separate, linear DNA molecules that overlap by 50 bp, including the *TEF1* promoter and terminator of *Ashbya gossypii*, and a nourseothricin-resistance (*Nat$^R$*) drug resistance ORF from *Streptomyces noursei* and targeted these molecules to the *URA3* locus.

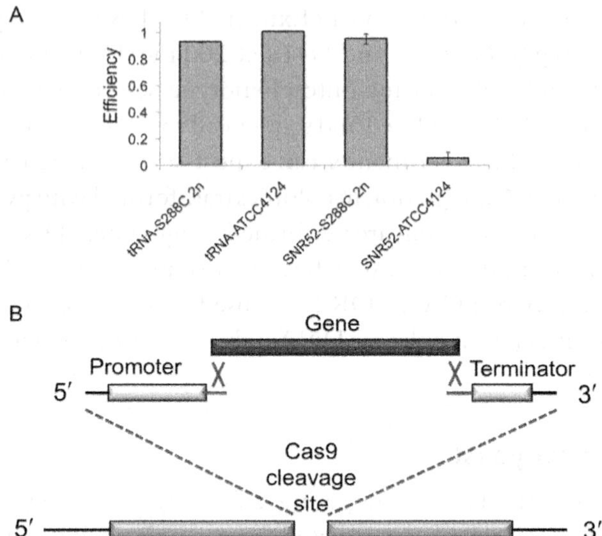


**Figure 22.5** Genome editing experiments. (A) Barcoding industrial yeast strains. The efficiency of barcode insertion, as determined by the percentage of auxotrophy, is shown by strain and RNA Pol III promoter tested. Barcode insertion was verified by PCR amplification and Sanger sequencing of the target site from each colony. (B) Assembling genes for chromosomal heterologous expression. CRISPRm creates a double-stranded break in the genome. Three linear DNA molecules generated by PCR contain 50 bp of homology to each other and 50 bp of homology to the genome (at the ends of the distal fragments). The homologous repair machinery of *S. cerevisiae* assembles the three fragments and integrates them into the CRISPRm-targeted chromosome.

The efficiency of Cas9-mediated integration and assembly of three DNA fragments to the correct locus was measured by a combination of 5-fluoroorotic acid resistance (5-FOA$^R$) and Nat$^R$. As observed with integrating barcoded oligos into the genomes of industrial isolates, the integration of functional expression modules was dependent on the tRNA promoter. We found 85% efficiency of correct homozygous targeting and assembly in diploid yeast S288C cells and 68% in yeast strain ATCC4124 using the tRNA$^{Phe}$ sequence as the sgRNA promoter (Ryan et al., 2014). Thus, CRISPRm enables the one-step markerless assembly of functional genes in the *S. cerevisiae* genome, including the genome of an industrial isolate. As an example of an application, we have exploited the high efficiency of gene assembly in the chromosome to evolve and select improved variants of a cellobiose transporter for lignocellulosic biofuel applications (Galazka et al., 2010).

## 6. CONCLUDING REMARKS

In this chapter, we describe methods for integrating linear DNA into yeast chromosomes for: (A) loss of function studies and (B) heterologous protein expression. Because CRISPRm does not require selectable markers to be integrated with the DNA, this method is fully scalable. We propose that by using iterations of CRISPRm, a synthetic chromosome (or genome) in theory could be reduced from a wild-type organism. This is particularly possible because of the multiplex function with minimal reduced efficiency in haploid cells as identified in our lab (Ryan et al., 2014).

The CRISPRm method of expressing Cas9 and multiple sgRNAs enables its broad utility for advanced genomic analysis and engineering of any yeast, including industrial isolates. First, the engineered sgRNA containing the HDV ribozyme enables the use of functional tRNAs as sgRNA promoters, which can be easily identified in any genome (Marck et al., 2006). Further, the HDV ribozyme significantly increases the cellular abundance of the sgRNA, removes the 5′ tRNA and leads to efficient multiplex targeting in diploid cells. The ability to remove 5′ sequences by HDV-like ribozymes without interfering with sgRNA functionality opens many sgRNA expression and multiplexing possibilities. For example, it should be possible to express many sgRNAs in eukaryotes using single or multiple tRNA promoters that can be posttranscriptionally converted into multiple high fidelity sgRNAs, each with a specific target encoded within it. This could be used to better regulate the coexpression and stoichiometry of sgRNAs for multiplexed editing.

Because the only requirements for CRISPRm are moderate expression of the Cas9 protein, an sgRNA expressed using a tRNA promoter, and unique target sequences, we anticipate that CRISPRm should be portable for genome engineering in nonmodel organisms, including fungal extremophiles used in the biotechnology industry and for novel drug target identification in fungal pathogens. The genetics of many of these organisms have not been studied in any depth due to the technological limitations of available genetic manipulation techniques. CRISPRm should serve as a rapid and high-throughput protocol for connecting the genotypes of these organisms to their phenotypes. For example, CRISPRm can be used to generate marker-free barcoded alleles for large-scale pooled fitness studies of loss-of-function mutants in these organisms. CRISPRm also facilitates high efficiency genome editing, synthetic biology, and protein engineering applications in industrial yeast strains and we have demonstrated that CRISPRm can be used to quickly engineer proteins for vastly improved metabolic activity. We anticipate that CRISPRm should be adaptable for engineering uses in any industrial fungus and will be used to accelerate advances in the production of commercially important chemicals and biofuels.

## ACKNOWLEDGMENTS

We thank J. Doudna for helpful discussions on the manuscript. This work was supported by funding from the Energy Biosciences Institute.

*Competing Financial Interests*: Some of the authors have filed a patent application related to the results presented here.

## REFERENCES

Boeke, J. D., LaCroute, F., & Fink, G. R. (1984). A positive selection for mutants lacking orotidine-5′-phosphate decarboxylase activity in yeast: 5-Fluoro-orotic acid resistance. *Molecular & General Genetics*, *197*(2), 345–346.

Cong, L., Ran, F. A., Cox, D., Lin, S., Barretto, R., Habib, N., et al. (2013). Multiplex genome engineering using CRISPR/Cas systems. *Science*, *339*(6121), 819–823.

Curran, K. A., Karim, A. S., Gupta, A., & Alper, H. S. (2013). Use of expression-enhancing terminators in Saccharomyces cerevisiae to increase mRNA half-life and improve gene expression control for metabolic engineering applications. *Metabolic Engineering*, *19*, 88–97.

DiCarlo, J. E., Norville, J. E., Mali, P., Rios, X., Aach, J., & Church, G. M. (2013). Genome engineering in Saccharomyces cerevisiae using CRISPR-Cas systems. *Nucleic Acids Research*, *41*(7), 4336–4343.

Dowell, R. D., Ryan, O., Jansen, A., Cheung, D., Agarwala, S., Danford, T., et al. (2010). Genotype to phenotype: A complex problem. *Science*, *328*(5977), 469.

Farrell, A. E., Plevin, R. J., Turner, B. T., Jones, A. D., O'Hare, M., & Kammen, D. M. (2006). Ethanol can contribute to energy and environmental goals. *Science*, *311*(5760), 506–508.

Galazka, J. M., Tian, C., Beeson, W. T., Martinez, B., Glass, N. L., & Cate, J. H. D. (2010). Cellodextrin transport in yeast for improved biofuel production. *Science*, *330*(6000), 84–86.

Giaever, G., Chu, A. M., et al. (2002). Functional profiling of the Saccharomyces cerevisiae genome. *Nature*, *418*(6896), 387–391.

Houseley, J., & Tollervey, D. (2009). The many pathways of RNA degradation. *Cell*, *136*(4), 763–776.

Hsu, P. D., Scott, D. A., Weinstein, J. A., Ran, F. A., Konermann, S., Agarwala, V., et al. (2013). DNA targeting specificity of RNA-guided Cas9 nucleases. *Nature Biotechnology*, *31*(9), 827–832.

Jinek, M., Chylinski, K., Fonfara, I., Hauer, M., Doudna, J. A., & Charpentier, E. (2012). A programmable dual-RNA-guided DNA endonuclease in adaptive bacterial immunity. *Science*, *337*(6096), 816–821.

Jinek, M., East, A., Cheng, A., Lin, S., Ma, E., & Doudna, J. A. (2013). RNA-programmed genome editing in human cells. *Elife*, *2*, e00471.

Jinek, M., Jiang, F., Taylor, D. W., Sternberg, S. H., Kaya, E., Ma, E., et al. (2014). Structures of Cas9 endonucleases reveal RNA-mediated conformational activation. *Science*, *343*(6176), 1247997.

Ke, A., Ding, F., Batchelor, J. D., & Doudna, J. A. (2007). Structural roles of monovalent cations in the HDV ribozyme. *Structure*, *15*(3), 281–287.

Kohler, A., & Hurt, E. (2007). Exporting RNA from the nucleus to the cytoplasm. *Nature Reviews. Molecular Cell Biology*, *8*(10), 761–773.

Krugel, H., Fiedler, G., Haut, I., Sarfert, E., & Simon, H. (1988). Analysis of the nourseothricin-resistance gene (nat) of Streptomyces noursei. *Gene*, *62*(2), 209–217.

Mali, P., Esvelt, K. M., & Church, G. M. (2013). Cas9 as a versatile tool for engineering biology. *Nature Methods*, *10*(10), 957–963.

Mali, P., Yang, L., Esvelt, K. M., Aach, J., Guell, M., DiCarlo, J. E., et al. (2013). RNA-guided human genome engineering via Cas9. *Science*, *339*(6121), 823–826.

Marck, C., Kachouri-Lafond, R., Lafontaine, I., Westhof, E., Dujon, B., & Grosjean, H. (2006). The RNA polymerase III-dependent family of genes in hemiascomycetes: Comparative RNomics, decoding strategies, transcription and evolutionary implications. *Nucleic Acids Research*, *34*(6), 1816–1835.

Parent, S. A., Fenimore, C. M., & Bostian, K. A. (1985). Vector systems for the expression, analysis and cloning of DNA sequences in S. cerevisiae. *Yeast*, *1*(2), 83–138.

Rubin, E. M. (2008). Genomics of cellulosic biofuels. *Nature*, *454*(7206), 841–845.

Ryan, O. W., Skerker, J. M., Maurer, M. J., Li, X., Tsai, J. C., Poddar, S., et al. (2014). Selection of chromosomal DNA libraries using a multiplex CRISPR system. *Elife*, e03703. http://dx.doi.org/10.7554/eLife.03703 [Epub ahead of print].

Solis-Escalante, D., Kuijpers, N. G., van der Linden, F. H., Pronk, J. T., & Daran-Lapujade, P. (2014). Efficient simultaneous excision of multiple selectable marker cassettes using I-SceI-induced double-strand DNA breaks in Saccharomyces cerevisiae. *FEMS Yeast Research*, *14*(5), 741–754.

Sopko, R., Huang, D., Preston, N., Chua, G., Papp, B., Kafadar, K., et al. (2006). Mapping pathways and phenotypes by systematic gene overexpression. *Molecular Cell*, *21*(3), 319–330.

Sternberg, S. H., Redding, S., Jinek, M., Greene, E. C., & Doudna, J. A. (2014). DNA interrogation by the CRISPR RNA-guided endonuclease Cas9. *Nature*, *507*(7490), 62–67.

van den Ent, F., & Lowe, J. (2006). RF cloning: A restriction-free method for inserting target genes into plasmids. *Journal of Biochemical and Biophysical Methods*, *67*(1), 67–74.

Webb, C. H., Riccitelli, N. J., Ruminski, D. J., & Luptak, A. (2009). Widespread occurrence of self-cleaving ribozymes. *Science*, *326*(5955), 953.

Wingler, L. M., & Cornish, V. W. (2011). Reiterative recombination for the in vivo assembly of libraries of multigene pathways. *Proceedings of the National Academy of Sciences of the United States of America*, *108*(37), 15135–15140.

CHAPTER TWENTY-THREE

# Protein Engineering of Cas9 for Enhanced Function

**Benjamin L. Oakes*, Dana C. Nadler†, David F. Savage*,‡,§,1**
*Department of Molecular & Cell Biology, University of California, Berkeley, California, USA
†Department of Chemical and Biomolecular Engineering, University of California, Berkeley, California, USA
‡Department of Chemistry, University of California, Berkeley, California, USA
§Energy Biosciences Institute, University of California, Berkeley, California, USA
[1]Corresponding author: e-mail address: savage@berkeley.edu

## Contents



## Abstract

CRISPR/Cas systems act to protect the cell from invading nucleic acids in many bacteria and archaea. The bacterial immune protein Cas9 is a component of one of these CRISPR/Cas systems and has recently been adapted as a tool for genome editing. Cas9 is easily targeted to bind and cleave a DNA sequence via a complementary RNA; this straightforward programmability has gained Cas9 rapid acceptance in the field of genetic engineering. While this technology has developed quickly, a number of challenges regarding Cas9 specificity, efficiency, fusion protein function, and spatiotemporal control within the cell remain. In this work, we develop a platform for constructing novel proteins to address these open questions. We demonstrate methods to either screen or select active Cas9 mutants and use the screening technique to isolate functional Cas9 variants with a heterologous PDZ domain inserted within the protein. As a proof of

*Methods in Enzymology*, Volume 546
ISSN 0076-6879
http://dx.doi.org/10.1016/B978-0-12-801185-0.00024-6

concept, these methods lay the groundwork for the future construction of diverse Cas9 proteins. Straightforward and accessible techniques for genetic editing are helping to elucidate biology in new and exciting ways; a platform to engineer new functionalities into Cas9 will help forge the next generation of genome-modifying tools.

## 1. INTRODUCTION

The manipulation of gene sequence and expression is fundamental to unraveling the complexity of biological systems. However, our inability to make such manipulations across different organisms and cell types has limited the power of recombinant DNA technology to a handful of model systems. Consequently, numerous strategies for genome engineering—the ability to programmably disrupt or replace genomic loci—have emerged in recent years, yet there remains no universal solution to the problem (Carroll, 2014).

Investigation of bacterial adaptive immunity, known as the "clustered regularly interspaced short palindromic repeats" (CRISPR) system, led to the discovery of the RNA-guided DNA nuclease Cas9, which has proven a particularly potent tool for genome engineering (Barrangou et al., 2007; Deltcheva et al., 2011; Jinek et al., 2012; Wiedenheft, Sternberg, & Doudna, 2012). In its biological context, Cas9 is part of a Type II CRISPR interference system that functions to degrade pathogenic phage or plasmid DNA. Targeting of Cas9 is enabled by host-encoded CRISPR-RNAs (crRNAs), which recognize, through RNA:DNA hybridization, 20 bp of complementary target DNA sequence (referred to as a protospacer) (Fig. 23.1A). The Cas9 protein itself also plays a role in target recognition by binding a short DNA sequence adjacent and opposite the protospacer, called the protospacer adjacent motif (PAM). Although there is significant variation in PAM specificity among Cas9 orthologs the commonly employed Cas9 from *Streptococcus pyogenes* (SpCas9) recognizes the PAM sequence 5′-NGG-3′. PAM binding is thought to prime Cas9 for target recognition by the crRNA sequence (Sternberg, Redding, Jinek, Greene, & Doudna, 2014). Upon target recognition, two nuclease domains, termed the RuvC and HNH domains because of their sequence similarity to other endonucleases, engage and cleave the separated strands of DNA between 3 and 4 bp upstream of the PAM site (Jinek et al., 2012). A second *trans*-activating RNA (tracrRNA), with partial complementarity to crRNA, is also required for crRNA maturation and activity. Doudna and colleagues

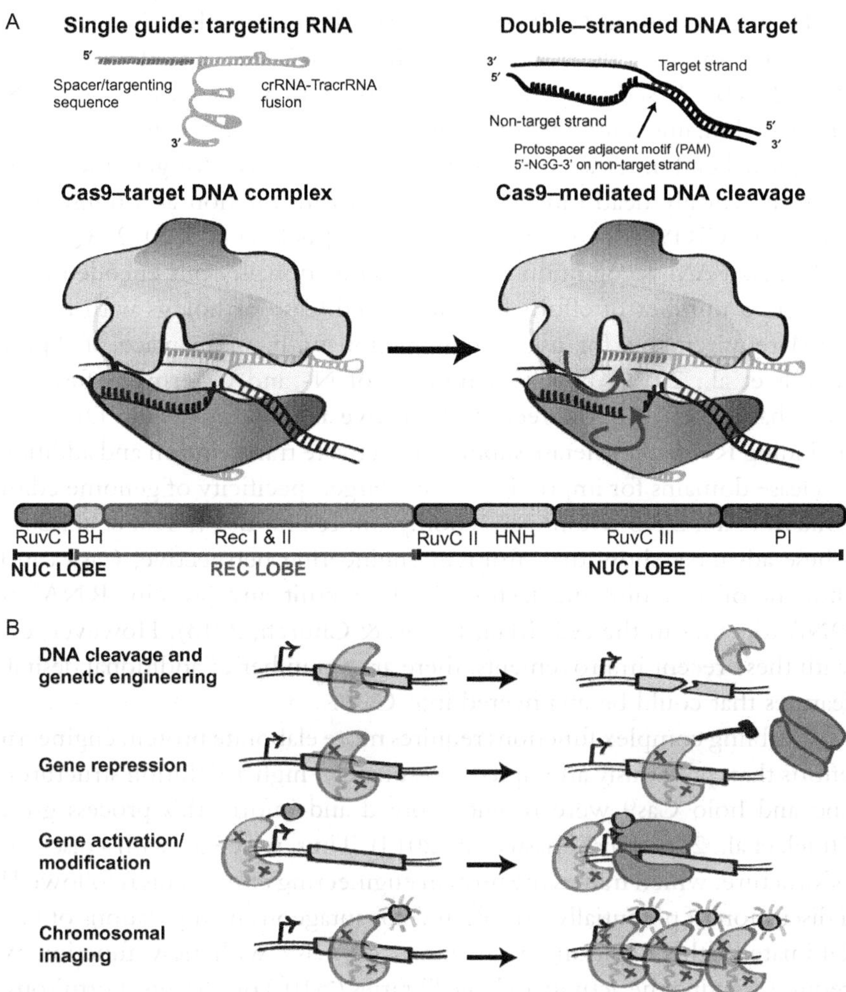


**Figure 23.1** Holo Cas9 model and its potential uses. (A) Single guide RNA, target dsDNA, and Cas9 are modeled. Domains of Cas9 are colored accordingly: RuvC-green, BH-pink, RecI-gray, RecII- dark gray, HNH-yellow PI-red. (B) Common uses of Cas9 as a tool. Red x's in Cas9 represent a nuclease dead variant. (See the color plate.)

have shown that the crRNA and tracrRNA can be fused together with a tetraloop insertion to form a single guide RNA (sgRNA or "guide") (Jinek et al., 2012). Expression of Cas9 and this sgRNA is both necessary and sufficient for targeting DNA. Therefore, the rapid success of Cas9-based engineering has been driven by programmability—Cas9 can be targeted to any DNA locus by simply changing the sgRNA sequence.

Intense interest in Cas9-based genetic engineering has already led to a number of directed alterations which change or improve Cas9 functionality (Fig. 23.1B). Based on sequence conservation of the RuvC and HNH nuclease domains, a number of point mutants were constructed to transform the normal endonuclease activity into either a nickase (for genome editing) or a catalytically dead mutant (dCas9) that can function as a transcription inhibitor (CRISPRi) (Cong et al., 2013; Jinek et al., 2012; Qi et al., 2013). As PAM recognition is critical to functionality but encoded by the protein, a number of efforts have identified Cas9 orthologs with minimal PAM requirements for use in conjunction with, or in place of, SpCas9 (Esvelt et al., 2013). Finally, a number of N- and C-terminal fusions to Cas9 have been used to recruit alternative factors to specific DNA loci, including RNA polymerase subunits to activate transcription and additional nuclease domains for improving the on-target specificity of genome editing (Bikard et al., 2013; Guilinger, Thompson, & Liu, 2014; Tsai et al., 2014). These advances show that, from an engineering perspective, Cas9 can be thought of as a unifying factor able to recruit any protein, RNA, and DNA together in the cell (Mali, Esvelt, & Church, 2013). However, even with these recent improvements, there are a number of additional desirable features that could be engineered into Cas9.

Enabling complex functions requires more elaborate protein engineering efforts than previously attempted. Fortunately, high-resolution structures of apo and holo Cas9 were recently solved and inform this process greatly (Jinek et al., 2014; Nishimasu et al., 2014). Thus, we begin with a discussion of structure, which frames the protein engineering effort. This is followed by a discussion of potentially feasible and advantageous manipulations of Cas9. Ultimately, the isolation of an enhanced Cas9 with new function will require assaying the activity of large libraries ($>10^6$) of variants. Fortuitously, the basic function of Cas9, disrupting gene sequence or gene expression, facilitates the construction of a genetic screen and means that directed evolution methods can be employed in the same conditions that functionality is ultimately desired. To this end, we present a detailed set of methods, employing either screening or selection, for the directed evolution of novel Cas9 proteins.

## 1.1. The structure of Cas9

Recently published high-resolution structures of Cas9 serve as a useful starting point to inform the protein engineering process. Doudna and colleagues reported the high-resolution structure of apoSpCas9 using X-ray crystallography and a holo complex using electron microscopy (Jinek et al., 2014). Concurrently, Zhang and colleagues solved the X-ray structure of SpCas9 bound to sgRNA and single-stranded target DNA (ssDNA) (Nishimasu et al., 2014). Here, we summarize these structures in the context of engineering novel Cas9 variants.

Cas9 possesses a hand-shaped structure of size 100 Å × 100 Å × 50 Å and is composed of two major lobes, an N-terminal recognition (REC) lobe, and a C-terminal lobe (NUC) possessing two endonuclease domains (Fig. 23.2). The REC lobe is composed of three segments: a small portion of the RuvC domain, an arginine-rich bridge helix (BH) which links the REC to the NUC lobe, and an α-helical recognition segment with two subdomains (RecI and II). The NUC lobe possesses three domains: the RuvC endonuclease, the HNH endonuclease, and a PAM-interacting (PI) C-terminal domain. The sgRNA–DNA complex lies at the interface of the two lobes, with the BH and RecI domains making many of the primary interactions. It should be noted that the sgRNA-target DNA heteroduplex generally makes nonspecific, sequence-independent interactions with the protein, while the sgRNA repeat:antirepeat region makes sequence-dependent interactions; this is consistent with the precise sgRNA recognition yet broad target flexibility of Cas9. Interestingly, although the sgRNA has extensive structural features—there are three stem loops

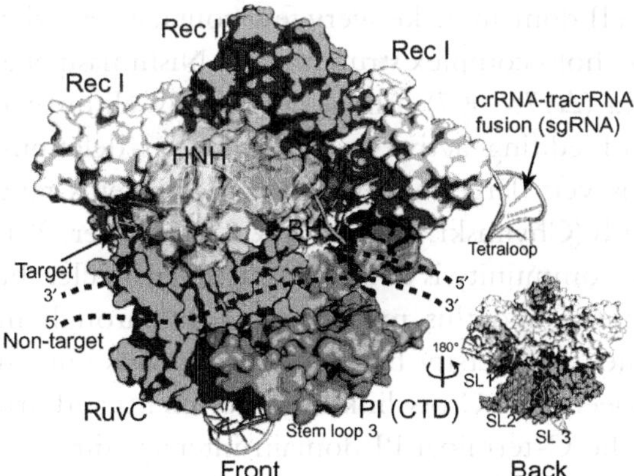


**Figure 23.2** The structure of Cas9 in complex with sgRNA and a single-stranded target DNA (ssDNA). The structure is shown in a surface representation and colored as in Fig. 23.1. Note the inset is rotated 180° and display the location of the sgRNA stem loops (SLs). *Adapted from Nishimasu et al. (2014) (PDB code 4OO8).* (See the color plate.)

in the tracrRNA sequence—stem loops 2 and 3 exit out the "back" of the structure and make few sequence-specific contacts with Cas9 (inset Fig. 23.2).

The structures inform a number of rational protein engineering design strategies. Many publications have employed N- and C-terminal fusions to Cas9 as a means of recruiting specific factors to a genomic locus, such as RNA polymerase for transcriptional activation (Bikard et al., 2013; Mali, Aach, et al., 2013). The structures suggest the protein N-terminus is adjacent to the 3′-end of the target DNA exiting Cas9, while the C-terminus is adjacent to the 5′-end of the same DNA. The close proximity of DNA to protein termini likely explains why many simple fusions have been successful. Intriguingly, the N- and C-termini are in the same lobe and roughly 50 Å apart, suggesting it might be possible to make circularly permuted forms of Cas9 for more elaborate protein engineering uses. Moreover, the bi-lobed nature of the apo complex indicates it could be possible to construct split variants of Cas9 that are only active under conditions when both halves are recruited together. Such variants would be useful in the construction of two-hybrid systems and for the engineering of allosterically controlled Cas9 derivatives, such as with optogenetic domains.

The sequence conservation of Cas9 orthologs mapped onto the structure also suggests domains that may prove malleable for engineering. Phylogenetic variation among Cas9 orthologs is significantly higher in the RecII domain of the REC lobe and the PI domain in the NUC lobe (Chylinski, Makarova, Charpentier, & Koonin, 2014; Nishimasu et al., 2014). The RecII domain makes very few contacts with the sgRNA or target DNA in the holo complex structure and Nishimasu et al. demonstrated that the domain (Δ175–307) can be completely eliminated yet still retain roughly 50% of editing activity. Such domain deletions are intriguing because Cas9 is very large; all known Cas9 proteins range from 984 to 1629 amino acids (Chylinski, Le Rhun, & Charpentier, 2013). Thus, a current goal in the community is to find or engineer smaller, but equally effective, Cas9 proteins and this may be possible through multiple domain deletions. Sequence diversity between PI domains also suggests a means to alter PAM specificity. Cross-linking experiments and structural evidence conclude that the C-terminal PI domain interacts directly with the PAM (Jinek et al., 2014). However, this specificity may in fact be modular. Domain swapping of the SpCas9 PI domain for that of the orthologous sequence from *S. thermophilus* imbues a preference for the *S. thermophilus* PAM sequence (TGGCG) (Nishimasu et al., 2014). Expanded domain swapping experiments, coupled with directed evolution, could therefore provide a means for creating orthogonal Cas9 variants from a single, well-characterized SpCas9 scaffold.

## 1.2. Current uses

Cas9 has rapidly established itself as a promising genome-engineering technology in widely used model organisms (Friedland et al., 2013; Gratz et al., 2013; Guilinger et al., 2014; Hou et al., 2013; Hsu et al., 2013; Hwang et al., 2013; Nishimasu et al., 2014; Niu et al., 2014; Shan et al., 2013; Tsai et al., 2014; Wang et al., 2013). In these systems, Cas9 has been used to create both small genomic insertions and deletions (indels) via nonhomologous end-joining and to facilitate larger sequence manipulations with homologous recombination. Cas9 also allows for multiplexed genome engineering, and has been used to create large knock-out libraries in human cells, a feat both surprising in its simplicity and impressive in its efficacy (Shalem et al., 2014; Zhou et al., 2014). Decoupling the DNA-binding activity of Cas9 proteins from cleavage activity has lead to a broader set of uses such as repression and activation of transcription (Gilbert et al., 2013). Finally, recent evidence suggests that Cas9 may be used to manipulate RNA (O'Connell et al., 2014). Although still nascent, the simple programmability and effectiveness of Cas9-based technology promises to democratize access to genome and now potentially transcriptome manipulation.

## 1.3. Initial engineering questions

As previously mentioned, there are a number of clear initial questions pertaining to Cas9 that are addressable using existing protein engineering tools. Namely, we believe that designing novel Cas9s with domain insertions and deletions will lead to the creation of a new family of synthetic orthologs whose outputs are manifold. For example, domain insertions could act to recruit additional protein partners with desired activity onto Cas9-associated nucleic acids; domain deletions will reduce Cas9's size and increase its versatility.

Alternatively, improving N- or C-terminal fusions with engineered linkers or creating Cas9s with new N- and C-termini altogether, may greatly increase the efficacy of fusions. For example, to address issues of Cas9 targeting specificity dCas9 has been fused to FokI, an obligate dimeric sequence-independent nuclease (Guilinger et al., 2014; Tsai et al., 2014). This system requires the mutual on-target activity of two different

FokI–dCas9 fusions to adjacent sites, a combined 40 bp of targeting, to catalyze a DNA cleavage event. Unfortunately, these FokI–dCas9 fusions are substantially less active than either WT Cas9 or the dual nickase strategy at inducing indels, rendering them a less attractive tool. Nevertheless, it is known that FokI protein fusions to other DNA-binding domains can achieve cleavage efficiencies similar to that of WT Cas9 (Hwang et al., 2013; Mali, Yang, et al., 2013). Therefore, lower activity of the current FokI–dCas9 is likely due to imperfect positioning of the FokI nuclease domain and further engineering the dCas9–FokI interface should yield an increase in activity.

Last, split proteins are known to function as switches or response elements in many different systems (Olson & Tabor, 2012). Splitting Cas9, as mentioned above, would be a simple method for engineering allosteric control and open the door to a number of uses including optogenetics, small-molecule dependence, or linking function to a cellular signal, such as a phosphorylation-dependent signal transduction cascade. Nevertheless, all of the previous engineered scenarios require that Cas9 is active despite the modifications introduced. Therefore, it is imperative that any engineering attempt start with an assay allowing for the separation of active mutant proteins.

# 2. METHODS

To advance some of the aforementioned goals related to Cas9 protein engineering, we have developed a suite of protocols allowing for the rapid isolation of functional Cas9 proteins. These techniques rely on the ability to screen or select an active Cas9 from a large pool of variants. As such, the methods may be applied equally to either rational or library-based approaches for engineering Cas9.

## 2.1. A note on applications

Although we present here a general method for directed evolution of Cas9, it is impossible to cover the myriad of nuances inherent to different applications. Instead, as a proof of concept, we demonstrate that Cas9 can be manipulated by domain insertion, a common event found in eukaryotic proteomes (Lander et al., 2001). Specifically, we created libraries of Cas9 in which the α-syntrophin PDZ domain was randomly inserted throughout the SpCas9 gene once per variant using *in vitro* transposition-based methods (Edwards, Busse, Allemann, & Jones, 2008). PDZ domains are small (~100 amino acids) proteins with adjacent N- and C-termini that mediate protein–protein interactions by specifically binding the C-terminal peptide of a cognate partner in the cell (Nourry, Grant, & Borg, 2003). These domains have been used extensively in synthetic biology applications as a tool for scaffolding proteins (Dueber, Mirsky, & Lim, 2007; Dueber, Yeh, Chak, & Lim, 2003). In the context of Cas9, PDZ insertion could recruit additional factors to a DNA-bound Cas9 in the cell, such as florescent tags, chromatin remodeling machinery, and nuclease domains.

## 2.2. Electrocompetent *E. coli* preparation for library construction

Construction of libraries containing $10^6$–$10^9$ diverse members is a basic step for engineering new functions into Cas9. A simple method for the creation of electrocompetent cells that consistently yields *E. coli* with transformation efficiencies of $\geq 10^8$ CFU per μg of plasmid is described. If necessary, cells can be pretransformed with additional screening/selection plasmids to maximize library transformation efficiency.

1. Begin with the desired strain grown as single colonies on appropriate plates. For example, we use plasmid 44251: pgRNA-bacteria (Addgene) with the RFP sgRNA from Qi et al. (2013) transformed and grown on carbenicillin plates (100 μg/mL) overnight at 37 °C to single colonies.
2. Pick a single colony and inoculate into 5 mL SOC (BD Difco 244310: Super Optimal Broth + 0.4% glucose) plus carbenicillin. Grow overnight at 37 °C.
3. Inoculate 1 L of SOC + carbenicillin with the 5 mL overnight culture. Grow at 37 °C for 2–4 h to an $OD_{600}$ between 0.55 and 0.65.
4. Rapidly cool the culture in an ice bath by swirling. Keep cells at 4 °C during all subsequent steps.
5. Centrifuge cells at 4000 × *g* for 10 min. Wash the cell culture with 500 mL of sterile ice-cold water by gentle resuspension. Centrifuge at 4000 × *g* for 10 min. Repeat wash step.
6. Centrifuge at 4000 × *g* for 10 min and wash cells with 500 mL of ice-cold 10% glycerol. Repeat.
7. Perform a final spin and discard the supernatant. Resuspend the pellet in 2.75 mL of 10% glycerol and aliquot into cold microcentrifuge tubes, 75 μL each. Flash freeze.
8. Transforming cells: Thaw on ice, add plasmid to 75 μL cells and vortex. Electroporate at 1800 V, 200 Ω, 25 μF in 0.1-cm cuvettes, resuspending in warm SOC immediately afterward. Recover for 1 h at 37 °C before adding antibiotics.

## 2.3. Discovery of functional, engineered, variants of Cas9 proteins

After generating any library of Cas9 variants, it is necessary to have a platform that can separate active variants with minimal effort. Two assays to probe this functionality are the coupling of dCas9 activity to either RFP expression or media-dependent cell growth. When devising these systems, it is important to keep in mind the First Law of Directed Evolution (Schmidt-Dannert & Arnold, 1999): "you get what you screen for."

## 2.4. Screening Cas9

The catalytically dead version of Cas9 has a functional output that can be tied directly to transcription in *E. coli*; namely, it can repress transcription of a desired gene (Qi et al., 2013). Qi et al. previously demonstrated that dCas9 with a guide sequence of 5′-AACUUUCAGUUUAGCGGUCU-3′ can target and repress a genome-encoded RFP while avoiding repression of a genome-encoded upstream GFP (Qi et al., 2013). In a screening context, this provides a simple output for assaying dCas9 functionality (i.e., RFP knock-down) while correcting for extrinsic noise in the population by monitoring GFP (Elowitz, Levine, Siggia, & Swain, 2002). The basic method of screening is schematized in Fig. 23.3A. Briefly, cells containing functional dCas9s will repress RFP and express GFP while those with nonfunctional dCas9s will express both fluorescent proteins. This signal is easily distinguished using flow cytometry and florescence imaging (Fig. 23.3B and C).

## 2.5. Selecting Cas9

To complement the screening method and for use with larger libraries of Cas9 mutants, we have also developed a technique to select functional dCas9s using cellular growth. We fashioned a derivative of the classic yeast counter-selection method, which takes advantage of the toxicity of 5-fluoroorotic acid (5-FOA) in cells with the *URA3* gene (Fig. 23.4A) (Boeke, Trueheart, Natsoulis, & Fink, 1987). In yeast, *URA3* encodes orotidine5′-phosphate decarboxylase, which catalyzes the conversion of 5-FOA into a highly toxic compound (Boeke, LaCroute, & Fink, 1984). The *E. coli* homolog of *URA3*, *pyrF*, is thought to act in a similar manner, and *pyrF* and the upstream gene *pyrE* are known to function as selectable markers in other Gram negative bacteria (Galvao & de Lorenzo, 2005; Yano, Sanders, Catalano, & Daldal, 2005). Nevertheless, it was unclear whether dCas9-based repression would mimic the effects of these full gene knockouts in an *E. coli* system. To this end,

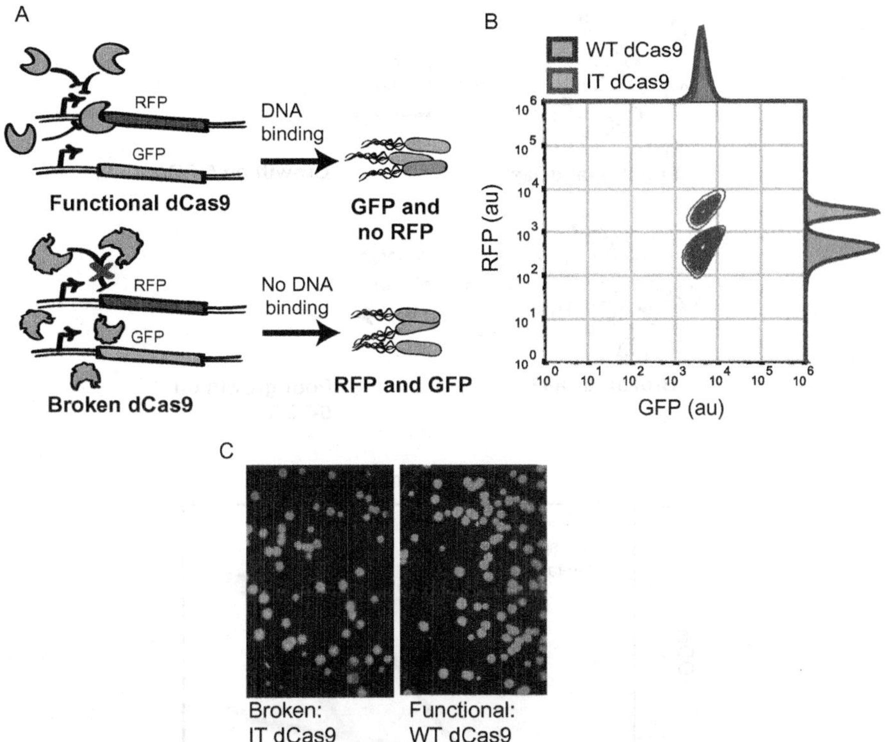


**Figure 23.3** Screen for functional Cas9s. (A) Schematic representation of the screen. (B) Flow cytometry data of the functional positive (WT dCas9) control in blue and negative "Inactive Truncation" Cas9 (IT dCas9) control in red. IT dCas9 contains only the C-terminal 250 amino acids. Both controls contain the sgRNA plasmid which targets RFP for repression. Samples were grown overnight in rich induction media. (C) Colony fluorescence of the functional (WT dCas9) and "Broken" negative (IT dCas9) controls. (See the color plate.)

we tested whether repression of either of *pyrF* and *pyrE* by dCas9 was sufficient to rescue a slow growth phenotype on 5-FOA. After creating a number of different sgRNAs targeted to the start of *pyrF* and *pyrE*, we determined that the guides 5′-ACCUUCUUGAUGAUGGGCAC-3′ for *pyrF* and 5′-UAAGCGCAAAUUCAAUAAAC-3′ for *pyrE* each rescued growth in 5 m*M* of 5-FOA (Fig. 23.4B).

Ultimately, it is important to decide which approach, screening or selection, will be used to enrich for functional engineered Cas9 mutants. A primary determining factor is the theoretical library size. Screening systems can effectively cover libraries of sizes up to ~$10^6$, which is roughly equivalent to the amount of *E. coli* that can be sorted 10× by a flow cytometer in one hour. On

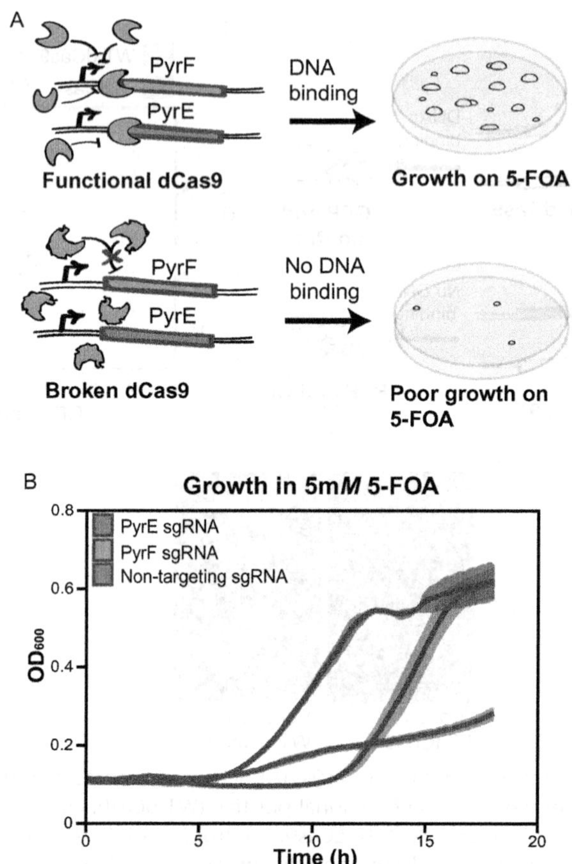


**Figure 23.4** Functional Cas9 selection overview. (A) Schematic representation of the selection system. (B) Growth rate of a functional dCas9+sgRNAs repressing the *pyrF* gene (purple), *pyrE* gene (green), and a no guide sequence control (red). Samples were grown in rich induction media+5 m*M* 5-fluoroorotic acid. All measurements represent the average (line) and standard deviation (shading) of three biological replicates. (See the color plate.)

the other hand, selection systems, which rely on repression/activation of a toxic/essential gene for growth, can screen libraries of random protein variants of up to $\sim 10^9$ in size (Persikov, Rowland, Oakes, Singh, & Noyes, 2014).

## 2.6. Screening for functional Cas9 variants

We have found that Fluorescence-Activated Cell Sorting (FACS) is a convenient method for isolating functional Cas9 variants. This approach is somewhat more flexible than a selection—a gating strategy in FACS is easily manipulated while the growth constraints of a selection are not—yet still provides reasonable throughput. As a proof of concept, we demonstrate the FACS-based screening of functional dCas9 variants possessing the insertion of an α-syntrophin PDZ domain.

1. Obtain a dCas9 library containing $\leq 10^6$ variants on an expression plasmid of choice. Here we used a tetracycline-inducible expression plasmid, plasmid 44249: pdCas9-bacteria from Addgene (Qi et al., 2013) to create a library which has a SNTA1 PDZ domain inserted across the whole dCas9 protein (Dueber et al., 2003). Based on the possible insertion sites and linkers, the size of this library is roughly equal to $10^6$.
2. Transform electrocompetent *E. coli* expressing GFP and RFP with 1 μg of the library plasmid and 1 μg of a sgRNA plasmid, if necessary. Here, the *E. coli* strain and guide RNA plasmid come from Qi et al. (2013) (plasmid is 44251: pgRNA-bacteria; Addgene).
3. To ensure adequate coverage of the library the transformation efficiency should be at least 5–10 × greater than the theoretical library size. To determine this, plate 5 μL aliquots of serially diluted transformants, and grow to colonies (overnight, 37 °C) on double-selection media (chloramphenicol (50 μg/mL) to maintain the engineered dCas9 plasmids and carbenicillin (100 μg/mL) for maintenance of the guide RNA plasmid). Store the remaining transformed cells at 4 °C overnight.
4. Determine the volume of the transformation mixture needed to cover the theoretical library size 5–10 × based on the results from step 3 and inoculate into 5 mL of rich induction media: SOC, chloramphenicol, carbenicillin and 2 μM anhydrotetracycline (aTC). Concurrently, inoculate tubes of rich induction media with controls WT dCas9 and IT dCas9 with the RFP sgRNA. Grow at 37 °C; we have found shaking ≥250 rpm is helpful for maximum RFP and GFP fluorescence.
5. After 8–12 h of growth, centrifuge 500 μL of each sample, wash 2 × with 1 mL of PBS and resuspend 1:20 in PBS for flow cytometry.
6. Run the controls on a FACS instrument to establish correct positive and negative gating (Fig 23.5).
7. Screen the library using FACS and collect the cell which fall within the previously determined positive gate in rich, non-selective media (Fig. 23.5). Screen at least 10 × the library size as cellular viability post-FACS is often substantially less than 100%.
8. Recover sorted cells for 2 h at 37 °C.
9. Depending on the library enrichment after a single round of sorting, repeating steps 4–8 may be necessary to further enrich for functional, engineered dCas9 clones.

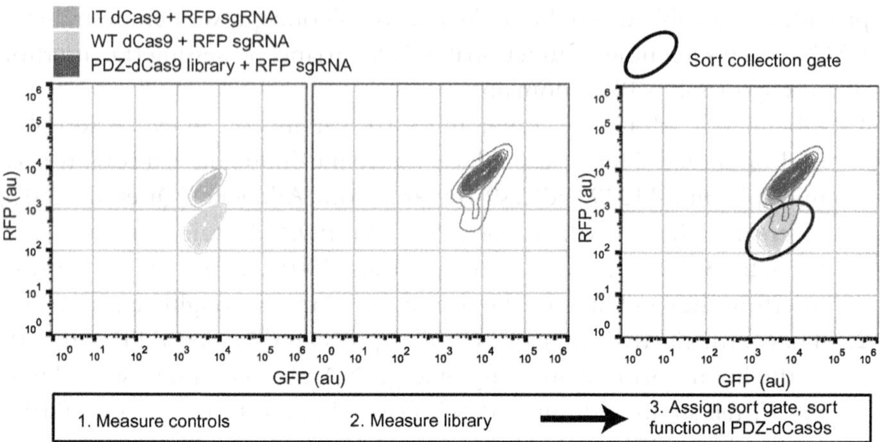


**Figure 23.5** Cell sorting data from the GFP-RFP screen. The first panel depicts the RFP versus GFP measurements of WT dCas9 (blue; light gray in the print version) and IT dCas9 (pink; dark gray in the print version) as run on a Sony SH800 cell sorter. The separation of these two controls into distinct populations is readily apparent. The second panel portrays the spread of the PDZ-Cas9 intercalation library (green; dark gray in the print version). The last panel shows the overlay of all three FACS plots. It is evident that there are populations of functional and nonfunctional proteins within the single PDZ-dCas9 library and the third panel also provides a demonstration of a gate for isolation of functional PDZ-dCas9 intercalations.

## 2.7. Determining screening enrichment of PDZ-dCas9 domain insertions

A successful round of screening with the PDZ-dCas9 library should enrich for functional PDZ-dCas9 insertion mutants (intercalations). A straightforward method to check for enrichment is to PCR with a primer specific to the inserted domain and one external to the engineered Cas9 (Fig 23.6A). An amplified smear indicates a relatively "naïve" library while specific bands indicate enriched library members. The following is a representative protocol for checking screening success.

1. Plate approximately 1000–10,000 of the sorted and recovered cells on rich induction plates with antibiotics and inducer. Add the remaining cells to 6 mL of liquid media with appropriate antibiotics.
2. Grow the induction plate(s) overnight at 37 °C and then allow 12 h at room temperature for RFP to fully mature. As described in Section 2.8 below, this plate will be used to pick colonies with functionally intercalated PDZ-dCas9s.

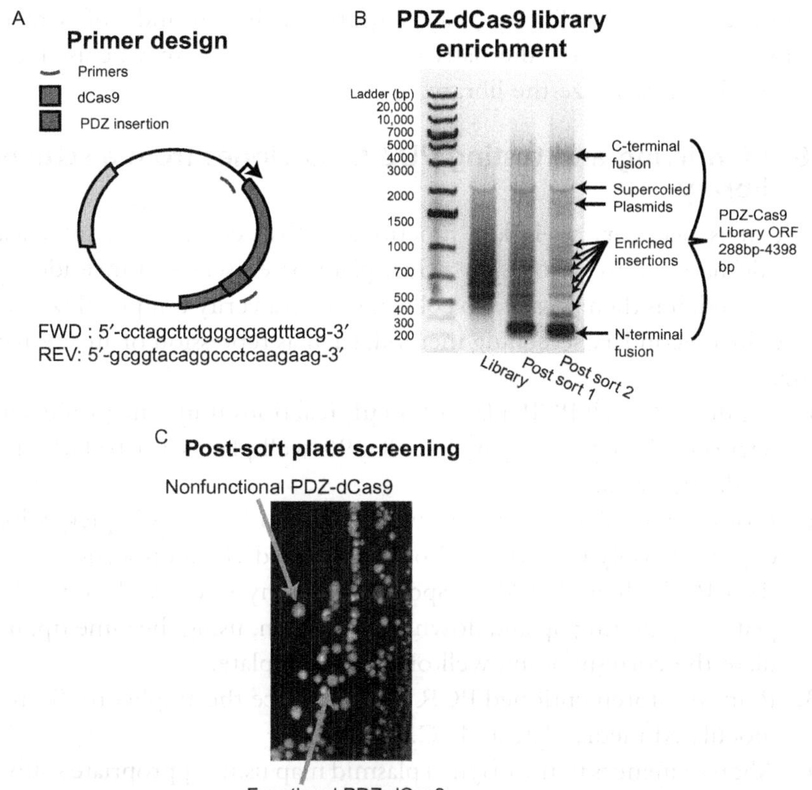


**Figure 23.6** Checking success of a screen and picking final clones. (A) An overview of primer design for the PDZ-dCas9 library. (B) Gel electrophoresis of PCRs run on the original PDZ-dCas9 library and the first and second round of screening. The banding patterns that appear after the first and second sorts are indicative of library enrichment, representing the insertion sites of a PDZ domain. It is also evident that the N- and C-termini fusions to PDZ are also enriched. Since these fusions are expected to be functional this serves as an internal control. (C) Fluorescent image of the on-plate "finishing" screen. Colonies that express only GFP are expected to have a functional PDZ-dCas9.

3. Grow the liquid culture from step 1 at 37 °C overnight and prepare a glycerol stock of the sorted cells for future use by mixing 800 μL of culture with 400 μL of 50% glycerol in lysogeny broth (LB).
4. Centrifuge the remaining liquid culture and miniprep to recover plasmid DNA (Qiagen).
5. Perform a PCR using plasmid DNA from the original and screened libraries with the primers described above (Fig. 23.6A). The screened library PCR should show enrichment bands (Fig. 23.6B). If bands are

not evident, the library may require further rounds of screening (Section 2.6, step 4). Alternatively, deep sequencing may be used to rigorously characterize the library.

## 2.8. Identifying and testing PDZ-Cas9 clones from a screened library

Next, it is necessary to isolate functional dCas9 clones with intercalated PDZ domains. This is done via a final plate-based screen. Once identified and isolated, it is then possible to collect, test, and verify unique PDZ-dCas9 clones in a secondary screening method, such as repression of an alternative gene.

1. Set up a 96-well PCR plate of 50 μL reactions using the primers from Section 2.7, step 5. In parallel, fill a 96-well plate with 100 μL of rich media per well.
2. From the induction plates grown in Section 2.7, step 2, pick colonies expressing only GFP (Fig. 23.6C), as assayed via fluorescence imaging (Bio-Rad Chemidoc MP). Spot each colony into a well of the PCR plate by pipetting up and down 5 × and then, using the same tip, inoculate the corresponding well of the media plate.
3. Run the aforementioned PCR and sequence the amplicons. Store the inoculated media plate at 4 °C.
4. Align sequences to the original plasmid map using appropriate software. Ensure variants are in-frame and determine unique clones.
5. Take 50 μL of the corresponding unique clones from the inoculated media plate and grow overnight in 5 mL of rich media with antibiotics. Miniprep DNA to obtain a mix of both the engineered PDZ-dCas9 plasmid and the RFP guide plasmid for each isolate.
6. Using a primer upstream of the dCas9 insertion site, sequence the plasmid to determine the insertion site and linkers (Fig. 23.7A).
7. Digest 5 μg of plasmid mixture with BsaI to remove the guide plasmid and clean up DNA (Qiagen) to remove restriction enzyme. (Digestion rxn occurs as follows: (1) 37 °C—60 min, (2) 50 °C—60 min, (3) 80 °C—10 min)
8. Transform the digested plasmid mixture with 200 ng of new guide plasmid to examine the function of the intercalated PDZ-dCas9 on other genes and endogenous loci. Specifically, we transformed one of our PDZ-dCas9 intercalations with guides for GFP and FtsZ, an essential cell division protein (sequences 5′-AUCUAAUUCAACAAGAAUU-3′, 5′-UCGGCGUCGGCGGCGG-3′, respectively).

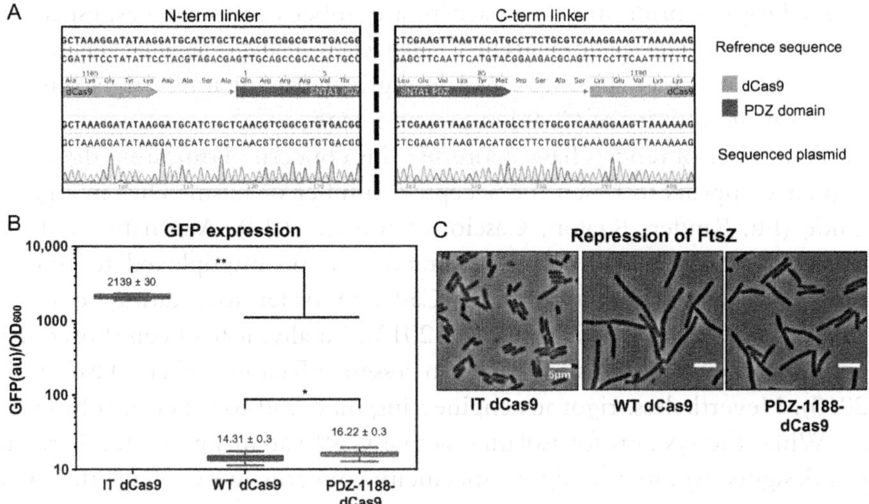


**Figure 23.7** Validating functionality of engineered dCas9. (A) Sequence validation of the site 1188 PDZ intercalation. Sequence alignment via SnapGene. (B) Quantitative repression of GFP by the PDZ-1188-dCas9 intercalation clone. Bulk florescence measurements of GFP expression levels over 5 h. Double asterisks represent a *p* value of <0.0001 in a one way ANOVA. Single asterisk represents *p* values of <0.0001 in an unpaired Student's *t*-test. (C) Qualitative repression of *ftsZ* gene by PDZ-dCas9 and controls. The scale bar is 5 μm.

9. Culture the bacteria with the PDZ-dCas9 intercalation isolates and the new guides. Grow the original dCas9 controls with these new guides, induce with 2 μ*M* aTC and measure the phenotypes accordingly.
10. Validate that the qualitative and quantitative phenotypes are within range of the WT Cas9 (Fig. 23.7B and C).

## 2.9. Expanding horizons

While the democratization of simple and multiplex genome engineering is central to the story of Cas9, the question of reliable specificity is paramount to its future use. Ideally, Cas9 would target and cleave one site in a complex genome yet leave other, similar, sites unmarred. Scarring the genome obscures the genotype–phenotype relationship, limiting basic science utility, and Cas9 cannot translate into the therapeutic arena if it is known to induce spurious mutations. Thus, how and when PAM and guide interactions integrate to provide specificity and activate Cas9-mediated cleavage is essential. Studies have shown that while the SpCas9 PAM 5′-NGG-3′ requirement is strict, only poorly tolerating one other target site (5′-NAG-3′), sgRNA:

target-DNA hybridization may accept a number of mismatches, especially toward the 5′-end of the sgRNA (Hsu et al., 2013; Mali, Aach, et al., 2013; Pattanayak et al., 2013). Accordingly, detrimental off-target binding and cleavage activity of Cas9 is a pressing issue.

A number of reports have addressed this concern. Truncating the guide sequence appears to lessen the accepted number of mismatches in a given guide (Fu, Sander, Reyon, Cascio, & Joung, 2014). Alternatively, Cas9 nickases, which cleave only one strand, can be multiplexed to require mutual on-target activity of two Cas9s in order for editing to occur (Mali, Aach, et al., 2013; Ran et al., 2013). Finally, it has been shown that lowering the expression of Cas9 can lessen off-target effects (Hsu et al., 2013). Nevertheless, rigorous engineering may lead to superior solutions.

While the systems for isolating active Cas9 variants presented here are not designed to directly address specificity concerns, we envisage that with small changes our selection and screening platforms could separate mutant Cas9s with less specificity from those with more. For example, it should be possible to introduce high affinity off-target binding-sites in front of the fluorescent reporter not actively targeted, such that any binding to these mock sites would act as an internal counterscreen. We are motivated by the likelihood that, in the future, screens and selections of this vein may be used to engineer synthetic Cas9 proteins that can tolerate few, if any mismatches, in the guide and/or PAM sequence.

## 3. CONCLUSION

Cas9 has fundamentally altered the genome-engineering landscape due to its simple programmability and overall effectiveness. Here, for the first time, we have delineated protein engineering-based methods for the directed evolution of Cas9 proteins with novel functions. We believe that such techniques will be critical for answering unresolved biochemical questions of protein structure and function. Moreover, directed evolution of Cas9 will allow for more refined improvement of this singular protein and the construction of next-generation tools for both interrogating the genome and biomedical therapies.

## REFERENCES

Barrangou, R., Fremaux, C., Deveau, H., Richards, M., Boyaval, P., Moineau, S., et al. (2007). CRISPR provides acquired resistance against viruses in prokaryotes. *Science (New York)*, *315*(5819), 1709–1712. http://dx.doi.org/10.1126/science.1138140.

Bikard, D., Jiang, W., Samai, P., Hochschild, A., Zhang, F., & Marraffini, L. A. (2013). Programmable repression and activation of bacterial gene expression using an engineered CRISPR-Cas system. *Nucleic Acids Research, 41*(15), 7429–7437. http://dx.doi.org/10.1093/nar/gkt520.

Boeke, J. D., LaCroute, F., & Fink, G. R. (1984). A positive selection for mutants lacking orotidine-5'-phosphate decarboxylase activity in yeast: 5-fluoro-orotic acid resistance. *Molecular and General Genetics, 197*(2), 345–346.

Boeke, J. D., Trueheart, J., Natsoulis, G., & Fink, G. R. (1987). 5-fluoroorotic acid as a selective agent in yeast molecular genetics. *Methods in Enzymology, 154*, 164–175.

Carroll, D. (2014). Genome engineering with targetable nucleases. *Annual Review of Biochemistry, 83*(1), 409–439. http://dx.doi.org/10.1146/annurev-biochem-060713-035418.

Chylinski, K., Le Rhun, A., & Charpentier, E. (2013). The tracrRNA and Cas9 families of type II CRISPR-Cas immunity systems. *RNA Biology, 10*, 726–737.

Chylinski, K., Makarova, K. S., Charpentier, E., & Koonin, E. V. (2014). Classification and evolution of type II CRISPR-Cas systems. *Nucleic Acids Research, 42*, 6091–6105.

Cong, L., Ran, F. A., Cox, D., Lin, S., Barretto, R., Habib, N., et al. (2013). Multiplex genome engineering using CRISPR/Cas systems. *Science (New York), 339*(6121), 819–823. http://dx.doi.org/10.1126/science.1231143.

Deltcheva, E., Chylinski, K., Sharma, C. M., Gonzales, K., Chao, Y., Pirzada, Z. A., et al. (2011). CRISPR RNA maturation by trans-encoded small RNA and host factor RNase III. *Nature, 471*(7340), 602–607. http://dx.doi.org/10.1038/nature09886.

Dueber, J. E., Mirsky, E. A., & Lim, W. A. (2007). Engineering synthetic signaling proteins with ultrasensitive input/output control. *Nature Biotechnology, 25*(6), 660–662. http://dx.doi.org/10.1038/nbt1308.

Dueber, J. E., Yeh, B. J., Chak, K., & Lim, W. A. (2003). Reprogramming control of an allosteric signaling switch through modular recombination. *Science (New York), 301*(5641), 1904–1908. http://dx.doi.org/10.1126/science.1085945.

Edwards, W. R., Busse, K., Allemann, R. K., & Jones, D. D. (2008). Linking the functions of unrelated proteins using a novel directed evolution domain insertion method. *Nucleic Acids Research, 36*(13), e78. http://dx.doi.org/10.1093/nar/gkn363.

Elowitz, M. B., Levine, A. J., Siggia, E. D., & Swain, P. S. (2002). Stochastic gene expression in a single cell. *Science (New York), 297*(5584), 1183–1186. http://dx.doi.org/10.1126/science.1070919.

Esvelt, K. M., Mali, P., Braff, J. L., Moosburner, M., Yaung, S. J., & Church, G. M. (2013). Orthogonal Cas9 proteins for RNA-guided gene regulation and editing. *Nature Methods, 10*(11), 1116–1121. http://dx.doi.org/10.1038/nmeth.2681.

Friedland, A. E., Tzur, Y. B., Esvelt, K. M., Colaiácovo, M. P., Church, G. M., & Calarco, J. A. (2013). Heritable genome editing in C. elegans via a CRISPR-Cas9 system. *Nature Methods, 10*(8), 741–743. http://dx.doi.org/10.1038/nmeth.2532.

Fu, Y., Sander, J. D., Reyon, D., Cascio, V. M., & Joung, J. K. (2014). Improving CRISPR-Cas nuclease specificity using truncated guide RNAs. *Nat Biotechnology, 32*, 279–284.

Galvao, T. C., & de Lorenzo, V. (2005). Adaptation of the yeast URA3 selection system to Gram-negative bacteria and generation of a betCDE Pseudomonas putida strain. *Applied and Environmental Microbiology, 71*(2), 883–892. http://dx.doi.org/10.1128/AEM.71.2.883-892.2005.

Gilbert, L. A., Larson, M. H., Morsut, L., Liu, Z., Brar, G. A., Torres, S. E., et al. (2013). CRISPR-mediated modular RNA-guided regulation of transcription in eukaryotes. *Cell, 154*(2), 442–451. http://dx.doi.org/10.1016/j.cell.2013.06.044.

Gratz, S. J., Cummings, A. M., Nguyen, J. N., Hamm, D. C., Donohue, L. K., Harrison, M. M., et al. (2013). Genome engineering of drosophila with the CRISPR RNA-guided Cas9 nuclease. *Genetics, 194*(4), 1029–1035. http://dx.doi.org/10.1534/genetics.113.152710.

Guilinger, J. P., Thompson, D. B., & Liu, D. R. (2014). Fusion of catalytically inactive Cas9 to FokI nuclease improves the specificity of genome modification. *Nature Biotechnology*, *32*(6), 577–582. http://dx.doi.org/10.1038/nbt.2909.

Hou, Z., Zhang, Y., Propson, N. E., Howden, S. E., Chu, L.-F., Sontheimer, E. J., et al. (2013). Efficient genome engineering in human pluripotent stem cells using Cas9 from Neisseria meningitidis. *Proceedings of the National Academy of Sciences of the United States of America*, *110*(39), 15644–15649. http://dx.doi.org/10.1073/pnas.1313587110.

Hsu, P. D., Scott, D. A., Weinstein, J. A., Ran, F. A., Konermann, S., Agarwala, V., et al. (2013). DNA targeting specificity of RNA-guided Cas9 nucleases. *Nature Biotechnology*, *31*(9), 827–832. http://dx.doi.org/10.1038/nbt.2647.

Hwang, W. Y., Fu, Y., Reyon, D., Maeder, M. L., Tsai, S. Q., Sander, J. D., et al. (2013). Efficient genome editing in zebrafish using a CRISPR-Cas system. *Nature Biotechnology*, *31*(3), 227–229. http://dx.doi.org/10.1038/nbt.2501.

Jinek, M., Chylinski, K., Fonfara, I., Hauer, M., Doudna, J. A., & Charpentier, E. (2012). A programmable dual-RNA-guided DNA endonuclease in adaptive bacterial immunity. *Science (New York)*, *337*(6096), 816–821. http://dx.doi.org/10.1126/science.1225829.

Jinek, M., Jiang, F., Taylor, D. W., Sternberg, S. H., Kaya, E., Ma, E., et al. (2014). Structures of Cas9 endonucleases reveal RNA-mediated conformational activation. *Science (New York)*, *343*(6176), 1247997. http://dx.doi.org/10.1126/science.1247997.

Lander, E. S., Linton, L. M., Birren, B., Nusbaum, C., Zody, M. C., Baldwin, J., et al. (2001). Initial sequencing and analysis of the human genome. *Nature*, *409*(6822), 860–921. http://dx.doi.org/10.1038/35057062.

Mali, P., Aach, J., Stranges, P. B., Esvelt, K. M., Moosburner, M., Kosuri, S., et al. (2013). CAS9 transcriptional activators for target specificity screening and paired nickases for cooperative genome engineering. *Nature Biotechnology*, *31*(9), 833–838. http://dx.doi.org/10.1038/nbt.2675.

Mali, P., Esvelt, K. M., & Church, G. M. (2013). Cas9 as a versatile tool for engineering biology. *Nature Methods*, *10*(10), 957–963. http://dx.doi.org/10.1038/nmeth.2649.

Mali, P., Yang, L., Esvelt, K. M., Aach, J., Guell, M., DiCarlo, J. E., et al. (2013). RNA-guided human genome engineering via Cas9. *Science (New York)*, *339*(6121), 823–826. http://dx.doi.org/10.1126/science.1232033.

Nishimasu, H., Ran, F. A., Hsu, P. D., Konermann, S., Shehata, S. I., Dohmae, N., et al. (2014). Crystal structure of Cas9 in complex with guide RNA and target DNA. *Cell*, *156*(5), 935–949. http://dx.doi.org/10.1016/j.cell.2014.02.001.

Niu, Y., Shen, Bin, Cui, Y., Chen, Y., Wang, J., Wang, L., et al. (2014). Generation of gene-modified cynomolgus monkey via Cas9/RNA-mediated gene targeting in one-cell embryos. *Cell*, *156*(4), 836–843. http://dx.doi.org/10.1016/j.cell.2014.01.027.

Nourry, C., Grant, S. G. N., & Borg, J.-P. (2003). PDZ domain proteins: Plug and play! *Science's STKE*, *2003*(179), RE7. http://dx.doi.org/10.1126/stke.2003.179.re7.

O'Connell, M. R., Oakes, B. L., Sternberg, S. H., East-Seletsky, A., Kaplan, M., & Doudna, J. A. (2014). Programmable RNA recognition and cleavage by CRISPR/Cas9. *Nature*, http://dx.doi.org/10.1038/nature13769.

Olson, E. J., & Tabor, J. J. (2012). Post-translational tools expand the scope of synthetic biology. *Current Opinion in Chemical Biology*, *16*(3–4), 300–306. http://dx.doi.org/10.1016/j.cbpa.2012.06.003.

Pattanayak, V., Lin, S., Guilinger, J. P., Ma, E., Doudna, J. A., & Liu, D. R. (2013). High-throughput profiling of off-target DNA cleavage reveals RNA-programmed Cas9 nuclease specificity. *Nat Biotechnology*, *31*, 839–843.

Persikov, A. V., Rowland, E. F., Oakes, B. L., Singh, M., & Noyes, M. B. (2014). Deep sequencing of large library selections allows computational discovery of diverse sets of zinc fingers that bind common targets. *Nucleic Acids Research*, *42*(3), 1497–1508. http://dx.doi.org/10.1093/nar/gkt1034.

Qi, L. S., Larson, M. H., Gilbert, L. A., Doudna, J. A., Weissman, J. S., Arkin, A. P., et al. (2013). Repurposing CRISPR as an RNA-guided platform for sequence-specific control of gene expression. *Cell*, *152*(5), 1173–1183. http://dx.doi.org/10.1016/j.cell.2013.02.022.

Ran, F. A., Hsu, P. D., Lin, C.-Y., Gootenberg, J. S., Konermann, S., Trevino, A. E., et al. (2013). Double nicking by RNA-guided CRISPR Cas9 for enhanced genome editing specificity. *Cell*, *154*, 1380–1389.

Schmidt-Dannert, C., & Arnold, F. H. (1999). Directed evolution of industrial enzymes. *Trends in Biotechnology*, *17*(4), 135–136.

Shalem, O., Sanjana, N. E., Hartenian, E., Shi, X., Scott, D. A., Mikkelsen, T. S., et al. (2014). Genome-scale CRISPR-Cas9 knockout screening in human cells. *Science (New York)*, *343*(6166), 84–87. http://dx.doi.org/10.1126/science.1247005.

Shan, Q., Wang, Y., Li, J., Zhang, Y., Chen, K., Liang, Z., et al. (2013). Multiplex and homologous recombination-mediated genome editing in Arabidopsis and Nicotiana benthamiana using guide RNA and Cas9. *Nature Biotechnology*, *31*(8), 688–691. http://dx.doi.org/10.1038/nbt.2654.

Sternberg, S. H., Redding, S., Jinek, M., Greene, E. C., & Doudna, J. A. (2014). DNA interrogation by the CRISPR RNA-guided endonuclease Cas9. *Nature*, *507*(7490), 62–67. http://dx.doi.org/10.1038/nature13011.

Tsai, S. Q., Wyvekens, N., Khayter, C., Foden, J. A., Thapar, V., Reyon, D., et al. (2014). Dimeric CRISPR RNA-guided FokI nucleases for highly specific genome editing. *Nature Biotechnology*, *32*(6), 569–576. http://dx.doi.org/10.1038/nbt.2908.

Wang, H., Yang, H., Shivalila, C. S., Dawlaty, M. M., Cheng, A. W., Zhang, F., et al. (2013). One-step generation of mice carrying mutations in multiple genes by CRISPR/Cas-mediated genome engineering. *Cell*, *153*(4), 910–918. http://dx.doi.org/10.1016/j.cell.2013.04.025.

Wiedenheft, B., Sternberg, S. H., & Doudna, J. A. (2012). RNA-guided genetic silencing systems in bacteria and archaea. *Nature*, *482*(7385), 331–338. http://dx.doi.org/10.1038/nature10886.

Yano, T., Sanders, C., Catalano, J., & Daldal, F. (2005). SacB-5-fluoroorotic acid-pyrE-based bidirectional selection for integration of unmarked alleles into the chromosome of Rhodobacter capsulatus. *Applied and Environmental Microbiology*, *71*(6), 3014–3024. http://dx.doi.org/10.1128/AEM.71.6.3014-3024.2005.

Zhou, Y., Zhu, S., Cai, C., Yuan, P., Li, C., Huang, Y., et al. (2014). High-throughput screening of a CRISPR/Cas9 library for functional genomics in human cells. *Nature*, *509*(7501), 487–491. http://dx.doi.org/10.1038/nature13166.

# AUTHOR INDEX

Note: Page numbers followed by "*f*" indicate figures and "*t*" indicate tables.

## C

## D

## E

## I



## J

## K

## L

## M

## N

## Q



## R

## S

## T

## U



## V

## X



## Y

# SUBJECT INDEX

Note: Page numbers followed by "*f*" indicate figures, "*t*" indicate tables, and "*b*" indicate boxes.

## D

## E



## F

## M



## N



## P

## V



## W



## X



## Z

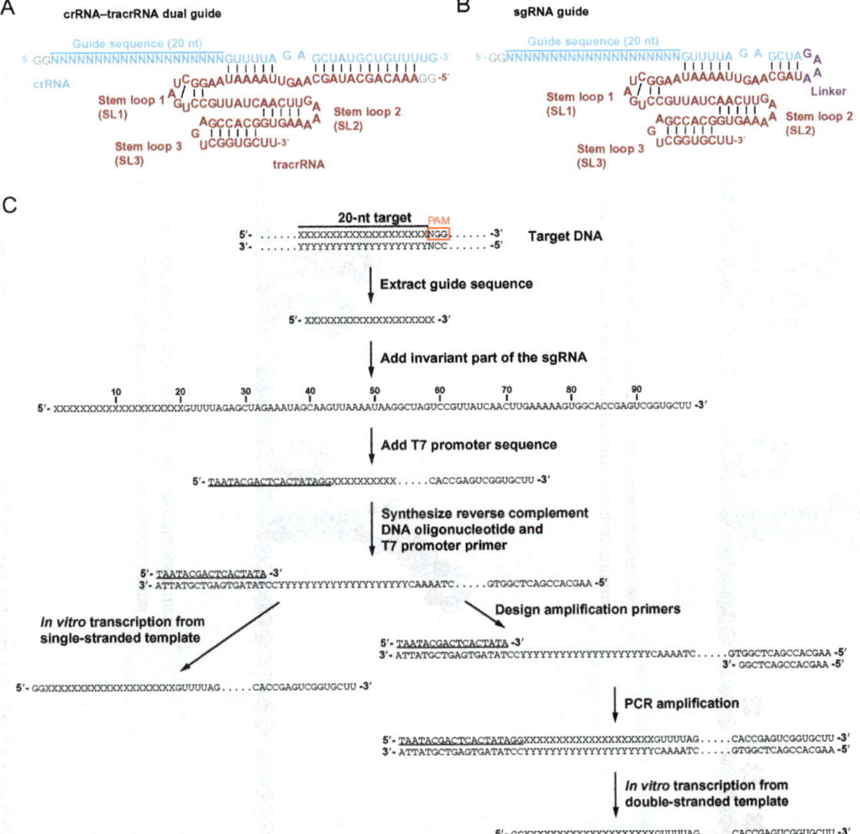


**Carolin Anders and Martin Jinek, Figure 1.1** (A) Schematic representation of the dual-RNA guide structure for programming SpyCas9. Blue: crRNA containing a 20-nt guide sequence and a 22-nt invariant sequence. The GG dinucleotide at the 5′ end is appended during *in vitro* transcription. Red: tracrRNA base-pairing with the invariant sequence of the crRNA. tracrRNA contains three stem-loops (SL1, SL2, and SL3) at its 3′ end. Although SL1 is sufficient for Cas9-mediated cleavage *in vitro* (Jinek et al., 2012), inclusion of both SL2 and SL3 increases cleavage efficiency by increasing the stability of the Cas9–crRNA–tracrRNA complex (Hsu et al., 2013; Nishimasu et al., 2014). (B) Schematic representation of chimeric single-guide RNA (sgRNA). crRNA- and tracrRNA-derived sequences are connected by a 5′-GAAA-3′ tetraloop linker. (C) Outline of the procedure for sgRNA guide design and preparation by *in vitro* transcription.

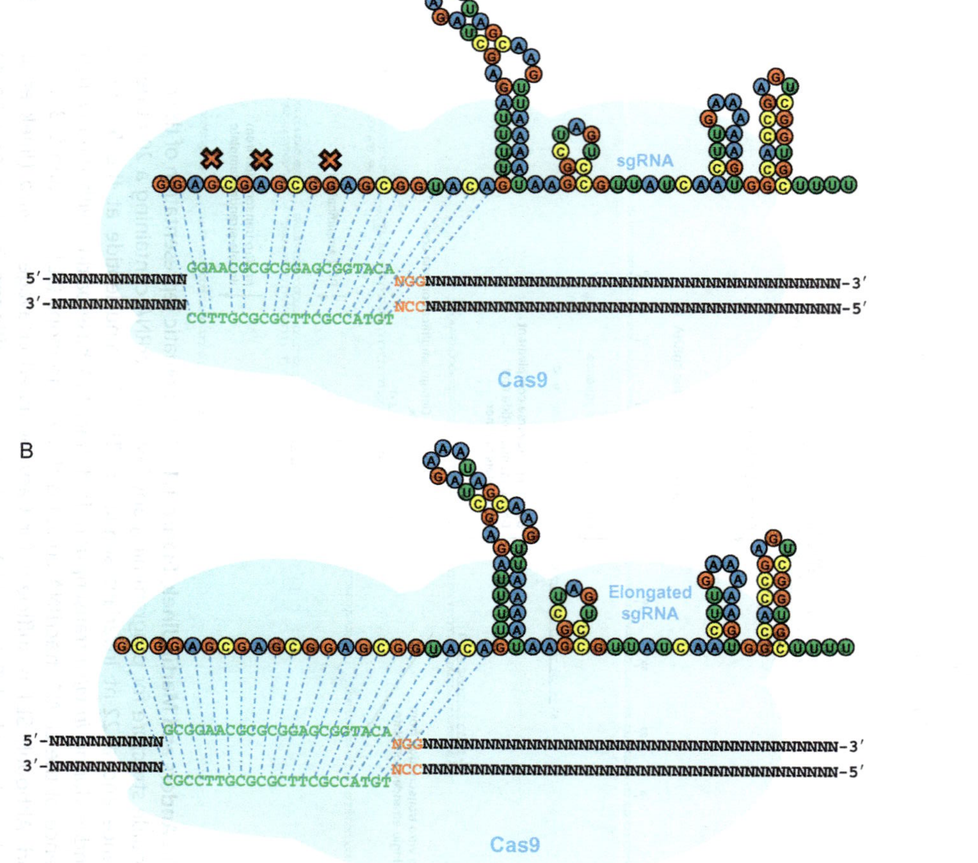


**Yanfang Fu *et al.*, Figure 2.1** Schematic overview of all the published methods on specificity improvement to CRISPR/Cas9 nucleases. (A) A first-generation CRISPR/Cas9 nuclease is directed to a target DNA site (green letters) by complementarity with the first 20 nucleotides of the guide RNA (gRNA) and the presence of a protospacer adjacent motif (PAM) sequence (red letters). Mismatches between the target DNA site and the sgRNA complementarity region (red Xs) can be tolerated, thereby leading to mutations at off-target sites. (B) Elongated guide RNAs bearing two additional nucleotides at their 5′-end may lead to recognition of longer target DNA sites, in some cases with improved specificity.

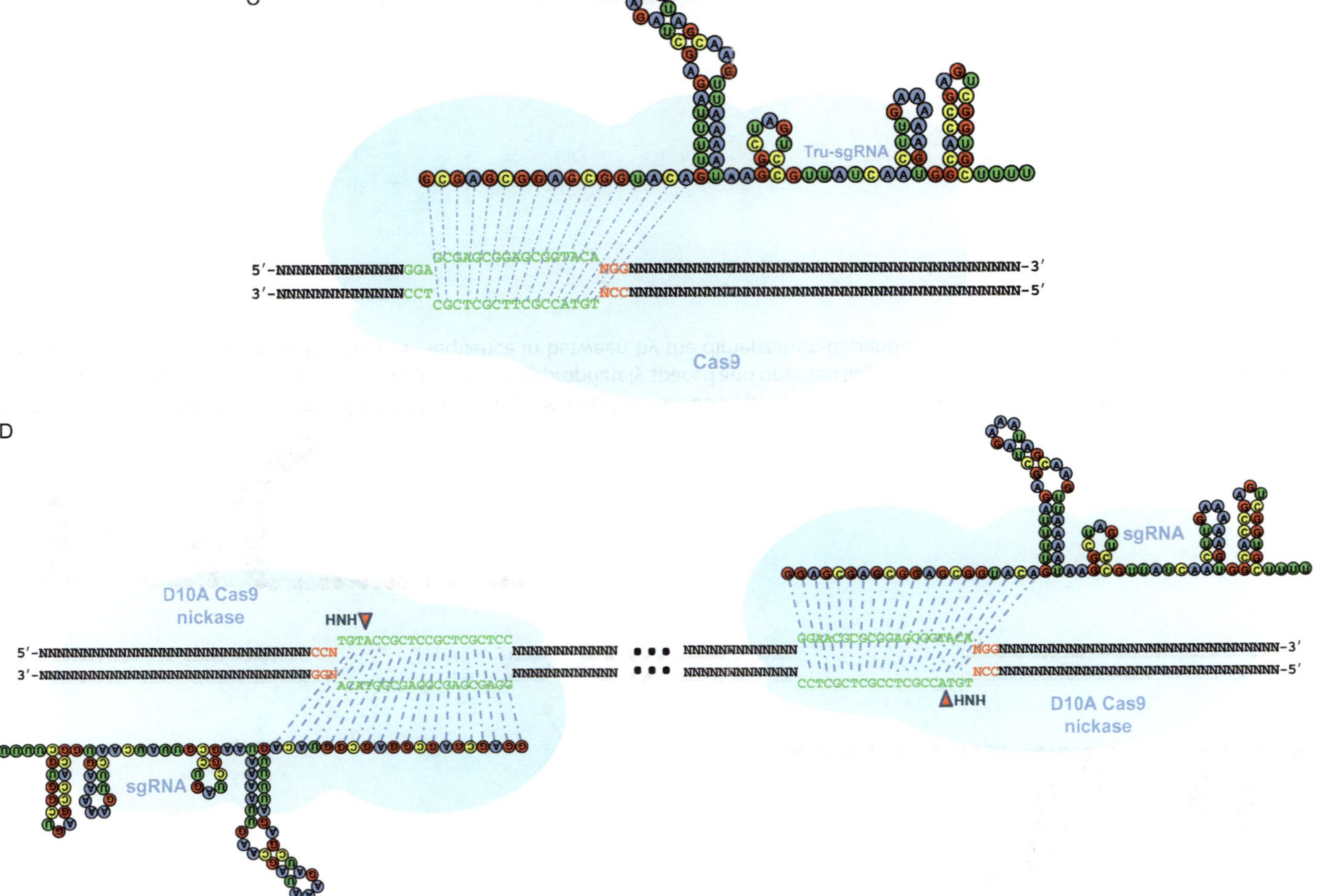


**Yanfang Fu *et al.*, Figure 2.1—Cont'd** (C) Truncated gRNAs (tru-gRNAs) are shortened on their 5′-end by two to three nucleotides. (D) A paired nickase strategy in which two sgRNAs direct a D10A Cas9 nickases to adjacent target sequences on opposite DNA strands.

E

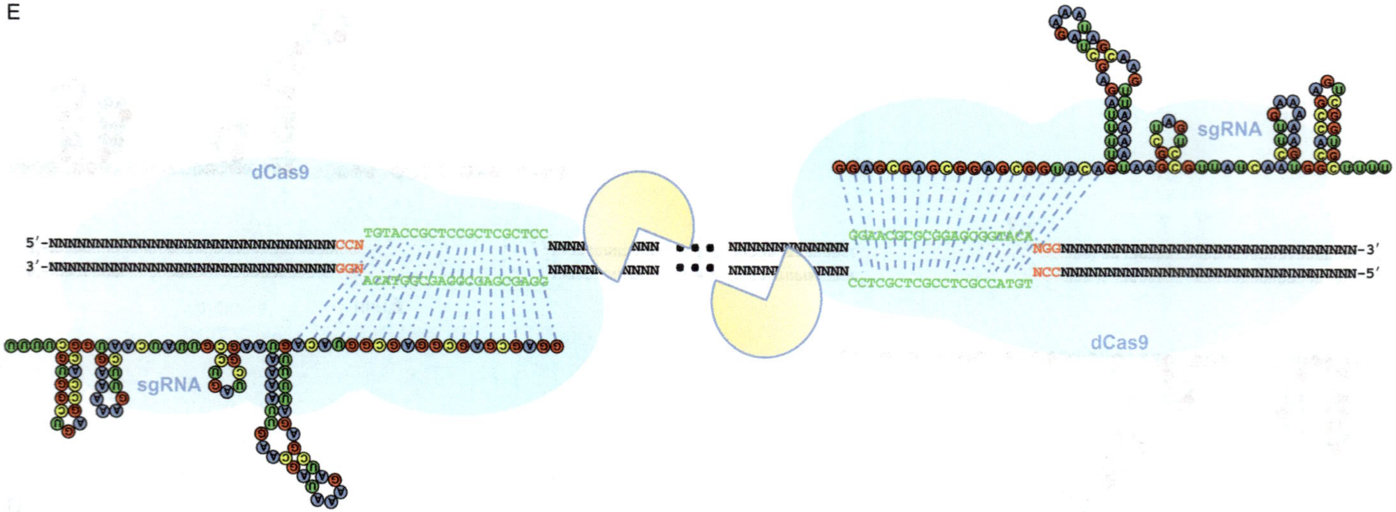

**Yanfang Fu *et al.*, Figure 2.1—Cont'd** (E) CRISPR RNA-guided nucleases. Fusion proteins consisting of the *Fok*I nuclease domain fused to a catalytically inactive Cas9 (dCas9) can be directed to two appropriately spaced and oriented adjacent sites on opposite strands of DNA by two sgRNAs, resulting in cleavage of the "spacer" sequence in between by the dimerization-dependent *Fok*I nuclease domains.

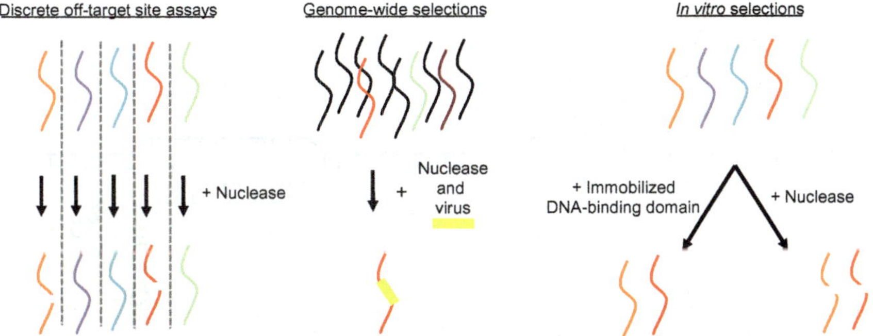


**Vikram Pattanayak *et al.*, Figure 3.1** Overview of methods to study the specificity of nucleases. Potential substrate sequences of interest (colored strands) are subjected to nuclease cleavage to identify cleaved sequences (broken red and orange strands). In discrete off-target site assays, sequences are individually subjected to nuclease cleavage in a low- or high-throughput manner. In genome-wide selections, a few potential off-target sites are cleaved within predominantly uncleaved genomic DNA (black strands) and detected by viral integration. Using *in vitro* selection, many potential off-target sites in a vast DNA library are selected for their ability to be bound or cleaved by site-specific nucleases *in vitro*.

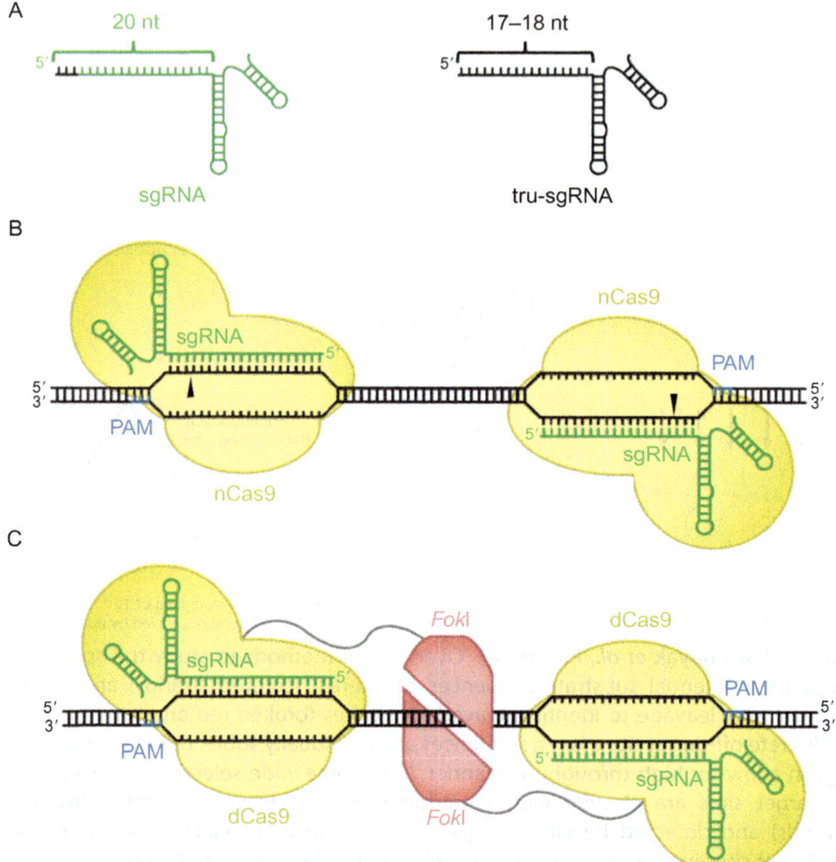


**Vikram Pattanayak *et al.*, Figure 3.4** Engineered Cas9 components with improved DNA cleavage specificity. (A) A truncated guide RNA (tru-gRNA, right), contains 17–18 base pairs of complementarity to its DNA target site, rather than 20 base pairs in a canonical sgRNA (left). The base pairs in the sgRNA that are not present in the tru-gRNA are colored black. (B) Mutant Cas9 proteins that cleave only a single strand of dsDNA (nCas9) can be targeted to opposite strands of adjacent sites as pairs to cause double-strand breaks. (C) Monomers of *Fok*I nuclease (red) fused to catalytically inactive Cas9 bind to separate sites within a target locus. Only adjacently bound *Fok*I–dCas9 monomers can assemble a catalytically active *Fok*I nuclease dimer, triggering dsDNA cleavage.

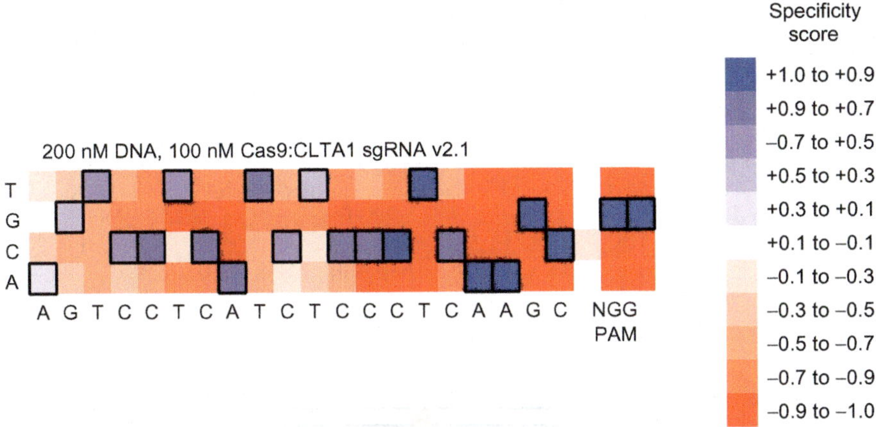


**Vikram Pattanayak *et al.*, Figure 3.6** An *in vitro* selection-derived specificity profile. The heat map shows the specificity profile resulting from a selection performed on Cas9: sgRNA targeting the human *CLTA* gene. Specificity scores of 1.0 (dark blue) and −1.0 (dark red) corresponds to 100% enrichment for and against, respectively, a particular base pair at a particular position. Black boxes denote the intended target nucleotides.

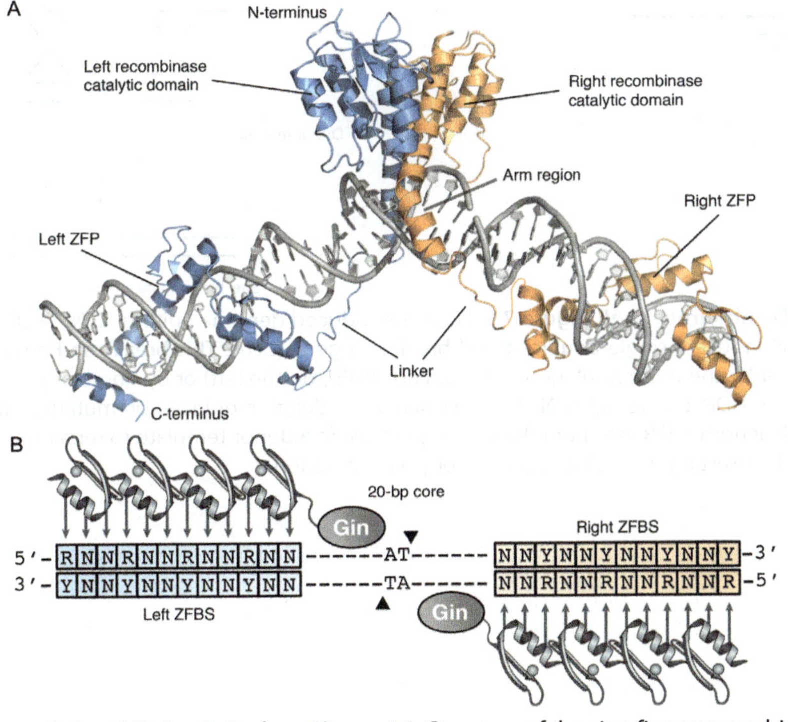


**Thomas Gaj and Carlos F. Barbas, Figure 4.1** Structure of the zinc-finger recombinase (ZFR) dimer bound to DNA. (A) Each ZFR monomer (blue or orange) consists of an activated serine recombinase catalytic domain fused to a custom-designed $Cys_2$-$His_2$ zinc-finger DNA-binding domain. (B) Cartoon of the ZFR dimer bound to DNA. ZFR target sites consist of two inverted zinc-finger binding sites flanking a central 20-bp core sequence recognized by the serine recombinase catalytic domain. Abbreviations are as follows: N = A, T, C, or G; R = G or A; and Y = C or T. *Adapted from Gaj, Mercer, et al. (2013).*

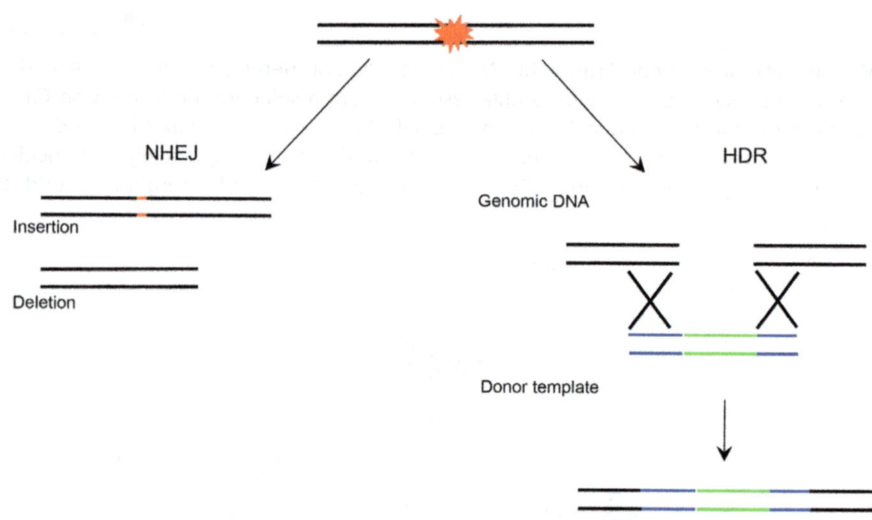


**D. Dambournet *et al.*, Figure 7.1** Nuclease-induced genome editing. ZFN, TALEN, and Cas9 induce a single double-strand break in a gene locus. This break can be repaired either by the nonhomologous end joining (NHEJ, on the left) or by homology-directed repair (HDR, on the right). NHEJ repair generates deletion or insertion mutations of variable length. HDR uses homologous sequences on a donor template to repair the break while inserting a specific sequence or point mutation.

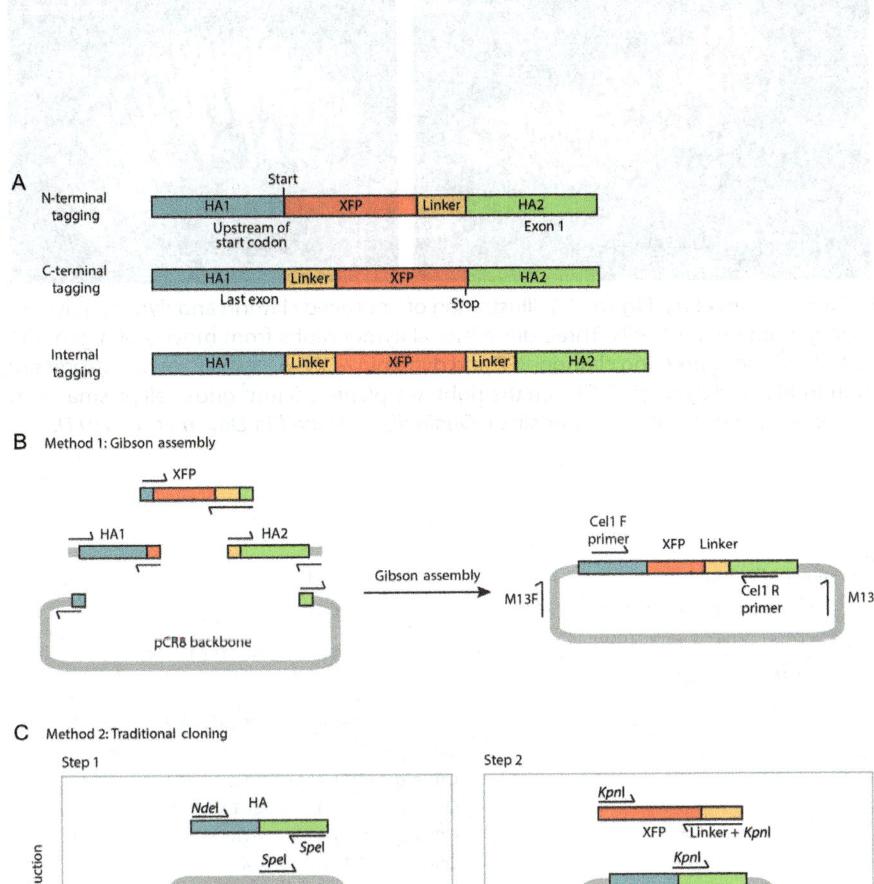


**D. Dambournet *et al.*, Figure 7.2** Donor plasmid construction. (A) Possible organization of donor plasmids for N-terminal tagging (top), C-terminal tagging (middle), and internal tagging (bottom). (B) Gibson assembly of the donor plasmid. The schematic depicts a donor for N-terminal tagging. (C) Traditional cloning of the donor plasmid.

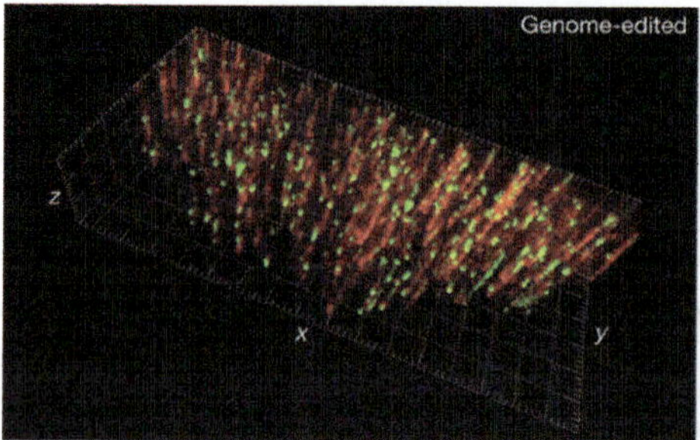


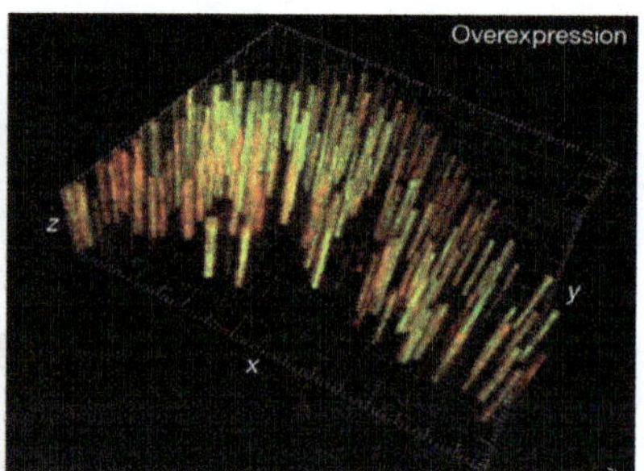


**D. Dambournet *et al.*, Figure 7.4** Illustration of improved clathrin and dynamin dynamics in genome-edited cells. Three-dimensional kymographs from movies of a genome-edited cell line expressing clathrin-RFP and dynamin2-GFP on the left, or overexpressing clathrin-RFP and dynamin2-GFP on the right. *x–y* plane (2.5 $\mu m^2$ grid), cell plasma membrane; *z*-axis, time (240 s, 2 s per slice). *Originally published in Doyon et al. (2011).*

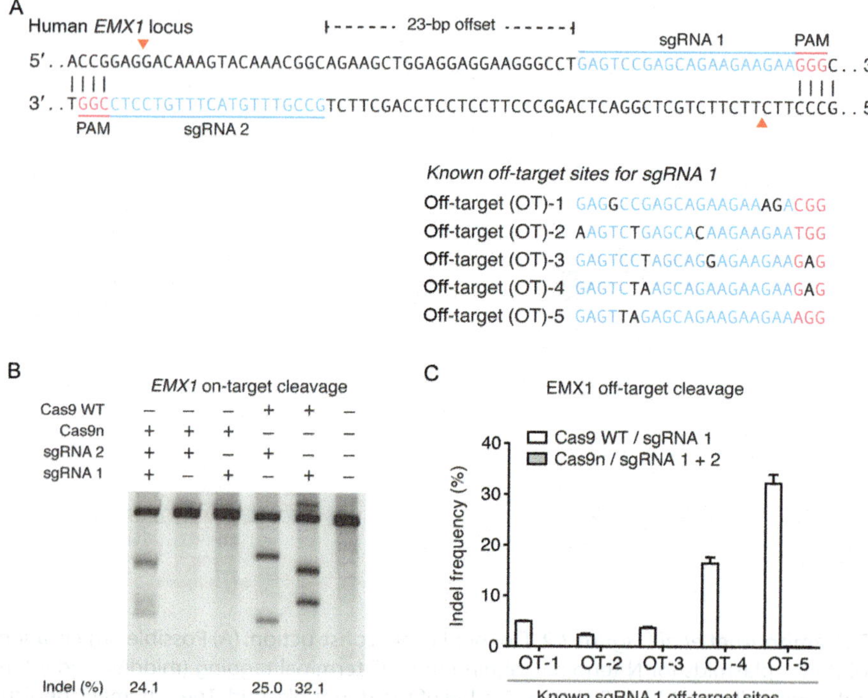


**Alexandro E. Trevino and Feng Zhang, Figure 8.2** Double nicking reduces off-target modification. (A) Diagram of a Cas9n D10A double-nicking sgRNA pair designed for the human EMX1 locus. Guide sequences are shown in blue, demonstrating a 23-bp offset. The PAM is shown in pink, and nicking sites are represented by red triangles. Five known genomic off-target sites (Hsu et al., 2013) for sgRNA 1 are listed. (B) Example SURVEYOR results showing modification of the EMX1 locus by Cas9 WT and Cas9n along with sgRNA 1 and/or 2. (C) Deep sequencing quantification of off-target modifications at five known off-target sites by Cas9 WT and sgRNA 1 or Cas9n with sgRNAs 1 and 2. *Adapted with permission from Ran et al. (2013).*

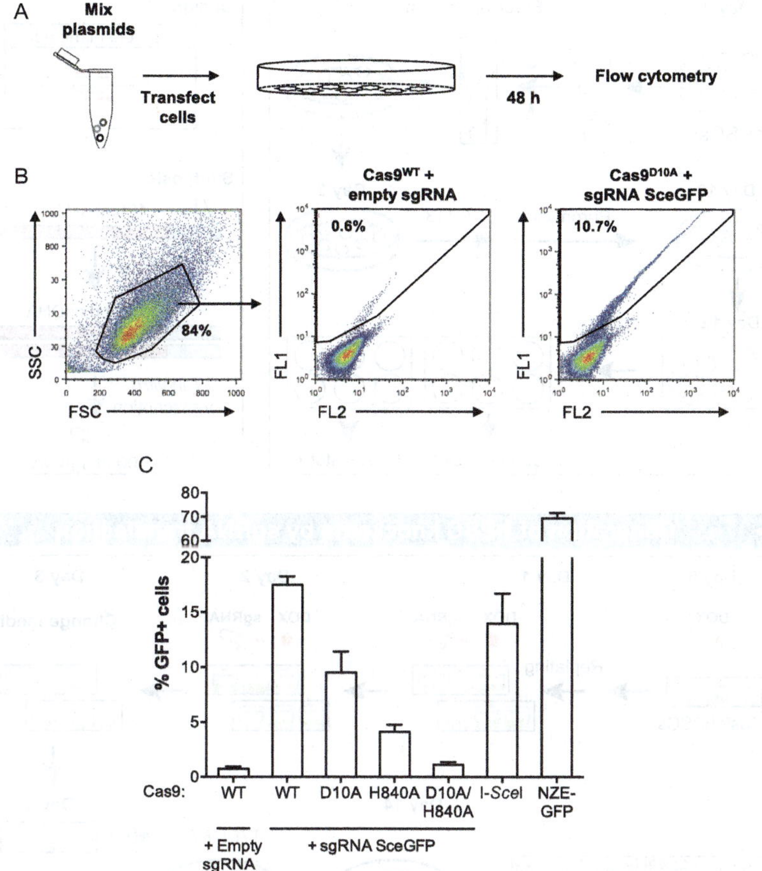


**Lianne E.M. Vriend *et al.*, Figure 9.2** Representative experiment testing different variants of Cas9 for HR in the DR-GFP reporter in HEK293T cells. (A) Schematic overview of the experiment. HEK293T cells were cotransfected with plasmid DNA (Table 9.2) using Nucleofector-2b (Lonza, program A-023). After 48 h, the percent GFP$^+$ cells, indicative of HR efficiency, was measured by flow cytometry. (B) FlowJo software was used to analyze the flow cytometry data. (Left) SSC versus FSC was used to set the gate for live cells. (Middle and Right) FL1 versus FL2 was used to determine the percent GFP$^+$ cells from a control sample (cells above diagonal line are GFP$^+$). (Middle) Note that there is a relatively high background when using the plasmid DR-GFP reporter. (Right) The percentage of GFP$^+$ cells is much higher when DNA damage is induced. (C) Results from three independent experiments. SSBs created by Cas9$^{D10A}$ and Cas9$^{H840A}$ are capable of inducing HR, albeit at reduced frequency, as compared to Cas9$^{WT}$. Error bars represent standard deviation from the mean.

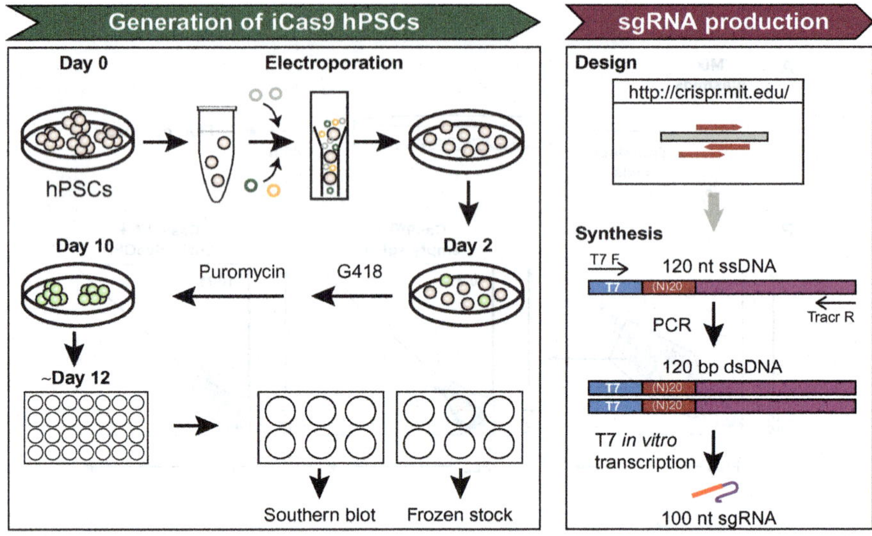


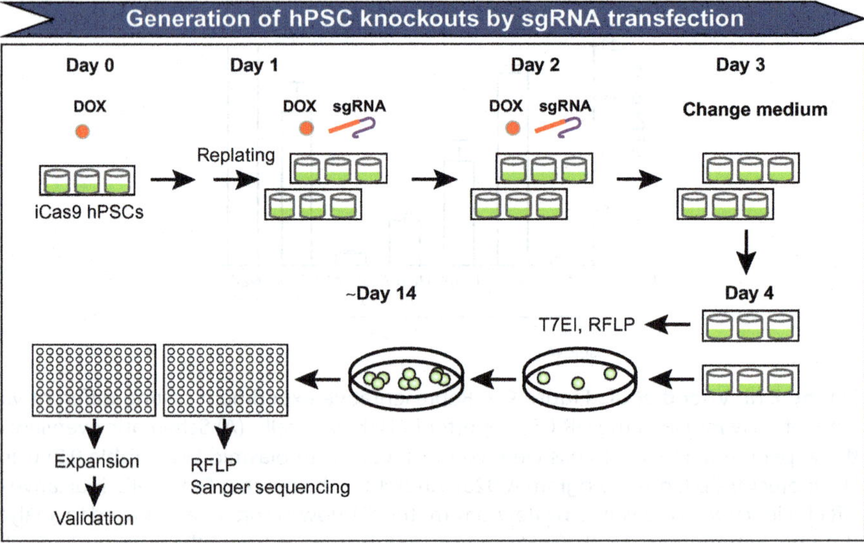


**Zengrong Zhu *et al.*, Figure 11.1** The iCRISPR platform for rapid genome editing in hPSCs. Instead of transient expression of Cas9 and sgRNAs from electroporated plasmids in hPSCs, targeted integration and inducible expression of Cas9 into the *AAVS1* locus provides a precise and reliable approach to express the invariable component of the CRISPR/Cas system (Section 2). Then, sgRNAs targeting specific loci are designed and synthesized by PCR amplification of *in vitro* transcription DNA templates (Sections 3.1 and 3.2). Due to their small size (~100 nucleotides), transfection of iCas9 hPSCs with sgRNAs is highly efficient, leading to reproducible and highly efficient gene knockout in hPSCs (Section 3.3).

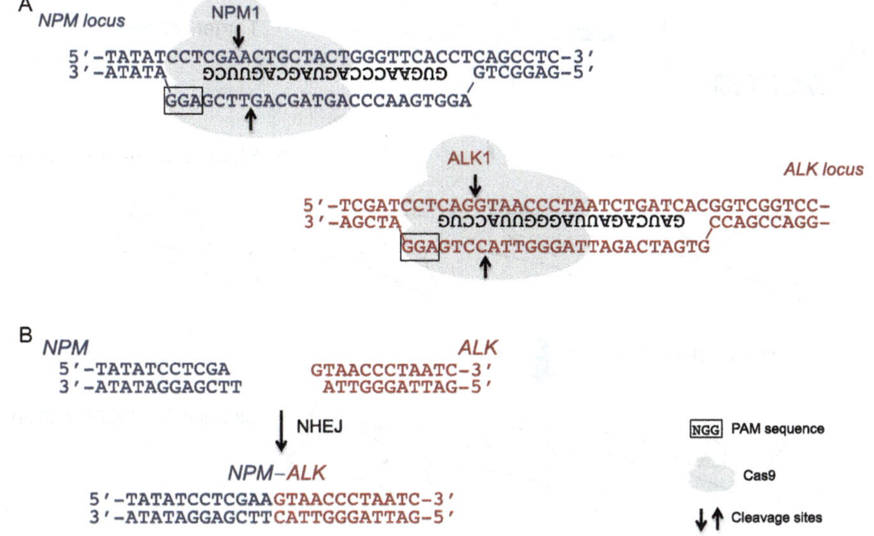


**Benjamin Renouf *et al.*, Figure 12.3** Cas9 cleavage at the *NPM* and *ALK* loci and example of a translocation junction sequence. (A) Cas9 cleavage at the *NPM* and *ALK* loci using the NPM1 and ALK1 sgRNAs, respectively, leads to DSBs. Only the part of the sgRNA that binds DNA is shown. PAM sequence, boxed; arrows, cleavage sites. (B) DNA ends from Cas9 cleavage that are relevant to Der5 formation are shown. NHEJ leads to a variety of Der5 translocation junctions, but a common junction (shown) is direct ligation of the DNA ends, presumably involving fill-in of the 1-base 5′ overhang. For additional junctions, see Ghezraoui et al. (2014).

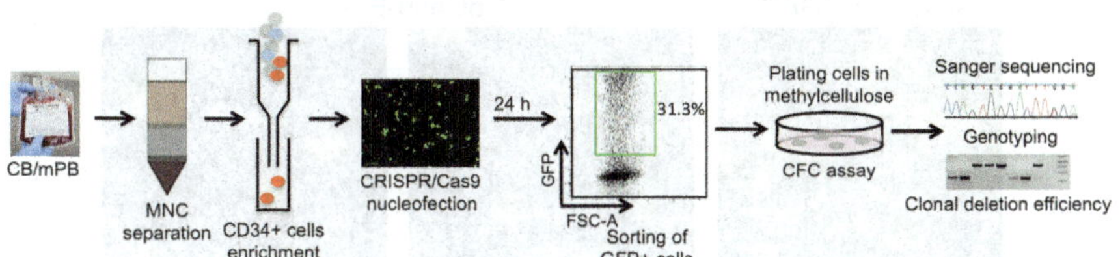


**Torsten B. Meissner *et al.*, Figure 13.7** Genome editing in CD34+ HSPCs. Schematic illustrating the workflow of genome editing in CD34+ HSPCs. Mononuclear cells (MNCs) are isolated from cord blood (CB) by Ficoll gradient centrifugation. Subsequently, CD34+ cells are isolated by CD34+ magnetic bead enrichment and are nucleofected with CRISPR/Cas9 plasmids. 24 h posttransfection, GFP-positive cells are sorted by FACS and cultivated in methylcellulose for 2 weeks. Colonies grown in methylcellulose are counted and scored for colony types. DNA isolated from individual colonies can be analyzed by either Sanger sequencing or gel electrophoresis.

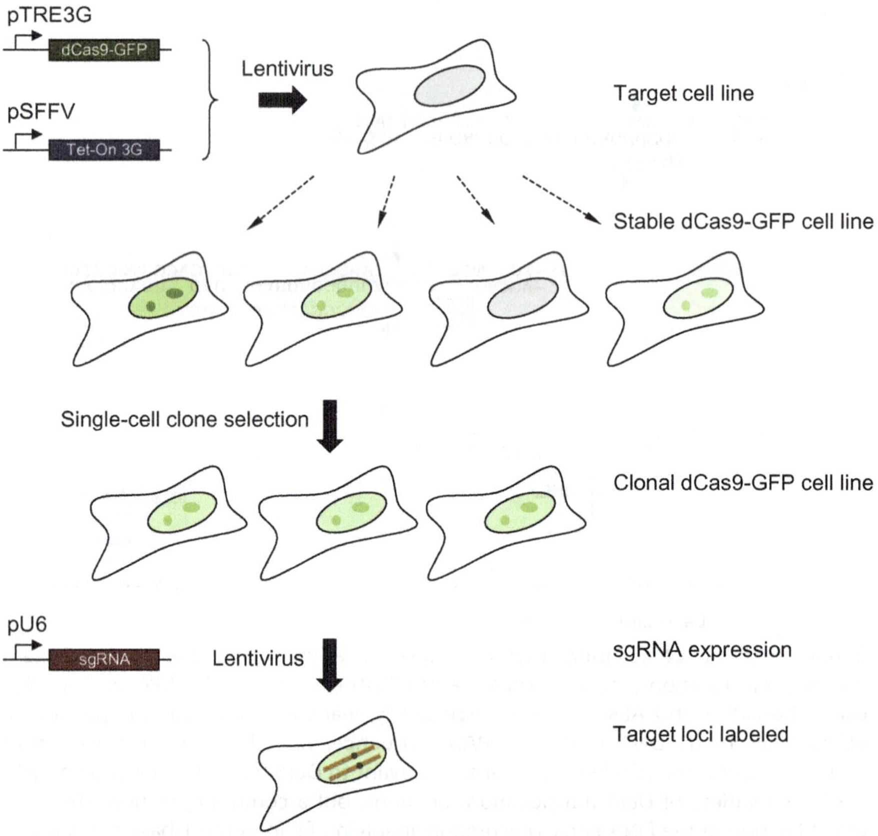

pTRE3G
dCas9-GFP
pSFFV
Tet-On 3G
Lentivirus
Target cell line
Stable dCas9-GFP cell line
Single-cell clone selection
Clonal dCas9-GFP cell line
pU6
sgRNA
Lentivirus
sgRNA expression
Target loci labeled

**Baohui Chen and Bo Huang, Figure 16.1** Schematic of the work flow for imaging genomic elements using the CRISPR/Cas9 system.

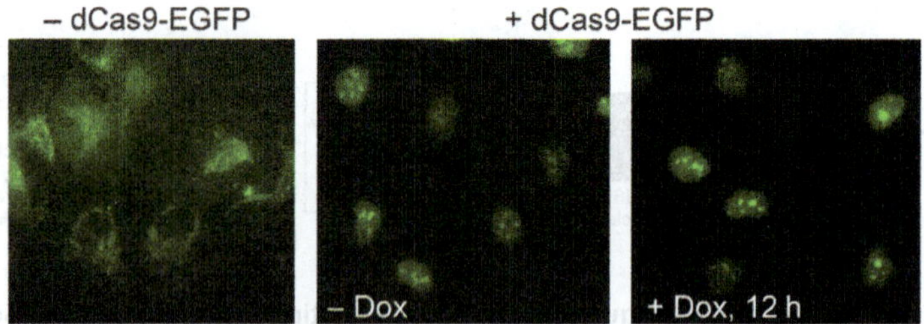

– dCas9-EGFP
+ dCas9-EGFP
– Dox
+ Dox, 12 h

**Baohui Chen and Bo Huang, Figure 16.2** Expression of dCas9-GFP in RPE cells without sgRNA. Only weak cytoplasmic autofluorescence can be detected in cells not expressing dCas9-GFP. For cells carrying a Tet-On 3G inducible expression system for dCas9-GFP, nuclear GFP signal can be observed at the basal expression level without doxycycline induction. At 12 h after doxycycline induction, much stronger (~50-fold) GFP signal can be recorded, with the nucleolar enrichment of dCas9-GFP clearly visible. The three panels are adjusted to different contrasts so that the weak signals are visible.

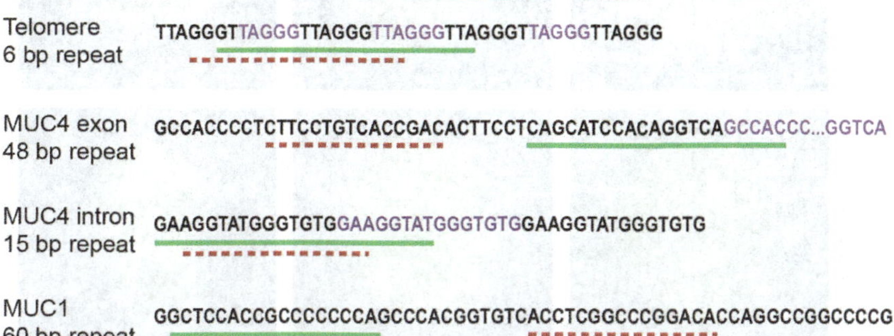


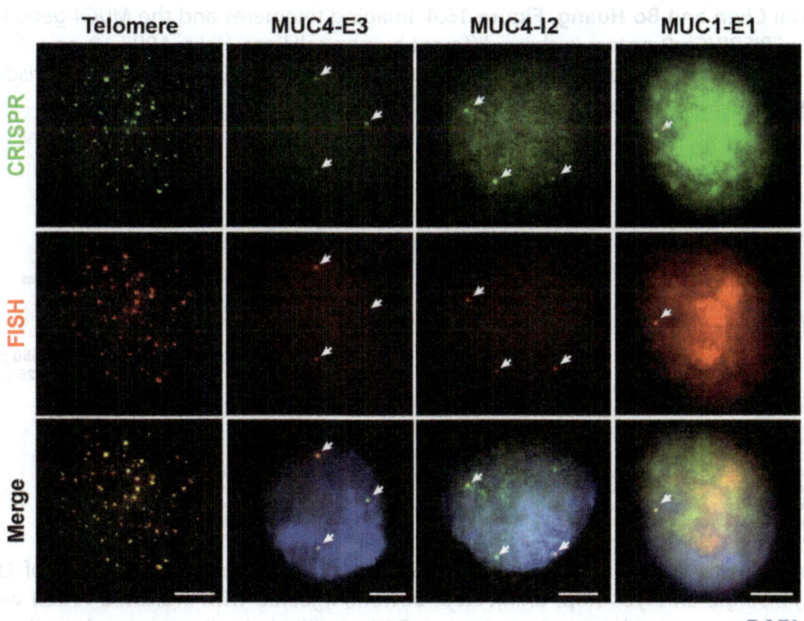


**Baohui Chen and Bo Huang, Figure 16.3** Verification of CRISPR signal specificity by FISH. Four target loci in RPE cells have been tested: telomere, the tandem repeats in exon 2 and intron 2 of the *MUC4* gene, and the tandem repeats in exon 2 of the *MUC1* gene. The recognition sequences for the sgRNAs and oligo-DNA FISH probes are underlined. Complete colocalization of the CRISPR and FISH signal is displayed with the nuclear DNA stained by DAPI. Scale bars: 5 μm.

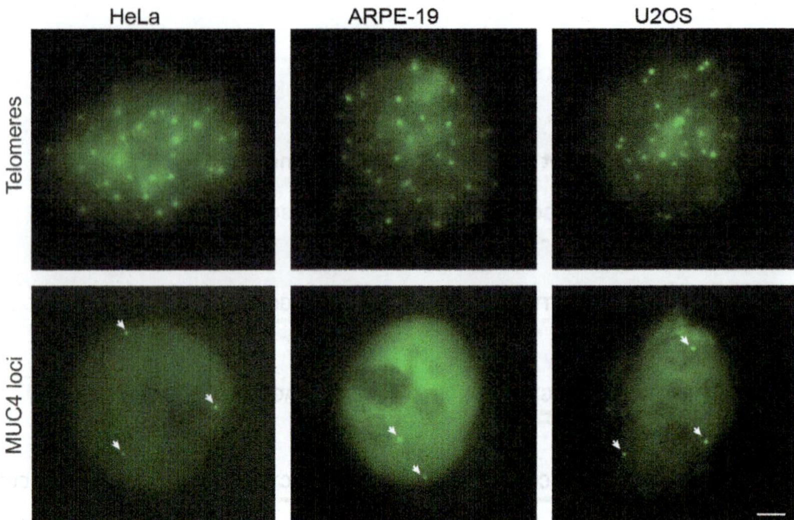


**Baohui Chen and Bo Huang, Figure 16.4** Imaging telomeres and the *MUC4* gene loci using CRISPR/Cas9 system in three different human cell lines: HeLa, ARPE-19, and U2OS. The HeLa and U2OS cell lines contain three copies of the *MUC4* gene due to trisomy chromosome 3, which is reflected in the CRISPR images. Scale bars: 5 μm.

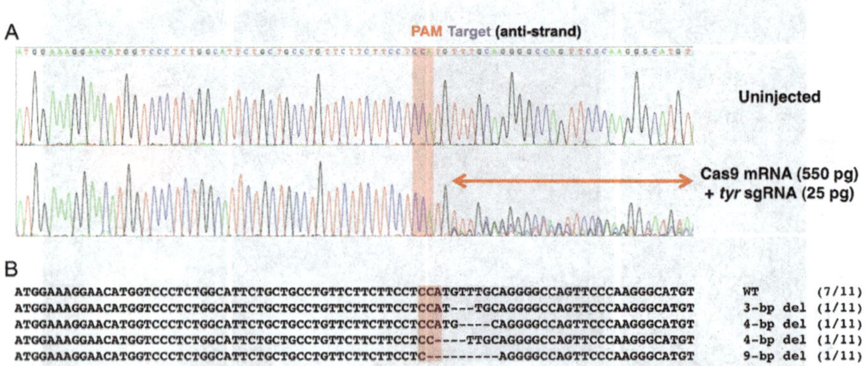


**Takuya Nakayama *et al.*, Figure 17.5** DSP assay. (A) Representative results of DSP assays. Single embryos (top, uninjected; bottom, injected with indicated RNAs) were lysed and the targeted genomic region was PCR amplified; amplicons were then directly sequenced. Perturbation of peaks (double-headed arrow) on the 3′ side of the PAM region (note that in this specific case, the target is on the antisense strand) seen in the sample of the injected embryo suggests indel events occurred between the target sequence (lightly shaded area where multiple peaks overlap) and the PAM region (darkly shaded). (B) The PCR amplicon from the injected embryo was recloned and sequenced to show the profile of individual mutations found in mosaic individuals (bottom sequence alignments). Dashes (-) indicate gaps. The numbers in parentheses indicate the frequency of each mutation pattern seen in the total number of sequenced clones, suggesting that a mutation frequency of 36% (i.e., 4/11) can be detected as a positive targeting event by the DSP assay. *The figure is adapted and modified from Nakayama et al. (2013).*

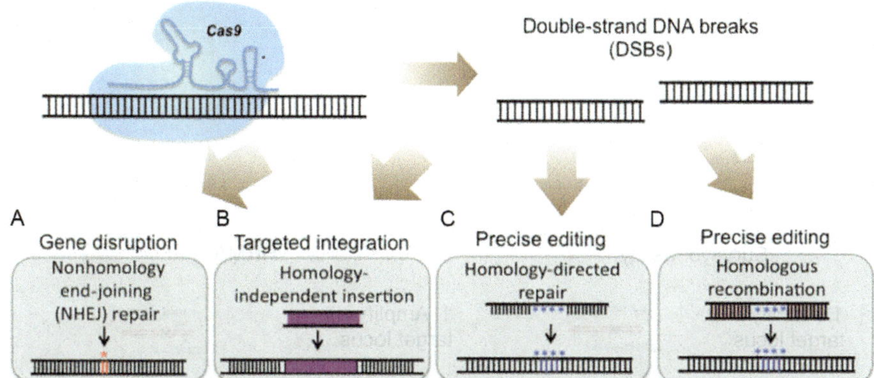


**Andrew P.W. Gonzales and Jing-Ruey Joanna Yeh, Figure 18.2** Cas9-mediated genome editing. The RNA-guided Cas9 endonuclease can induce DSBs at its genomic target site. Subsequently, a DSB may be repaired by a nonhomologous end-joining (NHEJ) repair mechanism. This mechanism may cause random-length insertion/deletion (indel) mutations (red asterisk) at the target site (A). Additional approaches can be used to create predetermined sequence modifications. Linear donor DNA fragments containing a desired functional cassette without any sequence homology to the target locus may be inserted into the target site during DNA repair (B). Moreover, donor DNA containing homologous sequences of the target locus, in the form of small single-stranded oligonucleotides (C) or plasmid DNA (D), may be recombined with the genomic DNA and replace the target site sequence.

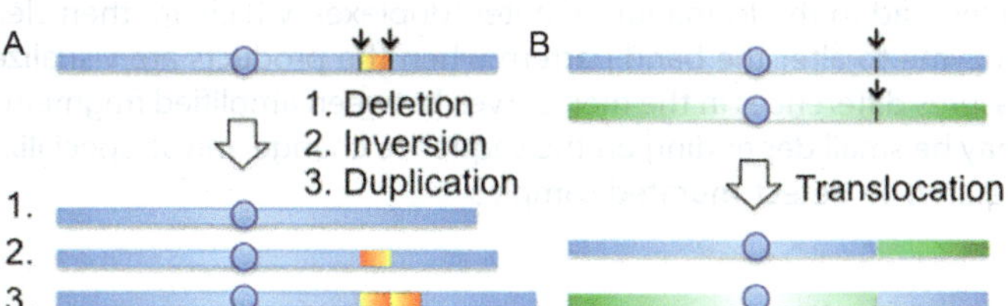


**Andrew P.W. Gonzales and Jing-Ruey Joanna Yeh, Figure 18.5** Cas9-mediated chromosomal rearrangements. (A) Chromosomal deletions, inversions, and duplications may be induced by two sgRNA–Cas9 complexes that target two distant sites on the same chromosome. (B) Chromosomal translocations may be induced by two sgRNA–Cas9 complexes that target two different chromosomes.

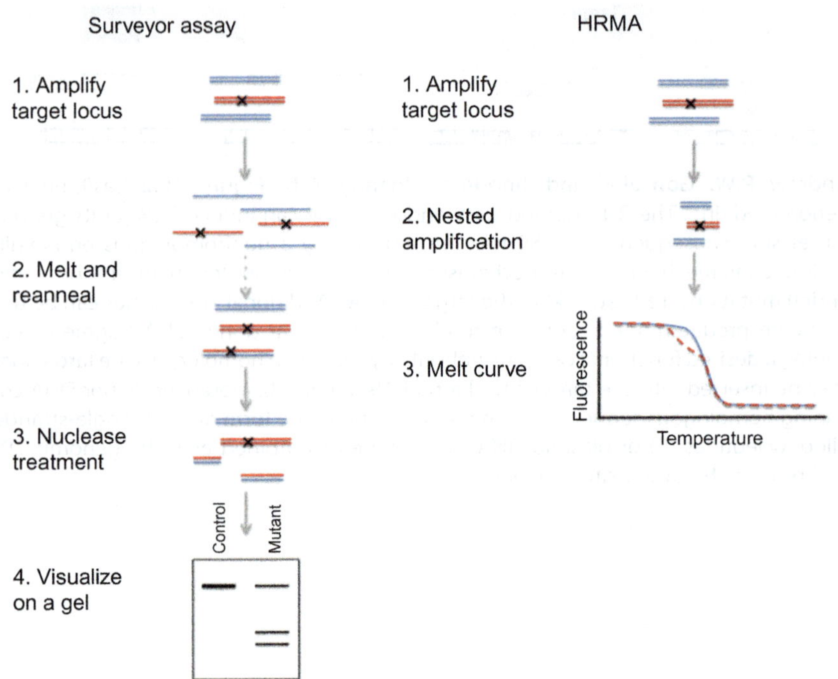


**Benjamin E. Housden *et al.*, Figure 19.2** Surveyor and HRMA screening protocols. Surveyor assays rely on the ability of the surveyor endonuclease to cleave mismatched DNA strands. Melting and reannealing fragments amplified from a mixture of wild-type and mutant alleles lead to the formation of heteroduplexes, which are then cleaved by the surveyor enzyme to alter the band pattern when the products are visualized on a gel. HRMA measures differences in the melt curves between amplified fragments. These differences may be small depending on the sequence change and so specialized software may be required to detect mutated samples.

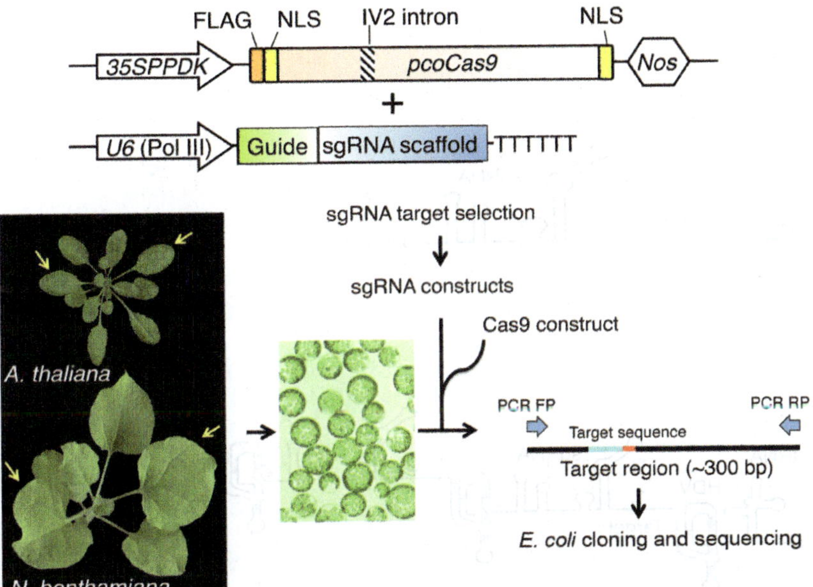


**Jian-Feng Li *et al.*, Figure 21.1** Unbiased sgRNA/Cas9-mediated genome editing in plant protoplasts. The expression cassettes of Cas9 and sgRNA are shown. Plant codon-optimized Cas9 (pcoCas9) is fused to dual nuclear localization sequences (NLSs) and FLAG tags. The constitutive *35SPPDK* promoter and the *Arabidopsis U6-1* promoter were used to express pcoCas9 and sgRNA, respectively, in protoplasts. NGG, the protospacer adjacent motif (PAM), in the target sequence is highlighted in red. The diagram illustrates the key procedure to generate and evaluate Cas9/sgRNA-mediated genome editing in *Arabidopsis* and tobacco protoplasts. Yellow arrows indicate the leaves at optimal developmental stage for protoplast isolation from 4-week-old plants. Scale bar = 2 cm. In the target region, the target sequence of $N_{20}$ and NGG (the PAM) are represented in cyan and red, respectively. Genomic DNA from protoplasts was PCR amplified and cloned into a sequencing vector. *E. coli* colonies were picked randomly for PCR amplification and sequencing.

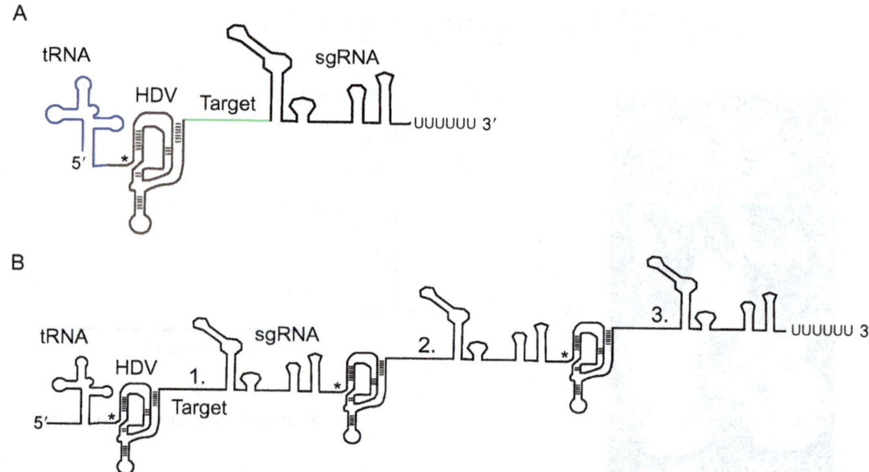


**Owen W. Ryan and Jamie H.D. Cate, Figure 22.3** The RNA structure of the sgRNA expression module. (A) The sgRNA is expressed using a tRNA as an RNA Pol III promoter. The HDV folds into its catalytically active form and removes the 5′ tRNA sequence from the mature sgRNA (cleavage site marked by an asterisk). The target sequence is protected between the HDV and sgRNA. RNA Pol III expression is terminated by a series of six or more uridine nucleotides. (B) Engineering sgRNAs for improved coexpression. Future polycistronic sgRNAs could be expressed using a single RNA PolIII promoter (tRNA) and processed internally by their catalytically active HDV (cleavage sites are marked by *). This panel contains three sgRNAs in tandem arrays.

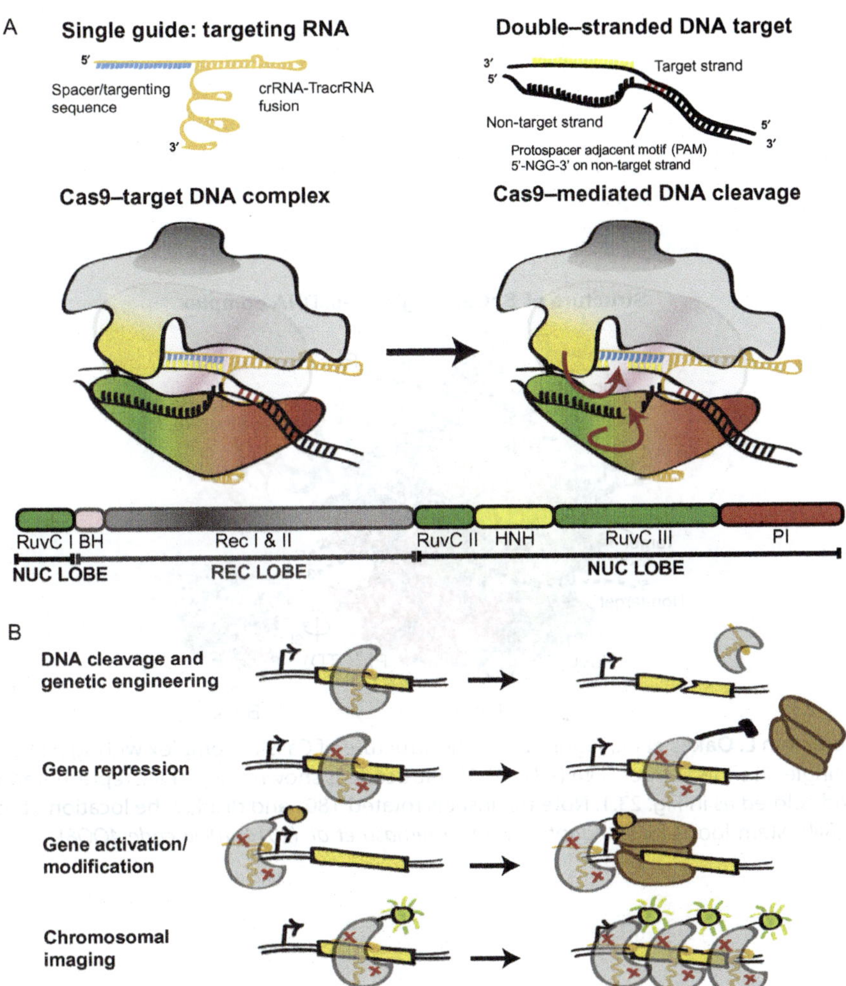


**Benjamin L. Oakes *et al.*, Figure 23.1** Holo Cas9 model and its potential uses. (A) Single guide RNA, target dsDNA, and Cas9 are modeled. Domains of Cas9 are colored accordingly: RuvC-green, BH-pink, RecI-gray, RecII- dark gray, HNH-yellow, PI-red. (B) Common uses of Cas9 as a tool. Red x's in Cas9 represent a nuclease dead variant.

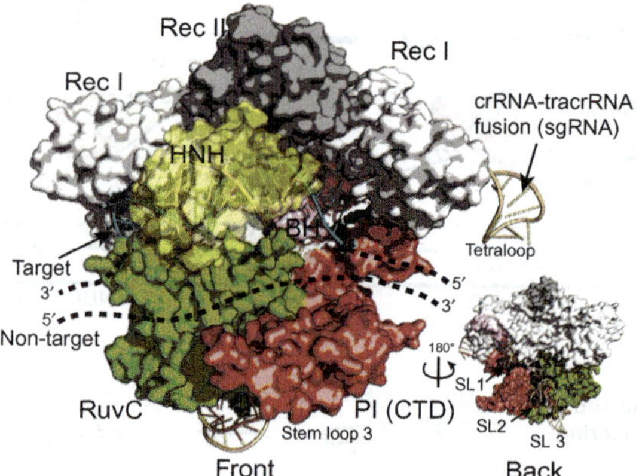


**Benjamin L. Oakes *et al.*, Figure 23.2** The structure of Cas9 in complex with sgRNA and a single-stranded target DNA (ssDNA). The structure is shown in a surface representation and colored as in Fig. 23.1. Note the inset is rotated 180° and display the location of the sgRNA stem loops (SLs). *Adapted from Nishimasu et al. (2014) (PDB code 4OO8).*

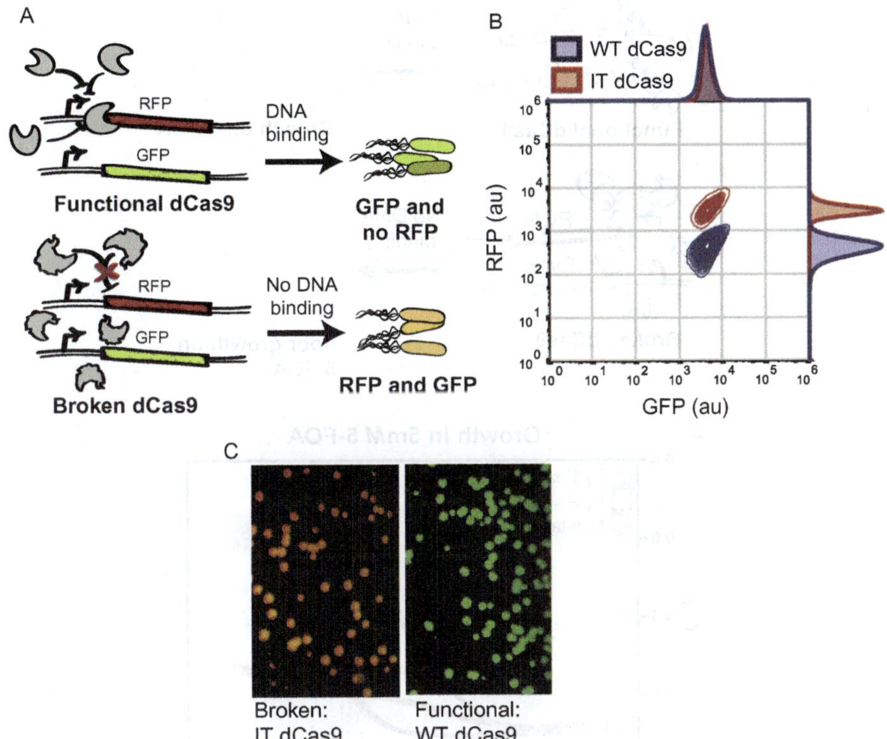


**Benjamin L. Oakes *et al.*, Figure 23.3** Screen for functional Cas9s. (A) Schematic representation of the screen. (B) Flow cytometry data of the functional positive (WT dCas9) control in blue and negative "Inactive Truncation" Cas9 (IT dCas9) control in red. IT dCas9 contains only the C-terminal 250 amino acids. Both controls contain the sgRNA plasmid which targets RFP for repression. Samples were grown overnight in rich induction media. (C) Colony fluorescence of the functional (WT dCas9) and "Broken" negative (IT dCas9) controls.

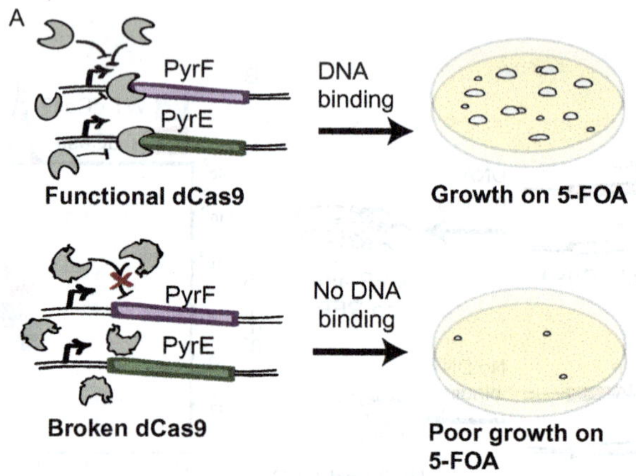


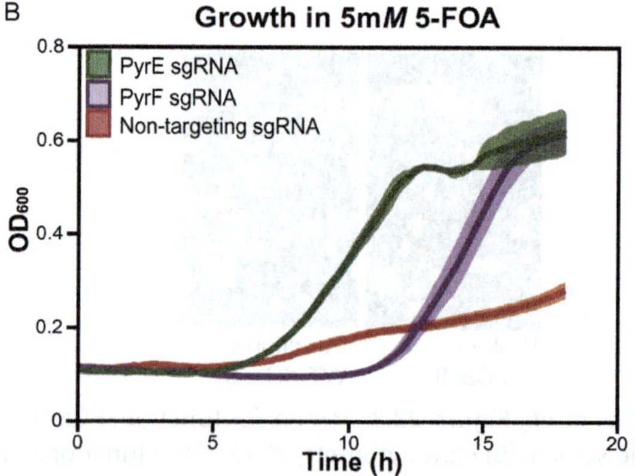


**Benjamin L. Oakes *et al.*, Figure 23.4** Functional Cas9 selection overview. (A) Schematic representation of the selection system. (B) Growth rate of a functional dCas9 + sgRNAs repressing the *pyrF* gene (purple), *pyrE* gene (green), and a no guide sequence control (red). Samples were grown in rich induction media + 5 m*M* 5-fluoroorotic acid. All measurements represent the average (line) and standard deviation (shading) of three biological replicates.

Printed by Printforce, the Netherlands